THE SOLAR CHROMOSPHERE

THE INTERNATIONAL ASTROPHYSICS SERIES

THE SOLAR CHROMOSPHERE

R. J. Bray and R. E. Loughhead

Commonwealth Scientific and Industrial Research Organization
National Standards Laboratory, Sydney, Australia

THE INTERNATIONAL ASTROPHYSICS SERIES

LONDON
CHAPMAN AND HALL

First published 1974
by Chapman and Hall Ltd
11 New Fetter Lane, London EC4P 4EE

Typeset by
Santype Ltd (Coldtype Division)
Salisbury, Wiltshire
Printed in Great Britain by
Butler & Tanner Ltd, Frome

SBN 412 10730 9

Distributed in the U.S.A.
by Halsted Press, a Division
of John Wiley & Sons, Inc., New York

Library of Congress Catalog Card Number 73–8101

Editors' Note

The aim of the International Astrophysics Series is to provide a collection of authoritative volumes dealing with the main branches of Astrophysics and Radio Astronomy. The need for such a series of books has arisen because of the great developments which have taken place in these fields of work during recent years.

The books will be suitable for both specialists and students. Some of the titles may have a wider and more popular appeal, but this will be secondary to their main purpose, which is to assist in the teaching of Astrophysics and Radio Astronomy and in the advancement of these subjects themselves.

Authors' Preface

Between the solar photosphere and corona there is a distinctive layer some 20 000 km deep having properties vastly different from the regions above and below. This layer – the chromosphere – displays a number of exotic characteristics which have aroused a widespread interest not only among solar physicists but also astrophysicists generally. From the bottom to the top of this layer the temperature rises steeply from less than 5000 K to about 1 000 000 K, indicating the presence of an efficient non-thermal source of heating. Moreover, high-resolution observations at the limb reveal a fine structure consisting mainly of high-speed jets ('spicules'), ceaselessly moving upward and downward and lasting, on the average, some 5 minutes. The chromosphere plays a central role in the study of the Sun as a whole, not least because modern observations suggest close links with such diverse phenomena as the solar wind and the deep-lying convection zone responsible for the supergranulation.

However, its significance goes well beyond such purely specialist interests: an understanding of the *heating* of the chromosphere carries us into the realm of hydrodynamic and magnetohydrodynamic shock wave theory and, indeed, the chromosphere provides a potentially valuable testing-ground for many aspects of hydrodynamics and magnetohydrodynamics, particularly in cases where radiation losses must be taken into account. (Some of these topics are of great interest in certain fields of physics and geophysics and, as we shall see in Chapter 6, extensive cross-fertilization of ideas has already taken place.) At the same time, the chromosphere remains our principal avenue for exploring the effects on the processes of spectral line formation of deviations from local thermodynamic equilibrium (LTE). In summary, therefore, it is fair to claim that both the hydrodynamic and radiative aspects of the physics of the chromosphere lie close to the core of current developments in the wider realm of theoretical astrophysics.

The last decade or so has witnessed a vast expansion in both our theoretical understanding and observational knowledge of the chromosphere. Our aim in this book is to present a detailed and comprehensive account of all this new work and thereby attempt to build up a unified and consistent picture. The book is confined almost entirely to the 'quiet' chromosphere, i.e. to locations far enough removed from any active regions which may be present on the Sun to remain apparently uninfluenced. An understanding of the quiet chromosphere is an essential aid in unravelling the much more diverse and complex phenomena associated with active regions. In some instances, however, there are reasons for believing that the converse may be true. Accordingly, a short chapter (Chapter 5) is devoted to examining a few selected aspects of the fine structure of active regions which may make some contribution to our understanding of the quiet chromosphere.

After a brief historical introduction (Chapter 1), we start by describing the morphological, statistical, and dynamical properties of spicules observed beyond the solar limb (Chapter 2). They were first described in 1877 and for many decades their observation provided virtually the sole means of investigating the inhomogeneous structure of the chromosphere. In recent years, however, the application of high-resolution telescopes equipped with narrow-band filters, tunable over the Hα line of hydrogen, have enabled observations of spicules to be carried out on the disk. Here they appear as the upper portions of structures known as *dark mottles*, which are best visible in the wings of Hα. In the centre of Hα fine structures of another type also appear, called *bright mottles*. They are closely related to the dark mottles but are located deeper down in the chromosphere; modern observations suggest that they may coincide with tubes of magnetic flux protruding up from the photosphere.

Chapter 3 opens with a description of these disk structures and continues with an account of the *chromospheric network*. This is a coarse network having a cell size of some 30 000 km, which is outlined by the bright and dark Hα mottles but is also well visible in other lines of chromospheric origin. For example, it appears as a prominent bright network in the H and K lines of Ca II, and it was in these lines that, in 1892, the network was first discovered. In stellar spectra, the presence of emission in the centres of the H and K lines may be taken as evidence for the existence of a stellar chromosphere; moreover, it has been established that the widths of the central emission peaks bear a close relationship to the absolute magnitude of the star, independently of spectral type (the so-called Wilson–Bappu effect). On the Sun the network is found to be closely correlated with the downward currents of the supergranule flow and with the relatively weak magnetic fields to be found outside active regions, both measured at the photospheric level; it must be regarded as a phenomenon of fundamental importance. Chapter 3 concludes with an account of velocities and magnetic fields observed in the chromosphere.

Our present knowledge of physical conditions such as temperature, density, etc., in the chromosphere and their variation with height has been derived from an almost bewildering array of spectroscopic data, pertaining to both the continuum and line radiation of the chromosphere. The relevant observations span a huge range of wavelengths, running from centimetric radio wavelengths through the far infrared and visible regions into the extreme ultraviolet. The observational techniques that have been employed are equally diverse, and both eclipse and non-eclipse observations have contributed to the overall picture.

These topics form the subject matter of Chapter 4, whose aim is threefold: (*a*) to summarize the relevant continuum and spectral line data; (*b*) to outline the astrophysical techniques which have been employed to extract the desired physical information; and (*c*) to present and compare models describing the run of physical conditions with height derived by various workers. Much of our most reliable information about physical conditions has been derived from continuum data. Spectral line data have not been exploited to the same extent, mainly because of the difficulties encountered in applying non-LTE theory to realistic situations with the necessary rigour. Nevertheless, application of this theory has achieved some notable successes such as, for example, the determination of the physical conditions in spicules; moreover, certain types of spectral line data have a promising potential for future analyses.

Knowledge only of the physical *conditions* in the chromosphere is evidently inadequate, by itself, to elucidate the physical *processes* which are occurring. In fact, solutions of many of the outstanding problems of chromospheric physics require the application of hydrodynamic or magnetohydrodynamic theory. Foremost amongst these are: (*a*) the elucidation of the temperature rise, and (*b*) the origin of spicules. These problems form the main subject matter of the concluding chapter of the book, Chapter 7.

Our chief aim in Chapter 7 is to confront theory with the results of observation wherever this is both possible and fruitful. In a general way, the observed temperature rise can reasonably be accounted for on the basis of dissipation by *shock waves*. These owe their origin to gravity-modified acoustic waves which start their upward journey at the photospheric level and, as a consequence of the steep decrease in density, degenerate into shocks at heights of a few hundred kilometres. Calculation shows that by an appropriate choice of wave parameters the rate of dissipation of the assumed shocks as a function of height can be brought into agreement with that of the known net radiative energy loss from the chromosphere, at least at the lower levels where the radiative loss can be evaluated with reasonable accuracy. However, it must be stressed that conclusive observational evidence for the existence of such shocks, or of the acoustic waves which give rise to them, is still lacking. In the upper chromosphere a full-scale confrontation between observation and theory is not feasible at the present time.

The basic theoretical background required for our discussion of the problems of the real chromosphere in Chapter 7 is widely scattered. In fact, material of relevance can be unearthed from a large number of books and papers dealing with a great diversity of applications spread over several quite separate disciplines. In Chapter 6, accordingly, we give a systematic introduction, relevant to the theoretical problems of the chromosphere, to the theory of hydrodynamic and magnetohydrodynamic waves. The topics of basic concern are the propagation of waves of various types, their degeneration into shock waves, and the subsequent dissipation of the shocks as they proceed upwards through a gravitationally-stratified atmosphere. The subject is developed largely from first principles, our starting-point being the basic equations of magnetohydrodynamics. Although Chapter 6 is orientated toward the problems of the chromosphere, it is nevertheless self-contained and may be read separately from the rest of the book. The authors hope that it will prove useful not only to solar physicists but also to astrophysicists with a general interest in wave phenomena in stellar atmospheres.

Commonwealth Scientific and Industrial Research Organization
National Standards Laboratory
Sydney, Australia

R. J. Bray
R. E. Loughhead

Contents

Plates

Acknowledgements

The authors wish to thank Miss Joan M. Cook, Librarian of the National Standards Laboratory, and her staff for their whole-hearted cooperation, and Mrs S. Williams and Mr R. M. Rattle for their skilful preparation of the Figures and Plates respectively. We have been sustained in a multitude of ways by a number of our close colleagues, especially Mr E. J. Tappere, Mr J. G. Winter, and Mr D. G. Norton.

For supplying illustrative material we thank:

Dr R. B. Dunn, Sacramento Peak Observatory (Plate 2.3b)
Dr E. M. Reeves, Harvard College Observatory (Fig. 4.12)
Dr G. D. Sandlin, U.S. Naval Research Laboratory (Plate 4.1)
Dr D. Vrabec, The Aerospace Corporation (Plate 3.13).

For permission to reproduce copyright material we are indebted to the Editors of the following journals: *Annales d'Astrophysique, Annual Review of Astronomy and Astrophysics, Astronomy and Astrophysics, Australian Journal of Physics, Bulletin of the Astronomical Institutes of the Netherlands, Comments on Astrophysics and Space Physics, Memoirs of the Faculty of Science of Kyoto University, Publications of the Astronomical Society of Japan, Smithsonian Contributions to Astrophysics,* and *Solar Physics*. The right to reproduce certain material from *The Astrophysical Journal* has been purchased from the University of Chicago Press. For permission to make use of other illustrative matter we thank Drs J. M. Beckers, R. B. Dunn, and A. Title, the General Secretary of the International Astronomical Union, the Director of the U.S. National Bureau of Standards, the Director of Sproul Observatory, and Interscience Publishers.

Our special thanks go to Professor H. H. Plaskett, F.R.S., for his timely encouragement and to a number of friends and colleagues in the U.S.A. for the benefit of valuable discussions and correspondence.

Finally, we thank the CSIRO Executive and Dr R. G. Giovanelli, Chief of the Division of Physics, for permission to publish this book.

R.J.B.

R.E.L.

CHAPTER 1

Historical Introduction

1.1 The First Photographic Observations of the Chromosphere

The total solar eclipse of 1860 may be regarded as marking the beginning of the serious study of the chromosphere. At this eclipse, the chromosphere – together with the corona and numerous prominences – was successfully photographed for the first time. This notable achievement was due to the independent efforts of the Italian astronomer Father Angelo Secchi (1818–1878) and the well-known British amateur astronomer Warren de la Rue (1815–1889), working at two sites in Spain separated by some 250 miles.

In his book *Le Soleil* Secchi carefully listed conclusions derived from examination of both his own and de la Rue's photographs. He concluded, amongst other things, that 'Outre les protubérances, il existe une couche de la même matière, et qui enveloppe le Soleil de toutes parts. Les protubérances proviennent de cette couche: ce sont des masses qui se soulèvent au-dessus de la surface générale et s'en détachent même quelquefois. . . .'[1] According to the American astronomer C. A. Young (1834–1908), the name 'chromosphere' for this layer was first proposed in 1869 by two Englishmen, the solar physicist J. N. Lockyer (1836–1920) and the chemist E. Frankland (1825–1899). The term derives from the vivid red colour of the chromosphere, which is due to the emission from the brightest line in its spectrum, the Hα line of hydrogen.

Although the eclipse of 1860 yielded the first photographic observations, the chromosphere had been observed visually on earlier occasions. For example, at the eclipse of 1851 several observers noticed an extended 'pink arc' surrounding the eclipsed Sun [Secchi, 1875, Fig. 133]. Similar appearances had also been noted during the eclipse of 1842, but less distinctly.[2]

[1] Translated this reads: 'Apart from protuberances [i.e. prominences], there exists a layer of the same material which surrounds the entire Sun. The protuberances originate from this layer: these rise up above the general surface and sometimes even become detached. . . .' [Secchi, 1875, p. 386].

[2] The interested reader will find descriptions of earlier eclipses, dating back to 1706, in some of the books cited at the end of the chapter, including those of Grant [1852], Clerke [1887], Young [1895], and Berry [1898]. In many cases these visual observations evidently refer to prominences or to the inner corona, but in others the true chromosphere may well have been distinguished.

1.2 Early Spectroscopic Observations: the Discovery of Spicules

The 1860's witnessed a marked growth of interest among astronomers and physicists in applying the relatively new science of spectroscopy to the Sun. In 1861 and 1862 the famous German physicist Gustave Kirchhoff (1824–1887) published the first detailed map of the solar disk spectrum, derived from measurements with a four-prism spectroscope. This was followed in 1869 by a map based on measurements made with a diffraction grating by A. J. Ångström (1814–1874), in which wavelengths were expressed in terms of the metre. The familiar spectroscopic unit of wavelength, 10^{-10} m, was later named after Ångström. Together, the maps of Kirchhoff and Ångström formed a solid foundation for the spectroscopic work that was to follow.

A number of astronomers who were interested in applying spectroscopic methods to examining the nature of prominences eagerly awaited the eclipse of 18 August 1868. The French astronomer Pierre Jules Janssen (1824–1907) was among the successful observers. Janssen was greatly impressed by the brightness of the prominence emission lines revealed by his spectroscope and wondered whether they might not remain visible outside of an eclipse. The next day he tried the experiment and, to his delight, found that the same lines were easily visible. An identical idea had previously occurred to Lockyer, who was not present at the 1868 eclipse; he too – after earlier unsuccessful attempts with inadequate equipment – succeeded on 20 October 1868 in observing the prominences spectroscopically.

By widening the spectroscope slit observers found that the whole extent of a prominence could easily be observed in light of one of the brighter lines, usually Hα or Hβ. Moreover, by rotating the spectroscope around the limb, using a tangential rather than a radial slit, they soon demonstrated that the chromosphere entirely surrounds the photosphere. Lockyer's spectroscope is shown in Plate 1.1; it employed seven prisms of dense flint glass and its slit could be positioned radially or tangentially on any point of the limb of the solar image provided by a 6¼-inch refractor.

Following Janssen's and Lockyer's discovery, it was no longer necessary to wait for the fleeting and comparatively rare moments of a total eclipse to study the properties of prominences and the chromosphere. The new technique was soon put into use at a number of observatories but, not surprisingly, much more effort was devoted to the more easily-observed prominences than to the chromosphere itself. Nevertheless, during the 1870's and 1880's several observers measured the 'height' of the chromosphere, giving values ranging from 8–15$''$ of arc (5800–11 000 km). (Young in 1869 had estimated.the height as 7–18$''$.) In addition, several observers noted the irregular appearance of its apparent upper 'boundary' (cf. Plate 2.3a), which varied from day to day. Most descriptions of the structure of the chromosphere observed with a tangential slit at this period are rather vague by modern standards. However, there are several references to little 'flames' projecting upwards from the chromospheric boundary which appeared to be distinguished from prominences by their smaller height and brightness. These could have been simply small prominences or, alternatively, larger examples of the *spicules* described in Chapter 2.

On the other hand, the clear account given by Secchi in 1877 leaves no doubt that this experienced observer did indeed recognize the spicules as distinct features in their own right. Plate 1.2 reproduces some of his drawings, based on observations made with a high magnification. Secchi describes the fine structure of the chromosphere portrayed in these drawings in the following terms: '[La chromosphère] est presque toujours recouverte de petits filets terminés en pointe et tout à fait semblables à des *poils*; on voit quelquefois des étendues considérables où ces filets sont tous inclinés dans le même sens (Fig. 1) [see our Plate 1.2]; souvent aussi ils sont inclinés dans des directions contraires, soit convergentes (Fig. 2), soit divergentes (Fig.3). . . . Les petits filets analogues à des *poils* se remarquent surtout dans les régions de calme. . .'[3] Secchi found the widths of the spicules to be only 100 or 200 km, whereas we now know that the true value is ~800 km. Nevertheless, the close resemblance between Secchi's drawings and modern high-resolution photographs of spicules (compare Plates 1.2 and 2.3b) bears striking testimony to the quality of his instruments and to his skill as an observer.

1.3 Early Spectroscopic Observations: the Flash Spectrum

The next major advance in the study of the chromosphere again resulted from observations at a total eclipse, this time the eclipse of 1870. Following the eclipse of 1868, the brighter lines of the chromospheric emission spectrum – due mainly to hydrogen, helium, and ionized calcium (cf. Table 4.8) – had been observed *outside* of an eclipse by a number of workers. However, any extensive emission line spectrum closely mimicking, in reverse, the absorption line spectrum of the photosphere had so far escaped detection. C. A. Young had looked for such a spectrum at the 1869 eclipse, but had been unsuccessful. At the 1870 eclipse he decided to use a tangential instead of a radial slit, which he carefully positioned at the point on the limb where the last ray of sunlight would disappear. When the moment of totality arrived, he was rewarded by the

[3] Translated this reads: '[The chromosphere] is nearly always covered by little filaments ending in a point and very similar to *hairs*; one sometimes sees large regions where these filaments are all inclined in the same sense (Fig. 1) [see our Plate 1.2]; frequently they are inclined in contrary directions, either convergent (Fig. 2), or divergent (Fig. 3). . . . The tiny filaments analogous to *hairs* are particularly noticeable in quiet regions. . . .' [Secchi, 1877, pp. 34–35].

sudden appearance ('flash') of numerous emission lines which, as far as he could judge, corresponded in position with the dark lines of the photospheric spectrum. The phenomenon observed by Young lasted only for a few seconds since the advancing Moon rapidly covered the lowest layers of the chromosphere, where the flash spectrum lines are brightest.

It was soon found that, under very good observing conditions, many of the emission lines were observable outside of an eclipse, particularly in the neighbourhood of active regions. Thus in a special study lasting several weeks made in 1872 Young was able to detect, in all, nearly 300 lines. In recent times, by comparison, no less than 11 500 lines have been catalogued from observations made outside of an eclipse (Section 4.3.2)! Nevertheless, especially when photometric measurements are required, eclipses remain the best method for observing the spectrum of the chromosphere.[4]

The flash spectrum was first successfully photographed at the 1883 eclipse, but only a few of the brighter lines were recorded. However, at the eclipse of 1898 the British astronomer J. Evershed (1864–1956), using a (slitless) prismatic camera of only 2¼-inch aperture, succeeded in photographing 313 lines extending from $\lambda3340$ to $\lambda6000$, including 28 lines of the Balmer series of hydrogen. In a paper published in 1901, Evershed carefully discussed differences and similarities between the flash spectrum and the photospheric spectrum. In addition to the emission lines, Evershed noticed in the spectrum of prominences (but not, apparently, in that of the chromosphere itself), a *continuum* extending from near the end of the hydrogen series into the ultraviolet. This was correctly identified by Evershed as due to hydrogen and is now known as the Balmer continuum.

The Balmer continuum of the chromosphere itself, together with the electron scattering continuum observable in the visible region of the spectrum, has been photographed at numerous subsequent eclipses. As we shall see in Chapter 4, owing to their relatively straightforward theoretical interpretation, these continua provide a sound basis for determining physical conditions in the chromosphere and, in this respect, they have so far proved more useful than the emission lines of the flash spectrum.

Following the eclipse of 1898, photography of the flash spectrum formed a major part of the observing programmes of numerous subsequent eclipse expeditions. Detailed accounts of the methods used and of the results obtained are to be found in the books by Dyson and Woolley and by Mitchell, and in the review article by van de Hulst, cited at the end of the chapter.

[4] In this book we shall follow modern usage in employing the term 'flash spectrum' to indicate the emission spectrum of the chromosphere observed at a total eclipse, regardless of the height to which the observations refer. Some writers restrict the term to the spectrum of the very lowest layers, e.g. below 1000 km.

1.4 Attempts to Explain Observations of the Chromosphere, 1860–1900

The four decades spanning the period between the first successful photography of the chromosphere in 1860 and the end of the century saw little real advance in understanding its nature and significance. It is true that by 1900 considerable progress had been made in cataloguing its spectral lines, many of which had been correctly identified as belonging to known terrestrial elements. Moreover, as early as the end of the 1860's it had become generally recognized that the entire Sun is gaseous, including the photosphere. That the chromosphere and prominences are also gaseous had been shown by their spectra of bright lines, similar to those of the laboratory spectrum of an incandescent gas and in agreement with Kirchhoff's laws. Finally, on the basis of an examination of hydrogen spectra in the laboratory carried out with the aid of Frankland, Lockyer had concluded in 1869 that the gas pressure in the chromosphere was far smaller than that in the Earth's atmosphere; however, he did not attempt to give numerical values.

Lockyer must be given credit for realizing the importance of a parallel study of solar and laboratory spectra, a policy later emulated with great success by Hale at the Mt. Wilson Observatory. On the other hand, his ideas on the nature of the chromosphere (often obscurely expressed) were strongly coloured by two explicit assumptions now known to be erroneous, namely, that (*a*) outwards from the photosphere, the temperature of the solar gases continuously decreases; and (*b*) the height attained in the atmosphere by any given element depends on its atomic weight. Similar assumptions were implicit in the interpretations of other leading workers of the time [see, for example, Young, 1895, pp. 334, 341]. Assumption (*a*) does not appear to have been based on specific observations. Assumption (*b*) arose from observations of the chromospheric spectrum (both inside and outside of an eclipse), which appeared to show that whereas hydrogen (atomic weight = 1) extended to great heights, the heavier elements were confined to the lower layers. One difficulty facing this assumption was early recognized, namely that the H and K lines of calcium (atomic weight = 40) extended to approximately the same height as the hydrogen lines. This anomaly was rather unsatisfactorily resolved by assuming that in the chromosphere the calcium atoms split up into some much lighter dissociation products.

We now know, of course, that differences in the distribution of emission with height among the various lines are not a true indication of any actual stratification of the elements involved. In fact, as we shall see in Section 4.3.2, for any selected line the volume emission at any given height depends not only on the total abundance of the element in question but also on (*a*) the intrinsic strength of the transition; and (*b*) the conditions of excitation and ionization, which depend in turn on the temperature and on the gas and electron densities.

Accordingly, with historical hindsight we are now able to see that the failure of the early workers to interpret properly the chromospheric spectrum can be attributed mainly to the absence of an adequate understanding of atomic spectra in general and of the laws governing the ionization and excitation of atoms in particular. Nearly three decades would elapse before advances in atomic physics made this information available.

1.5 Invention of the Spectroheliograph; Discovery of the Chromospheric Network

During the latter part of the nineteenth century a number of astronomers had considered possible ways of eliminating some of the inconveniences of the open-slit technique of observing prominences outside of an eclipse (Section 1.2). However, it was not until 1892 that the first successful spectroheliograph was constructed by the famous American astronomer George Ellery Hale (1868–1938), who later became the founder and first director of the Mt. Wilson Observatory. Hale's instrument was attached to the 12-inch refractor of the Kenwood Observatory near Chicago, a private observatory financed by his father which that year became part of the University of Chicago. Both the solar image and the photographic plate were fixed with respect to the telescope, while the first and second slits were made to move across them by means of a system of levers actuated by hydraulic power. Using the H and K lines of calcium, Hale was immediately successful in photographing not only prominences beyond the limb but also chromospheric faculae on the disk, together with – to quote his own words – '. . . a reticulated structure extending over the entire surface of the Sun.'[5]

The year 1892 thus marks both the invention of the spectroheliograph and the discovery of the chromospheric network, a quiet Sun phenomenon now known to be of fundamental importance (see Chapter 3). From then on it was possible to study, almost at will, both the quiet and active chromosphere on the disk.

The Kenwood telescope produced an image only 2 inches in diameter and the spectroheliograph itself was not well adapted to photographing the Sun's disk in lines other than the calcium H and K lines, for example the hydrogen lines. Accordingly, Hale decided to build a more powerful instrument designed to utilize the 7-inch image provided by the 40-inch refractor of the newly-established Yerkes Observatory. This instrument, the famous Rumford spectroheliograph, was completed in 1899. It was immediately put to good use in investigating, amongst other things, the structure of the disk chromosphere observed in various

[5] Illustrations of the Kenwood spectroheliograph have been published by Wright [1966, following p. 112] and Young [1895, Fig. 76]; the latter also reproduces spectroheliograms obtained with this instrument (see his Figs. 31*, 77–79).

parts of the H and K lines (cf. Plate 1.3), but it was not until 1903 that the first satisfactory photographs were obtained in the hydrogen lines Hβ, Hγ, and Hδ. For this work the spectroheliograph was equipped with two prisms and a plane grating. These observations resulted in the discovery of the *dark* chromospheric network, which is seen at all wavelengths in the lines Hβ–Hδ (cf. Section 3.3.4) and also in the wings, but not the core, of Hα. However, at this period it was not possible to observe in the Hα line since red-sensitive photographic plates had not yet become available.

Within a short time of Hale's first successful observations with the Kenwood spectroheliograph in 1892, other spectroheliographs were constructed by H. A. Deslandres (1853–1948) at Meudon, near Paris, and by Evershed in England. However, priority for the invention belongs to Hale who, writing to the English astronomer H. F. Newall in 1933, remarked: 'So far as I have ever known, Deslandres' first reference to the principle was in the *Comptes Rendus* of August 17, 1891, in which he mentions my papers. But he did not build and use a spectroheliograph until the summer of 1893. Meanwhile, Evershed had built one and confirmed my results.'[6]

After transferring the scene of his activities to Mt. Wilson in 1903–1904, Hale set out to design and construct a new 5-foot spectroheliograph which would incorporate all his extensive experience with the two earlier instruments. The new spectroheliograph, fed by the Snow horizontal telescope, was completed in 1905 and the following year was described by Hale in a joint paper with his colleague Ferdinand Ellerman (1869–1940). It was an immediate success, so much so that even by modern standards some of Hale's 1906 spectroheliograms are of excellent quality. (This is also true of some of the photographs taken with the earlier Rumford spectroheliograph, such as Plate 1.3.)

At this time red-sensitive photographic plates were not available commercially, but experiments carried out at Yerkes with emulsions treated with certain organic dyes had shown some success. In 1908, using this techinique at Mt. Wilson, Hale succeeded in photographing the disk in Hα for the first time, producing, in his own words, 'by far the most interesting photographs' he had ever taken.[7] The following year Hα spectroheliograms were also obtained at Meudon.

Much of the early work with the 5-foot and the later 13-foot spectroheliographs at Mt. Wilson was concerned with the *active* Sun, both on the disk and beyond the limb; this was also true of the large spectroheliograph completed at Meudon by Deslandres in 1909. However, in a paper published in 1916 entitled

[6] Quoted by Wright [1966, p. 85].

[7] One of Hale's 1908 Hα photographs, taken with the 5-foot spectroheliograph, has been reproduced by Abetti [1929, Fig. 68].

'The Minute Structure of the Solar Atmosphere' Hale and Ellerman described the appearance of the bright calcium flocculi visible in the H and K lines and the dark mottles visible in Hα. The smallest dark mottles they estimated to be only 2″ of arc (1450 km) in diameter, in good agreement with modern measurements (cf. Table 3.2). The French astronomer L. d'Azambuja (1884–1970) also showed interest in the structure of the quiet chromosphere and, during the years 1926–1928, succeeded in obtaining at Meudon an extensive series of spectroheliograms in a number of lines besides H and K and the hydrogen lines.

In modern times the spectroheliograph – as we shall see in Chapter 3 – has continued to play a vital role not only in photographing the disk chromosphere in various lines but also (with certain ingenious modifications) in measuring velocities and magnetic fields. On the other hand, it lacks the spatial resolution and speed of operation mandatory for properly observing such small, short-lived structures as the bright and dark mottles on the disk or the spicules at the limb. Further understanding of these basic chromospheric fine structures had to await the invention of the birefringent filter (Section 1.7) and the development of solar telescopes specifically designed to achieve high spatial resolution.

1.6 Progress in Theoretical Interpretation of the Chromosphere, 1920–1950

An important stage in the understanding of the flash spectrum was reached in 1920 when the Indian physicist M. N. Saha (1893–1956) published his well-known theory of gaseous ionization and applied it to the chromosphere. Saha showed that, in addition to temperature, pressure exercised a controlling influence on the degree of ionization of a gas. The prominence of lines of ionized metals ('enhanced' lines) in the flash spectrum compared with the Fraunhofer spectrum, which had been well known for a long time, thus received an explanation in terms of the lower pressure of the chromospheric layers. At this period, it should be noted, no one had envisaged the possibility of a large increase of *temperature* with height in the Sun's atmosphere.

However, a major difficulty remained. Many of the emission lines were observed to extend to heights h greater than 10 000 km, for example the Balmer and Paschen series of hydrogen, the H and K lines of Ca II, and several lines of He I and He II. How could these great heights be accounted for by an atmosphere in hydrostatic equilibrium at a temperature of the order of the 'boundary' temperature of the Sun? The density distribution in such an isothermal atmosphere follows the well-known law

$$\rho = \rho_0 \exp(-\beta^* h),$$

where $1/\beta^*$, the density scale height, is given by the expression

$$\frac{1}{\beta^*} = \frac{kT}{mg},$$

k being Boltzmann's constant, g the acceleration due to gravity, and m the atomic mass. Following the spirit of similar calculations carried out during the 1920's and 1930's, let us assume that the atmosphere consists entirely of neutral hydrogen and that $T = 5000$ K. Then we obtain a value for $1/\beta^*$ of only 150 km.

The practice was then to compare the calculated density scale height $1/\beta^*$ with the *emission* scale heights $1/\beta$ of a number of flash spectrum lines photographed at various eclipses. In making the comparison most authors, either explicitly or implicitly, identified emission and density scale heights with one another although, of course, this identification ignores the variation of excitation and ionization with height. Since the observed scale heights, particularly for hydrogen and ionized calcium, were much larger than the density scale height calculated for an isothermal hydrogen atmosphere, it was concluded that some mechanism for supporting the chromosphere was required.[8]

In 1924 the English astrophysicist E. A. Milne (1896–1950) proposed selective radiation pressure acting on the Ca II ions as a support mechanism and went on to construct a detailed theory.[9] These particular ions were chosen because their resonance lines – the H and K lines – lie in a fairly intense part of the solar spectrum; by contrast, the resonance lines of hydrogen (the Lyman series) lie in the far ultraviolet, where the radiant intensity is much weaker. One obvious difficulty facing Milne's theory was that it was too selective: in order to account for the support of atoms and ions other than ionized calcium, e.g. hydrogen, it would have been necessary to introduce viscous drag. In general, the theory gained little acceptance among the solar physicists of the day, at least in so far as its application to the solar chromosphere was concerned.

As an alternative mechanism, W. H. McCrea proposed in 1929 that the extra support needed was supplied by the turbulent motions of certain supposed volume elements of the chromospheric gases. The volume elements were not identified with any known chromospheric features but were supposed to possess a Maxwellian distribution of velocities characterized by a mean squared velocity U^2. This resulted in an increase in gas pressure from $p = N_H kT$ to an 'effective' value $p^* = N_H kT^*$, where

$$kT^* = kT + \frac{1}{3} m_H U^2.$$

Here, m_H is the mass of the hydrogen atom and N_H the number density of neutral hydrogen, assumed to be the only constituent. McCrea took U to be 18.4 km sec^{-1} on the basis of observations of the emission profiles of the H and K

[8] Modern values of emission scale heights are listed in Table 4.9.

[9] For references to Milne's series of papers, see Menzel [1931, p. 284] and Rosseland [1936, p. 269].

lines; he assumed T to be 5000 K. These figures yield an 'effective' temperature T^* of 18 700 K and a density scale height of ~560 km, i.e. nearly four times larger than in the absence of the McCrea mechanism and, accordingly, in better agreement with the observed emission scale heights.

McCrea's theory is of historical importance in that it represented the first attempt to introduce a *hydrodynamic* mechanism as a means of explaining the chromosphere's apparently anomalous extension in height. However, during the 1930's and early 1940's it gradually became apparent that the assumption of a constant temperature in the chromosphere of only 5000–6000 K could no longer be sustained in the face of mounting evidence for a very high temperature in the corona, of the order of a million degrees. Accordingly, by the late 1940's the problem of the *support* of the chromosphere can be said to have vanished, to be replaced by the problem of explaining the outward increase of temperature with height in the chromosphere and corona. This led a number of investigators to consider possible non-radiative heating mechanisms: these and subsequent developments in the theory of chromospheric heating are described in Chapter 7.

Meanwhile, the radiative aspects of the chromosphere were receiving an increasing amount of attention. In 1931 the American astronomer D. H. Menzel published the first detailed discussion to be thoroughly based on the quantum theory of the atom and on Saha's ionization theory. In Menzel's own words at the time: 'Physicists and astronomers are becoming "atom-minded".' The observational basis for Menzel's study was a series of 'moving-plate' spectrograms obtained by W. W. Campbell (1862–1938) of Lick Observatory at the eclipses of 1898, 1900, 1905, and 1908. The data were not adequate to allow Menzel to derive physical conditions as a function of height, but he was able to estimate the gas pressure at the base of the chromosphere. Menzel gave full recognition to possible and actual deviations from local thermodynamic equilibrium (cf. Section 4.3.2) and in this respect his paper was ahead of its time.

In 1935 the same author, in collaboration with G. G. Cillié, derived the electron density N_e in the low chromosphere from measurements of the Balmer continuum photographed at the 1932 eclipse. As we shall see in Chapter 4, the Balmer continuum, together with the electron scattering continuum in the visible region of the spectrum, has ever since proved to be of major importance in deriving reliable values for the physical conditions in the chromosphere.

Cillié and Menzel derived N_e only for a single height in the low chromosphere. It was not until 1950 that the Australian radio astronomer J. H. Piddington published for the first time a model expressing N_e and T as functions of height over a large range of heights. Piddington's model covered the region $h = 1500$ to $h = 16\,000$ km and was based largely on radio observations of the quiet Sun in the frequency range 600–24 000 MHz (50–1.25 cm) although use was also made of eclipse observations of the Balmer lines and continuum. In the same year, two pioneers of solar physics in Australia, C. W. Allen and R.v.d.R. Woolley, also published a chromospheric model based partly on radio data. Both models are in reasonable agreement with modern models in spite of the inadequacies of the radio data available at the time.

The year 1950 can therefore be taken as marking the beginning of the modern era in the search for a reliable knowledge of the physical conditions in the chromosphere. From that date to the present an enormous effort has been applied to this particular aspect of the physics of the chromosphere, using both continuum and spectral line data from widely-separated regions of the electromagnetic spectrum. Both the methods used and the results obtained are described at some length in Chapter 4.

1.7 Invention of the Birefringent Filter: Beginning of the Modern Era of High-Resolution Chromospheric Observations

In a short paper published in 1933 the French solar physicist Bernard Lyot (1897–1952) outlined constructional details of a new type of optical filter utilizing the phenomenon of birefringence. In the years ahead this device was destined to revolutionize observation of the fine detail of the chromosphere, both on the disk and at the limb, supplanting for these and several other applications the long-established spectroheliograph. Lyot proceeded to manufacture some of the optical elements according to the principles laid down in his paper. However, before finishing a complete filter he was brought to a halt by the difficulty of obtaining calcite crystals of adequate size and optical quality needed for the polarizers. Meanwhile, in 1937 the Swedish astronomer Y. Öhman took advantage of the circumstance that polaroid had recently become available to construct a simple Hα filter having a bandwidth of ~50 Å.

Öhman's filter represented the first concrete realization of a birefringent filter, but its bandwidth was far too large for serious observation of the chromosphere whether at the limb or on the disk. However, the following year Lyot, also using polaroid, constructed a temperature-controlled filter having a bandwidth of only 3 Å. In 1939 this filter was installed at the Pic-du-Midi Observatory, where it was used to photograph prominences in Hα and also, for the first time, the corona in the lines $\lambda\lambda$5303 and 6374. Further improvements by Lyot culminated in 1942 in a filter having a bandwidth of only 1.5 Å, which was successfully employed for cinematographic observations in Hα of both the limb and the disk.

One of the first fruits of the new development was improved observations of spicules, which had hitherto been photographed only during eclipses (cf. Section 2.3.1). Lyot himself noted in 1944 the presence on his 1942 films of chromospheric 'jets' having a lifetime of only a few minutes. However, the detailed study of spicules was inaugurated in 1943 by W. O. Roberts at Climax, Colorado, using a 4 Å filter designed and built by J. W. Evans. Roberts determined several of their basic properties and was the first to employ the term 'spicule', which derives from the Latin word *spiculum,* a dart. Later developments in the observation of spicules and other fine structures beyond the limb form the subject matter of the following chapter.

Lyot's films of 1942 were taken with a filter of 1.5 Å bandwidth, a value quite narrow enough for limb observations of spicules and prominences. On the other hand, such a bandwidth is too broad for the effective study of such disk features as the bright and dark mottles of the quiet chromosphere, although Lyot's films were successfully employed in an investigation of the fibrils in the neighbourhood of active regions (Section 5.2.2). However, by the late 1940's the Paris firm Optique et Précision de Levallois had undertaken – with Lyot's aid – the commercial production of birefringent filters having a bandwidth of only 0.65 Å. By the 1950's these filters had become quite widely distributed and in 1957 the German astronomer K. O. Kiepenheuer published the first systematic description of the chromospheric fine structure to be based on observations obtained with a birefringent filter. Observing at the Fraunhofer Institute's Observatory on the island of Capri, Kiepenheuer achieved a spatial resolution of 2–3″ of arc.

Subsequent years have seen the development of fully-tunable filters having bandwidths as small as 1/8 Å and the construction of specialized chromospheric telescopes capable of resolving better than 1″ of arc. These instrumental improvements have resulted in a large increase in our knowledge of the fine structure of the disk chromosphere and its relationship to features observed beyond the limb, an account of which is given in Chapter 3.

1.8 Chronological Summary

1860	Secchi and de la Rue successfully photograph the chromosphere during a total solar eclipse.
1868	Janssen and Lockyer observe prominences outside of eclipse.
1869	Open-slit method extensively applied to observe chromosphere outside of eclipse.
1869	Lockyer and Frankland propose name 'chromosphere'.
1869	Young measures apparent height of chromosphere.
1870	Young observes flash spectrum of low chromosphere during a total eclipse.
1872	Young observes numerous flash spectrum lines outside of eclipse.
1877	Secchi describes spicules in his book *Le Soleil.*
1883	Flash spectrum photographed during a total eclipse.
1892	Hale constructs first successful spectroheliograph at Kenwood Observatory, Chicago.
1892	Hale discovers chromospheric network on spectroheliograms taken in the H and K lines of Ca II.

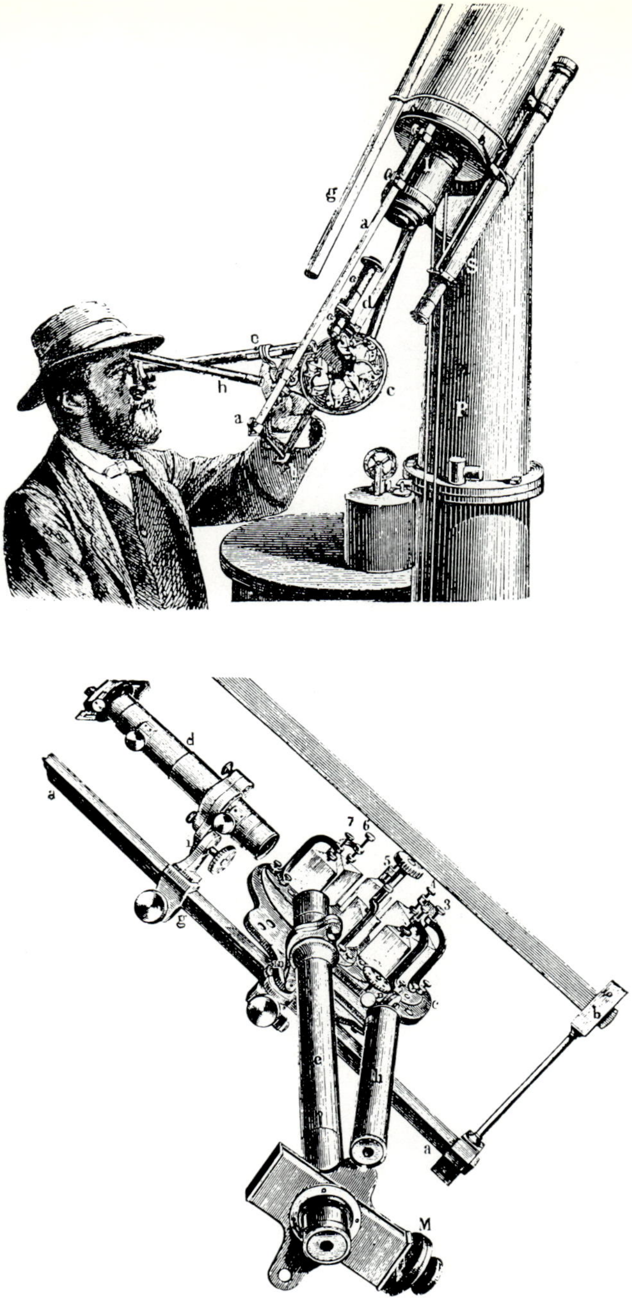

PLATE 1.1. Spectroscope built in 1868 for J. N. Lockyer by Browning of London for observation of prominences and the chromosphere outside of an eclipse. d is the collimator, e the viewing telescope, and h a small telescope for reading the prism assembly micrometre scale.

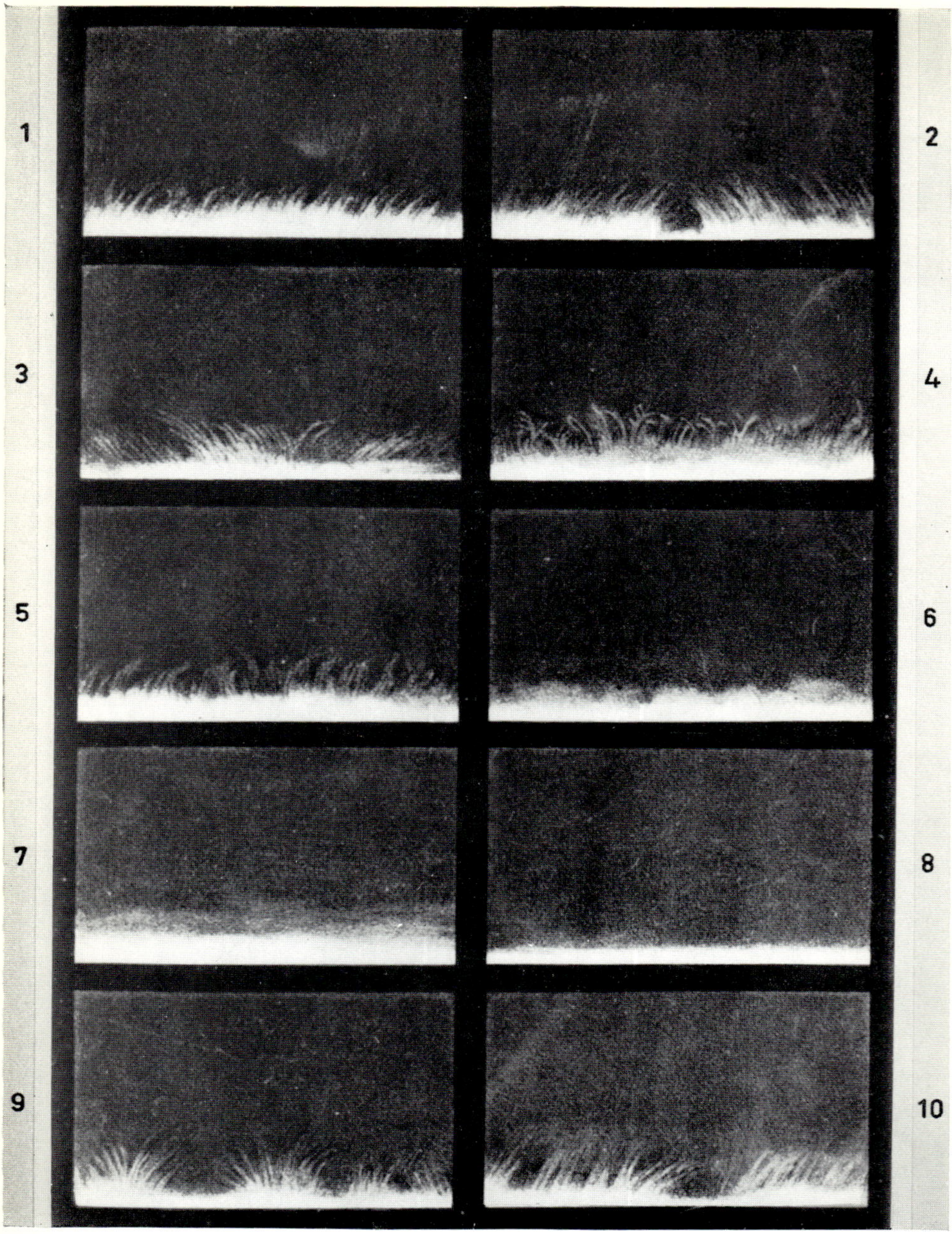

PLATE I.2. Secchi's drawings of spicules [Secchi, 1877, Plate A]. For description, see text.

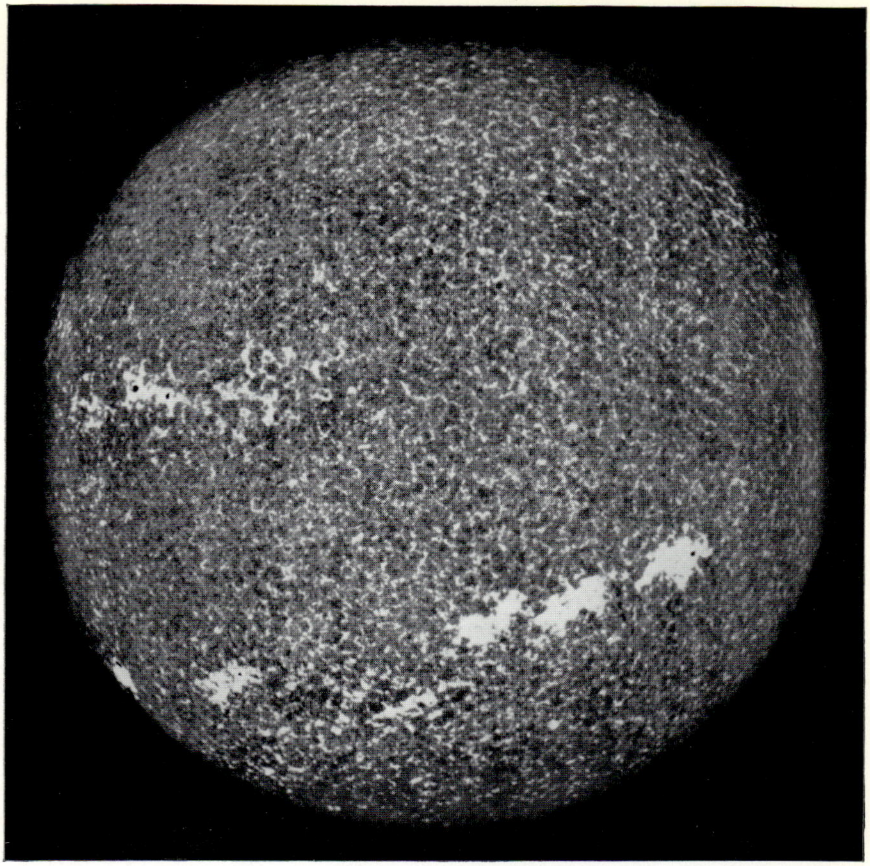

PLATE 1.3. Chromospheric network photographed in the H_2 portion of the H line of Ca II. This outstanding photograph was taken on 12 August 1903 by Hale and Ellerman with the Rumford spectroheliograph and gives a good representation of the bright flocculi which outline the network.

1898 313 lines in flash spectrum photographed by Evershed during a total eclipse.

1899 Completion of Rumford spectroheliograph at Yerkes Observatory.

1901 Evershed identifies Balmer continuum in spectrum of prominences.

1903 Hale obtains spectroheliograms in $H\beta$, $H\gamma$, and $H\delta$.

1905 Completion of 5-foot spectroheliograph at Mt. Wilson.

1908 Hale obtains spectroheliograms in $H\alpha$.

1909 Completion of large spectroheliograph at Meudon.

1916 Hale and Ellerman describe bright calcium flocculi and dark $H\alpha$ mottles.

1920 Saha publishes his ionization theory and applies it to the chromosphere.

1924 Milne formulates a theory of chromospheric support based on selective radiation pressure.

1929 McCrea proposes a theory of support based on 'turbulent' motions.

1930 Publication of d'Azambuja's collection of spectroheliograms obtained in various lines.

1931 Menzel publishes detailed discussion of radiative aspects of chromosphere.

1933 Lyot outlines design of a new type of optical filter.

1935 Cillié and Menzel derive electron density in low chromosphere from measurements of the Balmer continuum.

1942 Lyot obtains cinematographic observations of disk and limb chromosphere using a 1.5Å filter.

1945 The term 'spicule' introduced by Roberts.

1946 L. Biermann suggests shock waves as a mechanism for heating the chromosphere (cf. Section 7.3.1).

1949 E. Schatzman publishes a theory of shock wave heating of chromosphere and corona (cf. Section 7.4.1).

1950 Models of chromospheric physical conditions published by Piddington and by Woolley and Allen.

REFERENCES

ABETTI, G. [1929] 'Solar physics', *Handbuch der Astrophysik*, ed. G. Eberhard, A. Kohlschütter, and H. Ludendorff, vol. 4, p. 57; Springer, Berlin.

ATHAY, R. G. [1964] 'The challenge of chromospheric physics', *Science* **143**, 1129.

D'AZAMBUJA, L. [1930] 'Recherches sur la structure de la chromosphère solaire', *Ann. Obs. Paris-Meudon* 8, Part 2.

BERRY, A. [1898] *A Short History of Astronomy*, John Murray, London.

CILLIÉ, G. G., and MENZEL, D. H. [1935] 'The physical state of the solar chromosphere', *Harvard College Observatory Circular* No. 410.

CLERKE, A. M. [1887] *A Popular History of Astronomy during the Nineteenth Century*, 2nd Ed., A. and C. Black, Edinburgh.

DUNN, R. B. [1960] 'Photometry of the solar chromosphere', Thesis: Harvard University.

DYSON, F., and WOOLLEY, R. v. d. R. [1937] *Eclipses of the Sun and Moon*, Clarendon Press, Oxford.

EVERSHED, J. [1901] 'Wave-length determinations and general results obtained from a detailed examination of spectra photographed at the solar eclipse of January 22, 1898', *Proc. Roy. Soc.* **68**, 6.

GRANT, R. [1852] *History of Physical Astronomy,* Robert Baldwin, London.

HALE, G. E. [1892] 'Solar photography at the Kenwood Astro-physical Observatory', *Astronomy and Astro-physics* **11**, 407.

HALE, G. E., and ELLERMAN, F. [1903] 'The Rumford spectroheliograph of the Yerkes Observatory', *Publ. Yerkes Obs.* **3**(1), p. 3.

HALE, G. E., and ELLERMAN, F. [1906] 'The five-foot spectroheliograph of the solar observatory', *Astrophys. J.* **23**, 54.

HALE, G. E., and ELLERMAN, F. [1910] 'On the nature of the hydrogen flocculi and their structure at different levels in the solar atmosphere', *Proc. Roy. Soc.* A **83**, 177.

HALE, G. E., and ELLERMAN, F. [1916] 'The minute structure of the solar atmosphere', *Proc. Nat. Acad. Sci.* **2**, 102.

HULST, H. C. VAN DE [1953] 'The chromosphere and the corona', *The Sun,* ed. G. P. Kuiper, p. 207, Univ. Chicago Press.

KIEPENHEUER, K. O. [1957] 'Erfahrungen mit einem Lyotschen Hα-Filter sowie vorläufige Ergebnisse über die Struktur der Chromosphäre', *Z. Astrophys.* **42**, 209.

LINSKY, J. L., and AVRETT, E. H. [1970] 'The solar H and K lines', *Publ. Astron. Soc. Pac.* **82**, 169.

LYOT, B. [1933] 'Un monochromateur à grand champ utilisant les interférences en lumière polarisée', *C. R. Acad. Sci.* **197**, 1593.

LYOT, B. [1944] 'Le filtre monochromatique polarisant et ses applications en physique solaire', *Ann. Astrophys.* **7**, 31.

McCREA, W. H. [1929] 'The mechanics of the chromosphere', *Mon. Not. R.A.S.* **89**, 718.

MEADOWS, A. J. [1970a] *Early Solar Physics*, Pergamon, Oxford.

MEADOWS, A. J. [1970b] 'Lockyer as astronomer', *Nature* **225**, 230.

MENZEL, D. H. [1931] 'A study of the solar chromosphere', *Publ. Lick Obs.* **17**, Part 1.

MITCHELL, S. A. [1951] *Eclipses of the Sun,* 5th Ed., Columbia Univ. Press, New York.

ÖHMAN, Y. [1938] 'A new monochromator', *Nature* **141**, pp. 157, 291.

PAGEL, B. E. J. [1964] 'The structure of the solar chromosphere', *Ann. Rev. Astron. Astrophys.* **2**, 267.

PIDDINGTON, J. H. [1950] 'The derivation of a model solar chromosphere from radio data', *Proc. Roy. Soc.* A **203**, 417.

ROBERTS, W. O. [1945] 'A preliminary report on chromospheric spicules of extremely short lifetime', *Astrophys. J.* **101**, 136.

ROSSELAND, S. [1936] *Theoretical Astrophysics*, Clarendon Press, Oxford.

SAHA, M. N. [1920] 'Ionization in the solar chromosphere', *Phil. Mag.* **40**, 6th Series, p. 472.

SCHELLEN, H. [1872] *Spectrum Analysis*, Longmans, Green, & Co., London.

SECCHI, A. [1875–1877] *Le Soleil,* 2nd Ed., Gauthier-Villars, Paris.

WAYMAN, P. A. [1955] 'Optical systems employed in spectroheliographs', *Vistas in Astronomy,* ed. A. Beer, vol. 1, p. 422; Pergamon, London.

WOOLLEY, R. v. d. R., and ALLEN, C. W. [1950] 'Ultra-violet emission from the chromosphere', *Mon. Not. R.A.S.* **110**, 358.

WOOLLEY, R. v. d. R., and STIBBS, D. W. N. [1953] *The Outer Layers of a Star,* Clarendon Press, Oxford.

WRIGHT, H. [1966] *Explorer of the Universe – a Biography of George Ellery Hale,* Dutton, New York.

YOUNG, C. A. [1895] *The Sun,* Kegan Paul, London.

ZIRIN, H. [1968] 'George Ellery Hale, 1868–1938', *Solar Phys.* **5**, 435.

CHAPTER 2

Spicules and Other Fine Structures at the Solar Limb

2.1 Introduction

Ever since their discovery by Father Secchi (cf. Chapter 1), spicules have been a source of fascination to solar physicists and much effort has been devoted to their study. Yet, owing to their low brightness and narrow width, they remain among the most difficult of all solar fine structures to observe, while the elucidation of their origin continues to pose a major challenge to theoreticians. Until recently spicules were the only fine structures of the quiet chromosphere observed beyond the limb whose existence was known; nevertheless, by themselves, they have long borne graphic testimony to the inhomogeneous nature of the chromosphere. Moreover, the observation of ascending and descending motions in spicules, frequently faster than the local velocity of sound, focused attention on the probable existence of powerful hydrodynamical forces in the chromosphere. Through the agency of spicules these forces appear to convey large amounts of material from the low chromosphere to the base of the corona and, in so doing, may exert an important influence on the overall mass balance in the outer parts of the Sun's atmosphere.

It is appropriate, therefore, that we begin our account of the structure of the chromosphere with a chapter largely devoted to spicules and their properties. We begin in Section 2.3 with a consideration of the morphological properties of spicules, dealing successively with their appearance on photographs taken at different wavelengths in the Hα line, width, magnitude of Hα emission, height, abundance, inclination and clustering, and dependence on latitude and phase of solar cycle. Then, after a short digression to outline the appearance of spicular detail on spectrograms taken in Hα and other strong chromospheric emission lines (Section 2.3.9), we turn to a discussion of their dynamical properties (Section 2.4). These include lifetime and evolution, vertical and line-of-sight velocities, and possible internal motions; the question of the possible existence of spicule magnetic fields is also raised. Finally, we summarize in Section 2.5 the known quantitative data concerning the morphological and dynamical properties of spicules.

Several important topics relating to spicules are reserved for later chapters. The first is the important question of the identification of spicules with features observed on the disk. This question can be answered only after a detailed consideration and intercomparison of the properties of all the basic fine structures visible both on the disk and beyond the limb and is discussed in Chapter 3, Section 3.2.6. Secondly, a knowledge of the physical conditions in spicules, derived from measurements of the intensities and profiles of spicule emission lines, is essential for the construction of inhomogeneous models of the quiet chromosphere; both these topics are treated at length in Chapter 4. Finally, current theories of the origin of spicules are outlined in Chapter 7.

In 1969 observations made with the 12-inch refractor of the CSIRO Solar Observatory, Culgoora, revealed for the first time the presence at the centre of Hα of small bright features in the low chromosphere beyond the limb. These were found to show a close correspondence in properties with the *bright mottles* observed on the disk (Section 3.2.3) and, in fact, were unequivocally identified as the latter seen beyond the limb. The original observations, together with others made subsequently at Culgoora and at Athens, have provided a considerable amount of information on the properties of the bright mottles observed beyond the limb, and this is outlined in Section 2.6. In addition to bright mottles, the same observations revealed the existence of a narrow *dark band* lying immediately above the photospheric limb,[1] which is briefly described in Section 2.7. Observations of the dark band have been reported independently by workers in the Soviet Union, but its physical significance is not yet clear.

Photographs of the quiet chromosphere at the limb taken in Hα often show remarkable differences in appearance. These dissimilarities are due to differences in a number of instrumental parameters, for example, exposure time, amount of parasitic light present, etc.; an understanding of the influence of these parameters is essential to anyone wishing to compare chromospheric limb photographs taken under different instrumental conditions. Accordingly, before proceeding to describe the properties of spicules, we explain in Section 2.2 how the appearance of a limb photograph is affected by the choice of these parameters.

2.2 The Effect of Certain Instrumental Parameters on the Appearance of Limb Photographs

To date the best photographs of the quiet chromosphere at the solar limb have been obtained in the Hα line with the aid of high-resolution telescopes equipped

[1] According to White and Bhavilai [1970] this feature had been known to other observers prior to 1969 but had been dismissed as a purely instrumental artefact arising from the presence of parasitic light.

with birefringent filters of various bandwidths.[2] With the exception of Plates 2.1, 2.4, and 2.9, all the plates in this chapter give reproductions of Hα photographs of the quiet chromosphere at the limb, yet at first sight they show remarkable differences in appearance. The dissimilarities in appearance are due to differences in (*a*) exposure time; (*b*) the amount of parasitic light present; (*c*) the use or otherwise of an occulting cone, and its position with respect to the limb; (*d*) the wavelength setting of the filter; and (*e*) the bandwidth of the filter. We shall consider the main effects of each of these parameters in turn.

(1) *Exposure time.* In the case of photographs taken at the centre of the line the effect of increasing the exposure time is simply to burn out detail in the low chromosphere (and also on the disk, if an occulting cone is not employed), while bringing up spicular detail at the higher levels (cf. Plates 2.2, 2.3a, 2.5a). With lower exposures, on the other hand, the spicular detail disappears, while structure now becomes visible in the *low* chromosphere beyond the limb and also perhaps, in the absence of an occulting cone, on the disk (cf. Plates 2.5b, c; 2.10, 2.11).

(2) *Parasitic light.* It has long been recognized that the presence of a sharp photospheric limb at the base of the chromosphere is due, wholly or in part, to parasitic light coming from outside the Hα line as a result of imperfect filtering [Lyot, 1944]. However, this effect can be eliminated by inserting a sufficiently narrow-band pre-filter in tandem with the main filter [Bhavilai *et al.*, 1965; Simon and White, 1966; White and Simon, 1968; Loughhead and Tappere, 1971]. The presence of the photospheric limb on photographs taken at the centre of Hα is illustrated by Plates 2.4, 2.10, and 2.11, which are all affected by parasitic light. However, it is absent on Plate 2.5, taken with a 1/8 Å filter plus a 1 Å Halle filter in tandem.

The absence of the photospheric limb on narrow-band photographs free from parasitic light shows, incidentally, that in the *core* of Hα the chromosphere is optically thick along a tangential line-of-sight to the limb. This opaqueness was first pointed out by Lyot [1944] and Roberts [1950] and is shown even more directly by Plates 2.4a and 2.4c, where the lower parts of several small prominences projecting up from behind the limb are seen to be obscured by the chromosphere in the foreground.

Sometimes the presence of the photospheric limb is an advantage in that it

[2] Many of the illustrations in this and following chapters were obtained by the authors with the 12-inch refractor of the CSIRO Solar Observatory, Culgoora, situated some 370 miles north-west of Sydney, Australia. This instrument is equipped with a seeing monitor to select the moments of best definition. Detailed information on the design and performance of the telescope is given by Bray and Loughhead [1967, pp. 126–129]; Loughhead *et al.* [1968]; Bray and Winter [1970]; and Loughhead and Tappere [1971]. References to discussions of the principles underlying the optimization of telescopes and filters employed for chromospheric observations are given in Section 3.2.2.

can provide a convenient reference level from which to measure the heights of chromospheric features beyond the limb [Dunn, 1960, p. 11].

(3) *Occulting cone.* When parasitic light is present it is evidently desirable to mount an occulting cone at some appropriate intermediate focus of the telescope in order to reduce the effect of scattered light from the photosphere [see, for example, Dunn, 1960; Loughhead, 1969]. Examples of the use of an occulting cone are given in Plates 2.2, 2.3a, and 2.10; note that in the case of Plate 2.3a the edge of the cone was placed in approximate coincidence with the photospheric limb. The need for an occulting cone is much reduced when parasitic light is absent.

(4) *Wavelength.* For a given exposure time the apparent height of the chromosphere beyond the limb becomes smaller as the filter is tuned away from the centre of Hα (cf. Plate 2.2 and Section 2.3.2).

A second consequence of tuning the filter progressively further into the wings of the line is the eventual re-appearance of the photospheric limb, even in the absence of parasitic light [cf. Simon and White, 1966]. This is simply due to the fact that the tangential optical thickness of the chromosphere, while greater than unity in the *core* of Hα, must eventually drop below unity at some point in the wings. As we shall see in Section 2.3.2, high-resolution observations show that the photospheric limb first begins to re-appear at H$\alpha \pm 0.65$ Å. At these wavelengths the photospheric limb is visible along limited arcs at locations where, presumably, the chromospheric structures along the line-of-sight happen to be more transparent than elsewhere (Plate 2.7). As one moves further out from the line centre, the photospheric limb rapidly becomes more prominent and appears as a sharp regular boundary crossed at irregular intervals by isolated chromospheric features (Plate 2.8).

(5) *Filter bandwidth.* If the bandwidth of the filter is sufficiently broad it follows from (4) above that the photospheric limb will be present, regardless of the wavelength setting of the filter. In effect, portion of the image-forming light transmitted by the filter comes from wavelengths (beyond H$\alpha \pm 0.65$ Å) at which the chromosphere is partially transparent. The result is the same as that produced by parasitic light coming from wavelengths right outside the Hα line.

2.3 Chromospheric Spicules: Morphological Properties

2.3.1 INTRODUCTION

Owing to the small widths of the spicules, there are still relatively few good-quality photographs of these features. Prior to the general introduction of the birefringent filter (cf. Section 3.2.2), the available observational material was limited to isolated white-light photographs taken at various solar eclipses, notably by observers from the Lick and Sproul Observatories in the U.S.A. [Dunn, 1960, pp. 13–14]. The definition achieved in some of the eclipse photographs is illustrated in Plate 2.1, which gives two enlargements from one of the Sproul plates. The upper photograph shows the chromosphere projecting above the rugged outline of the Moon's limb with the much fainter corona beyond, while the lower photograph shows a further 6.4 times enlargement of the central portion of the upper photograph. Only a few individual spicules can be clearly distinguished in Plate 2.1b. Other reproductions of eclipse plates have been published by Menzel [1931, Plate 9; 1959, Fig. 106], van de Hulst [1953, Fig. 2], and Dunn [1960, Chapter 1, Figs. 4 and 5].

Eclipse photographs have the advantage of being free from scattered light from the photosphere and offer a way of studying the relationship between spicules and overlying features in the corona. However, it is purely a matter of chance whether or not the atmospheric seeing will be adequate at the required moments, while the brief duration of totality makes it impossible to obtain a sequence of photographs long enough to follow the evolution of individual spicules. Moreover, eclipse photographs cannot show the boundary between the photosphere and chromosphere and hence do not lend themselves to the accurate determination of heights. All of these disadvantages can be circumvented in observations made outside of an eclipse, in which the chromosphere is photographed through an Hα birefringent filter of suitable bandwidth incorporated in a telescope specifically designed to achieve high spatial resolution. The classic observations of this kind are those obtained by Dunn [1960] with a specially designed 15.5-inch chromospheric telescope at the Sacramento Peak Observatory. His films have provided the basis for a number of important studies of the morphological and dynamical properties of spicules by himself and other investigators [Lippincott, 1957; Athay, 1959; Mouradian, 1967].

Plate 2.2 shows four photographs of the same segment of the limb obtained simultaneously with the Sacramento Peak telescope through a filter of about 3.4 Å bandwidth at the centre of Hα (Plates 2.2b and d) and at H$\alpha \pm 2.5$ Å (i.e., double-peaked profile) (Plates 2.2a and c). Frames (c) and (d) were given ten times the exposure of (a) and (b), while the occulting cone was positioned about 2800 km lower in frames (a) and (b) than in (c) and (d). Reproductions of other spicule photographs obtained by Dunn are given in his thesis [loc. cit.: Chapter 5, Figs. 4–7]. Subsequently, high-quality Hα photographs of spicules have been obtained with the 12-inch refractor of the Culgoora Solar Observatory, employing both a 1 Å Halle filter [Loughhead, 1969] and CSIRO's tunable 1/8 Å filter plus a 1 Å pre-filter [Loughhead and Tappere, 1971]. An example of the former, taken at the centre of Hα, is given in Plate 2.3a, where the edge of the occulting cone is in approximate coincidence with the photospheric limb. Photographs obtained with the 1/8 Å filter are reproduced in Plates 2.5–2.8.

Finally, mention should be made of the exceptionally high-quality spicule photographs recently obtained by R. B. Dunn using a 1/4 Å Zeiss filter in conjunction with the new vacuum tower telescope of the Sacramento Peak Observatory [Dunn, 1969]. One of Dunn's photographs is reproduced in Plate 2.3b: it was taken at Hα + 0.9 Å with the telescope aperture diaphragmed down to 20 inches and shows the spicules with a clarity never before achieved. Other reproductions of the new observations are given in a recent review article by Beckers [1972, cf. Fig. 1].

It is of great interest to compare the spicular structure shown in Plates 2.2 and 2.3 with that visible on the Sproul eclipse photograph reproduced in Plate 2.1b. Dunn [1960, p. 15] notes that the Sproul photographs were taken on a blue-sensitive emulsion and, accordingly, show the chromosphere in the H and K lines of Ca II and in the higher lines of the Balmer series. For a given exposure time the H and K lines can be traced to greater heights than any other lines in the blue (cf. Section 4.3.2), so that Plate 2.1b effectively illustrates the spicules as seen in the H and K lines. The appearance of the spicules in these lines is similar to that in Hα, but a detailed comparison of simultaneous photographs made at different wavelengths has not yet been undertaken. Most of the results given in the sections that follow refer specifically to Hα spicules.

We must emphasize that, in examining spicule photographs and even more so spicule spectra, care must be taken to distinguish between spicules and small, spike-like *prominences.* These are of frequent occurrence, particularly in the neighbourhood of active regions. Plates 2.4b and 2.4c show examples of very thin prominences which, if the spatial resolution were slightly lower, could easily be mistaken for spicules. It must therefore be borne in mind that some of the results derived by investigators using limb observations of only medium resolution may well refer to small prominences and not to spicules at all.

2.3.2 GENERAL APPEARANCE ON Hα PHOTOGRAPHS: VARIATION WITH WAVELENGTH

The variation in the appearance of the spicules at the limb as one moves outwards from the line centre into the wings of the Hα line is well shown by the narrow-band observations of Loughhead and Tappere [1971], which were obtained with CSIRO's tunable 1/8 Å filter used in tandem with a 1 Å Halle filter to eliminate parasitic light (see Section 2.2). Plate 2.5a shows spicules at the centre of Hα projecting upwards from the low chromosphere, some distance above the (invisible) photospheric limb. The low chromosphere itself appears to be largely devoid of spicular detail (Plate 2.5b), a fact which is generally attributed to the overlapping of the much greater number of spicules believed to be present at these low heights [Beckers, 1968, pp. 369–370]. As far as one can judge from a mere comparison of photographs (e.g. Plates 2.2, 2.3a, and 2.5a), the shape and size of the individual spicules is much the same irrespective of whether they are photographed with a broad or narrow bandwidth. On the other hand, since many spicules display large Doppler shifts (Section 2.4.3), we may expect the number of spicules visible per unit arc length to be somewhat smaller on narrow-band photographs.

Over the wavelength range H$\alpha \pm 0.5$ Å the appearance of the spicules remains very similar to that at the line centre and, once again, little or no spicular structure is visible in the low chromosphere. These facts are well illustrated in Plate 2.6, taken at H$\alpha - 0.5$ Å. However, as one moves further out from the line centre, individual spicules become discernible in the *low* chromosphere and their apparent overall height seems to decrease.[3] At the same time the sharp photospheric limb begins to re-appear over limited arcs of the limb as a result of the decreasing optical thickness of the overlying chromosphere along the line-of-sight to the limb. This is well demonstrated in Plate 2.7, which shows a segment of the north limb photographed at H$\alpha - 0.65$ Å. As the wavelength shift $\Delta\lambda$ further increases, individual spicules extend right down to the photospheric limb, which rapidly becomes more prominent. Of course, this does not necessarily prove that individual spicules actually reach the photospheric limb since any spicule apparently observed to do so may be located some distance in front of or behind the limb, in which case its true base is unobservable [cf. Beckers, 1964, p. 35].

Plate 2.8 shows four segments of the north limb photographed at H$\alpha - 0.75$ Å (Plates 2.8a and c) and H$\alpha + 0.75$ Å (Plates 2.8b and d). At these wavelengths virtually all spicules extend down to the limb although, strictly speaking, quite a number are separated from the limb by the narrow dark band whose properties are described in Section 2.7. Moreover, in some instances individual spicules can actually be traced back onto the disk itself [see also Koutchmy and Macris, 1971], and this in turn raises the important question of the identification of spicules with features on the disk. However, limb observations alone are not adequate to settle this question, and the whole problem is discussed in Section 3.2.6 with reference to both limb and disk observations. Beyond H$\alpha \pm 0.75$ Å the number of spicules present, and their brightness, steadily decrease out to H$\alpha \pm$ 1.0 Å and presumably beyond. Observations outside this range cannot be made with the tandem filter system described by Loughhead and Tappere [1971].

More recently, Beckers [1972, Fig. 1] has published a montage of spicule photographs taken by R. B. Dunn through a 1/4 Å Zeiss filter at 11 wavelengths across the Hα line ranging from H$\alpha - 1.25$ Å to H$\alpha + 1.25$ Å. These confirm the virtual absence of spicular structure in the low chromosphere in the core of the line and, at greater distances from the line centre, show spicules extending further and further down towards the photospheric limb. Only a small number

[3] With lower spatial resolution these two effects combine to produce a decrease in the apparent height of the low chromosphere with increasing wavelength shift $\Delta\lambda$ [Bhavilai, 1965, Plate 10; Bhavilai *et al.*, 1965, Fig. 2; Simon and White, 1966, Fig. 9].

of spicules remain visible at Hα ± 1.25 Å. Dunn (private communication) finds that spicules are most visible around Hα ± 0.9 Å; this conclusion appears to disagree with the observations of Loughhead and Tappere [1971] described above.

Turning now to other properties of spicules, we find that most of our current knowledge is based on observations obtained with broad-band filters; for example, the photographs of Dunn [1960] were obtained with a bandwidth of 3.4 Å (Section 2.3.1). This should be borne in mind when interpreting the results given in Sections 2.3.3–2.3.8 below.

2.3.3 WIDTH

Examination of Plates 2.1–2.3 demonstrates the extreme narrowness of the individual spicules. A number of authors have given estimates of their widths based on direct measurements of photographs taken both during and outside of eclipses [Beckers, 1968, Table I], but these provide only a rough guide to the true dimensions of the spicules. In fact, the accurate determination of spicule widths poses a difficult photometric problem involving not only the photographic contrast but also the instrumental profile of the telescope and atmosphere. This task was undertaken by Dunn [1960, pp. 176–179] using spicule observations obtained with the 15.5-inch chromospheric telescope of the Sacramento Peak Observatory.

Dunn first determined the instrumental profile of the telescope and then made a total of 64 microphotometer tracings across 47 separate spicules. Each tracing was converted to an intensity curve and corrected to remove the intensity of the underlying background. The observed width of a spicule was taken to be the width of the intensity curve at an intensity level equal to half the peak intensity. From this the true width, i.e. corrected for the effect of the instrumental profile, was obtained by calculating the square root of the difference between the squares of the widths of the intensity curve and of the instrumental profile at the half intensity levels (cf. Section 4.3.3). Individual measurements were made at heights ranging from about 4200 to 10 400 km.

Figure 2.1 shows the distribution of the *corrected* widths of the 47 individual spicules, compiled from Dunn's results [loc. cit.: Chapter 6, Table 1]. Whenever an individual spicule was measured at more than one height, the mean width has been used. More than 80 per cent of the spicules have corrected widths lying in the range 300–1100 km ($0''.4$–$1''.5$ of arc), the average width being 815 km, or about $1''.1$. More recent measurements by Lynch *et al.* [1973] give the slightly higher figure of 950 km, more than 90 per cent of the spicules having widths between 500 and 1500 km.

Dunn's results reveal no obvious relationship between the width of a spicule and the (apparent) height to which the measurement refers. This conclusion is supported by the photometric measurements of Mouradian [1967] who, using one of Dunn's films, determined the widths (uncorrected for instrumental profile) of a number of individual spicules at two separate heights, 5000 and 7000 km. In most cases Mouradian found that, within the errors of measurement, the width of a spicule was the same at both heights. On the other hand, Lynch *et al.* [1973] claim to have detected a small decrease in width with height, amounting to some 50 km per 1000 km.

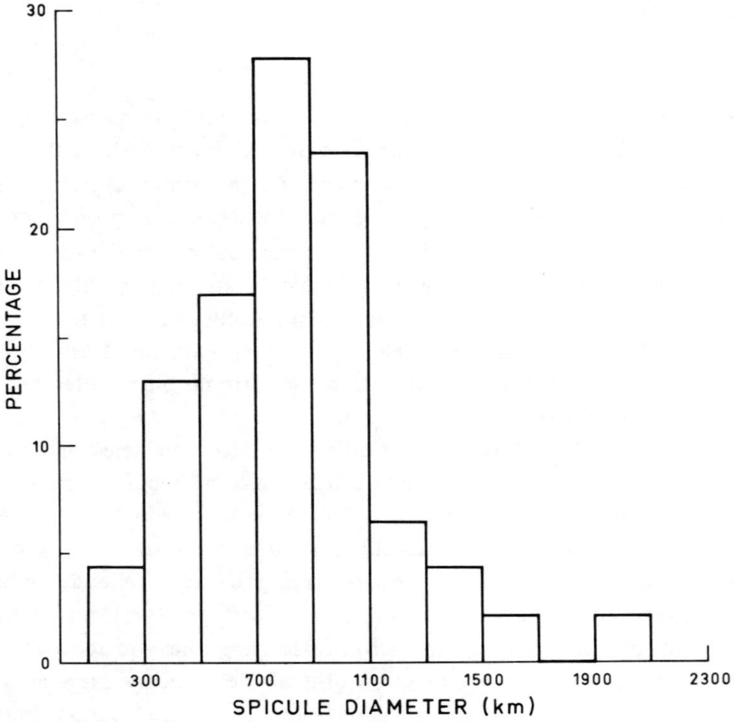

Fig. 2.1. Width of spicules. The histogram gives the distribution of spicule widths, corrected for instrumental profile, derived from the photometric measurements of Dunn [1960]. 85 per cent of the corrected widths lie in the range 300–1100 km ($0''.4$–$1''.5$ of arc), the average width being 815 km, or about $1''.1$.

However, the main aim of Mouradian's investigation was to study possible changes in the widths of spicules during their lifetimes. He succeeded in doing this for 16 separate spicules and found that, in all but two cases, there was a definite increase in the width of a spicule during its lifetime [Mouradian, 1967, cf. Fig. 2]. According to Mouradian the 'velocity' of expansion varies between 0 and 68 km sec^{-1}, with an average of 22 km sec^{-1}. The expansion usually starts after the spicule has reached its maximum intensity (see Section 2.4.1) and continues for 1–2 min. There is some indication that the duration of the expansion is longest for the brightest spicules. The observations are interpreted by Mouradian as implying the lateral diffusion of spicular material into the surrounding medium, and this, if true, would have important theoretical implications. However, considerable doubt has been thrown on Mouradian's results by the recent measurements of Lynch *et al.* [1973] which, with two exceptions, reveal no velocity of expansion or contraction exceeding 8 km sec^{-1}.

2.3.4 Hα EMISSION

Besides measuring spicule widths, Dunn [1960, p. 181] also measured total Hα intensities (i.e. integrated over wavelength) at the centres of 45 individual spicules and plotted the values as a function of height. After removal of the background intensity, he found a figure of approximately 2.5×10^4 erg cm^{-2} sec^{-1} $ster^{-1}$ for the mean value of the total Hα intensity at a height of 6000 km. Not unexpectedly, the spicule intensity decreases gradually with height, and from Dunn's measurements Beckers [1968, p. 378] has derived an emission scale height of 2200 km, this being the distance over which the spicule intensity drops by a factor of e. In general, therefore, a spicule does not have a sharp upper boundary. As we shall see in Section 4.3.4, Dunn's measurements have provided part of the observational basis for determining how the physical conditions in a spicule vary with height.

2.3.5 HEIGHT

As we have just seen, photometric measurements show that the Hα emission from a spicule decreases gradually with height, the effective emission scale height being 2200 km, or very nearly 3″ of arc. Consequently, a spicule does not have a sharply defined upper boundary and its apparent height on any particular photograph must depend on, amongst other things, the exposure time and the photographic contrast. Nevertheless, a number of investigations of the height distribution of spicules based on photographic observations have been undertaken, details of which can be found in Table 2.1. Spicule heights are conventionally measured from the photospheric limb, down to which the spicules appear to extend when observed in the wings of Hα. However, as we have previously emphasized in Section 2.3.2, this does not necessarily prove that individual spicules actually reach the photospheric limb since any spicule apparently observed to do so may be located some distance in front of or behind the limb, in which case its true base is unobservable (cf. Fig. 2.3).

Only three workers have given explicit values for the *average apparent height* of spicules. From an examination of films taken with the coronagraph at the Climax Observatory, Roberts [1945] concluded that the average apparent

TABLE 2.1

Measurements of Spicule Heights and Numbers

REFERENCE	OBSERVATIONS	TELESCOPE APERTURE (cm)	WAVE-LENGTH	BAND-WIDTH	NATURE OF INVESTIGATION
Roberts [1945]	Climax	–	Hα	4 Å	Average apparent height
Mohler [1951]	Eclipse [Marriott, 1931]	23	H & K (see Section 2.3.1)	–	Height distribution; counts of total number
Dizer [1952]	Pic-du-Midi [Lyot, 1944]	38	Hα	1.5 Å	Height distribution
Rush and Roberts [1954]	Climax	13	Hα	4 Å	Height distribution
Woltjer [1954]	Pic-du-Midi [Lyot, 1944]	38	Hα	1.5 Å	Height distribution; counts of total number
Lippincott [1957]	Sacramento Peak [Dunn, 1960]	39	Hα	3.4 Å	Height distribution; counts of total number
Athay [1959]	Sacramento Peak [Dunn, 1960]	39	Hα	3.4 Å	Counts of numbers at various heights

maximum height attained by polar spicules is about 7300 km, while a value of 6500 km was obtained by Dizer [1952] for spicules near the solar equator. Finally, Lippincott [1957] gave figures of 9500 km for polar spicules and 7300 km for equatorial spicules on the basis of an examination of films obtained by Dunn at the Sacramento Peak Observatory. The agreement among the various values is rather good in view of the uncertainty involved in defining the upper boundary of a spicule.

Information on the *distribution* of the apparent heights of spicules has been given by several of the authors listed in Table 2.1, the most extensive study being that of Lippincott [1957]. All of her measurements were made in a radial direction and no compensation was applied for spicule inclination (cf. Section 2.3.7). Using six separate sets of observations Lippincott measured the heights of the spicules within successive 12° arcs around the limb and averaged the results for each arc. Her results indicate that the vast majority of spicules have apparent heights in the range 5000–12 000 km, with very few exceeding 15 000 km.[4] This upper limit is somewhat greater than that given by Dizer [1952] who, using films obtained by Lyot at the Pic-du-Midi, found that few spicules exceed 9500 km, while those reaching a height of 11 000 km have the character of small prominences. According to Dizer spicules in the neighbourhood of active regions tend to be higher than those in quiet regions.

Lippincott's measurements [loc. cit.: Figs. 2 and 3] also reveal a steady decrease in the average apparent height of spicules from the poles to the equator of the Sun. The average height along a 12° arc centred on latitudes ± 84° was found to be 9500 km as compared to only 7300 km at the equator, a reduction of 23 per cent.

The decrease in the apparent height of spicules from the poles to the equator also emerges from a statistical investigation carried out by Athay [1959]. Using films taken at Sacramento Peak this author counted the numbers of spicules crossing various fixed heights in the chromosphere along successive 12° arcs of the limb; the counts were made at 1000-km height intervals over the range 3000 to 19 000 km. Fig. 2.2 shows the observed numbers of spicules, counted on films taken in October 1955, plotted on a height versus latitude grid. Examination of this figure shows that the number of spicules in the high chromosphere decreases from the pole to the equator, most of the reduction occurring within the first 40° from the pole. On the other hand, there is no discernible change from pole to equator at heights of 3000–4000 km. It should

[4] One has to remember, of course, that the height of any given spicule varies throughout its lifetime, so that the results presented above refer to the *instantaneous* distribution of spicule heights. Lippincott [1957] finds that the average apparent *maximum* height attained by the individual spicules is 11 600 km (see Section 2.4.1).

be noted that both Athay's and Lippincott's measurements were based on observations taken in 1955 near the time of solar minimum.

All the measurements so far discussed have dealt only with the apparent height of a spicule, i.e. the height above the photospheric limb measured in the plane normal to the line-of-sight. The geometrical situation in the case of a

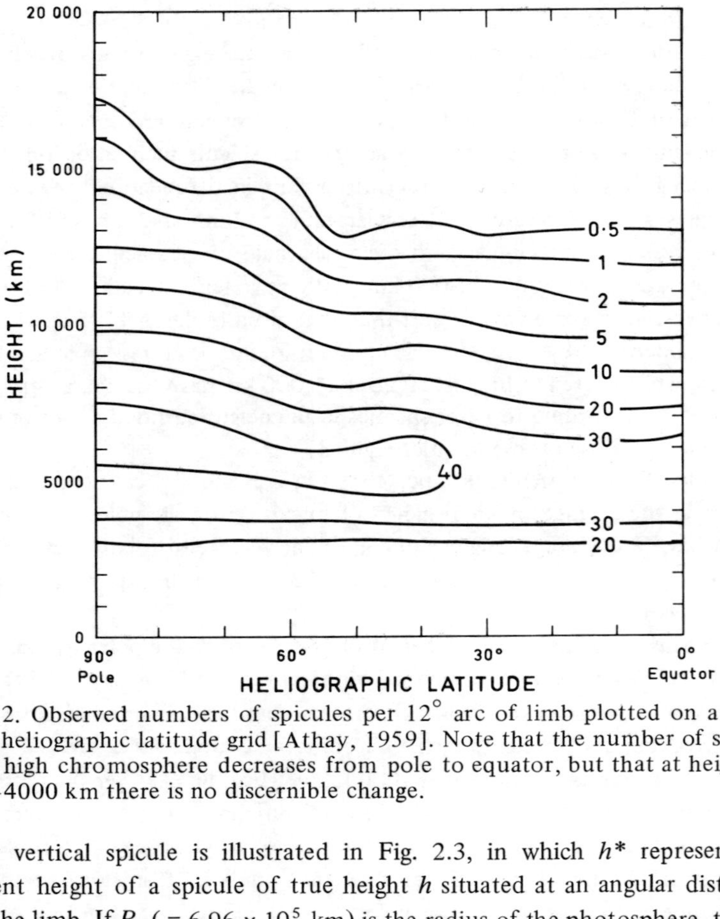

Fig. 2.2. Observed numbers of spicules per 12° arc of limb plotted on a height versus heliographic latitude grid [Athay, 1959]. Note that the number of spicules in the high chromosphere decreases from pole to equator, but that at heights of 3000–4000 km there is no discernible change.

purely vertical spicule is illustrated in Fig. 2.3, in which h^* represents the apparent height of a spicule of true height h situated at an angular distance α from the limb. If $R_\odot(= 6.96 \times 10^5$ km) is the radius of the photosphere, then

$$h^* = h \cos \alpha - R_\odot(1 - \cos \alpha), \tag{2.1}$$

whence, expressing h^* and h in km and α in degrees, we have for small α

$$h^* = h - 106\, \alpha^2 . \tag{2.2}$$

Thus the apparent heights h^* of a group of spicules all of a given actual height h will cover the range 0 to h. Moreover, it follows from equation (2.2) that even an exceptionally tall spicule having an actual height of, say, 15 000 km will be completely invisible when its distance from the limb is 12° or greater. At the centre of the Hα line, of course, this is an overestimate of the zone of visibility since, as we have already remarked in Section 2.3.2, spicular structure is not discernible in the low chromosphere at this wavelength.

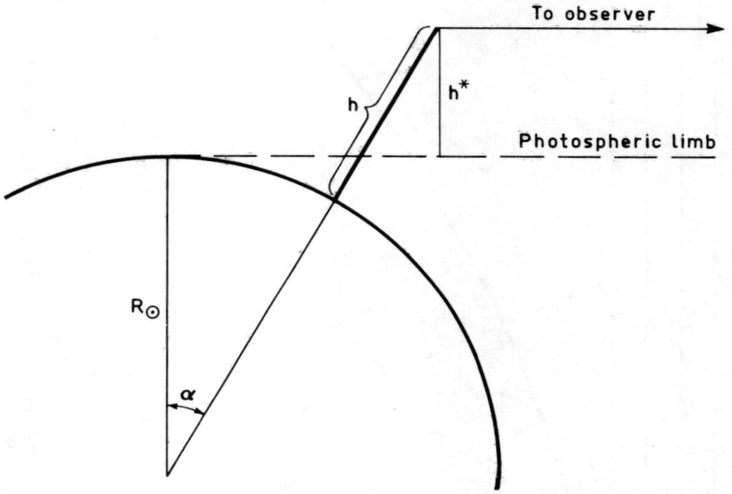

Fig. 2.3. Effect of limb truncation on the measured height of a vertical spicule situated at an angular distance α from the limb. h is the true height of the spicule and h^* the apparent height seen by an observer.

It is evident, therefore, that geometrical truncation must have an important influence on the form of the observed height distribution of spicules. Since the apparent height of a spicule steadily decreases as its angular distance from the limb increases, the observed distribution must be biased towards low spicule heights. However, there is a simple geometrical technique devised by Rush and Roberts [1954] which enables one to approximately reconstruct the *true* height distribution from the observed distribution. The procedure may be described by reference to Fig. 2.4, in which *a, b, c, d,* and *e* represent an arbitrary series of equal ranges Δh of true spicule heights, extending upwards from the photospheric limb L to some maximum height LM = H, say. The angles A, B, C, D, and E are the successive angular increments from the limb over which the radial displacement of the limb away from the tangent L is equal to $\Delta h/2$, $3\Delta h/2$, $5\Delta h/2$, $7\Delta h/2$, and $9\Delta h/2$ respectively. Since these angles are small the reduction of h due to foreshortening may be neglected. Moreover, the ratio $H/R_\odot$ is sufficiently small to allow us to take the radial displacement from the tangent for any angle considered to be the same for all radii from $R_\odot$ to $R_\odot + H$.

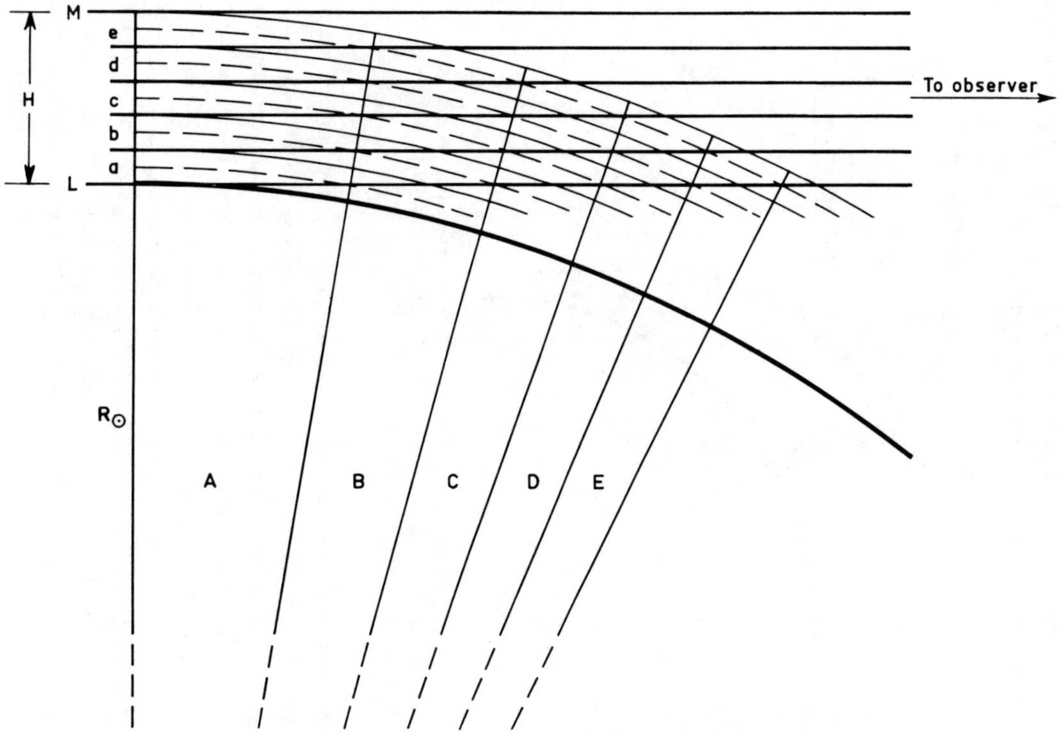

Fig. 2.4. Correction of observed distribution of spicule heights for the effect of geometrical truncation by the limb [Rush and Roberts, 1954]. Notation is defined in text.

An observer whose line-of-sight is parallel to the tangent to the limb L will see spicules apparently in range a arising throughout the angular range A + B + C + D + E, although it is evident from Fig. 2.4 that this class of apparent heights is actually a composite derived from all values of h. Within the degree of approximation adopted, heights actually in range a are observed in this range over the angle A; heights actually in range b are observed as being in range a throughout the angle B; those actually in c appear in a throughout angle C; and so on. Hence the heights appearing in range a are derived in reality from all height ranges. However, the apparent range b receives no contribution from true heights over angle E; c receives none in angles D or E; and so on. It is evident, therefore, that if spicule heights are randomly distributed over the solar surface, the distribution observed in the line-of-sight will be a distortion of the true distribution and will show disproportionately large numbers of spicules in the lesser height ranges.

To begin the correction procedure the observed distribution curve $N^*(h^*)$ is examined and a choice made, albeit somewhat arbitrarily, of the maximum height that occurs with any significant frequency; this height is taken as the maximum true height H attained by the spicules. Fortunately, because of the relatively small numbers of spicules in this part of the distribution, the correction is not very sensitive to the choice of H. One next computes the angles A, B, C, D, and E such that the mean height of range e falls in the arbitrary height ranges e, d, c, b, and a respectively. Since the maximum range e can receive no apparent contributions from the lower ranges, all of the N_e^* spicules appearing in range e are assumed to be located in the angular range $-$A to +A, remembering that spicules may be located behind as well as in front of the limb. Thus the true number of spicules in this range is $N_e = N_e^*$. Furthermore, the N_d^* spicules appearing in range d include (2B/2A) N_e spicules from true range e distributed over angle B on the two sides of the limb, and the difference between this term and N_d^* gives the true number N_d in the range $-$A to +A, i.e. $N_d = N_d^* - (B/A)\,N_e$. Similarly we have $N_c = N_c^* - (B/A)\,N_d - (C/A)\,N_e$, and so on. These equations suffice to determine the numbers $N_a, N_b, \ldots$, of spicules occurring in the range $-$A to +A, which is the required approximation to the true height distribution. Naturally, as pointed out by Athay [1959], any errors in the observed distribution N^* are magnified in the true distribution derived in this manner.

True height distributions obtained by this procedure have been given by Rush and Roberts [1954] and Lippincott [1957]. The latter, in the course of her extensive statistical study mentioned above, obtained the apparent height distribution of a group of 6231 spicules, considered without regard to their heliographic latitude. Taking H = 14 500 km (see above), Lippincott found that 37 per cent of all the spicules lay within ±2° of the limb, and for these she obtained the true height distribution given in Fig. 2.5. The true heights all lie between 6000 and 17 000 km, the most probable value being about 9000 km. Lippincott does not give a figure for the average true height of the spicules, but working directly from Fig. 2.5 we obtain an approximate value of 9800 km. The sharp decline in numbers at low heights reflects the apparent absence of spicule structure in the low chromosphere when observed over the central region of the

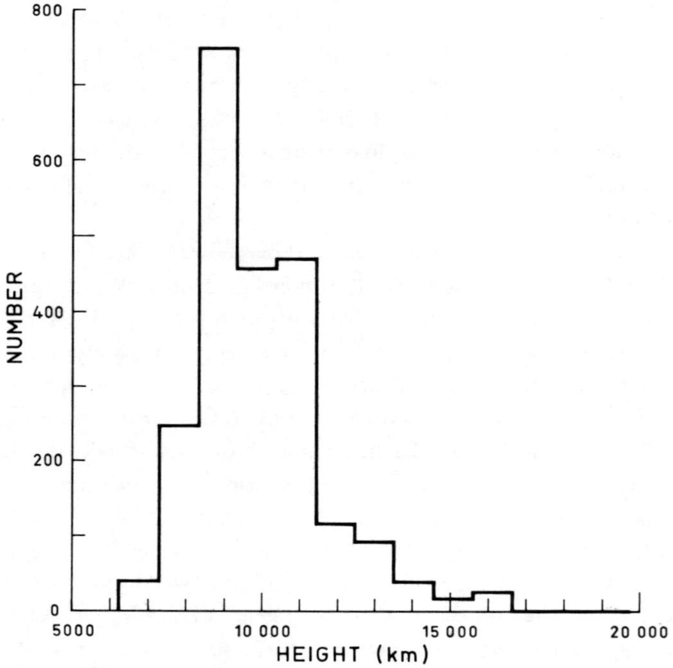

Fig. 2.5. True heights of spicules. The histogram shows the distribution of spicule heights obtained by Lippincott [1957] after correction for the effect of limb truncation. The true heights all lie between 6000 and 17 000 km, the average being about 9800 km.

Hα line. In reality, however, it is generally believed that spicules are actually present in the low chromosphere in such large numbers that overlapping – and possibly other factors – obscures their presence. This conjecture is supported by the fact that spicules are nevertheless observed in the low chromosphere in the wings of Hα (cf. Section 2.3.2).

At the present time the data presented in Fig. 2.5 represent the most reliable information available on the true height distribution of the spicules.

2.3.6 ABUNDANCE

A number of investigators have counted the total number of spicules per unit arc length visible at the solar limb, the different measurements referring to a variety of heliographic latitudes [see Lippincott, 1957, Table 2]. The extensive counts made by Lippincott herself [loc. cit.: Fig. 1] indicate that the number of spicules per 12° arc decreases steadily from the poles down to latitudes ±36° and then increases slightly towards the equator; the number per unit arc length at the equator is less than 80 per cent of that at the poles. A decrease in the frequency of spicules from the poles to the equator is also revealed by the statistical analysis of Athay [1959] who, like Lippincott, utilized films obtained by R. B. Dunn at Sacramento Peak. In contradiction to Lippincott, however, Athay found that the spicule distribution showed, in addition to the maximum at the poles, a second *maximum* at latitudes ±35° (cf. Fig. 2.2), but remarked that this might be spurious. The contradiction between the results of the two workers suggests that the frequency of spicules may, in reality, simply decrease monotonically from the poles to the equator, but further observations will be required to finally resolve the question.[5]

From her spicule counts Lippincott [1957] obtained a figure for the average total number of spicules visible around the solar limb at any one time. Then, using the fact that 37 per cent of this number lie within an angular distance of 2° of the limb on either side (see Section 2.3.5), she derived a value of 11 000 for the total number of spicules over the entire solar surface. It should be noted that this result gives the abundance of spicules in the middle and upper chromosphere ($h > 5000$ km) since the true heights of the spicules observed all exceed 6000 km (cf. Fig. 2.5). Another point to emphasize is that the number of spicules counted depends very strongly on the quality of the observations. In the course of nine separate determinations, Lippincott found that the number of spicules counted around the solar limb – disregarding possible obscuration by prominences – varied from a minimum of 845 to a maximum of 1154. One may expect, therefore, that future spicule observations of higher spatial resolution will lead to an upward revision of the figure for the total number of spicules over the solar surface.

We also require information on the way in which the spicule abundance varies with height.[6] This problem has been considered in some detail by Athay [1959] who, using Dunn's observations, counted the numbers of spicules crossing various fixed heights in the chromosphere along successive 12° arcs of the limb. The counts, averaged over latitude, are given in the second column of Table 2.2.

[5] Lippincott [1957] comments that the peak of prominence activity at the limb at the times of the observations used in her study also occurred around latitudes ±36°. Her adopted spicule count for any particular 12° arc was the average of nine separate determinations, but care was taken to exclude the values for regions obviously affected by prominences.

[6] This problem is closely related to the determination of the height distribution of spicules, discussed in the previous section. As we shall see in Section 4.5, a knowledge of the height variation of the spicule abundance is required for the construction of inhomogeneous models of the chromosphere.

It is evident that the observed number rises rapidly to a peak at 5000–6000 km and then gradually declines.

The drop in the observed spicule abundance below 6000 km was attributed by Athay to a greater crowding of the spicules at the lower heights, leading to their mutual obscuration [see also Woltjer, 1954; de Jager, 1959, pp. 136–138]. Accordingly, he corrected the observed numbers of spicules for the effect of

TABLE 2.2

Spicule Abundance as Function of Height [Athay, 1959]

HEIGHT (km)	OBSERVED NUMBER	CORRECTED NUMBER	REDUCED NUMBER	TOTAL ON SUN
1000	–	–	–	–
2000	–	–	–	–
3000	19.6	292	167	9.3×10^4
4000	35.0	176	123	6.8
5000	38.8	128	100	5.6
6000	38.2	88	76	4.2
7000	32.5	59	52	2.9
8000	23.8	36	32	1.8
9000	16.5	22	20	1.1
10 000	9.6	11	10	5.6×10^3
11 000	5.5	6.1	5.7	3.2
12 000	2.8	2.8	2.7	1.5
13 000	1.5	1.5	1.4	7.8×10^2
14 000	0.8	0.8	0.8	4.4
15 000	0.4	0.4	0.4	2.2
16 000	0.2	0.2	0.2	1.1
17 000	0.1	0.1	0.1	5.6×10

Note: the entries in columns 2 and 3 refer to a 12° arc of the limb, while those in column 4 give the actual number of spicules in an area 12° × 6.°2 of the solar surface; all values are averaged over latitude.

overlapping. The validity of the method employed depends on a number of basic assumptions, viz.,

(1) the spicules are randomly distributed over the solar surface;

(2) they all have the same brightness;

(3) they all have the same diameter, which is independent of height (cf. Section 2.3.3);

(4) the disappearance of spicules at low heights is due solely to the effect of overlapping; and

(5) all spicules are optically thick in Hα.

On the other hand, the possibility of interspicular absorption is ignored and no account is taken of the fact that spicules are often inclined at quite large angles to the vertical, thus reducing the effect of overlapping. The corrected numbers of spicules obtained by Athay are given in the third column of Table 2.2. One can see that the correction for obscuration is insignificant in the upper chromosphere but becomes very large in the low chromosphere, increasing from a factor of 3.3 at a height of 5000 km to a factor of nearly 15 at 3000 km.

The entries in the third column of Table 2.2 remain uncorrected for the effect of limb truncation. To perform this correction Athay used the method of Rush and Roberts outlined in the previous section. In this way he reduced the spicule counts to include only those spicules situated no more than $3\overset{\circ}{.}1$ from the limb on either side. The reduced numbers given in the fourth column of Table 2.2 thus refer to an area $12^\circ \times 6\overset{\circ}{.}2$ of the solar surface. Finally, the total numbers of spicules on the Sun at the specified heights can be obtained by multiplying the corresponding entries in the fourth column by the factor $30/\sin 3\overset{\circ}{.}1 = 555$, yielding the values set out in the last column.

It is probably safe to conclude that the spicule abundances given in Table 2.2 are reasonably reliable for the upper chromosphere ($h > 10\,000$ km). However, whether they are valid for the low chromosphere ($h < 5000$ km) depends on the extent to which the assumptions underlying Athay's analysis are themselves justified, particularly the conjecture that overlapping is the sole cause of their disappearance at low heights. Further clarification of this point will depend on gaining a better understanding of the morphological and radiative properties of the low chromosphere and, until this is achieved, the derived abundances at low heights should be treated with caution.

A determination of the abundance of spicules as a function of height had earlier been made by Coates [1958], who used data on the distribution of observed spicule heights given by Lippincott [1957]. Like Athay, Coates attempted to correct for the effect of mutual obscuration of spicules by one another, but the abundances so derived (see his Table 1) differ greatly from those given in Table 2.2. More recently, a determination of the height abundance of spicules has been published by Beckers [1968, Table IV], based on spicule counts made by Hiei [1963] on well-resolved flash spectra. In addition, Mouradian [1965] has counted the number of spicules resolved at different heights on Hα spectrograms taken with a 9-metre spectrograph at the Pic-du-Midi Observatory, the measurements being made both at the line centre and in the wings (at about H$\alpha \pm 0.6$ Å). The results show that the number of spicules per unit arc length visible in the wings exceeds that at the line centre below a height of about 3500 km, whereas above this height the numbers are approximately the same for all three wavelengths (see also Section 2.3.9). Mouradian interprets this behaviour as evidence for the existence of interspicular material whose absorption is high at the centre of Hα but decreases as one moves out into the wings of the line.

The high-quality observations of spicules now being made with the new vacuum tower telescope at Sacramento Peak (Section 2.3.1) should eventually lead to a more accurate determination of their abundance. However, according to Beckers [1972, p. 76] a preliminary examination of the new data indicates that the existing statistics (Table 2.2) appear to be fairly reliable.

2.3.7 INCLINATION AND CLUSTERING

Although the majority of spicules are directed more or less radially outwards from the solar limb, individual spicules may be inclined at quite large angles to the vertical. An example of a spicule inclined at more than 45° is shown in Plate

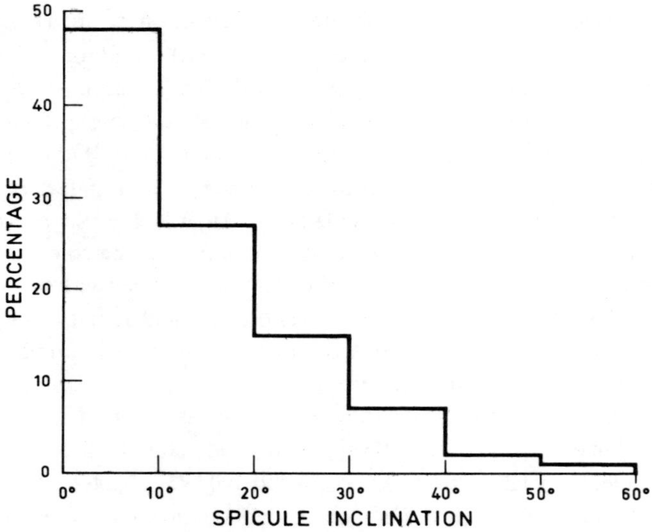

Fig. 2.6. Inclination of spicules to the vertical. The histogram shows the distribution of inclinations derived by Coates [1958] for a sample of 129 spicules. The mean inclination inferred from the histogram is about 14°.

2.2d slightly to the left of centre, while Beckers [1964] has observed spicules in the wings of the Hα line with slopes of up to 70°. Determinations of spicule inclinations have been made by van de Hulst [1953, p. 214], Coates [1958], and Beckers [1964, pp. 56–57]. Van de Hulst states that half the spicules have slopes exceeding 20°, while Beckers finds from measurements made on photographs taken by R. B. Dunn an average slope of 19° for spicules around heliographic latitude 60°. However, Coates' results lead to a somewhat lower value, although he too used Dunn's observations. Fig. 2.6 shows the distribution of spicule inclinations derived by Coates for a sample of 129 spicules; the mean value inferred from the histogram is about 14°.

However, it should be pointed out that since spicules are visible on photographs taken at the centre of Hα only above a height of some 4000 km, some of those inclined at large angles will not rise high enough to be seen and hence the observed distribution of spicule inclinations will be biased towards small values. On the average, therefore, the spicules will appear to be more nearly radial than is actually the case. Moreover, in measuring the inclination of a spicule we have to remember that we actually see a spicule as its projection onto the plane normal to the line-of-sight, and this will of course contribute to an underestimate of its true inclination.

Using Dunn's observations Lippincott [1957, cf. Fig. 4] has found some evidence that spicules tend to slope towards the equator. Other observers have concluded from examination of eclipse photographs that the inclinations of the spicules tend to follow the direction of the *coronal rays* above them [for references see van de Hulst, 1953, p. 215]. Rush and Roberts [1954] have noted several instances of spicules appearing to be superimposed on coronal arches, their directions being tangential to the contours of the arches, while Waldmeier [1960] has claimed that he was able to photograph a spicule extending to a height of at least 60 000 km among the coronal rays visible at the eclipse of February 1952! Bugoslavskaya [1950, p. 142] has reported that, out of 22 chromospheric 'clusters' observed near the north pole at the solar eclipse of September 1941, 20 coincided with overlying coronal rays (cf. her Fig. 19).[7] However, Beckers [1968, p. 380] has failed to find such a detailed correspondence between the spicules and polar rays visible on good-quality eclipse photographs obtained by the Sproul Observatory. The hypothesis has been advanced [cf. van de Hulst, loc. cit.] that the tilt of both the chromospheric spicules and the coronal rays is determined by weak magnetic fields, but direct evidence to confirm this view is so far lacking.

Despite the tendency of spicules to follow the alignment of overlying coronal rays, their general appearance in quiet regions usually seems to imply a more or less random distribution of inclinations, heights, and intensities [Dunn, 1960, p. 31]. However, Lippincott [1957] has claimed that there are occasions on which the inclinations of the spicules follow more orderly arrangements. One of these, the so-called 'porcupine' pattern, is described as a dome-like rise in the general level of the chromosphere, averaging 18 000 km across, from which spicules protrude in an approximately radial fashion, their apparent point of intersection being located from 7000 to 30 000 km inside the limb. Illustrations of 'porcupine' patterns are given by Dunn in Chapter 5 of his thesis [Dunn, 1960, see Fig. 5, right of centre; and Fig. 6, centre]. The other type of spicule

[7] The chromospheric 'clusters' described by Bugoslavskaya presumably represent groups of spicules, since their mean cross-section is given as 10″ of arc.

configuration described by Lippincott is the so-called 'wheat field' pattern in which the spicules display similar inclinations, heights, and intensities along arcs of average length 140 000 km. According to Lippincott all the spicules within a 'wheat field' pattern grow and disappear *simultaneously*. Both patterns are observed only within 30° of the poles and have lifetimes of the order of 5 min, a figure comparable to the lifetime of the individual spicules (see Section 2.4.1).

Further evidence of the clustering tendency of spicules has recently been put forward by Platov and Shilova [1971]. Using a sequence of Hα photographs extending over a period of some 42 min, these authors determined the height of the apparent chromospheric 'boundary' at 101 points spaced some 1250 km apart as a function of time and found that time changes in the height of the boundary were coherent over regions of widths $\gtrsim$4000 km. At the same time, their analysis revealed a strong correlation between the height changes at points on the limb separated by distances of the order of 25 000 km. This distance is comparable to the observed separation between the boundaries of a typical cell in the supergranulation (see Section 3.4.3) and, accordingly, Platov and Shilova interpret their results as implying the existence of clusters of spicules of widths $\gtrsim$4000 km located on the boundaries of supergranules.

2.3.8 DEPENDENCE ON LATITUDE AND PHASE OF SOLAR CYCLE

In the course of the preceding sections we have noted a number of cases in which the properties of the spicules show a clear dependence on heliographic latitude. Firstly, there is a steady decrease in the average apparent height of the spicules from the poles to the equator (Section 2.3.5). According to Lippincott [1957] the average apparent height at the equator is 7300 km as compared to 9500 km at latitudes ±84°, a reduction of 23 per cent. Secondly, the number of spicules per unit arc length drops from the poles to the equator, the abundance at the equator found by Lippincott being less than 80 per cent of that at the poles (Section 2.3.6). Thirdly, there is evidence that the inclinations of the spicules near the poles follow the direction of the overlying coronal rays, but tend to be more randomly inclined in the equatorial region (Section 2.3.7). Finally, we shall see in Section 2.4.3 that the average Doppler (i.e. line-of-sight) velocity of polar spicules seems to be somewhat lower than that of equatorial spicules.

The tendency for spicules to align themselves in the direction of the overlying coronal rays is particularly interesting in view of the well known fact that the gross form of the corona varies markedly with the phase of the solar cycle [see, for example, Billings, 1966, p. 58 et seq.]. This raises the question of whether the properties of the spicules also change with the phase of the solar cycle. No systematic study of this problem has yet been undertaken, although high-quality films of spicules spanning the period between solar maximum and minimum have been available at the Sacramento Peak Observatory for over a decade [Dunn, 1960, p. 28]. It is to be regretted that they have not so far been utilized for a systematic investigation of any possible dependence of the properties of the spicules on the phase of the solar cycle.

2.3.9 APPEARANCE ON SPECTROGRAMS TAKEN IN Hα AND OTHER CHROMOSPHERIC LINES

Almost all the information on the properties of spicules presented so far has been derived from high-quality direct photographs obtained with the aid of Hα birefringent filters of various bandwidths. Nevertheless, the effort devoted to direct photography has been small in comparison to that expended on *spectroscopic* observations of spicules in Hα and, as we shall see below, other strong chromospheric emission lines. Plate 2.9 shows a series of spectra illustrating the spicule structure visible in the Hα line at various heights above the limb; the spectra were obtained by Zirker [1962] with the 13-metre Littrow spectrograph of the Sacramento Peak Observatory. Only at heights above about 8000 km are the spicules sufficiently well separated to be seen as completely distinct emission streaks crossing the line. As we have seen above (Table 2.2), at heights between 4000 and 8000 km much greater numbers of spicules are present and, with the spatial resolution achieved, one can no longer be sure that a given emission streak refers to just a single spicule. Below about 4000 km a central reversal appears in the core of the Hα line,[8] within which spicule structure becomes increasingly difficult to discern. At these heights spicules are best visible in the wings of the line, a fact which accords with the apparent absence of spicules in the low chromosphere on photographs taken with medium- or narrow-band filters tuned to the centre of Hα (Section 2.3.2). On the other hand, spicules appear to extend right down to the photospheric limb on photographs taken in one or the other wing of the line.

Besides Hα, the presence of spicules or unresolved clusters of spicules is conspicuous in the spectra of other strong chromospheric emission lines, notably the early members of the Balmer series, the D_3 line of He I, and the H and K resonance lines of Ca II. Spicules can also be observed – albeit usually with greater difficulty – in the λ3888.65 and λ10830 lines of He I, the infrared triplet of O I ($\lambda\lambda$7772, 7774, and 7775), and the infrared triplet of Ca II ($\lambda\lambda$8498, 8542, and 8662). Among the thousands of fainter emission lines in the visible spectrum, arising principally from singly-ionized rare earth metals, some appear to show no structure, whereas others show small-scale inhomogeneities.

[8] This feature is also characteristic of Hα profiles observed with a spatial resolution inadequate to resolve the spicules (see Fig. 4.9).

These features resemble spicules, but conclusive proof of their identity is lacking [cf. Beckers, 1972, pp. 79–80].[9]

Of particular importance are spectroscopic observations made simultaneously in two or more lines, since they provide answers to the crucial question of whether spicules seen in one line coincide in position with those seen in other lines or, in other words, whether a given spicule radiates simultaneously in all the strong chromospheric lines listed above. Illustrations of simultaneous spectra have been published by a number of authors, including Pasachoff *et al.* [1968, Fig. 2] (K, H, Hϵ, and D_3), Krat and Krat [1971, Fig. 1] (Hα, Hβ, D_3, H, and K), and Beckers [1972, Fig. 2] (Hα, Hβ, Hϵ, H, and K). However, as emphasized by Beckers [1968, p. 391], the comparison of features in different lines is fraught with difficulty, especially where the lines are of very different strengths. Thus at *low* heights spicules tend to disappear in the cores of strong lines with central reversals, whereas this is not true of lines without self-absorption. Examples of the former are Hα, Hβ, and the H and K lines, and of the latter Hϵ and D_3 (Section 4.3.3). Moreover, the height at which spicules are most prominent varies from line to line. Finally, care must be taken to allow for the effect of differential atmospheric refraction [cf. Pasachoff *et al.*, loc. cit.].

The close relationship between the spicule structures observed simultaneously in different lines is illustrated in Fig. 2.7, which shows four microphotometer tracings made by Pasachoff *et al.* at K + 0.57 Å, H + 0.51 Å, Hϵ centre, and D_3 centre. Not surprisingly, there is very close agreement between the positions and intensities of features in the H and K lines: in fact, the values of cross-correlation coefficients computed between tracings made in the wings of these two lines – where the spicules are most clearly visible – are almost unity. On the other hand, Pasachoff *et al.* find that, although there are fewer features present in Hϵ and D_3 than in H and K, these can almost always be identified with spicules in one or the other of the H and K line wings. The values of the cross-correlation coefficients calculated between a wing of H or K and the Hϵ line centre are about 0.6, while those between a wing of H or K and the D_3 centre are rather less, namely about 0.3. In both cases, however, the coefficients are artificially reduced by considering only one wing of H or K at a time, thus tending to ignore the presence of spicules with profiles Doppler-shifted towards the other side of the line (see Section 2.4.3). Finally, Fig. 2.7 shows the reasonable agreement between the positions of corresponding features in the Hϵ and D_3 spectra, although the relative *intensities* of the individual spicules are not always the same in the two lines. For example, the feature marked with an arrow

[9] Further references to spectroscopic observations of spicule emission lines are to be found in Sections 4.3.3 and 4.3.4, with particular emphasis on the deduction of information about physical conditions in spicules from line profile and intensity measurements.

near the top centre of Fig. 2.7 is very bright with respect to other features in the Hε, H, and K lines, but is scarcely visible in D_3.

Similar results have been obtained by a number of other observers and, in addition, a very strong correspondence appears to have been established between

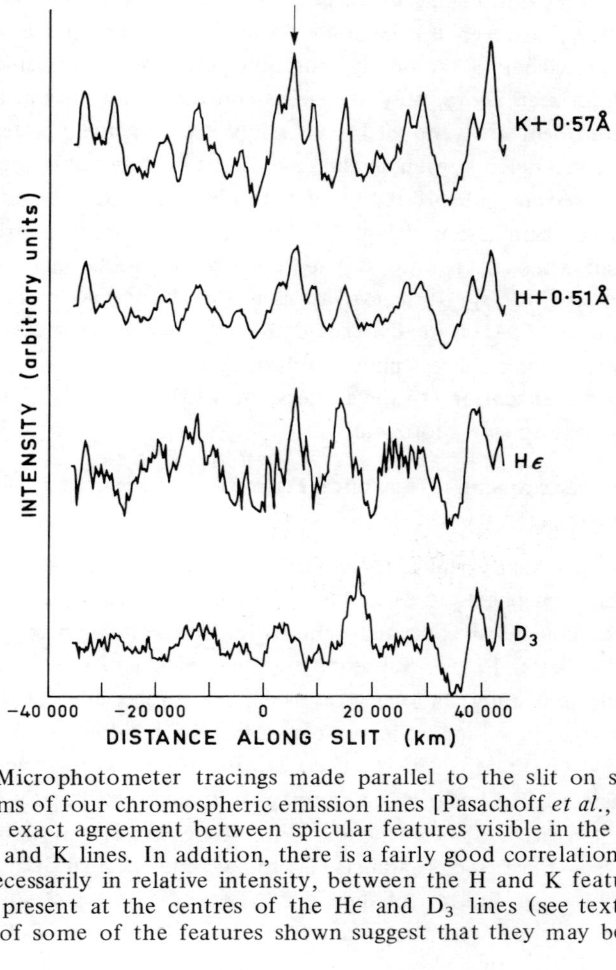

Fig. 2.7. Microphotometer tracings made parallel to the slit on simultaneous spectrograms of four chromospheric emission lines [Pasachoff *et al.*, 1968]. Note the almost exact agreement between spicular features visible in the wings of the calcium H and K lines. In addition, there is a fairly good correlation in position, but not necessarily in relative intensity, between the H and K features and the structures present at the centres of the Hε and D_3 lines (see text). The large diameters of some of the features shown suggest that they may be clusters of spicules.

spicules seen in Hα and in other members of the Balmer series [cf. Beckers, 1972, Fig. 2]. Moreover, as one would expect, a high degree of correlation has been found between Hα spicules and those observed in D_3 and the H and K lines [see, for example, Athay, 1958; Beckers, 1968, pp. 391–392; 1972, Fig. 2; Mouradian, 1969].

Summarizing all the available evidence, Beckers [1968, p. 392] and Pasachoff *et al.* [1968] conclude that most spicules radiate simultaneously in the Balmer lines, the calcium H and K lines, and the helium D_3 and λ3888.65 lines. In addition, it is highly probable that the infrared triplets of O I and Ca II and the helium λ10830 line are also emitted by each spicule. (Incidentally, this assumption is implicit in the discussions of Sections 4.3.3 and 4.3.4.)

Unfortunately, however, this large collection of spectroscopic data adds very little to our knowledge of the purely morphological properties of spicules.[10] In fact, as we have seen above, only at heights above about 8000 km can one be reasonably confident that the visible emission streaks actually refer to single spicules. At lower heights, even in the case of the best available spectrograms, the features observed probably represent unresolved clumps of spicules. This is borne out by an examination of Fig. 2.7, derived from a spectrogram taken with a tangential slit whose centre was situated some 4000–5000 km above the limb [Pasachoff *et al.*, 1968]. It is evident that the diameters of many of the prominent features present greatly exceed the average spicule width of 815 km derived from high-quality direct photographs (Section 2.3.3).

With only one exception (to the authors' knowledge), only Hα filters have so far been applied to spicule photography.

2.4 Chromospheric Spicules: Dynamical Properties

2.4.1 LIFETIME AND EVOLUTION

Observations show that a typical spicule first appears as a barely detectable lump in the low chromosphere. It then rapidly elongates upwards and increases in brightness, reaching its maximum height within a minute or two of its initial appearance [Roberts, 1945]. Subsequently, the spicule may either fade from visibility while remaining stationary or else appear to descend back to the low chromosphere with a velocity comparable to that of its initial ascent [Rush and Roberts, 1954; Lippincott, 1957]. The majority of descending spicules appear to retrace their original paths but, according to Beckers [1972, p. 74], there are many which fall back along different paths, often forming an arch. The time interval between the first appearance of a spicule and its final disappearance is taken as the observed lifetime of the spicule. Several determinations have been made, the results of which are summarized in Table 2.3. Details of the observations used by the various investigators are given in Table 2.1, with the exception of Alissandrakis and Macris [1971]. Their observations were obtained at Athens with a 40-cm refractor equipped with a 0.5 Å filter set at the centre of Hα.

[10] Krat and Krat [1961, 1971] and Nikolsky [1965] have concluded that spicules are fuzzier and have larger widths in D_3 than in Hα, while Beckers [1972, p. 75] finds them somewhat larger in H and K.

Judged by the quality of the observations employed the most reliable determination is presumably that of Lippincott [1957], who estimated the lifetimes of 77 individual spicules from films obtained by R. B. Dunn at Sacramento Peak; the interval between successive frames generally did not exceed 30 seconds. Examination of the resulting lifetime distribution, given in

TABLE 2.3

Lifetime of Spicules

REFERENCE	AVERAGE LIFETIME (min)	REMARKS
Roberts [1945]	4–5	–
Dizer [1952]	2	–
Rush and Roberts [1954]	3.5	–
Lippincott [1957]	5.1 6.3	Mean for all spicules Mean for tallest 25 per cent
Alissandrakis and Macris [1971]	6	–

Fig. 2.8, indicates that the observed lifetimes range from 1 to 10 min, the average value being 5.1 min.[11] Dizer [1952] claims that the spicule lifetimes in active regions are slightly smaller than those in quiet regions, but it is not clear whether or not this result is statistically significant.

According to Lippincott, spicules with very short lifetimes are extremely faint, appearing and disappearing alternately from frame to frame. She suggests that these may represent the tops of tall spicules actually located some distance from the limb. In fact, as we have already seen in Section 2.3.5, only 37 per cent of the spicules observed lie within $\pm 2^\circ$ of the limb, and hence one observes the true lifetimes of only a relatively small fraction of the total number. In agreement with this conclusion Lippincott finds that when only the tallest 25 per cent of the spicules are considered, the mean lifetime is increased to 6.3 min. Unfortunately, unlike the case of the distribution of spicule heights, there appears to be no obvious method for correcting the observed lifetimes to obtain the true lifetime distribution [cf. Rush and Roberts, loc. cit.].

[11] Despite the contrary impression given by her histogram (Fig. 2.8), Lippincott explicitly states that none of the spicules studied had lifetimes of less than 1 min.

Although spicules are not visible in the low chromosphere on photographs taken at the centre of Hα, the possibility remains that they may in fact extend right down to the bottom of the chromosphere (Section 2.3.6). If they do originate at this level, then their total lifetime will obviously be considerably

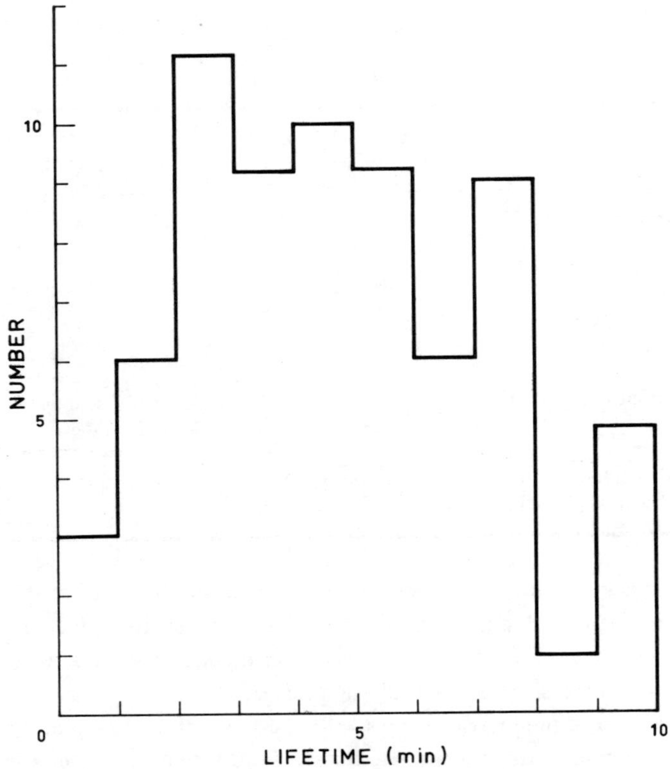

Fig. 2.8. Lifetimes of spicules. The histogram shows the distribution of the lifetimes of 77 spicules determined by Lippincott [1957]. The observed lifetimes range from 1 to 10 min (see footnote 11), the average being 5.1 min.

greater than that observed. Assuming that the velocity of ascent through the first 5000 km is the same as that actually observed above this height, Lippincott [1957] calculates that the total average lifetime is increased by 4 min if the spicule fades at maximum; the increase is doubled if it descends back to the bottom of the chromosphere at the same speed. Consequently, the total lifetime of spicules originating at the bottom of the chromosphere would be of the order of 10 to 15 min. However, we must again emphasize that it is not known with certainty at the present time whether or not spicules do in fact originate in the very low chromosphere (cf. Section 2.3.6).

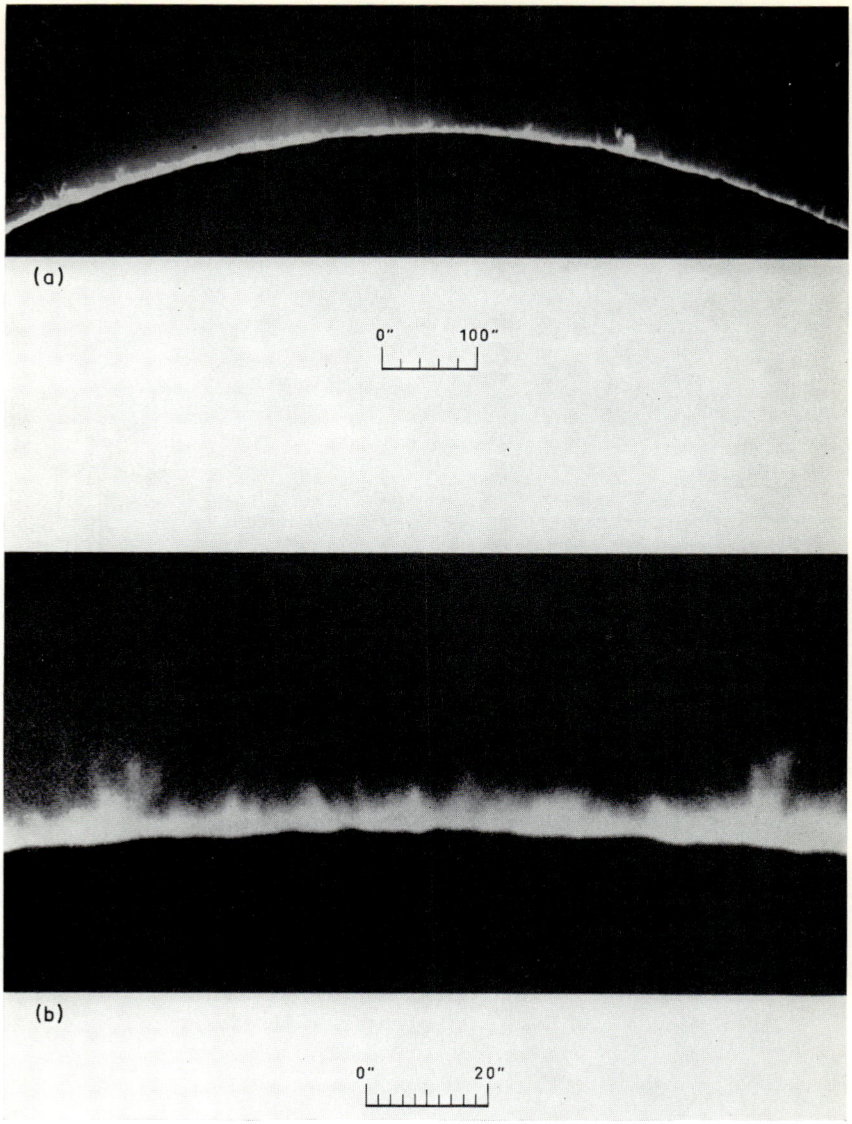

PLATE 2.1. Enlargements from a photograph taken at the 1930 solar eclipse. (a) shows the chromosphere projecting above the rugged outline of the Moon's limb with the much fainter corona beyond, while (b) shows a further 6·4 times enlargement of the central portion of (a). A few individual spicules are clearly delineated in both (a) and (b). (*By courtesy of Sproul Observatory.*)

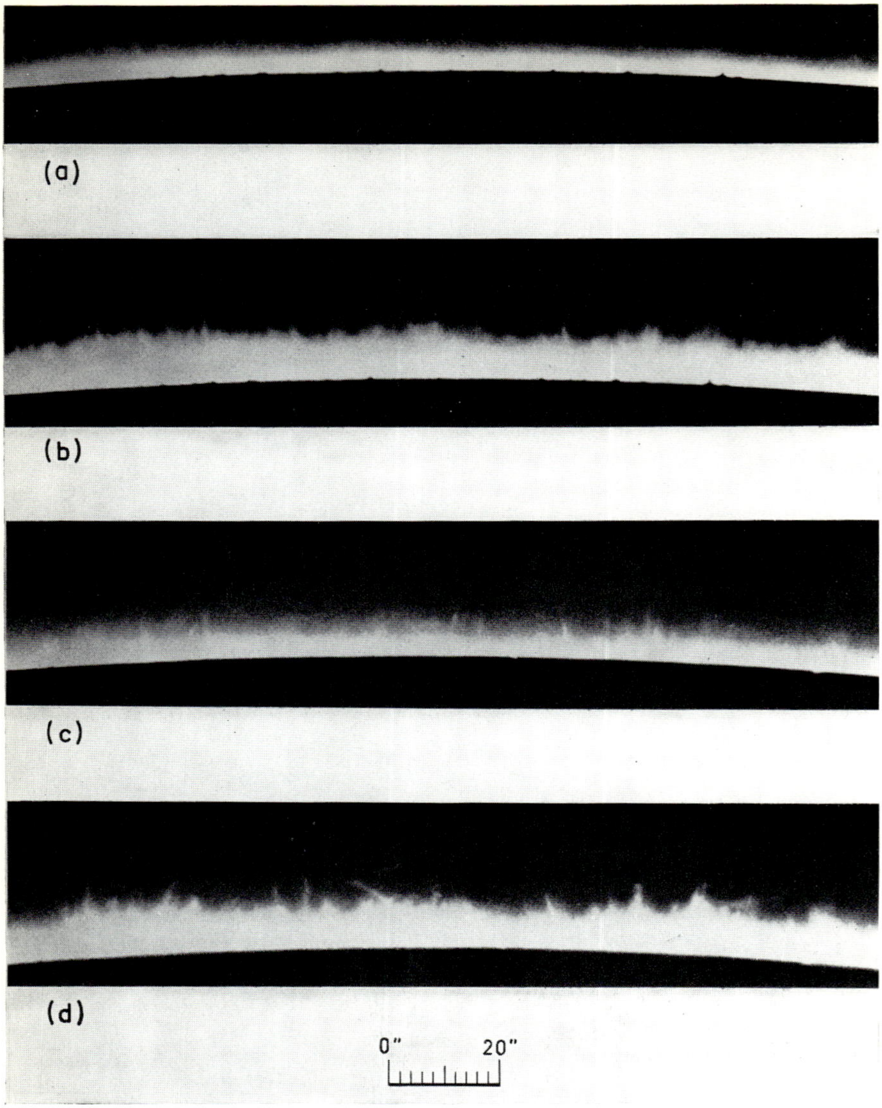

PLATE 2.2. Four photographs of the same segment of the chromosphere beyond the limb obtained simultaneously by R. B. Dunn with the 15·5-inch chromospheric telescope of the Sacramento Peak Observatory. They were taken through a filter of about 3·4 Å bandwidth at the centre of Hα (images b and d) and at Hα ± 2·5 Å (i.e. double-peaked profile) (images a and c). Frames (c) and (d) were given ten times the exposure of (a) and (b); the occulting cone was positioned about 2800 km lower in frames (a) and (b) than in (c) and (d). (*By courtesy of Sacramento Peak Observatory.*)

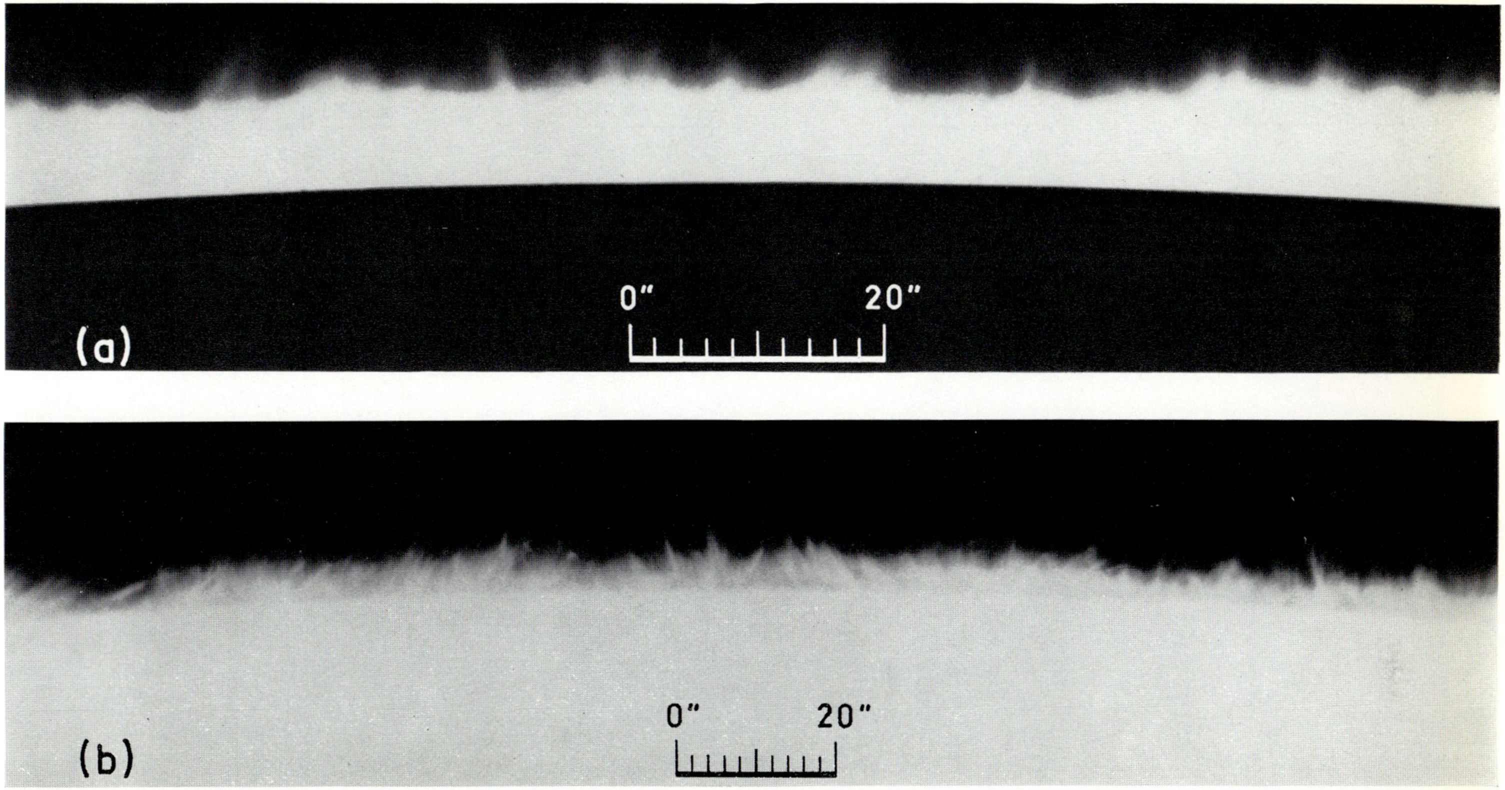

PLATE 2.3. Chromospheric spicules photographed (a) at the centre of Hα through a 1 Å Halle filter and (b) at Hα + 0·9 Å through a $\frac{1}{4}$ Å Zeiss filter. (a) was obtained with the 12-inch refractor of the Culgoora Solar Observatory and (b) with the new vacuum tower telescope of the Sacramento Peak Observatory, diaphragmed down to 20 inches. (*Photograph* (b) *by courtesy of Sacramento Peak Observatory.*)

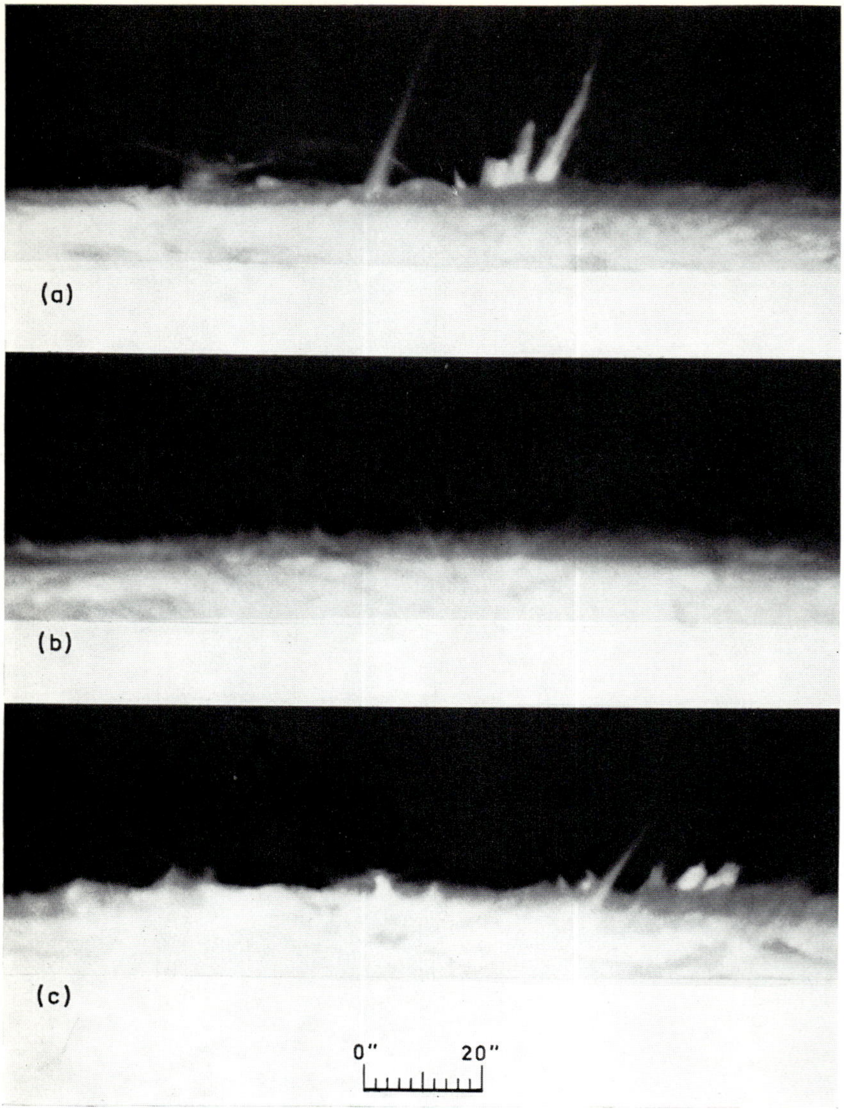

PLATE 2.4. Low chromosphere beyond west limb in the vicinity of an active region photographed on three separate occasions near the centre of Hα with a bandwidth of 0·5 Å (frames a and b) and 1·0 Å (frame c). (a) and (c) show the low chromosphere blocking out the light of several small bright prominences projecting up from behind the limb, demonstrating that at the centre of Hα the chromosphere is optically thick along a tangential line-of-sight. (b) and (c) show examples of small, spike-like prominences which, in observations of lower spatial resolution, could easily be mistaken for spicules.

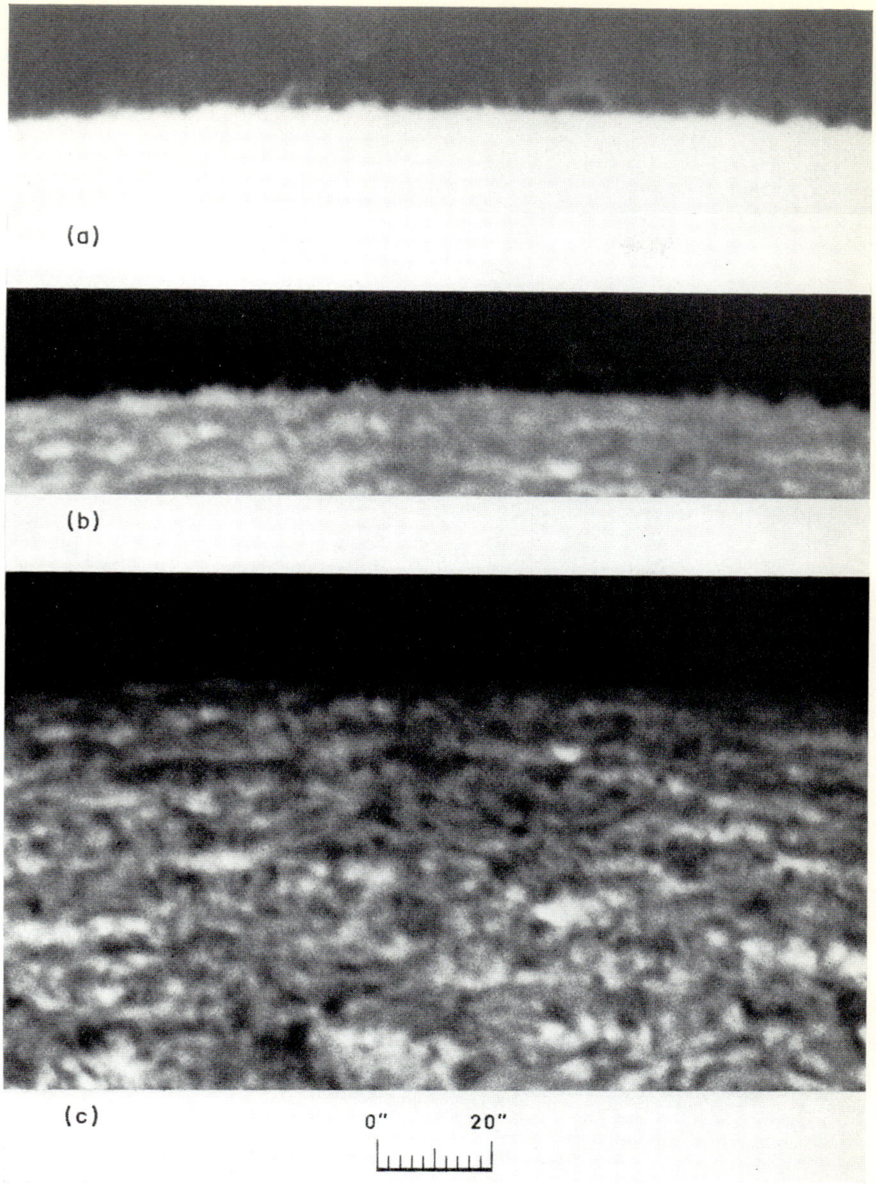

PLATE 2.5. Quiet chromosphere bordering the north limb photographed at the centre of Hα through a $\frac{1}{8}$ Å filter plus 1 Å Halle filter. All three photographs have been printed from the same negative: (a) shows spicules protruding upwards from the low chromosphere; (b) brings up detail at the extreme limb; and (c) reveals structure on the disk bordering the limb. Note that no trace of the photospheric limb is discernible.

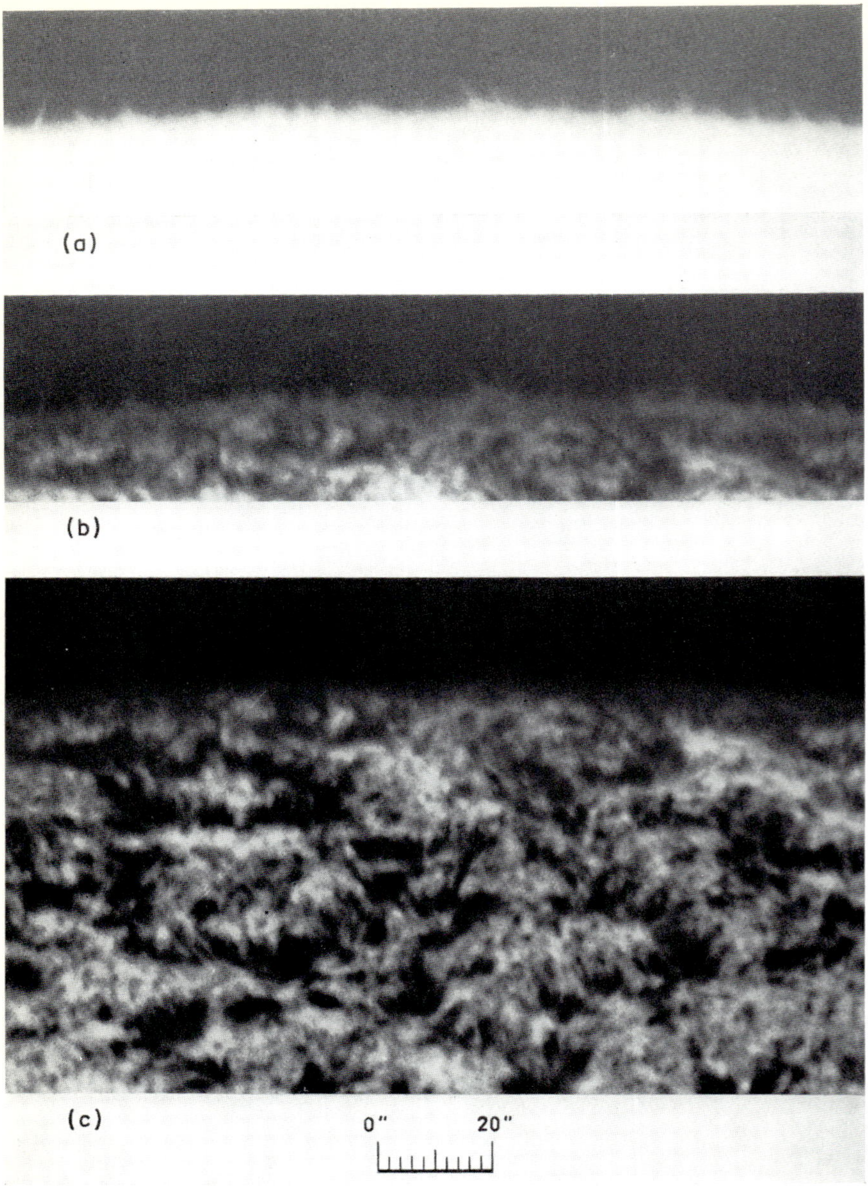

PLATE 2.6. Quiet chromosphere bordering the north limb photographed at Hα − 0·5 Å through a $\frac{1}{8}$ Å filter plus 1 Å Halle filter. All three photographs have been printed from the same negative: (a) shows spicules protruding upwards from the low chromosphere; (b) brings up detail at the extreme limb; and (c) reveals structure on the disk bordering the limb. Note that no trace of the photospheric limb is discernible.

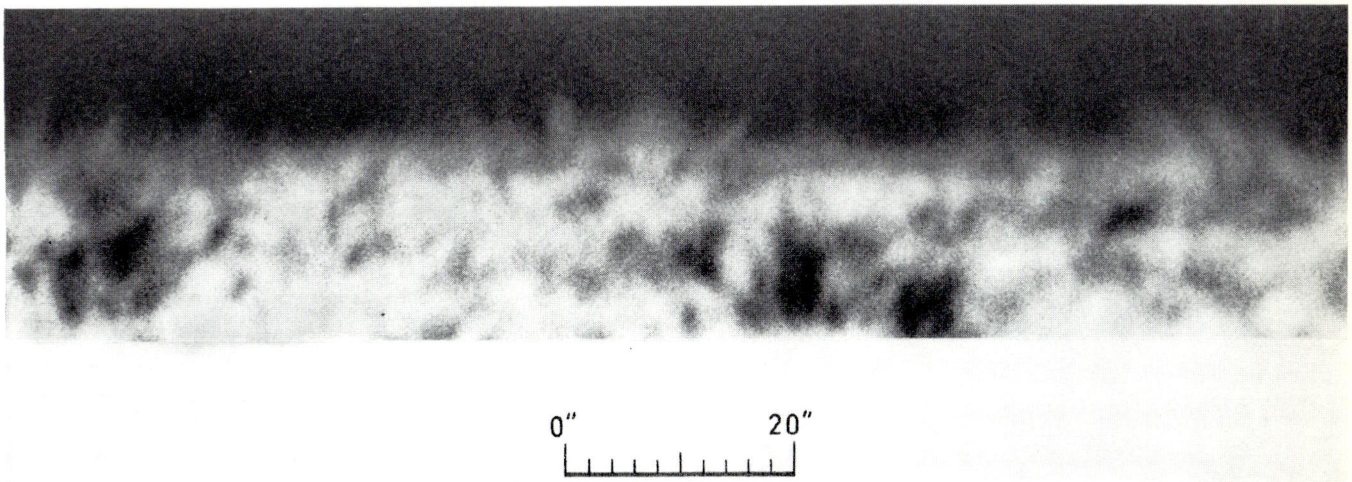

PLATE 2.7. Quiet chromosphere bordering the north limb photographed at Hα − 0·65 Å through a $\frac{1}{8}$ Å filter plus 1 Å Halle filter. Note the presence of the photospheric limb along a limited arc near the centre, where the transparency of structures in the foreground happens to be greater than on the sides.

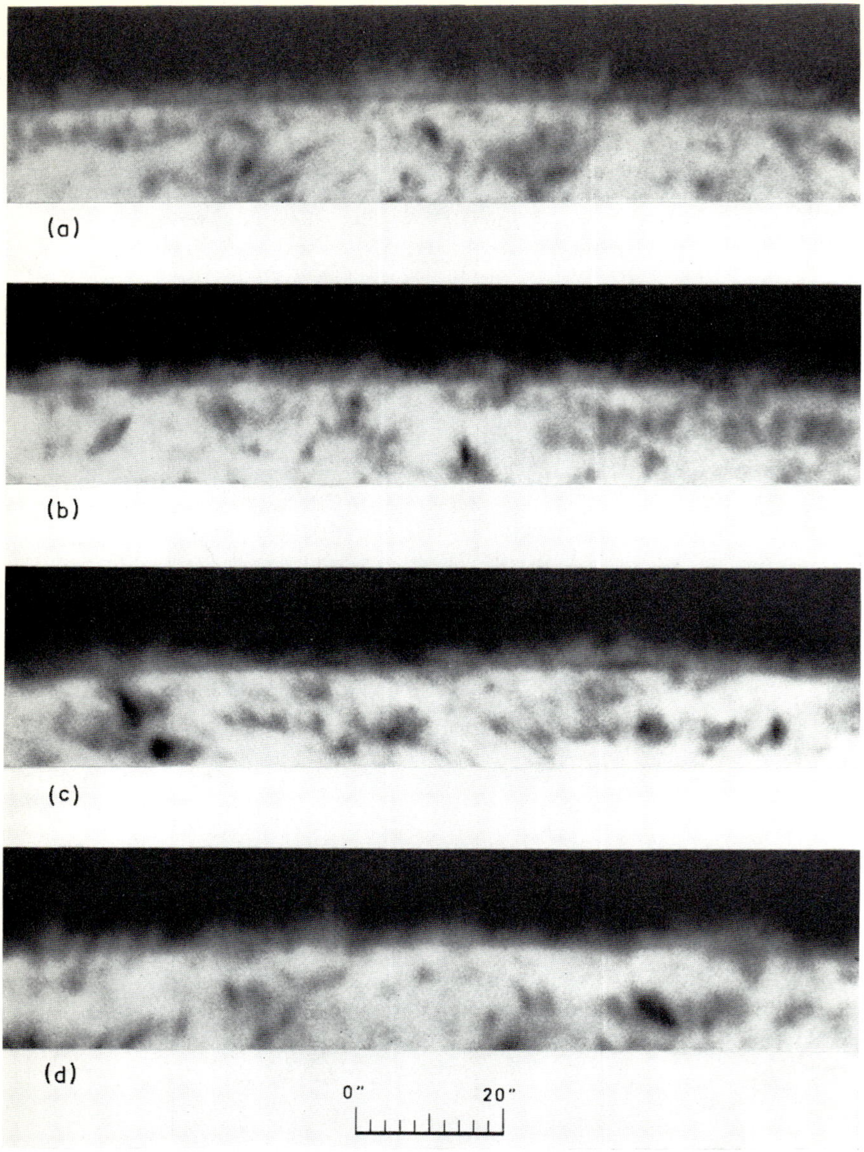

PLATE 2.8. Quiet chromosphere bordering the north limb photographed on separate occasions at Hα − 0·75 Å (a and c) and Hα + 0·75 Å (b and d) through a $\frac{1}{8}$ Å filter plus a 1 Å Halle filter. Note the narrow dark band lying immediately above the photospheric limb, together with isolated examples of dark mottles and spicules crossing the limb.

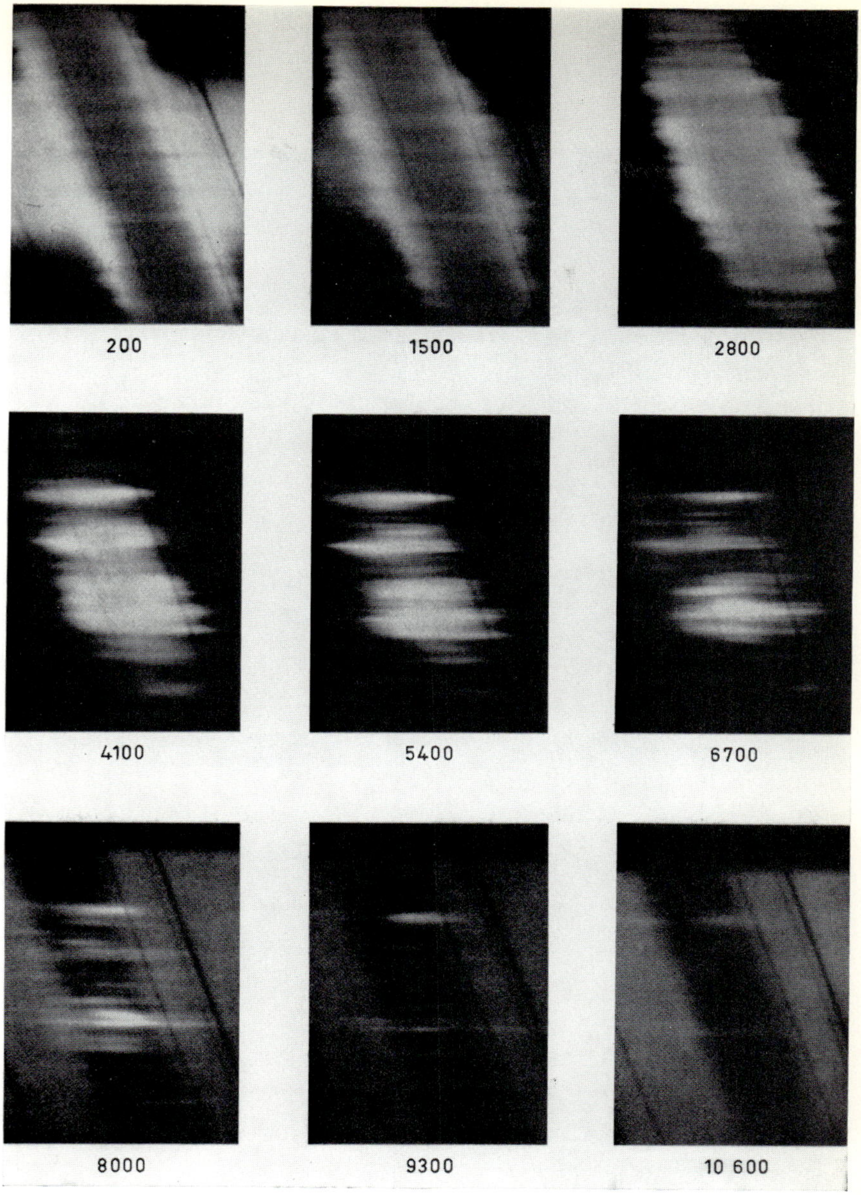

PLATE 2.9. Series of good-quality spectrograms obtained by Zirker [1962] showing the spicule structure visible in Hα at various heights above the photospheric limb. Note the large Doppler shifts of some of the structures and the characteristic central reversal of Hα at low heights (cf. Fig. 4.9). The spectrograph slit was aligned parallel to the limb, its minimum height in km being indicated by the figures.

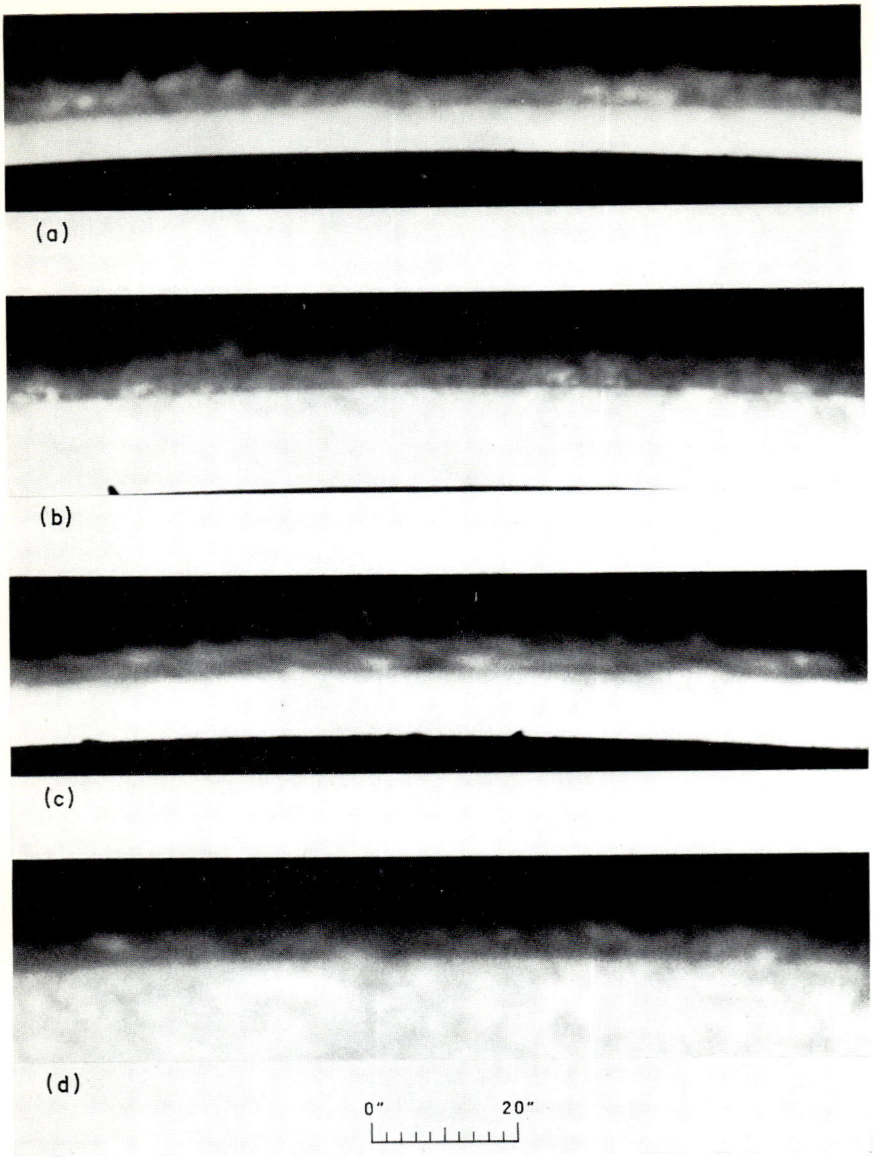

PLATE 2.10. Low chromosphere beyond limb photographed on four separate occasions at the centre of Hα through a 1 Å Halle filter. The conspicuous small bright blobs are identified as the bright mottles seen on the disk (Section 3.2.3). Their apparent heights range from about 1″ to 4″·5 of arc (725–3300 km). (a), (b), (c), and (d) refer to the south, north, west, and north limbs respectively.

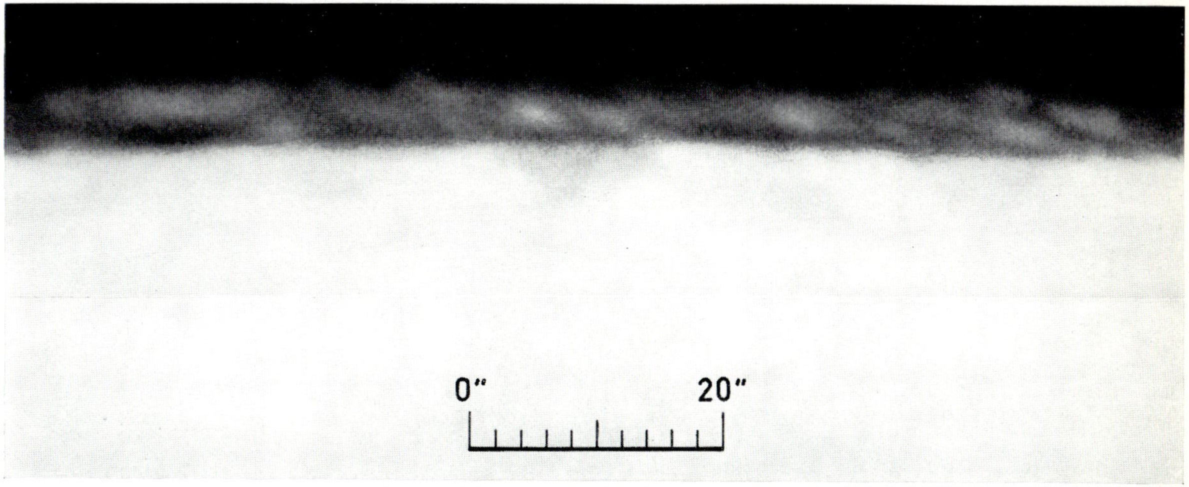

PLATE 2.11. Low chromosphere beyond west limb photographed at the centre of Hα through a 1 Å Halle filter. It shows the bright mottles and, in addition, the dark band lying immediately above the photospheric limb, whose width and intensity vary markedly from point to point along the limb (see also Plate 2.8).

Sometimes, after a spicule has descended or has shrunk to a mound in the low chromosphere, another spicule may appear to rise from the same position [Lippincott, loc. cit.]. The lifetime of such spicule-producing locations may be several times the observed lifetime of an individual spicule. Platov and Shilova [1969] and Krat and Krat [1971] have also commented on the tendency for spicules to recur at approximately the same locations.

Finally, from the viewpoint of spicule evolution, it is interesting to note that out of the 77 spicules studied by Lippincott [1957], 32 (or 42 per cent) faded over their entire lengths at maximum extension, while the remaining 45 (or 58 per cent) appeared to descend or to fade from the top (the observations do not allow one to distinguish between these possibilities). Of the latter, 21 remained at their maximum height for one minute or more.[12] The maximum heights attained by the individual spicules ranged from about 7000 to 18 000 km [see Lippincott, loc. cit.: Fig. 5], the average maximum height being 11 600 km. Dizer [1952] found that the maximum height reached by a spicule increases with the duration of its ascent, whereas Rush and Roberts [1954] established a strong correlation between the maximum height and the velocity of ascent. The velocities of spicules are discussed in the following sections.

2.4.2 VERTICAL VELOCITIES DERIVED FROM PHOTOGRAPHIC MEASUREMENTS

A number of determinations of the apparent upward and downward velocities of spicules have been made based on the measurement of their heights on successive photographs. The upward motion is usually fairly regular and continuous, the spicule rising with constant velocity and stopping abruptly at its maximum height [Rush and Roberts, 1954; Lippincott, 1957]. However, it is not uncommon for a spicule to rise in a series of jerks. In these cases Lippincott finds that the velocity is roughly the same during each separate movement, although Rush and Roberts report that large changes in velocity do occasionally occur. According to the latter authors the downward velocities of the spicules – in the instances where they are observed (see Section 2.4.1) – may be smooth but are more often abrupt and irregular. The results of all available determinations of the upward and downward velocities of spicules are summarized in Table 2.4.

In comparing the various values for the average velocities given in Table 2.4, one has to remember that some of the determinations may be affected by observational selection. For example, Roberts and his collaborators seem to have concentrated their attention on the larger, longer-lived spicules, whereas

[12] This last statement contradicts the observation of Dizer [1952] that all spicules fall back immediately.

Lippincott analysed a sample of 77 spicules covering a wide range of maximum heights and lifetimes. Fig. 2.9 shows the distributions of upward and downward velocities found by Lippincott. The upward velocities cover the range 5 to 60 km sec^{-1} and have a mean value of 24 km sec^{-1}, while the downward velocities cover the range 0 to 70 km sec^{-1} and have a mean value of 28 km sec^{-1}. The distribution of upward velocities has also been investigated by Rush and Roberts

TABLE 2.4

Vertical Velocities of Spicules Derived from Photographic Measurements

REFERENCE	AVERAGE VELOCITY (km sec^{-1})		REMARKS
	UPWARD	DOWNWARD	
Roberts [1945]	30	–	–
Roberts *et al.* [1949]	38, 72	–	Values for two separate determinations
Dizer [1952]	20 40	20 40	Quiet regions Active regions
Rush and Roberts [1954]	31	–	–
Lippincott [1957]	24	28	–

[1954, cf. Fig. 3], using a sample of 59 spicules. Taking the results of all observers into account it is perhaps reasonable to adopt a figure of 25 km sec^{-1} as a representative value for the average velocity, upward or downward, of the spicules.

According to Rush and Roberts [1954] the upward velocities of the individual spicules are independent of their lifetime but show a strong correlation (coefficient of correlation = 0.70) with their maximum heights. The relationship appears to be essentially linear [cf. Rush and Roberts, loc. cit.: Fig. 4]. These results, however, contradict those of Dizer [1952], who concluded that the velocity of ascent (and descent) is much the same for all normal spicules and that the maximum height attained by a spicule depends on its time of ascent. These aspects of spicule behaviour clearly deserve further study.

2.4.3 LINE-OF-SIGHT VELOCITIES DERIVED FROM SPECTROSCOPIC MEASUREMENTS

It has to be emphasized that the velocities deduced from direct observations of the changes in spicules with time are, strictly speaking, only *apparent* velocities, since the observed changes could equally well result from temporal variations in the Hα emission of the spicules. Fortunately, however, one can discriminate between the two alternatives by appealing to spectroscopic observations, which

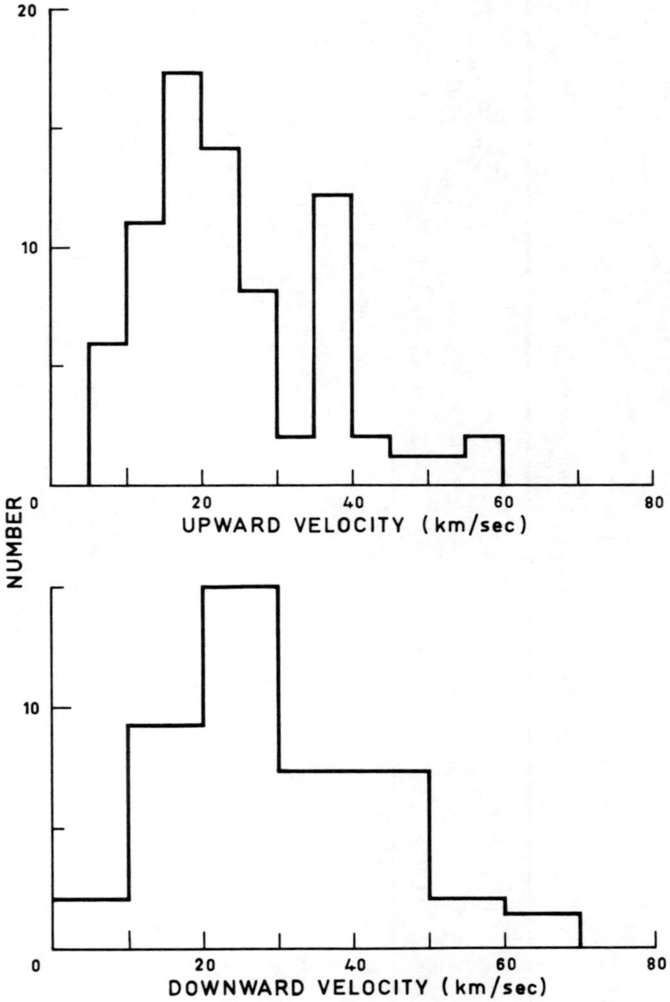

Fig. 2.9. Vertical velocities of spicules. The histograms show the distributions of the upward and downward spicule velocities determined by Lippincott [1957] from direct photographs. The mean upward and downward velocities are 24 and 28 km sec^{-1} respectively.

indicate the existence of Doppler shifts corresponding to velocities of the same order as those described above.

Nevertheless, in comparing the velocity information provided by the two

TABLE 2.5

Measurements of Spicule Doppler Shifts

REFERENCE	OBSERVATIONS	SPECTRAL LINE	LATITUDE	HEIGHT (km)	R.M.S. LINE-OF-SIGHT VELOCITY (km sec^{-1})	AVERAGE LINE-OF-SIGHT VELOCITY (km sec^{-1})
Michard [1954]	Arcetri	Hα	Pole	3000–6500	13.7	–
Michard and Laborde [1955]	Arcetri	Hα	–	4200–6150	20.5	–
Michard [1956]	Arcetri	Hα	–	4000–9000	19.5	14.4
Michard [1959]	Meudon	Hα	–	–	6	–
Athay and Bessey [1964]	Climax	Hα	Equator	–	9.2	3.7
		Hα	Pole	–	4.6	2.8
		D_3	Equator	–	5.6	4.1
		H	Equator	–	3.8	2.5

Mouradian [1965]	Pic-du-Midi	Hα	Equator	5500	15.0	–
				6500	13.2	–
				7500	13.8	–
				8500	10.3	–
				9500	8.9	–
Zirker [1967]	Sacramento Peak	Hα	Equator	5000–6000	21.9	16.2
Pasachoff *et al.* [1968]	Sacramento Peak	Hα	–	5000	15.4	12.2
				6800	16.1	12.2
Krat and Krat [1971]	Pulkovo	Hα	Equator	5000–9000	–	14.5*
		D_3	Equator	5000–9000	–	10.9
		K	Equator	5000–9000	–	10.6

* Values inferred from the velocity measurements given in Table V of the authors' paper.

Note: The measurements of Nikolsky and Sazanov [1966] have been omitted since these authors give neither the average nor r.m.s. values of the observed velocities.

different techniques, one has to bear in mind two important differences between them. In the first place, whereas photographic observations give the vertical velocities of the spicules in the plane of the limb, spectroscopic observations give the velocity component along the line-of-sight and hence in the direction normal to the plane of the limb. (Thus the fact that large Doppler shifts are observed at all supplies further evidence that many of the spicules must be inclined at appreciable angles to the vertical: see Section 2.3.7.) Secondly, the spatial resolution of spicule spectrograms is generally low compared to that of high-quality direct photographs (Section 2.3.9) and hence the observed Doppler shifts must refer mainly to the larger or brighter spicules or sometimes perhaps, to unresolved clumps of spicules [see, for example, Zirker, 1962; Giovanelli *et al.*, 1965]. In addition, it is clear that spicules with large Doppler velocities will be more easily distinguished than those with small velocities. Finally, there is always the possibility of mistaking small, spike-like prominences for spicules (cf. Section 2.3.1 and Plate 2.4).

As we shall see in Chapter 4, considerable effort has been directed towards spectroscopic investigations of spicules with a view to determining their physical conditions and, in fact, such investigations far outnumber those based on direct photographs. Nevertheless, spectrograms of the quality needed for determining the Doppler shifts of spicules are rather difficult to obtain [cf. Athay and Bessey, 1964], and only a relatively small number of such measurements have so far been made. These are summarized in Table 2.5.

In comparing the various measurements of the Doppler shifts presented in this table we are confronted with the immediate difficulty that some authors quote their results only in the form of an r.m.s. spicule velocity, whereas what we really want to know are the *average* velocities towards and away from the observer, or at least their mean magnitudes.[13] Certain other points also have to be noted, viz.,

(1) Michard [1959] states that not too much weight should be attached to his earlier velocity values, which have probably been overestimated;

(2) both the r.m.s. and average values found by Athay and Bessey [1964] and Zirker [1967] are quoted from Beckers [1968, Table XII]; and

(3) the features measured by Zirker were located in a region bordering an old active centre and it is not certain that they were all spicules.

Taking all factors into account it seems reasonable, following Athay and Bessey [1964] and Beckers [1968, p. 393; 1972, p. 81], to adopt a value of

[13] The magnitude V_D of the line-of-sight velocity corresponding to a shift $\Delta\lambda$ of a line of wavelength λ is given by the usual Doppler formula

$$V_D = \frac{\Delta\lambda}{\lambda} \cdot c \ \ (\text{km sec}^{-1}),$$

where $c = 2.998 \times 10^5$ km sec^{-1} is the velocity of light.

about 6 km sec^{-1} for the magnitude of the average line-of-sight velocity of spicules observed in Hα.

We will now show that the adopted line-of-sight velocity of 6 km sec^{-1} is consistent with the figure of 25 km sec^{-1} for the average vertical velocity derived from direct photographs (see Section 2.4.2), provided due account is taken of the known inclinations of the spicules to the vertical. The vertical velocity of 25 km sec^{-1} is measured in the plane of the limb and, in order to obtain the actual velocity along the axis of the spicule, allowance should be made for the inclination of the spicule to this plane. However, since this correction is probably small in most cases, we may with sufficient accuracy take $V = 25$ km sec^{-1} as the true velocity along the spicule axis. Then, following Athay and Bessey [1964], one can show that the magnitude of the average Doppler velocity of spicules lying in a cone whose axis is normal to the line-of-sight and whose apex angle is 2β is given by the formula

$$\bar{V}_{\mathrm{D}} = \frac{V \sin\beta \int_0^{\pi/2} \sin\phi \; \mathrm{d}\phi}{\int_0^{\pi/2} \mathrm{d}\phi} = \frac{2}{\pi} V \sin\beta, \tag{2.3}$$

where ϕ is the azimuthal angle of the projection of the spicule velocity vector onto a plane tangent to the limb (cf. Fig. 2.10). Hence, putting $V = 25$ km sec^{-1}, we have

$$\bar{V}_{\mathrm{D}} = 16 \sin\beta \quad (\mathrm{km\ sec^{-1}}). \tag{2.4}$$

Thus for $\bar{V}_{\mathrm{D}} = 6$ km sec^{-1} we get $\beta \simeq 22°$, which is close to the average inclination of the spicules determined from direct photographs (Section 2.3.7).

Among the results presented in Table 2.5 those of Athay and Bessey [1964] are especially interesting, because they give the average line-of-sight velocities measured not only in Hα but also in the helium D_3 and Ca II H lines. Furthermore, their Hα measurements indicate that the average Doppler velocity of polar spicules is less than that of equatorial spicules, thus providing yet another example of the dependence of spicule properties on latitude (Section 2.3.8).

The distribution of spicule Doppler velocities has been considered by Michard [1956], Athay and Bessey [loc. cit.], and Zirker [1967]. Fig. 2.11, taken from Zirker's paper, shows the distributions derived from the data of Athay and Bessey and of Mouradian [1965]. (We have omitted from the diagram the distribution found by Zirker himself because not all of the features studied could conclusively be identified as spicules.) It should be noted that Mouradian's data refer only to equatorial spicules, whereas Athay and Bessey's data refer to a sample containing both equatorial and polar spicules. In both cases the observed line-of-sight velocities are less than 30 km sec^{-1}.

Michard [1956] studied the distribution of the Doppler velocities of 35 spicules having heights in the range 4000–9000 km and found almost as many velocities directed towards as away from the observer. On the other hand, Mouradian [1965] concluded that this is true only for spicules above 6000 km. He measured the velocities of 10 spicules during the course of their lives and found that, below 6000 km, almost all the velocities are directed *towards* the

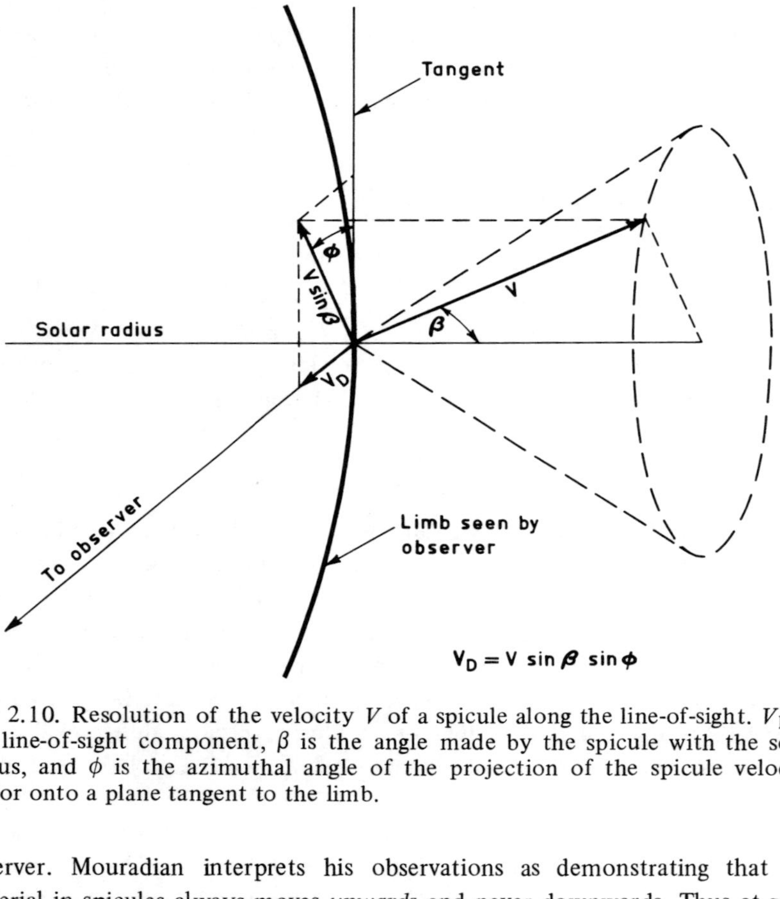

Fig. 2.10. Resolution of the velocity V of a spicule along the line-of-sight. V_D is the line-of-sight component, β is the angle made by the spicule with the solar radius, and ϕ is the azimuthal angle of the projection of the spicule velocity vector onto a plane tangent to the limb.

observer. Mouradian interprets his observations as demonstrating that the material in spicules always moves *upwards* and never downwards. Thus at great heights, where we presumably see only spicules located close to the limb, the observed Doppler shifts will be positive (red) or negative (blue) depending on whether the spicule is inclined away from or towards the observer. At low heights, on the other hand, Mouradian assumes that we see only foreground spicules and hence the velocities will predominantly be directed towards the observer.

However, it must be emphasized that a predominance of blue shifts among low-lying spicules has not so far been reported by any other observers, with the possible exception of Krat and Krat [1971]. They have published the results of 16 spicule velocity measurements (referring to a total of 14 separate spicules) made in Hα (see their Table V) which yield an *algebraic* mean value of -3.5 km sec^{-1}, the negative sign indicating a Doppler displacement to the blue. However,

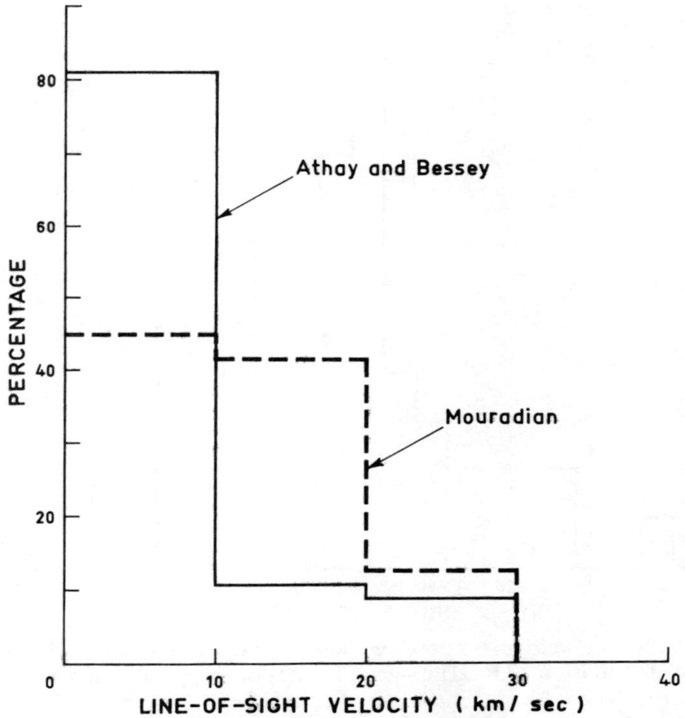

Fig. 2.11. Spicule velocities: observed line-of-sight components. The histograms show the distributions of spicule Doppler velocities plotted by Zirker [1967] from the data of Athay and Bessey [1964] and Mouradian [1965].

it is not clear how much significance should be attached to this result in the present context, owing to the fact that Krat and Krat obtained their spectrograms at heights ranging from 5000 to as high as 9000 km.

On the other hand, Nikolsky and Sazanov [1966] found that the 'centre of gravity' of curves showing the Doppler velocities of 11 spicules as functions of time was displaced to the *red* by an amount corresponding to a velocity of 5 km sec^{-1}, although they admit that the displacement might be due simply to an inaccuracy in the determination of the centre of the unshifted spectral line; the measurements were made in Hα at a height of about 6000 km. Moreover, their result is supported by Doppler measurements made in Hα by Pasachoff *et al.* [1968], who found that the algebraic means of the velocities of 64 spicules were + 2.7 km sec^{-1} at 5000 km and + 3.9 km sec^{-1} at 6800 km, the positive sign indicating a displacement to the red. The actual distribution of the positive and negative spicule velocities measured by these authors at 6800 km is shown in Fig. 2.12, which is taken from their paper.

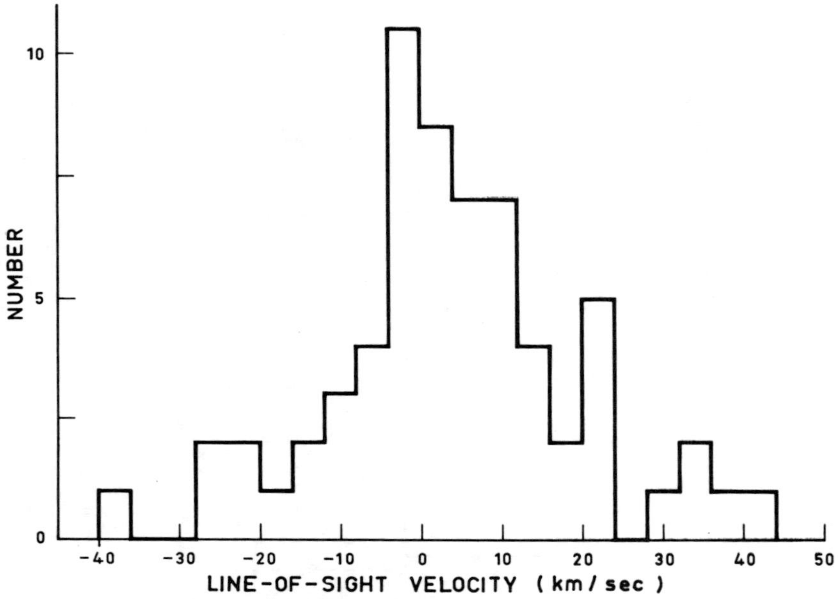

Fig. 2.12. Spicule velocities: observed line-of-sight components. The histogram shows the distribution of the Doppler velocities of 64 spicules measured in Hα at a height of 6800 km by Pasachoff *et al.* [1968]. Positive velocities correspond to red shifts and negative velocities to blue shifts; the algebraic mean velocity is +3.9 km sec^{-1}.

In conclusion, we should mention that Pasachoff *et al.* [1968], Nikolsky [1970], Weart [1970], and Nikolsky and Platova [1971] have recently found evidence from sequences of Hα spectra taken with the slit tangential to the limb that some spicules change their position along the slit relative to the average positions of other features, thus apparently showing a real *horizontal* component of velocity. According to Pasachoff *et al.* the direction of motion parallel to the limb remains the same for a given spicule, but Weart has noted a few exceptions to this rule. On the other hand, Nikolsky and Platova have concluded that the spicules execute quasi-periodic *oscillations* parallel to the limb (see their Fig. 2). They found that the oscillations are random in character, showing no definite relationship between velocity and amplitude. However, it must not be forgotten that, even under good observing conditions, atmospheric seeing will tend to make spicules appear to execute random oscillations over distances of the order of 1″ of arc or greater. At the present time it is not clear to what extent the observations of the various authors cited above – and hence their conclusions – may have been influenced by purely seeing effects.

2.4.4 VELOCITY VARIATION WITHIN INDIVIDUAL SPICULES

When we turn to the question of possible spatial and temporal variations of the motions within individual spicules we find that our present information is very scanty and often quite contradictory.

(1) *Variation with height.* The observations pertaining to this aspect of spicule behaviour could hardly be more contradictory. Evidence suggesting a decrease in velocity with height in a spicule has been given by Michard [1956] and Mouradian [1965], although the latter found that during the later part of a spicule's life its velocity becomes uniform. On the other hand, Michard [1959] observed two spicules with large Doppler velocities which increased with height, while Pasachoff *et al.* [1968] found that, out of a sample of 64 features measured in Hα, 27 per cent showed an increase in velocity with height, 28 per cent showed a decrease, and the remaining 45 per cent displayed virtually no change. Summarizing the results of their rather extensive Doppler measurements both in Hα and in the H line of Ca II, Pasachoff *et al.* have concluded that there is no systematic trend in the variation of spicule velocities with height. However, they have pointed out the existence of a class of features on their spectrograms (perhaps some 10 per cent of the total) characterized by large velocities of the order of 50 km sec^{-1}, which do show an increase in velocity with height. Nikolsky and Sazanov [1966] found no change in the velocity of individual spicules over a wide range of heights.

(2) *Variation with time.* Curves illustrating the time variation of the Doppler velocity of a spicule during its lifetime have been given by Mouradian [1962, 1965], Nikolsky and Sazanov [1966], Zirker [1967], and Pasachoff *et al.* [1968]. According to Mouradian the velocity of a spicule rises to a maximum and then declines to zero, but never changes sign during the lifetime of the spicule (cf. Fig. 2.13). The latter conclusion was also reached by Bhavilai [1965] from a comparison of pairs of spicule photographs obtained simultaneously at Hα ± 0.75 Å. These observations imply that the direction of motion is either always upwards or always downwards and, in fact, as we have seen in Section 2.4.3, Mouradian has adduced further evidence to show that the motion is always directed upwards. On the other hand, examination of spicule spectra taken at the Sacramento Peak Observatory has led Pasachoff *et al.* to the conclusion that, although the majority of spicules show the behaviour described

by Mouradian, a certain number (of the order of 20 per cent) do undergo a reversal in their Doppler velocity. By comparison, direct photographs indicate that more than 50 per cent of spicules appear to descend or fade from the top (Section 2.4.1).

Completely different results have been obtained by Nikolsky and Sazanov [1966], using spectra obtained with the coronagraph of the Institute of

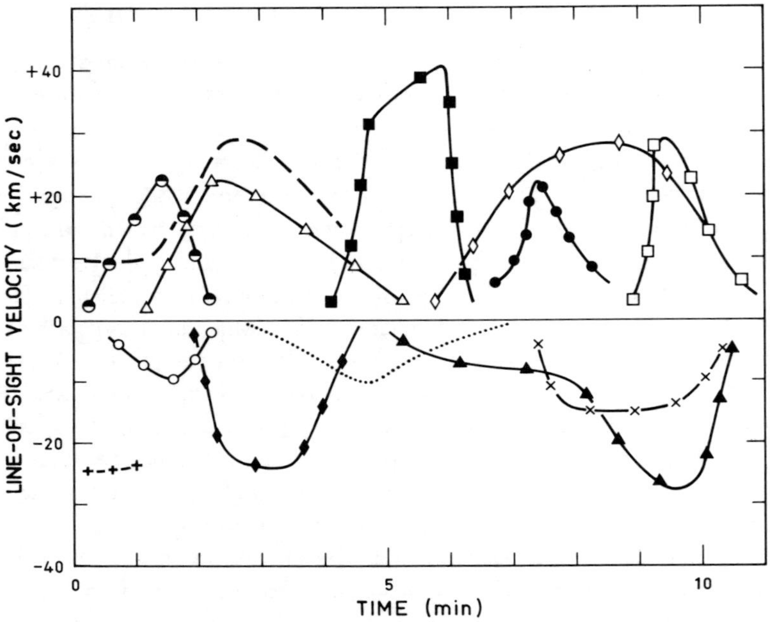

Fig. 2.13. Time variation of the line-of-sight velocities of 13 spicules during the course of their lives [Mouradian, 1965]. Note that these observations imply that the direction of motion for a given spicule is either always upwards or always downwards (but see discussion in text). Positive velocities correspond to red shifts and negative velocities to blue shifts.

Terrestrial Magnetism near Moscow (NIZMIR). They find that the line-of-sight velocities of individual spicules vary irregularly with a mean period of about one minute and often change sign [see also Nikolsky, 1970]. Apparently similar changes have also been reported by Zirker [1967], while Pasachoff *et al.* [1968] have mentioned instances of spicules displaying apparently oscillatory motions with a period measured in minutes (cf. their Fig. 6). Nevertheless, at the present time the balance of the evidence seems to indicate that most spicules show a constant direction of motion along their axes throughout their lifetimes. However, a reversal of velocity does occur in some cases, and this supports the conclusion that the downward velocities inferred from direct photographs must at least partly correspond to true downward motions and are not due solely to changing conditions of excitation. On the other hand, Nikolsky [1970] and Nikolsky and Platova [1971] have drawn attention to the alternative possibility that the velocity oscillations described above may be due to bodily movement of the spicules backwards and forwards along the line-of-sight.

(3) *Time difference between velocity changes at different heights.* Pasachoff *et al.* [1968] have studied the difference between the times of occurrence of the maximum Doppler velocity in individual spicules at two different heights 1800 km apart, using sequences of Hα spectra obtained at 5-second intervals with the aid of an image slicer. Both positive and negative time lags were recorded and, out of 21 cases in which the times of maximum velocity could be accurately determined at both heights, 60 per cent showed time lags within the range −10 sec to +10 sec and 70 per cent within the range −5 sec to +15 sec. Pasachoff *et al.* have concluded that the existence of a phase lag of more than 20 sec of either sign can certainly be ruled out; the true values are likely to be considerably less.

Accordingly, if the short time interval between events at the two heights is attributed to a disturbance propagating upward or downward along the spicule, then the phase velocities involved must be very high. As Pasachoff *et al.* point out, a time delay of less than 20 sec over a height interval of 1800 km implies a phase velocity greater than 90 km sec^{-1}.[14] This is five times greater than the velocity of sound in spicules (~18 km sec^{-1}) but is comparable to the Alfvén wave velocity for a magnetic field of some 17 gauss, if we assume a spicule density of 2.8×10^{-13} gm cm^{-3} at a height of 6000 km (cf. Chapter 4, Table 4.23 and footnote 10). Beckers [1966] suggests that the presence of a magnetic field may be essential for the existence of a spicule (cf. Sections 2.4.5 and 7.6).

Pasachoff *et al.* also remark that they have never seen the appearance of a spicule at the lower level followed by a brightening at the upper level seconds or minutes later. It is difficult to reconcile this result with the characteristic ascent of new spicules shown by direct photographs (Section 2.4.1), but perhaps it may simply reflect an effect of some as yet unrecognised observational selection.

(4) *Velocities measured in different lines.* An interesting observation was reported by Krat and Krat [1961] of two spicules appearing to have a considerably larger Doppler velocity in Hα than in the helium D_3 line. Subsequently, the same workers determined the velocities of 11 spicules in the Hα, D_3, and K lines simultaneously [Krat and Krat, 1971, Table V], two of the features being measured on spectrograms taken 15 seconds apart at somewhat different heights. Only in a minority of cases is there good agreement among the

[14] Since we observe a Doppler shift the spicule must be inclined to the vertical, and hence the actual distance between the two points along the length of the spicule exceeds the apparent height difference of 1800 km.

velocities measured in the three lines. The majority show substantial differences and, in fact, the velocity of one feature appears to change from +2.3 km sec^{-1} in Hα to −13.3 km sec^{-1} in D_3! However, it may well be that the origin of the apparent discrepancies lies in the difficulty of inferring *true* spicule velocities from the highly distorted and asymmetric line profiles shown by most of the features studied by Krat and Krat (see their Figs. 2 and 3). In fact, as these authors themselves point out, the observed profiles probably result from the blending of the separate profiles – each with its own Doppler shift – of two or more closely adjacent spicules which, as a consequence of inadequate spatial resolution, appear as a single feature on the spectrogram (cf. Section 2.3.9).

At the present time, we must conclude that there is no convincing evidence of any large systematic difference in the velocities of spicules inferred from different spectral lines. This conclusion is supported by the velocity measurements of Athay and Bessey [1964], given in Table 2.5, which reveal no substantial differences among velocities measured in Hα, D_3, and H.

(5) *Tilt of spicule emission streaks.* A number of workers have noticed that the emission streaks of spicules observed on spectrograms (cf. Plate 2.9) do not always lie parallel to the direction of dispersion but are sometimes inclined at a small angle to it [Michard, 1959; Athay and Bessey, 1964; Beckers, 1966; Livshitz, 1966; Pasachoff *et al.*, 1968]. Attempts to measure the actual angles by Pasachoff *et al.* on both K and Hα spectra suggest that the tilts are of the order of 1–2° and are as likely to be in one direction as the other. One obvious explanation would be that in these cases the feature observed actually consists of two adjacent, but separate, spicules with different Doppler velocities which are seen as one because of inadequate spatial resolution. However, Beckers and his co-workers have pointed out that tilts are seen up to heights where the number of spicules present is so low that there is very little possibility of such near-coincidence. Moreover, some inclined features appear to move as single entities. Against this, however, Weart [1970] has observed several features which in the course of time have each separated into two distinct components, one red-shifted and the other blue-shifted, moving apart along the direction of the slit. In most cases, nevertheless, we seem to have no alternative but to attribute the observed tilt to an actual difference in the line-of-sight velocity between the two sides of the spicule. The possibility of detecting such a difference clearly depends both on the spatial resolution of the spectrogram and on the magnitude of the velocity differential [cf. Beckers, 1968, p. 397].

A change in the Doppler shift across a spicule may be due either to a difference in the upward or downward velocity of the (inclined) spicule between its two sides or to a change in the line-of-sight velocity due to rotation about its axis (see Fig. 2.14). If the first explanation is correct, one would expect the degree of tilt of the emission streak to be related to the overall Doppler velocity of the spicule: in particular, no tilt should be observed in the case of spicules moving in the plane normal to the line-of-sight and therefore showing no Doppler shift. No such correlation would be anticipated if the tilt is due to spicule rotation. In an attempt to resolve this question Beckers and his co-workers [Beckers, 1968, p. 397; Pasachoff *et al.*, 1968] examined filtergrams obtained simultaneously in the red and blue wings of the Ca II K line (at

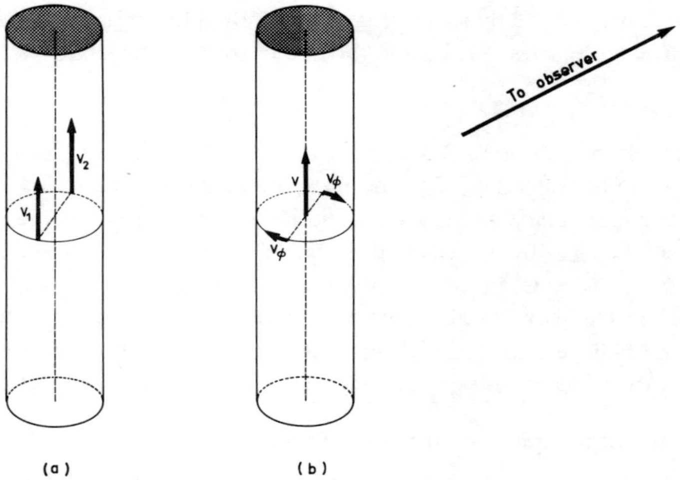

Fig. 2.14. Schematic drawing showing possible models of the velocity field inside a spicule proposed to explain the observed tilts of emission streaks. In (a) the magnitudes of the upward velocities V_1 and V_2 on the two sides are different, while in (b) the upward velocity V is constant but the spicule rotates around its own axis with a peripheral velocity V_ϕ.

K ± 0.3 Å) and found no evidence of a correlation, implying that the observed tilt is due to rotation. On filtergrams the presence of a velocity change across an individual spicule is revealed by a difference in its position on photographs taken on the two sides of the line; more than half of the 20 spicules examined showed a definite difference in position, the average shift being 450 km.

Taking account of the observed widths of the K and Hα lines Pasachoff *et al.* conclude that, if spicules rotate, their velocity of rotation cannot exceed about 30 km sec^{-1}. They further point out that a column of gas 1000 km in diameter rotating with this peripheral speed would have a velocity gradient of 0.06 km sec^{-1} per km across it – sufficient to produce a tilt of about 1°.9 in K and 2°.6 in Hα. Such a peripheral speed implies a centripetal acceleration approximately six times greater than the value of solar gravity! According to Pasachoff *et al.* the tilts of some spicule emission streaks are occasionally observed to change with time and even, in at least four observed cases, apparently reverse in direction. A clear illustration of such a reversal has been given by Noyes [see Avery, 1970, Fig. 3]. This behaviour would be consistent with material rising in a helix and then falling back along a similar path.

A final decision as to the exact mechanism responsible for producing the observed tilts of spicule emission streaks must await further observations. In particular, one may hope that continuing advances in high-resolution observing techniques may in future enable spectra to be obtained with sufficient spatial resolution to permit the direct measurement of Doppler shifts at several points across individual spicules – at least in the case of some of the larger ones.

2.4.5 MAGNETIC FIELD

As we have seen in Section 2.3.7, the spicules in the polar regions of the Sun tend to follow the direction of the overlying coronal rays, and this has led some authors to suggest that the alignment of both phenomena is determined by local magnetic fields [see, for example, Bugoslavskaya, 1950, p. 144; van de Hulst, 1953, p. 215]. To date, however, no direct measurements of magnetic fields in individual spicules have been published, although G. Brückner is quoted by Wentzel and Solinger [1967] as having observed fields in spicules with strengths not greater than 4 gauss, using a scanning aperture of 2″ × 2″ of arc.

2.5 Chromospheric Spicules: Summary of Properties

To complete our account of spicules we summarize in Table 2.6 the known quantitative data concerning their morphological and dynamical properties. This table provides a useful bird's-eye view, but for more comprehensive information the reader should, of course, refer to the appropriate section of the chapter. For convenience, the relevant section numbers are given in the penultimate column of the table. The physical conditions obtaining in spicules are summarized in Chapter 4, Table 4.23.

2.6 Bright Mottles in the Low Chromosphere

The presence of bright mottles in the low chromosphere beyond the photospheric limb was first pointed out by Loughhead [1969] on the basis of observations obtained during the course of a systematic programme of limb photography undertaken with the 12-inch Culgoora refractor.[15] For this work an occulting cone was mounted at the prime focus of the telescope and the photoelectric guider was equipped with motor-controlled fine motions to enable the solar image to be positioned precisely with respect to the cone. In most instances the edge of the occulting cone was placed a short distance within the

[15] Nikolsky [1970] independently reported the existence of small bright structures in the low chromosphere beyond the limb but did not identify them with any disk features.

limb in order to permit observations of features actually crossing the limb and to facilitate the measurement of the height of features above the limb. All photographs were taken through a Halle filter either at the centre or in the red wing of the Hα line; a bandwidth of 1 Å was generally used in order to keep the exposure times as short as possible. The spatial resolution achieved is illustrated in Plates 2.10 and 2.11, which show segments of the limb photographed at the centre of Hα on five separate occasions. Bright mottles are visible in the low chromosphere above the photospheric limb on every photograph; they are very conspicuous, although their number varies from frame to frame.

Subsequently, Loughhead and Tappere [1971] obtained photographs showing the bright mottles at a number of wavelengths in the Hα line with CSIRO's 1/8 Å filter, used in tandem with a 1 Å Halle filter to eliminate parasitic light. One such photograph, taken at the line centre, is reproduced in Plate 2.5. Due to the absence of parasitic light no trace of the photospheric limb is visible on Plate 2.5 (cf. Section 2.2) and bright mottles can be seen both on the disk and in the low chromosphere beyond the (now invisible) photospheric limb. In the latter region the appearance of the bright mottles does not differ essentially from that shown by the broad bandwidth photographs reproduced in Plates 2.10 and 2.11. The Culgoora observations, together with those made more recently by Alissandrakis and Macris [1971] through a 0.5 Å Halle filter, provide a considerable amount of information on the properties of the bright mottles observed beyond the limb. This information may be summarized as follows:

(1) *Shape and size.* The smaller bright mottles are roughly circular in shape, whereas many of the larger ones show a definite elongation parallel to the limb. As illustrated on the left-hand side of Plate 2.5b, some of the higher bright mottles show cone-shaped tops. Care must be taken not to confuse these tops with spicules: the latter are fainter and project much higher into the chromosphere (cf. Plate 2.3a).[16] According to Loughhead [1969] the horizontal dimensions of the bright mottles range from about 2″ to 5″.5 of arc (1450–4000 km) and their vertical dimensions from 1″ to 3″ (725–2200 km). The corresponding ranges found by Alissandrakis and Macris [1971] are 0″.6 to 8″.6 (450–6250 km) and 1″.2 to 4″.2 (850–3050 km) respectively.

On the very best photographs some of the bright mottles lose their diffuse, uniform appearance and begin to show traces of fine structure. Sub-structures smaller than 1″ can be detected on Plate 2.10b.

(2) *Lifetime.* The determination of the lifetimes of the bright mottles beyond the limb poses some difficulty because of their relative paucity and the stringent requirements on the quality of the seeing. Nevertheless, Loughhead [1969] was

[16] This conflicts with the claim of Alissandrakis and Macris [1971] that, at the centre of Hα, most bright mottles beyond the limb show upward extensions in the form of spicules.

TABLE 2.6

Summary of Properties of Spicules

PROPERTY	LATITUDE	HEIGHT (km)	ADOPTED AVERAGE VALUE	SECTION	REMARKS
Total Hα intensity	–	6000	2.5×10^4 erg cm^{-2} sec^{-1} $ster^{-1}$	2.3.4	Integrated over wavelength
Width	–	–	815 km	2.3.3	Corrected for instrumental profile
Apparent height	Pole Equator	– –	9500 km 7300 km	2.3.5	Measured at any given time
True height	Averaged	–	9800 km	2.3.5	Measured at any given time
Max. apparent height	Pole	–	11 600 km	2.4.1	Max. height attained during lifetime

Number at various heights	–	3000 6000 9000 12 000 15 000	(9.3×10^4) (4.2×10^4) 1.1×10^4 1.5×10^3 2.2×10^2	2.3.6	First two values are rather uncertain
Inclination	–	–	~20°	2.3.7	–
Lifetime	–	–	5 min	2.4.1	Greater for taller spicules
Vertical velocity (upward or downward)	–	–	25 km sec^{-1}	2.4.2	–
Line-of-sight velocity	–	–	6 km sec^{-1}	2.4.3	Value still rather uncertain
Magnetic field	–	–	⩽4 gauss	2.4.5	Value still very uncertain

able to obtain estimates of the lifetimes of 12 bright mottles. Four of these had lifetimes of 2¾, 6½, 11¼, and 12¾ min. For the remaining eight only lower limits could be assigned; these range from 4 to 7 min. Subsequently, Alissandrakis and Macris [1971] derived a mean lifetime of 12 min from measurements of a larger sample of 50 bright mottles beyond the limb.

(3) *Height.* According to Loughhead [1969] the apparent heights of the centres of the bright mottles above the photospheric limb range from about 1″ to 4″.5 of arc (725–3300 km), while Alissandrakis and Macris [1971] find some apparent heights as high as 6″ (4350 km). Of course, the measured values represent only lower limits since we do not know whether any given feature lies directly above the limb or is situated some distance in front or behind.

(4) *Change in visibility with wavelength.* Observations made with the 1/8 Å filter show that the bright mottles become less conspicuous as one moves outwards from the centre into the wings of the Hα line. In fact, at Hα ± 0.5 Å only a few remain (cf. Plate 2.6), and at Hα ± 0.65 Å they are almost entirely absent (cf. Plate 2.7).

(5) *Observation of bright mottles crossing the photospheric limb.* The properties of the bright mottles beyond the photospheric limb, listed above, are in good accord with those found for the bright mottles on the disk, as demonstrated in Table 3.1. These similarities suggest that they are the same structures. Conclusive proof of their identity is afforded by direct observations: as pointed out by Bhavilai [1965] and Bray [1969], the bright mottles on the disk observed at the centre of Hα remain visible right up to the photospheric limb and are actually seen to cross it. Plate 2.10d shows two bright mottles crossing the limb. The first, slightly to the left of the centre, is about halfway over the limb, while the second, on the right-hand side, has almost completed its passage.

Equally impressive evidence of the identity of limb and disk bright mottles is afforded by 1/8 Å photographs taken at the centre of Hα, such as that reproduced in Plate 2.5. Here one can see that, apart from the effects of foreshortening and limb darkening, the bright mottles remain identical in appearance on both sides of the (now invisible) photospheric limb. Taken together, the broad- and narrow-band observations leave no doubt as to the identity of the limb and disk phenomena.

The available information on the properties of the bright mottles – determined from both limb and disk observations – is summarized in Table 3.1.

2.7 Dark Band at Base of Chromosphere

The first published observations of the dark band at the base of the chromosphere are those of Loughhead [1969] and Nikolsky [1970]. They were taken at the centre of Hα through Halle filters having bandwidths of 0.5 or 1 Å. The appearance of the dark band on such broad-band photographs is well illustrated in Plate 2.11, where one can see that its width and intensity vary markedly from point to point along the limb; at some places it is entirely absent. Loughhead [1969, Fig. 4] has published a microphotometer tracing made across the dark band shown in Plate 2.11 in a direction radially outwards from the limb. On the tracing the dark band appears as a distinct depression immediately above the photospheric limb, whose apparent width is close to $1''.5$ of arc (1100 km).

In contradiction to the earlier observations, however, the more recent work of Loughhead and Tappere [1971] has shown that the dark band is not truly a phenomenon of the line centre. On the contrary, spectrally-pure observations taken through CSIRO's tunable 1/8 Å filter show that it occurs over only short ranges of wavelengths roughly centred on H$\alpha \pm 0.75$ Å. Plate 2.8 shows four segments of the north limb photographed through the 1/8 Å filter at H$\alpha - 0.75$ Å (Plates 2.8a and c) and H$\alpha + 0.75$ Å (Plates 2.8b and d). In each case the dark band is visible immediately above the photospheric limb. Its appearance on the broad-band photograph of Plate 2.11 and the spectrally-pure photographs of Plate 2.8 is closely similar, indicating that its presence cannot be attributed in any way to parasitic light. In agreement with the earlier observations, Plate 2.8 demonstrates that the dark band is more prominent at some locations than at others and is not continuous even over comparatively short segments of the limb.

Examination of Plate 2.8 also indicates that, at H$\alpha \pm 0.75$ Å, the dark mottles on the disk, whose properties are described in Section 3.2.4, not only extend right out to the photospheric limb but are often seen to actually *cross* it. Furthermore, some of the dark mottles crossing the limb appear to link up with the dark band (e.g. on the left-hand side of Plate 2.8b and the right-hand side of Plate 2.8c). This may be entirely coincidental but, if real, would suggest that the dark band, when observed, could be due to the presence in the foreground of chains of dark mottles which are optically thick in their lower parts and hence block out the light from the material behind them.

In any event, one should note that the dark band straddles the range of heights over which there is a pronounced minimum in the chromospheric temperature distribution (see Section 4.2.5). The question of whether the two phenomena are related merits investigation, but the interpretation of the dark band poses a very complex problem whose solution might require, among other things, the application of spherical radiative transfer theory to a non-uniform medium.

REFERENCES

ALISSANDRAKIS, C. E., and MACRIS, C. J. [1971] 'A study of the fine structure of the solar chromosphere at the limb', *Solar Phys.* **20**, 47.

ATHAY, R. G. [1958] 'Hα and D_3 profiles in chromospheric spicules', *Ann. Astrophys.* **21**, 98.

ATHAY, R. G. [1959] 'The number of spicules in the middle chromosphere', *Astrophys. J.* **129**, 164.

ATHAY, R. G., and BESSEY, R. J. [1964] 'Doppler shifts and line broadening in spicules', *Astrophys. J.* **140**, 1174.

AVERY, L. W. [1970] 'The formation of the Ca II K line in a spinning spicule', *Solar Phys.* **13**, 301.

BECKERS, J. M. [1964] 'A study of the fine structures in the solar chromosphere', AFCRL–Environmental Research Paper No. 49.

BECKERS, J. M. [1966] 'Observations of chromospheric spicules', *The Fine Structure of the Solar Atmosphere*, ed. K. O. Kiepenheuer, p. 80; Franz Steiner Verlag, Wiesbaden.

BECKERS, J. M. [1968] 'Solar spicules', *Solar Phys.* **3**, 367.

BECKERS, J. M. [1972] 'Solar spicules', *Ann. Rev. Astron. Astrophys.* **10**, 73.

BHAVILAI, R. [1965] 'The structure of the solar chromosphere. I. Identification of spicules on the disk', *Mon. Not. R.A.S.* **130**, 411.

BHAVILAI, R., NORTON, D. G., and GIOVANELLI, R. G. [1965] 'The "double limb" in Hα', *Astrophys. J.* **141**, 274.

BILLINGS, D. E. [1966] *A Guide to the Solar Corona*, Academic Press, New York.

BRAY, R. J. [1969] 'High-resolution photography of the solar chromosphere. VI. Properties of the bright mottles', *Solar Phys.* **10**, 63.

BRAY, R. J., and LOUGHHEAD, R. E. [1964] *Sunspots*, Chapman and Hall, London.

BRAY, R. J., and LOUGHHEAD, R. E. [1967] *The Solar Granulation*, Chapman and Hall, London.

BRAY, R. J., and WINTER, J. G. [1970] 'High-resolution photography of the solar chromosphere. VIII. Computer control of a tunable 1/8 Å filter', *Solar Phys.* **15**, 309.

BUGOSLAVSKAYA, E. Y. [1950] 'Structure of the solar corona', *Publ. Sternberg Astron. Inst.*, vol. 19.

COATES, R. J. [1958] 'A model of the chromosphere from millimeter wavelength observations', *Astrophys. J.* **128**, 83.

DIZER, M. [1952] 'Étude des films cinématographiques de la chromosphère solaire au bord du disque', *C.R. Acad. Sci.* **235**, 1016.

DUNN, R. B. [1960] 'Photometry of the solar chromosphere', Thesis, Harvard University.

DUNN, R. B. [1969] 'Sacramento Peak's new solar telescope', *Sky and Telescope* **38**, No. 6, p. 368.

GIOVANELLI, R. G., MICHARD, R., and MOURADIAN, Z. [1965] 'Some remarks on spectroscopic observations of chromospheric limb structures', *Ann. Astrophys.* **28**, 871.

HIEI, E. [1963] 'Continuous spectrum in the chromosphere', *Publ. Astron. Soc. Japan* **15**, 277.

HULST, H. C. VAN DE [1953] 'The chromosphere and the corona', *The Sun*, ed. G. P. Kuiper, p. 207, Univ. Chicago Press.

JAGER, C. DE [1959] 'Structure and dynamics of the solar atmosphere', *Handbuch der Physik*, ed. S. Flügge, vol. 52, p. 80; Springer, Berlin.

KOUTCHMY, S., and MACRIS, C. [1971] 'Observation in the wing of the Hα line and identification of the spicular structure near the solar limb', *Solar Phys.* **20**, 295.

KRAT, V. A., and KRAT, T. V. [1961] 'On the physics of the solar chromosphere', *Izv. Pulkovo Astron. Obs.* **22**, No. 167, p. 6.

KRAT, V. A., and KRAT, T. V. [1971] 'On physical properties of solar spicules', *Solar Phys.* **17**, 355.

LIPPINCOTT, S. L. [1957] 'Chromospheric spicules', *Smiths. Contr. Astrophys.* **2**, 15.

LIVSHITZ, M. A. [1966] 'Line widths in spicules', *Astron. J. U.S.S.R.* **43**, 718 (*Sov. Astron. AJ* **10**, 570).

LOUGHHEAD, R. E. [1969] 'High-resolution photography of the solar chromosphere. VII. Structure of the low chromosphere', *Solar Phys.* **10**, 71.

LOUGHHEAD, R. E., BRAY, R. J., TAPPERE, E. J., and WINTER, J. G. [1968] 'High-resolution photography of the solar chromosphere. I. The 30-cm refractor of the C.S.I.R.O. Solar Observatory', *Solar Phys.* **4**, 185.

LOUGHHEAD, R. E., and TAPPERE, E. J. [1971] 'High-resolution photography of the solar chromosphere. IX. Limb observations of high spectral purity', *Solar Phys.* **19**, 44.

LYNCH, D. K., BECKERS, J. M., and DUNN, R. B. [1973] *Solar Phys.* (in press).

LYOT, B. [1944] 'Le filtre monochromatique polarisant et ses applications en physique solaire', *Ann. Astrophys.* **7**, 31.

MARRIOTT, R. W. [1931] 'The United States Naval Observatory eclipse expedition to Niuafoou', *Popular Astron.* **39**, No. 5.

MENZEL, D. H. [1931] 'A study of the solar chromosphere', *Publ. Lick Obs.* **17**, 1.

MENZEL, D. H. [1959] *Our Sun*, revised Ed., Harvard Univ. Press.

MICHARD, R. [1954] 'Radial velocities of spicules and the width of chromospheric lines', *Observatory* **74**, 209.

MICHARD, R. [1956] 'Spectrophotométrie des raies d'émission intenses de la chromosphère. II. La raie Hα des spicules', *Ann. Astrophys.* **19**, 1.

MICHARD, R. [1959] 'La raie Hα dans la chromosphère', *Ann. Astrophys.* **22**, 547.

MICHARD, R., and LABORDE, G. [1955] 'Spectrophotométrie des raies d'émission intenses de la chromosphère solaire', *Atti Accad. Naz. Lincei*, Ser. 8, vol. 19, p. 283.

MOHLER, O. C. [1951] 'Eclipse observations of the structure of the chromosphere', *Mon. Not. R.A.S.* **111**, 630.

MOURADIAN, Z. [1962] 'Sur la cinématique des spicules de la chromosphère solaire', *C.R. Acad. Sci.* **255**, 3372.

MOURADIAN, Z. [1965] 'Contribution à l'étude du bord solaire et de la structure chromosphérique', *Ann. Astrophys.* **28**, 805.

MOURADIAN, Z. [1967] 'La diffusion des spicules dans la couronne solaire', *Solar Phys.* **2**, 258.

MOURADIAN, Z. [1969] 'Observations of spicules in the spectral lines of H and

Ca II', *Chromosphere–Corona Transition Region*, Proc. Conf. in Boulder, August 1969, p. 295; National Center Atmospheric Research, Boulder, Colorado.

NIKOLSKY, G. M. [1965] 'Observations of the spectrum of spicules and the fine structure of the solar chromosphere', *Astron. J. U.S.S.R.* **42**, 86 (*Sov. Astron. AJ* **9**, 64).

NIKOLSKY, G. M. [1970] 'The observation of the chromospheric fine structure by the 53-cm Lyot coronagraph', *Solar Phys.* **12**, 379.

NIKOLSKY, G. M., and PLATOVA, A. G. [1971] 'Motions of Hα-spicules along the solar limb', *Solar Phys.* **18**, 403.

NIKOLSKY, G. M., and SAZANOV, A. A. [1966] 'The motion and nature of Hα spicules in the solar chromosphere', *Astron. J. U.S.S.R.* **43**, 928 (*Sov. Astron. AJ* **10**, 744).

PASACHOFF, J. M., NOYES, R. W., and BECKERS, J. M. [1968] 'Spectral observations of spicules at two heights in the solar chromosphere', *Solar Phys.* **5**, 131.

PLATOV, J. V., and SHILOVA, N. S. [1969] 'Regularity in the distribution of spicular groups and their time characteristics', *Solnechnye Dannye 1969*, No. 12, p. 102.

PLATOV, J. V., and SHILOVA, N. S. [1971] 'Oscillations of visible chromosphere boundary and regularity in position of spicule groups along the limb', *Solar Phys.* **19**, 52.

ROBERTS, W. O. [1945] 'A preliminary report on chromospheric spicules of extremely short lifetime', *Astrophys. J.* **101**, 136.

ROBERTS, W. O. [1950] 'Observation of chromospheric opacity', *Publ. Astron. Soc. Pac.* **62**, 219.

ROBERTS, W. O., BRENTON, V. K., SHAPLEY, M. B., and KOPAL, Z. [1949] 'Further measures of chromospheric spicules', *Publ. Astron. Soc. Pac.* **61**, 160.

RUSH, J. H., and ROBERTS, W. O. [1954] 'Recent studies of chromospheric spicules', *Aust. J. Phys.* **7**, 230.

SIMON, G. W., and WHITE, O. R. [1966] 'The H-alpha double limb: a scattered-light phenomenon', *Astrophys. J.* **143**, 38.

WALDMEIER, M. [1960] 'Die Struktur der aktiven Chromosphäre', *Z. Astrophys.* **50**, 225.

WEART, S. R. [1970] 'The horizontal component of spicule motion', *Solar Phys.* **14**, 310.

WENTZEL, D. G., and SOLINGER, A. B. [1967] 'On the motion of shocks through the solar chromosphere', *Astrophys. J.* **148**, 877.

WHITE, O. R., and BHAVILAI, R. [1970] 'Comments on the discovery of the dark band in the Hα solar chromosphere', *Astrophys. Letters* **5**, 137.

WHITE, O. R., and SIMON, G. W. [1968] 'Resolution of the Hα double-limb controversy', *Solar Phys.* **3**, 269.

WOLTJER, L. [1954] 'A photometric investigation of the spicules and the structure of the chromosphere', *Bull. Astron. Inst. Netherlands* **12**, 165.

ZIRKER, J. B. [1962] 'Spectral observations of solar chromospheric spicules', *Astrophys. J.* **136**, 250.

ZIRKER, J. B. [1967] 'On the motions of chromospheric fine-structure in a weak plage', *Solar Phys.* **1**, 204.

CHAPTER 3

The Morphology and Dynamics of the Quiet Chromosphere Observed on the Disk

3.1 Introduction

In the previous chapter we have described the fine structure of the quiet chromosphere as revealed by observations made beyond the limb, paying particular attention to the properties of spicules. We now turn our attention to the structures seen on the disk, dealing with their morphology, velocities, and magnetic fields, but postponing to Chapter 4 any consideration of the physical conditions within them. An account of the basic fine structures observed in *active regions* is reserved for Chapter 5.

As in the case of features visible beyond the limb, most observations of the *fine* structures on the disk have been made in the Hα line with the aid of birefringent filters of various bandwidths. At first sight a high-definition photograph of the quiet chromosphere taken at the centre of Hα gives the bewildering impression of a more or less random conglomeration of bright and dark features of diverse shapes and sizes, inextricably mixed together. However, further observations at a variety of wavelengths make it clear that, despite its apparent complexity, the quiet chromosphere is composed basically of just two fine structures. One is the *bright mottles*, whose properties observed beyond the limb have already been described in Section 2.6, while the other is the *dark mottles* which, on the basis of substantial indirect evidence, are believed to be identified with spicules.

We begin the present chapter by describing certain features of the performance of birefringent filters and problems connected with their use (Section 3.2.2), and then proceed to outline the properties of the bright and dark mottles (Sections 3.2.3 and 3.2.4) and the spatial relationship between them (Section 3.2.5). The important question of the identification of spicules is discussed at some length in Section 3.2.6. After a brief description of other fine structures observed in Hα (Section 3.2.7), we conclude Section 3.2 by summarizing the somewhat scanty information available concerning the fine detail observed in the K line of Ca II and its correlation with Hα features (Section 3.2.8).

The bright and dark mottles seen in Hα are not randomly distributed over the solar surface but are mostly strung out along the boundaries of a large-scale cellular pattern known as the *chromospheric network*. The network is best observed with *medium* spatial resolution and is more prominent in the K line than in Hα. An account of its appearance and properties in the K line, Hα, and other chromospheric lines in the visible and infrared regions of the spectrum is given in Sections 3.3.2–3.3.4. Observations in these lines reveal the structure of the network in the low and middle chromosphere, i.e. up to about 10 000 km. In principle, by observing in selected ultraviolet lines we could extend our knowledge of the properties of the network to the upper chromosphere. In practice, however, most existing observations lack even the medium resolution – say, 6–8″ of arc – necessary to properly observe the network (Section 3.3.5). On the other hand, owing to rapid developments in rocket and satellite techniques this situation is unlikely to continue for long.

Essential clues to the origin and significance of the chromospheric network are provided by an examination of its relationships with two *photospheric* phenomena, namely, the supergranulation (Section 3.4.3) and the photospheric magnetic field distribution (Section 3.4.4). The supergranulation emerges as the fundamental agency: horizontal and vertical motions associated with the individual supergranules concentrate photospheric magnetic fields at the cell boundaries. These photospheric fields are presumed to extend upwards into the chromosphere, and together the photospheric and chromospheric fields are believed to be responsible for the excess heating which manifests itself as a bright network in the K line, in the centre of Hα, and also in various photospheric lines (Section 3.4.2).

The observed correlation between the network and the photospheric magnetic field is so close that a photograph of the network may be regarded as essentially a map of the magnitude of the field or, more precisely, of its *longitudinal* component. Thus, during the dissolution of an active region the network serves as a tracer for the ultimate dispersal of its field over the surface of the 'quiet' Sun.

Attempts to determine the velocities of the chromospheric fine structures from disk observations encounter a number of intrinsic difficulties (Sections 3.5.2 and 3.5.3) and, for the time being, our knowledge of the velocities of these features must continue to rest mainly on the limb observations already described in Sections 2.4.2–2.4.4. On the other hand, we now have a detailed knowledge of the oscillatory velocity field in the upper levels of the underlying photosphere (Section 3.5.1). Observations in lines such as Na D_1 and Mg b_2 show that this field extends up at least into the low chromosphere, where the oscillations are probably a manifestation of the existence of progressive gravity-type waves (Section 7.3.4). Finally, we conclude the chapter with a brief account of the serious difficulties facing attempts to obtain and interpret observations of magnetic fields in the chromosphere; the existing meagre data are summarized in Section 3.5.4.

3.2 Chromospheric Fine Structures Observed in the Hα and K Lines

3.2.1 INTRODUCTION

Our knowledge of the fine structure of the disk chromosphere, described in the sections that follow, is of relatively recent origin. Two factors have been responsible for the rapid growth in the last decade or so: a steady improvement in the performance of solar telescopes, and the general availability of birefringent filters transmitting the Hα line. Earlier studies, of necessity, employed the spectroheliograph. This is a versatile instrument which, as we shall see in Sections 3.3–3.5, is still of great importance; however, it lacks the spatial resolution and speed of operation mandatory for properly observing small, short-lived structures such as the chromospheric bright and dark mottles. Filters for the K line have also been constructed, but so far they have not been widely used. Accordingly, our knowledge of the K line fine structure is meagre in comparison to that of the Hα structure and is still based on spectroheliograms (Section 3.2.8).

3.2.2 SOME INSTRUMENTAL CONSIDERATIONS[1]

Following the pioneering work of Bernard Lyot in 1942 (Section 1.7), the first systematic observations of the chromospheric fine structure through an Hα filter were undertaken by Kiepenheuer in 1957. He used an 11-cm refractor installed at the Fraunhofer Institute's Observatory on the island of Capri [Kiepenheuer, 1957; 1960a,b] and achieved a spatial resolution of 2–3″ of arc. The years that followed saw the attainment of similar or slightly better results by a number of other observers, including Moreton, Harvey, and Ramsey at the Lockheed Solar Observatory, California, who used an 18-cm refractor; Beckers [1964a] (Sydney, 13-cm refractor; Sacramento Peak Observatory, 30-cm coelostat telescope); and Zirin [1966] (High Altitude Observatory, 41-cm coronagraph). Chromospheric photographs obtained by some of these workers have been reproduced by Zirin [op. cit.: see Figs. 9.6, 12.16, 13.2–13.7]. The resolution achieved by Beckers at Sydney and Sacramento Peak was 1.″8 and 1.″3 respectively, figures which may be taken as typical for the early 1960's.

The next significant improvement in performance followed the completion of the 35-cm domeless coudé refractor at Capri. The best chromospheric

[1] References included in this section have been selected mainly to illustrate certain features of the performance of birefringent filters and problems connected with their use. General principles underlying the optimization of solar telescope design have been discussed in detail elsewhere [Bray and Loughhead, 1964, Chap. 2; 1967, Chap. 5].

photographs obtained with this instrument show a resolution of 0″.7 of arc [Kiepenheuer, 1966; 1967–private communication], although the number of occasions on which the seeing conditions at the site permit this resolution to be achieved has proved to be disappointingly small. Shortly afterwards, the authors and their colleagues brought into operation the 30-cm refractor of the CSIRO Solar Observatory, Culgoora, Australia [Loughhead *et al.*, 1968]. This telescope was specifically designed for high-resolution photography of the chromosphere and, in use, has systematically attained a resolution of 0″.7–0″.75, a figure which compares favourably with a theoretical resolution limit (= 1.22 λ/D) at Hα of 0″.55. Many of the photographs in this chapter were taken with this telescope, equipped either with a 0.5 Å Halle filter or CSIRO's 1/8 Å filter plus a Halle filter in tandem (see below).

Since 1968, equivalent performance has been achieved at a number of other observatories, including the Big Bear and Aerospace Observatories in California, the Athens Observatory [Banos and Macris, 1970], and Sacramento Peak Observatory [Dunn, 1969].

The first commercial birefringent filters were made by the Paris firm Optique et Précision de Levallois with the assistance of B. Lyot and, later, A. Dollfus, both of Meudon Observatory;[2] they have a bandwidth of about 0.65 Å. At the time of writing, the most widely-distributed commercial filters are those of the Berlin firm Bernhard Halle Nachfl., which have a bandwidth of ~0.5 Å.

For many applications, the relatively broad bandwidth and absence of full tunability of the commercial filters are not serious limitations. However, as shown by Plate 3.8, the fine structures of the quiet chromosphere show significant differences between wavelengths only 0.25 Å apart. Such differences may hold important clues to the interpretation of the structures and, to avoid concealing them, one evidently has to use a bandwidth not exceeding 1/8 Å; indeed, even a 1/8 Å bandwidth may be too broad for complete discrimination. On the other hand, for Hα this bandwidth appears to represent a practical lower limit at the present time for a filter of the birefringent type [Steel *et al.*, 1961].

The CSIRO's 1/8 Å filter was the first narrow-band, fully-tunable filter. Details of its optical design have been given by Steel *et al.* [loc. cit.], who also describe certain features of its performance. Further aspects of its performance have been discussed by Beckers [1964a] who, together with Bhavilai [1965], employed it in conjunction with a 13-cm telescope. The filter was subsequently fitted with a computer-controlled tuning system and installed in the 30-cm telescope of the CSIRO Solar Observatory [Bray and Winter, 1970]. This system enables the observer to obtain photographs automatically at seven pre-selected

[2] The birefringent filter was invented independently by Lyot and Y. Öhman (see Section 1.7).

wavelengths in the Hα line, the whole sequence occupying a time short compared with the lifetimes of the chromospheric fine structures under study, or the times over which they undergo real changes (see Plate 3.8). Such a system is close to ideal because it combines the narrow bandwidth and tunability of the spectroheliograph with the high spatial resolution and short exposure times possible with the birefringent filter. Observations with this new system have led to a further understanding of the properties of the bright and dark mottles, described in Sections 3.2.3–3.2.6.

A tunable Hα filter having a bandwidth of 1/4 Å is currently available from the firm of Carl Zeiss, Oberkochen, West Germany. In addition, the latest catalogue from Bernhard Halle Nachfl. lists tunable Hα and K filters with bandwidths of only 0.14 and 0.15 Å respectively.

Somewhat surprisingly in view of the large number of precision optical elements comprising a birefringent filter, the measured width of the central passband often turns out to be close to the theoretical value [see, for example, van Griethuysen and Houtgast, 1959; Beckers, 1964a; Krafft, 1968]. For example, the CSIRO 1/8 Å filter has a nominal bandwidth of 0.112 Å and a measured bandwidth of 0.126 Å [Beckers, loc. cit.]. On the other hand, a 'clean' profile is achieved only when the filter is properly assembled and adjusted: several workers have in fact reported the detection and correction of serious errors in newly-delivered filters [Krafft, 1968; Ramsay *et al.*, 1968]. Such errors can lead to a broadened, asymmetric, or even a double-peaked profile which, in turn, can lead to otherwise inexplicable discrepancies with the results of other observers.[3] When the full profile of a filter has been accurately determined, e.g. with the aid of a large spectrograph, photometric investigations become feasible; suitable methods and applications have been described by van Griethuysen and Houtgast [1959], Stenflo [1966], and Krafft [1968]. In general, however, birefringent filters have found their greatest use up to the present time in morphological and dynamical studies.

One final problem connected with filters and their use is the presence of *parasitic light*, i.e. light of unwanted wavelengths. At the limb, the effect of this is to produce a sharp photospheric 'inner' limb at the base of the chromosphere (Section 2.2). On the disk, the unwanted light reduces the contrast of the chromospheric features and superimposes a picture of the photospheric granulation; this is especially noticeable in regions where dark mottles are absent (see

[3] Similar discrepancies can arise from wavelength shifts due to temperature changes: all birefringent filters, particularly those of narrow bandwidth, therefore require elaborate temperature controllers. With such a system, wavelength shifts in the CSIRO 1/8 Å filter are found not to exceed 0.01–0.02 Å when the ambient conditions are relatively constant, but may amount to 0.04–0.07 Å, or possibly slightly more, when conditions are rapidly changing [Bray and Winter, 1970]. Temperature-cycling effects in a Zeiss 1/4 Å filter are referred to by Janssens [1970].

Section 3.2.7). Spectrograms graphically illustrating parasitic bands in a commercial 0.5 Å filter have been published by Zirin [1966, Fig. 2.16]. All birefringent filters seem to suffer from this defect, which is particularly serious in the case of CSIRO's 1/8 Å filter [Beckers, 1964a; Bray and Winter, 1970] even when it is preceded by a relatively narrow-band pre-filter of the interference type (e.g., 30 Å bandwidth). As mentioned in Section 2.2, the solution adopted by the authors has been to place a 1 Å Halle filter in tandem with the 1/8 Å filter, the two filters being simultaneously tuned by the same computer-controlled stepping motor [Loughhead and Tappere, 1971]. This expedient increases the exposure time 15-fold but no loss of spatial resolution is apparent on the best photographs, and the contrast, of course, is greatly improved. On the other hand, the increase in exposure time evidently makes greater demands on the quality of the atmospheric seeing.

3.2.3 PROPERTIES OF THE BRIGHT MOTTLES

Although the bright mottles have been less extensively studied than the dark mottles described in the following section, we nevertheless have an adequate knowledge of such basic properties as their lifetime, size, and height. Our description is based mainly on observations obtained with a 0.5 Å Halle filter [Bray, 1969], but use is also made of confirmatory data obtained with CSIRO's tunable 1/8 Å filter used in tandem with a 1 Å Halle filter (cf. Section 3.2.2). Limb observations of the bright mottles have already been discussed in Section 2.6; a summary of their properties, derived from both disk and limb observations, is given in Table 3.1 below.

(1) *Variation with wavelength.* Observations with the 1/8 Å filter show that, irrespective of heliocentric angle, the bright mottles in quiet regions are most prominent at or near the line centre and are still well visible at H$\alpha \pm 0.25$ Å. However, at H$\alpha \pm 0.5$ Å only a few remain and at H$\alpha \pm 0.65$ Å they are almost entirely absent. Plate 3.1 illustrates the appearance, at the centre of Hα, of the bright mottles in the central region of the disk.

Near the limb, as we shall see below, the bright mottles become more easily visible and their appearance on 1/8 Å photographs closely resembles that shown by Plate 3.2, taken with a 0.5 Å Halle filter.

Since the bright mottles are most prominent at or near the line centre for all heliocentric angles, our description from now on refers solely to observations taken at that wavelength.

(2) *Variation with heliocentric angle.* Plate 3.1, taken at the centre of Hα, shows that at the disk centre bright and dark mottles form a rather confused pattern. The bright mottles occupy a smaller area than the dark mottles and occur mainly in small clusters. Careful intercomparison of photographs taken simultaneously at the centre and in the wings of Hα shows very clearly that

individual bright mottles in a cluster occupy the space *between* prominent dark mottles (cf. Plate 3.8).

At intermediate heliocentric angles ($\theta \simeq 60$–$70°$: Plate 3.2a) the pattern becomes noticeably different in three respects: (*a*) it is somewhat less confused than at the disk centre; (*b*) the bright mottles now appear to be more or less equally numerous as the dark mottles; and (*c*) there now appears to be a definite spatial relationship between bright and dark mottles, bright and dark features often lying in close juxtaposition, with the latter being displaced towards the limb. (This effect is discussed in detail in Section 3.2.5.)

Finally, as the extreme limb is approached ($75° \leqslant \theta \leqslant 90°$: Plate 3.2b) the bright mottles stand out distinctly as individual structures against a darker background, whereas the dark mottles are now much less prominent. The bright mottles are visible right up to and indeed actually crossing the limb, as first pointed out by Bhavilai [1965]. Their appearance beyond the limb has already been described and illustrated in Section 2.6.

(3) *Size, shape, and foreshortening.* Measurements of the bright mottles on Plates 3.2a and 3.2b in a direction parallel to the limb indicate sizes ranging from 2″ to about 6″ of arc (1450–4400 km), many of the mottles having irregular shapes.

Very close to the limb a perfectly flat structure would, owing to foreshortening, appear elongated parallel to the limb, although the elongation could be masked by the effect of finite resolution. A close examination of Plate 3.2b demonstrates that individual bright mottles appear to show *no* foreshortening and, in fact, a few of them seem to have their longest dimension in a direction *perpendicular* to the limb [cf. Bhavilai, 1965]. This suggests that the absence of foreshortening cannot be due wholly to finite resolution, but that the vertical dimension of some of the mottles is comparable with their horizontal dimension. Measurement of Plate 3.2b indicates a typical vertical dimension for mottles near the extreme limb of about 2–3″ (1450–2200 km), in good agreement with the range 725–3000 km derived from limb observations (Section 2.6).

(4) *Lifetime.* Bray [1969] obtained two sequences for determining the lifetime, one lasting 7 min and the other 1 h. Plate 3.3 shows selected pairs of photographs from these films, separated by intervals ranging from 1 min to 17 min. Each pair shows the same region of the disk, but different pairs correspond to different regions; all photographs were obtained in the general neighbourhood of the north limb. Referring to Plate 3.3, we can describe the changes occurring over various intervals Δt as follows:

$\Delta t = 1$ min: over this interval any change is almost imperceptible;

$\Delta t = 2.5$ min: several of the bright mottles undergo noticeable changes, the more prominent of which are the birth of a small mottle near the centre of the upper edge of the photograph and the increase in size of the large mottle in the lower left-hand corner;

Δt = 4.3 min: a detailed examination of the individual mottles shows that about one-half are no longer recognizable;

Δt = 7 min: a majority of the mottles are no longer recognizable;

Δt = 10.5 min: nearly all the individual mottles have changed so markedly in size, shape, and brightness that taken by themselves they are unrecognizable. However, the general pattern of their arrangement retains a well-marked similarity;

Δt = 17 min: after this interval even the pattern itself is hard to identify in detail.

TABLE 3.1

Properties of the Bright Mottles

PROPERTY	VALUE	REFERENCE
Horizontal dimension (km)	1450–8700	Bhavilai [1965]
	3000–6500	Beckers [1968]
	1450–4400	Bray [1969]
	1450–4000	Loughhead [1969] *
	1000–2000	Banos and Macris [1970]
	500–6000	Alissandrakis and Macris [1971] *
Vertical dimension (km)	1450–2200	Bray [1969]
	725–2200	Loughhead [1969] *
	1000–3000	Alissandrakis and Macris [1971] *
Height (km)	725–3300	Loughhead [1969] *
(lower limits)	0–4000	Alissandrakis and Macris [1971] *
Lifetime (min)	11	Bray [1969]
	2¾–12¾	Loughhead [1969] *
	12	Macris and Alissandrakis [1970]
	12	Alissandrakis and Macris [1971] *
Total no. on Sun	7.7×10^4	Bhavilai [1965]

*Derived from observations beyond the limb (see Section 2.6).

In summary, after only 2.5 min changes can be detected in individual bright mottles, and about one-half are no longer recognizable after 4.3 min. After 7 min most of the mottles are unrecognizable. On the other hand, the general pattern of their arrangement retains a well-marked similarity even after 10.5 min, but becomes hard to identify in detail after 17 min. Since more than one-half the mottles become unrecognizable after an interval of 4.3–7 min, Bray [1969] concluded that a reasonable estimate for the average lifetime is ~11 min.

This figure agrees closely with the value of 12 min derived independently by Macris and Alissandrakis [1970].

Table 3.1 summarizes the basic properties of the bright mottles as determined by various workers from observations both on the disk and beyond the limb.

3.2.4 PROPERTIES OF THE DARK MOTTLES

While the importance of the bright mottles has been realized only recently, the dark mottles have long been recognized to be one of the basic elements constituting the inhomogeneous structure of the chromosphere. Accordingly, they have been subjected to intensive observation, especially since 1957, when high-quality Hα photographs first became available (Section 3.2.2).

(1) *Variation with wavelength*. Let us first consider the appearance of the dark mottles when observed with the 1/8 Å filter in the central region of the disk. At line centre the bright and dark mottles form a rather confused pattern (cf. Plate 3.1); the pattern becomes less confused as one moves out into the wings of the line, but the more prominent dark mottles visible in the wings are nevertheless easily distinguishable at the line centre. At H$\alpha \pm 0.25$ Å the dark mottles become more distinct while the bright mottles decline in contrast and apparent size. However, the most dramatic changes occur when the filter is tuned out a further 1/4 Å to H$\alpha \pm 0.5$ Å: only a few bright mottles remain whereas the dark mottles are now greatly enhanced in contrast (Plate 3.4). At or near these wavelengths the dark mottles show their maximum visibility and further tuning, out to H$\alpha \pm 0.75$ Å, reduces their number and apparent sizes (Plate 3.5). Only a few are still present at H$\alpha \pm 1.0$ Å.

Near the disk centre the dark mottles appear somewhat coarser at H$\alpha + 0.5$ Å than at H$\alpha - 0.5$ Å. This is well illustrated by 1/8 Å photographs published by Beckers [1964a, compare Figs. 7 and 8]. Near the limb, on the other hand, the difference in appearance at wavelengths equally spaced on the two sides of Hα is inappreciable. For a given bright or dark mottle, the change in appearance (e.g., contrast) as the wavelength is altered evidently depends on the line profile of the structure; this topic is discussed in Section 3.5.3.

Since the dark mottles are best seen in the line wings, our description from now on refers to such wavelengths.

(2) *Variation with heliocentric angle.* A noteworthy property of the dark mottles is that as the limb is approached many of them appear elongated in the direction of the limb (Plate 3.6). Such elongation has been pointed out by numerous authors, including Kiepenheuer [1957, 1960a], Bruzek [1959], Beckers [1963], and Bhavilai [1965, 1966]. At the same time the mottles appear reduced in size, as though their lower portions are now obscured and only their radial extensions remain. Measurements of Plate 3.6 indicate an apparent width of only $\sim 1''$ of arc (725 km) for some of the narrow radial extensions, a figure

comparable with the observed widths of spicules beyond the limb (Section 2.3.3).

Viewed in the line wings, the dark mottles extend right out to the limb and are sometimes seen to actually cross it. Their appearance in the neighbourhood of the limb and their possible relationship to the dark band at the base of the chromosphere are discussed in Section 2.7.

(3) *Size and shape.* Viewed at the disk centre in the wings of Hα (Plates 3.4 and 3.7), the dark mottles show a wide diversity in size and shape [Bray, 1968]. The smallest, with apparent diameters of slightly under $2''$ of arc (1450 km), appear more or less circular (perhaps owing to limited resolution), but the largest, which attain dimensions of up to $11''$ (8000 km), may be of any shape. Although there are many elongated mottles among the larger ones, others have quite irregular shapes. Measurements of mottle sizes by other workers at a variety of wavelengths are listed in Table 3.2.

Some mottles have exceedingly sharp boundaries while others have diffuse boundaries which make the mottle appear to merge into its surroundings. Some mottles are sharp on one side and diffuse on the other.

Some of the mottles form part of a 'rosette', a characteristic grouping of bright and dark mottles which is discussed in the next section. Others form only a partially-completed rosette or are isolated structures.

In many cases, the larger mottles of Plates 3.4 and 3.7 appear to be distinct structures in their own right and not merely conglomerations of small mottles. On the other hand, the right-hand pairs of photographs of Plate 3.7 provide several clear examples of large mottles which are single structures at one stage but multiple at earlier or later times.

(4) *Evolution and lifetime.* The evolution of the dark mottles is illustrated in Plate 3.7, which shows selected pairs of photographs separated by intervals ranging from 0.5 to 15 min. Each pair shows the same region of the disk, but different pairs correspond to different, although neighbouring, regions. A detailed examination of the photographs [see Bray, 1968] reveals that even after only 1 min small changes can be detected in individual mottles; most of them show well-marked changes after 2.5 min. After 5–9 min most of the mottles become unrecognizable, but the general pattern of their arrangement remains similar even after 15 min. These observations lead to an estimated average lifetime of ~5 min. This figure agrees closely with the values obtained at various wavelengths by the majority of the other workers, although a few authors have reported values 2 to 3 times as large (see Table 3.2). If we accept the two most recent determinations as being the most reliable, then the best available estimate of the lifetime of the dark mottles is ~5 min, a value significantly smaller than that for the bright mottles (~11–12 min).

Only meagre information is available concerning the details of the mottles' mode of evolution. Bray [1968] has concluded that individual mottles seem to undergo continual changes in size, shape, and contrast. Howard and Harvey [1964] have characterized time changes in the mottle pattern as a 'seething' effect, which is best seen in the blue wing of Hα but is also visible at the line

TABLE 3.2

Properties of the Dark Mottles

PROPERTY	VALUE	Δλ(Å)	REFERENCE
Horizontal dimension (km)*	1700–10 000	–	de Jager [1957]
	2000	±0.5	Kiepenheuer [1957]
	<1450–8700	–	Macris [1957]
	2000–5500	0	Bruzek [1959]
	725–7250	–	Kiepenheuer [1960a]
	<1300–3300	+0.5	Beckers [1963]
	1450; 3600–7250	0	Howard and Harvey [1964]
	2200–2900	+0.75	Bhavilai [1966]
	<1450–8000	±0.3†	Bray [1968]
Vertical dimension (km)‡	8000–12 000	0	Bruzek [1959]
Lifetime (min)	5	–	de Jager [1957]
	5½	–	Macris [1957]
	10	–	Bruzek [1959]
	3–5	–	Kiepenheuer [1960a]
	16	±0.5	Beckers [1963]
	10–15	0	Howard and Harvey [1964]
	5	±0.3†	Bray [1968]
	5½	−0.6	Macris and Alissandrakis [1970]
Total no. on Sun	$<5 \times 10^4$	–	Macris [1957]
	1.5×10^4	0	Bruzek [1959]
	3.5×10^5	±0.5	Beckers [1963]

*Derived from measurements near the disk centre.
†Observations obtained with a filter having a double-peaked profile with red and blue peaks at ±0.3 Å.
‡Derived from measurements near the limb.

centre. According to these authors, the seething appearance is due not merely to changes in size and shape but also to *random horizontal motions*. The maximum displacement due to these motions was observed on photographs taken about 30 sec apart and amounts, on the average, to 3000 km (4″ of arc), corresponding to a velocity of 100 km sec^{-1}. The authors' own observations, however, do not confirm the existence of random displacements of such magnitude (compare, e.g., the first pair of photographs of Plate 3.7).

Table 3.2 summarizes the basic properties of the dark mottles as determined by various workers; the third column gives the wavelength of observation, when this is known.

3.2.5 RELATIONSHIP BETWEEN BRIGHT AND DARK MOTTLES

Disk observations of the bright and dark mottles described in the two previous sections, together with the limb observations of Section 2.6, show unmistakably that the two types of mottles are different structures with different properties. Nevertheless, there exists a close spatial relationship between them, two aspects of which are discussed below.

(1) *Rosettes.* Bright and dark mottles show a strong tendency to cluster together in a larger structure called a 'rosette', a term introduced by Beckers [1963] and now generally adopted. Plate 3.8 shows 1/8 Å filter photographs of a particularly clearly-defined rosette, obtained nearly simultaneously at seven wavelengths across the Hα line ranging from Hα – 0.75 Å to Hα + 0.75 Å. The radial orientation of the dark mottles is well shown at Hα + 0.75 Å: it is this appearance that gave rise to the term 'rosette'. Note that the visibility of both the bright and dark mottles changes with wavelength in the ways described in Sections 3.2.3 and 3.2.4, the former being most prominent in the line centre, the latter in the line wings.

The bright material evidently concentrates either at the *centre* of the rosette, or, away from the centre, *between* the dark mottles. Moreover, a close examination shows that as the wavelength is changed dark mottles do *not* become bright mottles. In general, the dark mottles often appear to penetrate somewhat further towards the centre of a rosette at Hα + 0.5 Å than at Hα – 0.5 Å.

It is important to point out that not all dark mottles are clustered together in well-defined rosettes such as the one shown in Plate 3.8. Many occur in broken or partially-completed rosettes [cf. Janssens, 1970], while others exist in isolation. Nevertheless, the rosette is by no means merely a chance grouping of mottles and, accordingly, we are led to ask, what is its physical significance? One possible answer to this question is suggested in Section 5.1, namely that a rosette should be regarded as an *active region seen in microcosm.* This suggestion is supported by the fact that prominent rosettes tend to be located along the boundaries of the chromospheric network, where the photospheric and chromospheric magnetic fields of the quiet Sun are known to be concentrated (Sections 3.4.4 and 3.5.4). Further support comes from a direct comparison of the rosette of Plate 3.8 and the incipient active region of Plate 5.1: the resemblance is quite striking, although the latter differs from the rosette in size and complexity and in showing more bright material and an (apparently) bipolar field configuration.

Two consequences of the hypothesis are, in principle, capable of observational verification. Firstly, on occasion one might hope to observe the evolution of an active region in its earliest stages from a rosette. Secondly, with sufficient spatial resolution one might expect to see small magnetic fields concentrated at the *centre* of a rosette and generally coincident, perhaps, with the bright material.

Similar descriptions of rosettes have been published by other authors, including Beckers [1963; 1964a,b], Noyes [1967], and Janssens [1970]. The latter has also studied time changes in rosettes; these are hard to follow and difficult to describe. Cragg *et al.* [1963] have observed what they call 'bushes' in the neighbourhood of the limb. These are simply rosettes seen partially side-on, and we shall now discuss the additional information provided by such observations.

(2) *Height difference between bright and dark mottles.* Plate 3.9a shows several clusters of bright and dark mottles strung out along a strip of the chromosphere roughly parallel to the limb and located at a heliocentric angle of about 58°. This photograph, taken with a 0.5 Å Halle filter at the centre of Hα, strikingly demonstrates the close relationship between bright and dark mottles when observed at *intermediate* heliocentric angles. (Note that such an observation is not possible very close to the limb, because there the dark mottles lose their prominence when viewed at the Hα line centre.) It will be noticed that the dark material lies predominantly on the limb side of the bright material; interpreted as a vertical displacement, the measured shift implies a height difference between dark and bright material (measured between the centres of adjacent dark and bright mottles) amounting, on the average, to about 4300 km [Bray, 1969]. Since the bright mottles themselves are found to lie at least 725–3300 km above the photospheric limb (Section 2.6), it follows that the centres of the dark mottles measured on Plate 3.9a are located at least 5000–7600 km above the base of the chromosphere;[4] their tops must be even higher. Purely geometrical considerations therefore lead us to the important conclusion that whereas the bright mottles are located in the low chromosphere, the dark mottles extend to heights comparable with those of spicules (see Table 2.6).

Bhavilai [1965], Beckers [1968], and Banos and Macris [1970] have also remarked on the effect described above. Bhavilai compared photographs of bright mottles at the Hα line centre and of dark mottles at Hα ± 0.75 Å, but did not evaluate the displacement. Beckers gives a figure of ~2000 km. Zirin [1966, p. 220] has also concluded that the dark mottles lie above bright material.

[4] These figures are evidently subject to considerable uncertainty and, accordingly, have not been included in Table 3.2.

A careful one-by-one examination of the mottles in the left-hand cluster of Plate 3.9a leads to the conclusion that the bright and dark mottles are distinct entities and are *not* merely parts of a single structure which is bright and dark in its lower and upper portions respectively [see Bray, 1969]. This observation is consistent with the distribution of bright and dark material in a rosette, described above. On the other hand, Banos and Macris [1970], on the basis of photographs similar to Plate 3.9a and of comparable quality (see their Figs. 5b and 6), have come to the opposite conclusion! However, in our opinion the balance of the evidence, taken as a whole, strongly favours the view that although bright and dark mottles are spatially related, they are nevertheless *distinct structures*. A similar view has recently been expressed by Alissandrakis and Macris [1972].

Unfortunately, existing observations provide little information bearing on the *physical*, as opposed to the spatial, relationship between bright and dark mottles.

3.2.6 IDENTIFICATION OF SPICULES

As we have seen in the previous three sections, observations of the fine structure of the quiet chromosphere at the centre of the disk reveal just two basic elements, the bright and dark mottles, which are separate and distinct entities with different properties. Similarly, as we have seen in Chapter 2, just two distinct fine structures are observed beyond the limb, the bright mottles and the spicules. Moreover, the identity of the bright mottles on the disk with the bright mottles beyond the limb is firmly established by (*a*) the close correspondence in their properties (cf. Table 3.1); and (*b*) the observation of bright mottles at all stages of crossing the limb (Section 2.6). On purely logical grounds, therefore, we are led to identify spicules with either (*a*) dark mottles or (*b*) some other fine structure on the disk, not as yet properly identified or observed [Loughhead, 1969].

At the present time the first alternative seems by far the more likely. It receives strong support from (*a*) the predominantly radial elongation of many of the dark mottles as the limb is approached (Section 3.2.4); and (*b*) the height difference between bright and dark mottles, which places the former in the low chromosphere and the latter at heights to which spicules are known to extend (Section 3.2.5).

Table 3.3 shows that the identification of spicules with dark mottles is consistent with the known properties of the two features. Lifetime, height, and vertical dimension are seen to be in good agreement, as is the *minimum* horizontal dimension. Total numbers do not agree, but these are difficult quantities to measure accurately. In general, however, there is a substantially better agreement between the basic properties of dark mottles and spicules than of bright mottles and spicules. On the other hand, the reader must bear in mind the considerable uncertainty attached to some of the measurements included in the table. Earlier workers who identified spicules with dark mottles on the basis of similarity in properties and of the radial elongation of dark mottles near the limb include de Jager [1957], Macris [1957], Kiepenheuer [1957, 1960a], Bruzek [1959], and Beckers [1963, 1964a].[5]

TABLE 3.3

*Comparison of Basic Properties of Spicules, Dark Mottles, and Bright Mottles**

PROPERTY	SPICULE	DARK MOTTLE	BRIGHT MOTTLE
Horizontal dimension (km)	815	1000–8000	1450–4000
Vertical dimension (km)	9800	8000–12 000	1000–2200
Height (km)†	4900	5000–7600 (lower limit)	725–3300 (lower limit)
Lifetime (min)	5	5	11–12
Total no. on Sun	9.3×10^4 (at h = 3000 km)	3.5×10^5 (Beckers' figure)	7.7×10^4

*Extracted from Tables 2.6, 3.1, and 3.2.
†Measured at the midpoint of the structure.

One argument *against* the identification is the large size and irregular shapes of many of the dark mottles when viewed at the disk centre. However, this probably implies that the identification refers not to the whole volume of a mottle but only to its thin elongated upper portions, which on the disk only become clearly visible as the limb is approached (Plate 3.6).

The above arguments are all indirect: to establish the identification directly one would have to show that spicules actually cross the limb as dark mottles and vice versa. Attempts to do this have been made by Beckers [1966], Loughhead [1969], Alissandrakis and Macris [1971], and Loughhead and Tappere [1971] but, unfortunately, with inconclusive results. The latter observers noticed cases of spicules apparently crossing the limb both as dark and *bright* material, but a large number of photographs produced only isolated examples. Beckers observed dark mottles crossing the limb, beyond which they showed up in absorption.

Owing to the presence of a number of complicating observational factors, this direct approach to the problem of identification of spicules is less attractive than, at first sight, it might appear [Beckers, 1968, p. 427; Loughhead and Tappere, loc. cit.]. In the first place, foreshortening near the extreme limb may make it hard to decide whether two apparently neighbouring features are

[5] In the authors' opinion, Bhavilai's [1965] published photographs fail to substantiate his conclusion that spicules are to be identified with *bright* mottles (see his Plates 7–10).

actually close together or far apart: the horizontal distance between a point on the limb and another point just 1″ of arc (725 km) within the limb on the same solar disk radius exceeds 30 000 km! Secondly, depending on the bandwidth of the filter system employed (e.g., 1/8 Å), only some of the spicules actually present will be visible on photographs taken in one of the wings of Hα, namely those whose Doppler shift happens to be roughly comparable to the wavelength setting of the filter.

Thirdly, although both spicules and dark mottles have lifetimes of ~5 min, a spicule may be observable in one of the wings of Hα during only part of its lifetime. This is because, after an initial rapid ascent, a spicule either fades from visibility while remaining stationary or else appears to descend back to the lower chromosphere at a velocity comparable to that of its ascent (cf. Section 2.4.1); in either case the resulting change in its Doppler shift may cause the spicule to disappear on a wing photograph. Finally, the question of the appearance of an emitting and absorbing (and moving) structure at the extreme limb poses a complex theoretical problem whose proper solution would involve a full non-LTE theory of line formation and spherical radiative transfer theory.

Summarizing, one can see that the possibility of observing a direct relationship between a dark mottle and a spicule at the limb is determined by the product of a number of separate probabilities, some of which may be quite low. The latter include the probability that both the spicule and the dark mottle are favourably positioned on the observer's side of the limb; the probability that the Doppler shift of the spicule is roughly comparable to the wavelength setting of the filter; the probability that the spicule is observed at a favourable moment in its lifetime; and so on. Consequently, even if spicules and dark mottles represent essentially the same phenomenon, only in rare, isolated cases would one expect to find direct evidence of such a relationship. Accordingly, for the time being, support for this view must continue to rest on the (very substantial) weight of the indirect evidence outlined above.

3.2.7 OTHER FINE STRUCTURES OBSERVED IN Hα

The bright and dark mottles described in previous sections appear to be the basic fine structures observed in Hα on the disk. However, the presence of other fine structures has been reported. Beckers [1964a, 1968] has pointed out the existence in the blue wing of Hα of small dark dots which he calls 'grains'. The grains are so small that very good-quality photographs are needed to show them. Such a photograph, taken by the authors at Hα – 0.5 Å, is reproduced in Plate 3.9b. The grains are well visible in the interiors of the cells of the chromospheric network described in Section 3.3.3 and are easily distinguishable, by virtue of their smaller size, from the dark mottles which outline the cell walls. Measurement of this photograph indicates apparent diameters ranging from only

1″.0 to just over 2″.0. Virtually no further information is available about these structures. Beckers has suggested that they may be identical to the *bright* grains observed in K_{2V} and K_3 but, as pointed out in Section 3.2.8, positive identification is lacking.

A second reported structure which many observers have noticed at various wavelengths is a bright pattern resembling the photospheric granulation and best visible in regions free from dark mottles [see, for example, Bray, 1968, Fig. 1; 1969, Fig. 1a; Loughhead *et al.*, 1968; Bray and Winter, 1970, Fig. 3; Janssens, 1970; Rogers, 1970]. The present authors have expressed the view that this is not a true chromospheric structure but simply reflects the presence of parasitic light which, in effect, superimposes a picture of the photosphere onto that of the chromosphere. However, Banos and Macris [1970] have disputed this conclusion, claiming that their measured cell size of the bright pattern (4000–5000 km) exceeds that of the photospheric granulation.

The observations of the bright pattern – obtained with a variety of filters – are all influenced to some degree by the presence of parasitic light. To eliminate this influence, the authors used the 1/8 Å filter plus a 1 Å filter in tandem to obtain a number of good-quality, spectrally-pure photographs at Hα + 0.65 Å: at this wavelength the dark mottles are relatively sparse, so that any structure within the network cells should be well visible. However, these photographs show no trace of a bright cellular structure resembling the photospheric granulation. Instead, they show within the network cells what appears to be a very low contrast field of brightness variations, devoid of any measurable scale. The significance of this structure, if real, is unknown.

3.2.8 FINE STRUCTURES OBSERVED IN THE K LINE OF Ca II

Information about the K line fine structures and their correlation with Hα features has been provided by good-quality spectroheliograms obtained by Beckers [1964a], which have an effective resolution of about 2″ of arc.

The most prominent structures visible in the K line are the well known *bright flocculi.* It is the distribution of these flocculi which outlines the chromospheric network, described in Section 3.3. Each bright flocculus coincides in position with an Hα rosette (Section 3.2.5). Its diameter is somewhat smaller than the overall diameter of the rosette but greater than that of the bright central portion of the rosette, as seen at the centre of Hα. When the two features overlap, each Hα dark mottle appears to coincide with a local weakening of the brightness of any coincident flocculus; moreover, outside of a flocculus there is often a pronounced darkening in K at the position of a dark Hα mottle. The spatial relationship between an Hα rosette and a K line flocculus and its immediate surroundings is schematically illustrated in Fig. 3.1, which is taken from Beckers [loc. cit.].

Near the limb, the bright K flocculi again coincide with the bright mottles seen at the centre of Hα; thus they appear to coincide with the bases of the Hα 'bushes', such as those shown in Plate 3.9a [Cragg *et al.*, 1963; Beckers, loc. cit.].

A second prominent structure is visible *within* the cells of the chromospheric network. This structure is composed of small round bright features, called 'grains' by Beckers. Good photographs of these structures have been published by Beckers [1964a, Fig. 43], Leighton [1965, Fig. 6], Title [1966a, Plates 12–14, 41], and Bappu and Sivaraman [1971, Fig. 5]. Beckers remarks that the grains are best visible on the violet side of the K line; this is confirmed by comparing Title's Plate 41 (K_{2V}) with his Plate 43 (K_{2R}). They are also well visible in K_3, the centre of the K line (Title's Plates 14 and 42).

Fig. 3.1. Spatial relationship between an Hα rosette and a K line flocculus, as depicted by Beckers [1964a]. The dotted area on the right represents bright material. The diameter of a typical rosette is ~30″ of arc.

According to Beckers and to Bappu and Sivaraman, the bright K line grains have a lifetime of ~3–4 min. Beckers suggests that they may be identical to the *dark* grains seen at Hα – 0.5 Å and discussed in the previous section; however, a positive identification has not yet been established. The bright grains appear more or less round at all parts of the disk and presumably, therefore, cannot be identified with spicules. Their significance remains to be explained.

Owing to the absence of observations having a spatial resolution comparable to that of the best Hα photographs (e.g., 0″.7–1″.0 of arc), our knowledge of fine structures observed in the K line is by comparison rather limited. Moreover, the complicated profile of the K line core makes the task of interpreting K line spectroheliograms more difficult than that of interpreting Hα photographs. With low spatial resolution, the core appears to consist of two emission peaks (K_{2V} and K_{2R}) separated by a central self-reversal (K_3) (Fig. 3.2). With high spatial resolution, on the other hand, only a single peak (red or violet) is observed at most locations [Zirin, 1966, p. 230; Linsky and Avrett, 1970; Pasachoff, 1970]. Apparent Doppler shifts measured in the K line core are similarly hard to interpret; this topic is discussed in Section 3.5.2.

3.3 The Chromospheric Network: Basic Properties

3.3.1 INTRODUCTION

In contrast to the fine structures described in previous sections of the chapter, the chromospheric network is best studied on medium-resolution photographs reproduced on a relatively modest scale. With high resolution and a large scale the network as such is not readily discernible although, as we have seen, such observations are mandatory for elucidating the properties of the fine structures composing it. Accordingly, our account of the properties of the network in the sections that follow is based on spectroheliograms obtained in various lines of chromospheric origin having a resolution of about 6″ of arc, or perhaps slightly better.

We start with the bright network observed in the K line of Ca II (Section 3.3.2), the Fraunhofer line in which the chromospheric network is most clearly defined. Closely related to the K network are the bright and dark networks observed in the centre and wings respectively of Hα (Section 3.3.3); these are made up of the bright and dark mottles whose properties have been described in previous sections. Other lines of chromospheric origin in the visible and infrared regions of the spectrum reveal networks similar to those observed at various wavelengths in either Hα or the K line (Section 3.3.4): for example, the centre of Hβ shows a *dark* network similar to that observed in the *wings* of Hα.

All the lines referred to above reveal the network either in the *low* ($h < 5000$ km) or *middle* ($5000 \leqslant h \leqslant 10\,000$ km) chromosphere (for example, see the height estimates of bright and dark Hα mottles given in Table 3.3). To establish what happens to the network in the upper chromosphere we require spectroheliograms in a number of ultraviolet emission lines carefully chosen to cover the relevant range of temperatures, say, $10^5 - 10^6$ K (cf. Tables 4.12 and 4.18). Unfortunately, most existing observations in this region of the spectrum lack even the minimum spatial resolution needed to delineate the network cell boundaries, say, 6–8″ of arc. However, observations in Lyman α of He II (λ304) have revealed some evidence of structure, and programmes planned for the immediate future should soon extend our knowledge of the network to the upper chromosphere (Section 3.3.5).

The chromospheric network bears a close relationship to two photospheric phenomena, namely, the supergranulation and the photospheric magnetic field distribution. However, a discussion of these important relationships – for which a knowledge of the basic properties of the network is an essential prerequisite – is postponed to Section 3.4.

3.3.2 PROPERTIES OF THE K LINE NETWORK

Before proceeding to outline the basic properties of the chromospheric network as observed in the K line of Ca II, we shall find it convenient to describe the profile of the central emission core of the K line. This is illustrated in Fig. 3.2, which is due to Sheeley [1967b]; the four separate traces refer to a bright facular area ('plage'), a weaker facular area, a bright flocculus, and a 'quiet' region, which we may assume to be located within one of the network cells. In each case there are two emission peaks, designated K_{2V} and K_{2R}, separated by a central self-reversal K_3. The violet peak is usually somewhat higher than the red, a circumstance which has been pointed out by many observers [e.g. McMath *et al.*, 1956; de Jager, 1959, p. 129; Linsky and Avrett, 1970].[6]

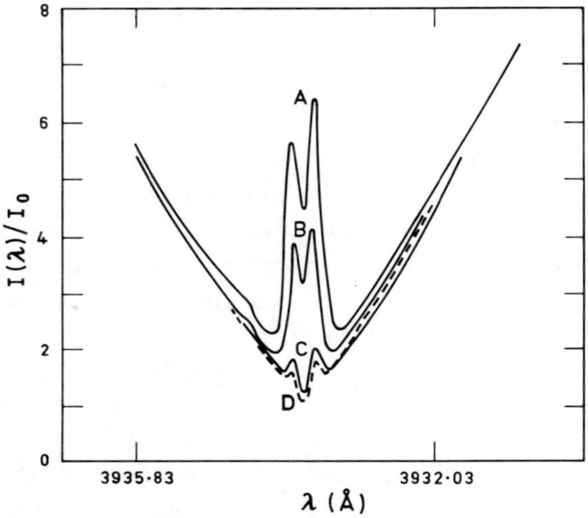

Fig. 3.2. Profiles of the central portion of the K line for various solar features, obtained with medium or low spatial resolution [Sheeley, 1967b]. A refers to a bright facular area, B to a region near its edge, C to a bright flocculus, and D to a 'quiet' region, which we may assume to be located within a network cell. (I_0 is approximately equal to 4 per cent of the continuum in the neighbourhood of λ4000.)

If we now imagine the second slit of a spectroheliograph to be placed at K_3 or at either of the K_2 peaks then we shall obtain a picture of the distribution of the bright flocculi – i.e. a picture of the chromospheric network – and, if active regions are present, of the K line chromospheric faculae.[7] On the other hand, in the troughs beyond the central emission (K_{1V} and K_{1R}) and in the rising portions of the profile beyond, we would evidently expect to see the various structures only with reduced contrast, an expectation which examination of K_1

[6] It is important to recall, however, that the above description applies only to observations of *medium* or *low* spatial resolution: with high spatial resolution only a single peak (red or violet) is observed at most locations (Section 3.2.8).

[7] Called chromospheric or calcium 'plages' by some observers.

spectroheliograms quickly confirms [see, for example, Title, 1966a, Plate 14].

K line spectroheliograms have been obtained regularly at a number of observatories since the end of the nineteenth century (Section 1.5); the interested reader will find references to much of the earlier work in the following sources: Abetti [1929]; d'Azambuja [1930]; Unsöld [1955, p. 703]; Macris [1962]; and Linsky and Avrett [1970]. Over the years the spectroheliograph itself has undergone a substantial improvement in performance and good examples of modern medium-resolution spectroheliograms obtained in the K line have been published by Kiepenheuer [1960b, Fig. 7]; Athay [1964, Fig. 4]; Leighton [1965, Figs. 4 and 6]; Title [1966a]; and Zirin [1966, Figs. 9.8 and 9.10]. Title's atlas, in particular, is typical of the best results obtained with modern techniques and forms a homogeneous collection of spectroheliograms taken at various wavelengths in some of the more important chromospheric lines.

Plate 3.10, reproduced from this atlas [Title, 1966a, Plate 43], illustrates the chromospheric network as seen in the K_{2R} emission peak. Individual cells of the network are outlined by the bright flocculi, whose appearance under *higher* spatial resolution has been described in Section 3.2.8. In spite of occasional gaps in the bright cell boundaries, the cells in general are well defined. However, this is not so near the limb, where the cellular appearance is largely lost although the brighter flocculi are still well evident. Note that as the prominent active region is approached the cell boundaries appear to become brighter and thicker until, eventually, each cell is filled with bright material, thus forming part of the chromospheric faculae surrounding the spot group.

The appearance of the network in K_{2V} and K_3 is similar to that in K_{2R} (compare Title's Plates 41, 42, 43). However, within the network cells the small bright grains referred to in Section 3.2.8 are more prominent in K_{2V} than in K_{2R}.

The great diversity in the sizes and shapes of individual network cells is well illustrated in Plate 3.10. According to measurements made by Rogerson [1955] the cell sizes range from 25 000–70 000 km; Kiepenheuer [1960b] quotes a range of 25 000–50 000 km. From a careful autocorrelation analysis, Simon and Leighton [1964] derived a *mean* cell size of 30 000–35 000 km. Sýkora [1970] confirmed this value, using a similar procedure: he found mean cell sizes of 34 600 and 32 800 km measured at solar minimum parallel and perpendicular to the equator respectively. Combining Simon and Leighton's and Sýkora's values, we obtain for the mean cell size of the K line network a figure of 33 000 km. The apparent elongation found by Sýkora is small compared to the intrinsic spread of cell sizes and, in the absence of confirmation by other workers, should perhaps be viewed with caution.

A second property of the network cells, also measurable in principle, is their lifetime. Macris [1962] determined the lifetimes of individual bright flocculi and found that 85 per cent persist for at least 9 h and 15 per cent persist for at least 26–30 h. These measurements, however, do not really tell us the lifetime of the network cells themselves. Simon and Leighton [1964] measured the correlation as a function of time of several large areas of the quiet chromosphere containing up to 244 cells. Since their correlation machine responded to all intensity fluctuations in its field of view, their measured lifetime is evidently an average value associated with both the flocculi and the network which the flocculi outline. They derived a mean lifetime – defined as the interval required for the correlation to fall to $1/e$ of its initial value – of 19 to 21 h. This figure is in good agreement with the lifetime of the network cells observed in the wings of Hα, described in the next section.

3.3.3 PROPERTIES OF THE Hα NETWORK

Both in the core and the wings of Hα, the chromospheric network is less prominent than in the K line. Plate 3.11 shows a medium-resolution spectroheliogram taken at the centre of Hα, which is reproduced from Title's atlas of Mt. Wilson spectroheliograms [Title, 1966a, Plate 6]. On this photograph the bright Hα network is well visible over a large area in the neighbourhood of the disk centre; nevertheless, cells with completely defined boundaries are the exception rather than the rule. Nearer the limb network cells as such are no longer apparent, although the bright mottles (or, more accurately, *groups* of bright mottles) are visible with good contrast.

It would be misleading, however, to imply that the appearance of the bright Hα network shown on Plate 3.11 is necessarily typical. In fact, on other spectroheliograms taken at the Hα centre published by Title the network is far less prominent (see his Plates 21 and 38). When we compare these last two photographs with Title's Plates 1,[8] 6 (our Plate 3.11), 9c, and 33, an important conclusion emerges: the bright network becomes prominent only in those regions which lie in the general neighbourhood of existing activity centres or which show other residual signs of activity revealed, for example, by the presence of dark filaments. Where these signs of activity are absent, the bright Hα network may become virtually invisible. We may conclude, therefore, that the bright Hα network is a fairly sensitive indicator of the presence of magnetic fields; this is also true of the K line network (see Section 3.4.4).

The bright network observed at the centre of Hα is made up essentially of bright mottles and groups of bright mottles, whose properties have already been outlined in Sections 3.2.3 and 3.2.5. Since the bright K line flocculi, observed with high resolution, coincide with the bright Hα mottles (Section 3.2.8), then evidently we must expect a close correlation between the bright K network and

[8] Portion of this spectroheliogram has been reproduced by Zirin [1966, Fig. 9.4a].

the bright Hα network. Such a correlation is well-established and has been known for a long time [cf. Noyes, 1967]. It can be demonstrated by comparing spectroheliograms in K and in the Hα centre taken within a time short compared with the lifetime of the network cells, such as those reproduced by Title [1966a] : compare his Plates 1 and 5, 21 and 22, and 38 and 40. A good way of effecting the comparison is to superimpose transparencies of spectroheliograms taken in the two lines. A moiré pattern is observed when one of the transparencies is shifted slightly relative to the other, despite the fact that the Hα cells are far less completely delineated than the K cells. No quantitative measurements of the cross-correlation coefficient have yet been published.

When the spectroheliograph or filter is tuned into one of the wings of Hα, the bright network disappears and is replaced by a network of dark mottles; the dark network is best visible on medium-resolution photographs taken at some wavelength within the ranges Hα ± 0.5 Å to Hα ± 0.7 Å. The change of the network with wavelength is, of course, exactly what one would expect from the wavelength variation of the bright and dark mottles described in Sections 3.2.3 and 3.2.4.

Plate 3.12b shows a spectroheliogram taken at Hα + 0.56 Å, which is reproduced from Title's atlas [Title, 1966a, Plate 9a]. Once again, the description of the mottle pattern as a 'network' is somewhat less convincing than in the case of the K network. Nevertheless, except in active regions, it is easy to pick out clearly delineated cells. Other illustrations of the dark network have been published by Zirin [1966, Fig. 9.4c]; Kiepenheuer [1967, Fig. 2b]; Janssens [1970, Figs. 2 and 3]; and Rogers [1970, Fig. 1].

Using *negative* transparencies of the dark Hα network one can readily show by the moiré method outlined above that the dark Hα network coincides approximately with the bright K network (compare Title's Plates 3 and 5, and 39 and 40). Once again the correlation has been known for a long time and has been pointed out by numerous authors [see, for example, Kiepenheuer, 1960a, 1967; Noyes, 1967]. It is important to realize, however, that the coincidence is only approximate: observed with higher resolution, the bright flocculi outlining the K network coincide with the *bright,* not the *dark* mottles (Section 3.2.8). However, since the bright and dark mottles occur in close juxtaposition (see Section 3.2.5 and Plate 3.8), on medium-resolution photographs the K network necessarily appears to correlate closely with the dark Hα network.

Observed with high resolution in the wings of Hα the dark network is seen to be composed of isolated dark mottles, rosettes, and partially completed rosettes. In effect, the dark mottles constitute the 'walls' of the network cells. This is strikingly illustrated in Plate 3.12a, a high-resolution photograph of a single network cell located at a heliocentric angle of 63° and thus viewed obliquely. Note how the dark mottles appear to enclose the cell, individual mottles standing up like the slats of a picket fence. This photograph, taken by the authors with the 1/8 Å filter tuned to Hα – 0.5 Å, is reproduced on about six times the scale of the medium-resolution spectroheliogram of the dark network shown in Plate 3.12b.

Plate 3.12b demonstrates that the dark Hα network forms a rather irregular pattern with many incomplete cell boundaries and, like the K network, a large diversity in cell sizes and shapes. Kiepenheuer [1960a] quotes a size range of 25 000–50 000 km, while Janssens [1970] gives an average figure of 30 000 km, in agreement with an earlier estimate of Jefferies and Loughhead [1956]. Rogers [1970] attempted to use an autocorrelation method similar to that successfully applied by Simon and Leighton [1964] and Sýkora [1970] to measuring the cell size of the K network (Section 3.3.2). However, in the case of the dark Hα network, many of the autocorrelation curves showed no significant secondary maxima, although some exhibited weak secondary maxima in the range 20 000–40 000 km. No secondary maxima appeared in Rogers' average curve, which is indicative of the irregularity of the pattern and the large spread in cell sizes evident from direct inspection of his photographs.

Rogers' paper is of particular interest for his clear exposition of the measuring technique involved in a correlation analysis of this kind, which he also applied to a determination of the lifetime of the cells. The observational material consisted of a nearly continuous 62 h sequence of photographs taken at Hα + 0.65 Å through a Zeiss 1/4 Å filter. The telescope was situated at Thule, Greenland, where the Sun lies continuously above the horizon during summer. Measurement of the correlation as a function of time interval Δt revealed an interesting behaviour [Rogers, 1970, Figs. 7 and 8]: initially, there was a rapid decay lasting for $\Delta t \simeq 7$ min, attributed by Rogers to the appearance and disappearance of the dark mottles. This was followed by a slower decline lasting from $\Delta t = 7$ min to ~4 h, presumably due to changes in rosettes and similar groupings of mottles. Finally, for $\Delta t > 4$ h there was a more gradual decay due to changes in the network pattern as a whole. This portion of the curve gave a mean $(1/e)$ lifetime of 25.0 ± 1.6 h. Using the same observational material but following the history of individual cells instead of making a correlation analysis, Janssens [1970] derived a lifetime of 21 h. Janssens describes in detail the complicated evolutionary changes undergone by the cells during their lifetime.

Within the uncertainty limits of the measurements the average lifetimes and cell sizes of the Hα and K networks are therefore the same. This agreement, together with the close spatial correspondence between the two networks, leaves little doubt that both owe their origin to the same underlying agency. This topic is discussed in Section 3.4.5.

3.3.4 THE NETWORK OBSERVED IN OTHER CHROMOSPHERIC LINES

The stronger Fraunhofer lines are formed in the chromosphere (Section 4.3.5). Spectroheliograms obtained in any of these lines may be expected to show structures similar to those observed in Hα and K and, in particular, some analogue of the Hα and K networks.[9] Earlier discussions of spectroheliograms in lines other than Hα and K have been based on the extensive pioneering observations of d'Azambuja [1930] [see, for example, Reichel, 1953; Unsöld, 1955, pp. 707–713; de Jager, 1959, pp. 126–129; Pagel, 1964; Dubov, 1969]. However, examination of the published reproductions of d'Azambuja's spectroheliograms shows them to be markedly inferior to modern examples in terms of spatial resolution. In fact, except on his Ca II spectroheliograms, even the chromospheric network is scarcely visible, let alone finer structures. Accordingly, we shall base our description on the spectroheliograms in Title's [1966a] atlas, which are of much better spatial and spectral resolution.

(1) *λλ8498, 8542, and 8662 of Ca II.* These three infrared lines share an upper atomic level with the H and K lines of Ca II, namely the level 4^2P [Moore and Merrill, 1968]. Their lower level is 3^2D, whereas the lower level for H and K is the ground state 4^2S. All five lines are among the strongest lines in the flash spectrum (Table 4.8).

Spectroheliograms in the three infrared lines present a similar appearance. At the centres of the lines ($\Delta\lambda = 0$) a bright network is well visible [Title, loc. cit.: Plates 17 and 35a], although it is not quite so prominent as in K_2 and K_3. The network is still faintly present at $\Delta\lambda = \pm 0.036$ Å (Title's Plates 19 and 23) but has disappeared at $\Delta\lambda = -0.06$ and -0.072 Å; at these wavelengths the bright faculae are still well visible, together with a coarse mottled structure of unknown origin.

(2) *λλ4861 (Hβ) and 4340 (Hγ) of H I.* An intriguing aspect of the behaviour of Hβ is that even at line *centre* it shows a network of dark mottles characteristic of the *wings* of Hα (Title's Plates 25a and 26c). The approximate correspondence can be written Hβ + 0 Å ≡ Hα + 0.5 Å (compare Title's Plates 25a, 8a, and 9a).[10] At Hβ ± 0.16 Å the appearance is similar to that at the line centre; tuning out to Hβ ± 0.44 Å results in a reduction in the number of mottles, as in the case of Hα (Plates 30 and 31). At this wavelength we have the correspondence Hβ + 0.44 Å ≡ Hα + 0.7 Å (compare Plates 31c, 10a, 3, 39).

At such corresponding wavelengths the dark Hβ network closely resembles the dark Hα network described in the previous section, although the contrast appears somewhat reduced. At no wavelength in Hβ do we see the bright

[9] The network observed in the weaker lines is called the 'photospheric network' and is discussed in Section 3.4.2.

[10] This result disagrees with de Jager's [1957] findings.

network visible at the centre of Hα. The appearance at the centre of Hγ resembles that at the centre of Hβ (compare Title's Plates 25a and b).

(3) *λ5896 (D_1) of Na I and λ5173 (b_2) of Mg I.* The D_1 line shows a well-defined bright network similar to that visible on a K_1 spectroheliogram (compare Title's Plates 35b, 14b and c). However, the network is only faintly discernible on the particular magnesium spectroheliogram published by Title (Plate 32).

(4) *λλ5876 (D_3) and 10830 of He I.* The D_3 line is strongly present in the flash spectrum, where its strength is exceeded only by Hα, Hβ, and the K and H lines of Ca II (Table 4.8). On the disk it appears in emission and absorption in flares and in absorption (weakly) in faculae and filaments [see Bray, 1964, Table 2]. There is a possibility, therefore, that spectroheliograms in this line would show a very weak dark network.

However, a better candidate among the helium lines for revealing the network is λ10830. The lower level of this transition is the metastable triplet level 2^3S, which behaves as a sort of ground state for the triplet series. Consequently, it is easier to excite than the lower level of the λ5876 transition. Modern techniques have brought the necessary observations within the realm of possibility and the network in λ10830 has recently been successfully observed for the first time by Giovanelli *et al.* [1972].

3.3.5 ULTRAVIOLET OBSERVATIONS: HEIGHT VARIATION OF CHROMOSPHERIC NETWORK

Spectroheliograms obtained in chromospheric lines located in the visible and infrared regions of the spectrum – described in Sections 3.3.2 to 3.3.4 above – reveal the appearance and properties of the network in the *low* ($h < 5000$ km) and middle ($5000 \leqslant h \leqslant 10\,000$ km) chromosphere. The effective height corresponding to any given line is hard to estimate without a detailed application of the theory of line formation. In the case of Hα, however, direct measurement of the displacement of dark mottles with respect to bright mottles, as seen at intermediate heliocentric angles, has given us an estimate of at least 5000–7600 km for the height of dark mottles above the base of the chromosphere (Section 3.2.5). Presumably, therefore, the dark network visible in the wings of Hα is located in the middle chromosphere and the bright network visible at Hα centre in the low chromosphere. The middle chromosphere is probably also the site of the network observed in Hβ, Hγ, K_2, K_3, and in the centres of the infrared Ca II lines, while the network visible in K_1, D_1 of Na I, and the b_2 line of Mg I refers to the low chromosphere.

What happens to the network at greater heights, say, in the range $10\,000 < h < 20\,000$ km? Does it become modified in some way or, alternatively, does it disappear altogether at some particular height? To answer these important and fundamental questions we need to obtain spectroheliograms of adequate spatial resolution in a number of ultraviolet emission lines chosen to cover the range of temperatures known to occur in the upper chromosphere, viz. 10^5-10^6 K (Table 4.18, Chapter 4). Some suitable lines are listed in Table 4.12, including emission lines of carbon, nitrogen, oxygen, sulphur, neon, magnesium, and helium in various stages of ionization; the last column gives the 'effective' temperatures, $\overline{T}_e$, of the chromospheric levels emitting the lines in question.

Unfortunately, however, most existing ultraviolet spectroheliograms lack the necessary spatial resolution. Examination of Plates 3.10 and 3.12a and b shows that the *walls* of the network cells are approximately 6–8″ of arc wide, so that if we assume the network remains largely unchanged with height, a resolution of this order is needed.

One ultraviolet spectroheliogram which comes close to satisfying this criterion has been reproduced by Austin *et al*. [1967, Fig. 7]. This is portion of a slitless spectrogram obtained from an Aerobee rocket on 28 April 1966, and refers to Lyman α of He II (λ304). An enlarged print has been published by Tousey [1967, Fig. 29]. The resolution claimed is ~10″, and a *network* as such is not apparent. Nevertheless, quiet regions on the photograph show a great deal of detail, which takes the form of a confused pattern of coarse bright and dark 'mottles'. This detail is even more prominent on a λ304 spectroheliogram recently published by Tousey [1971, Fig. 11], which has a claimed resolution of ~5″. The presence of this structure has been interpreted by Austin *et al.* [loc. cit.], Goldberg [1967], and Tousey [loc. cit.] as implying the existence, at the relevant height, of the network itself. Referring to Table 4.12, one finds that the value of $\overline{T}_e$ for λ304 of He II is 85 000 K; this temperature is reached at $h \cong 9000$ km according to the adopted homogeneous model of Table 4.18.[11]

An apparently similar structure is visible in λ584 of He I [Goldberg, 1967, Fig. 13]. For this line $\overline{T}_e = 32\,000$ K and $h \simeq 8000$ km.[11] On the other hand, the above authors have remarked on the apparent *absence* of structure in lines of O IV and O V (and in lines of still higher ionization potential), for which $\overline{T}_e$ ranges from 125 000–160 000 K ($h \simeq 10\,000$ km),[11] and have concluded that the network disappears at or near this level. However, it should be pointed out that the quality of the images upon which this conclusion appears to be based leaves much to be desired [see Goldberg, 1967, Fig. 13; Tousey, 1967, Fig. 27]. Perhaps judgement should be reserved until spectroheliograms with higher spatial resolution and less overlapping of images are available.

Rocket photographs in the K line of Mg II (λ2795.5) have been obtained by Fredga [1969]. In quiet regions these show bright flocculi very similar to those

[11] If we use instead the interspicular component of our adopted *inhomogeneous* model, each of these estimated heights has to be increased by ~4000 km (cf. Fig. 4.14). The spicule component is too cool to excite the ultraviolet lines under discussion.

observed in the K line of Ca II (see Fredga's Figs. 3–5). The resolution claimed is about 8″ and the network itself is not apparent. According to Linsky and Avrett [1970] the H and K lines of Mg II are formed higher in the chromosphere than the H and K lines of Ca II.

Somewhat surprisingly, the spatial resolution achieved from rockets exceeds that so far achieved from satellites. An extensive atlas of ultraviolet spectroheliograms obtained with the fourth Orbiting Solar Observatory (OSO IV) has been published by Reeves and Parkinson [1970]. It contains many lines of interest in the region 300–1400 Å, corresponding to a number of elements in various stages of ionization. However, the resolution is no better than 1′ of arc, so these observations unfortunately tell us nothing about the network in the higher levels of the chromosphere. This situation, however, is unlikely to continue for long. Later members of the OSO series have given improved resolution and the necessity for space observations of the Sun having a resolution of at least 5″ is well recognized [cf. Doyle, 1969, pp. 149–176]. The most immediate prospect of achieving this is the planned launching in 1973 of 'Skylab', a three-man satellite which will include a solar observatory [Watts, 1970]. Among the instruments will be an ultraviolet spectroheliograph developed by Harvard College Observatory, which is designed to cover the range 300–1310 Å with a spectral resolution of 1.3 Å and a spatial resolution of 5″. Each picture will be 5′ of arc square, made up of 60 scan lines; for comparison purposes, Hα photographs will also be taken from the satellite. Observations with this and other future instruments should ultimately extend our knowledge of the properties of the network to the upper chromosphere.

3.4 The Chromospheric Network: Relationship to Photospheric Phenomena

3.4.1 INTRODUCTION

In the preceding four sections we have outlined the properties of the network as observed in the K, Hα, and other chromospheric lines and have related the network to the bright K flocculi and the bright and dark Hα mottles. In the sections that follow we shall describe the close relationships that exist between the chromospheric network and three underlying *photospheric* phenomena, viz., the photospheric network, the supergranulation, and the photospheric magnetic field distribution (Sections 3.4.2–3.4.4).

The photospheric network can be regarded simply as a downward extension of its chromospheric counterpart and, accordingly, does not greatly add to our knowledge of the latter. On the other hand, the relationships between the chromospheric network and the other two photospheric phenomena listed above are of great theoretical importance because they yield essential clues to the origin and significance of the network. Essentially, the fundamental agency is the *supergranulation.* The horizontal and vertical motions associated with the supergranules concentrate magnetic fields at the cell boundaries, a process whose feasibility can be tested by comparing the magnetic and kinetic energy densities at the photospheric level (Section 3.4.5). The photospheric magnetic fields are presumed to extend upwards into the chromosphere, and together the photospheric and chromospheric fields are believed to be responsible for the excess heating which manifests itself as a bright network in the K line, in the centre of Hα, in various photospheric lines, and also (very weakly) in the continuum.

This picture implies a close correlation between the network and the field and, in fact, the observed correlation is so close that a photograph of the network may be regarded as essentially a map of the magnitude of the field or, more precisely, of its longitudinal component (Section 3.4.4). The network is thus the poor man's magnetograph. During both the formation and dissolution of active regions, the network – together with the enhanced network in their neighbourhood and the chromospheric faculae within them – serves as a tracer for the magnetic field, showing in detail its initial concentration and its final dispersal over the entire surface of the Sun.

3.4.2 THE PHOTOSPHERIC NETWORK

The photospheric network may be regarded as a downward extension of the more prominent chromospheric network described in Section 3.3. It takes the form of a *bright* network visible in some of the weaker Fraunhofer lines and, for any given line, is produced by a local *weakening* of the line. Since the network is found to be closely correlated with the photospheric magnetic field distribution (Section 3.4.4), it might be conjectured that the line weakening is solely a consequence of Zeeman splitting. However, this explanation is ruled out by the circumstance that a number of Zeeman-*insensitive* lines are found to show local weakenings in (non-spot) magnetic regions. Accordingly, Chapman and Sheeley [1968] have sought an explanation in terms of a *temperature* effect: appealing to the Saha–Boltzmann equation they have calculated that the observed weakenings may be accounted for by a temperature excess of some 100–200 K.

The visibility of the photospheric network in various lines has been investigated by Chapman and Sheeley [loc. cit.], who find it to be most prominent in low-excitation lines of neutral metals and in certain molecular lines. Thus it is quite prominent in the Fe I line λ5250 but weak in the Fe II line λ5265 [Chapman and Sheeley, Fig. 8]. Figures 2–6, 8, and 11 of their paper illustrate the network visible in the lines Fe I λλ6302.5, 6241, 5269.5, 5250, 5131, 5124; Fe II λλ6248, 5265; Ni I λ4855; Ca I λ6103; and CH λ3144, together with the stronger lines Ca I λ4227 and b_1 of Mg I. Spectroheliograms in Fe I λ4071 and near the CN band head at λ3883 have been reproduced by Sheeley and Engvold [1970, Figs. 1 and 2] and Sheeley [1969, Figs. 3 and 4] respectively. The general appearance of the photospheric network on these photographs is somewhat reminiscent of the network observed in the K_1 portion of the K line of Ca II (Section 3.3.2). It is most prominent in the neighbourhood of sunspots and active regions, resembling in this respect both the K network

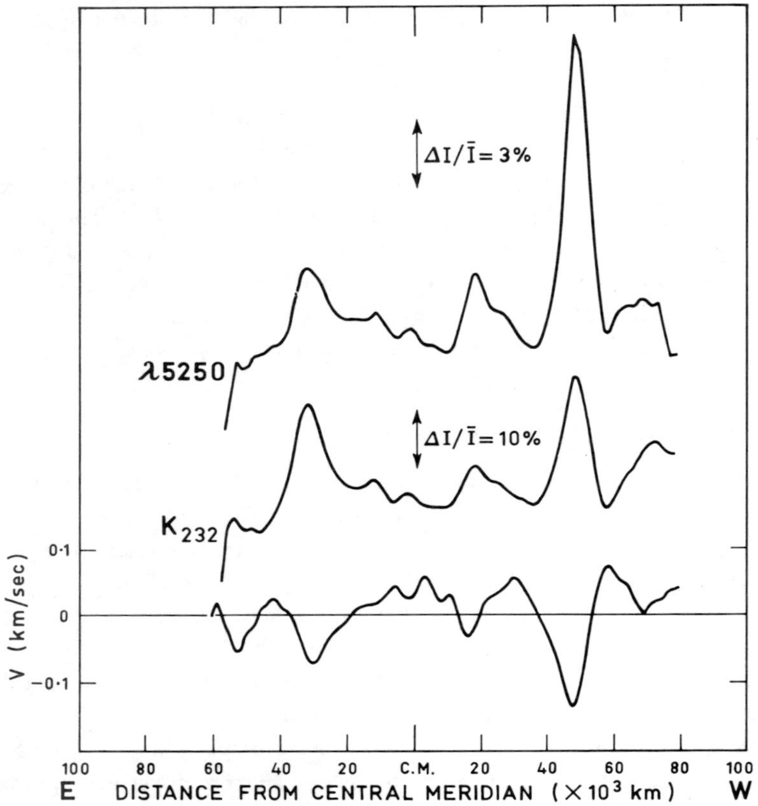

Fig. 3.3. Correlation between the photospheric and chromospheric networks and the supergranulation [Tanenbaum *et al.*, 1969, Fig. 11]. The upper and middle curves show the brightness in Fe λ5250 and $K_{2\,32}$ respectively, the lower curve the velocity measured in Fe λ5233. Each curve is the average of 300 scans made with an aperture of 2″.4 along a line crossing the disk centre. Note the close correlation between the maxima of the two upper curves and the minima (downward velocities) of the lower; these points are located on the boundaries of a network or supergranule cell.

and the bright network observed at the centre of Hα (Section 3.3.3). It appears to be made up of small bright points approximately 1000 km in diameter [Sheeley, 1969].

The observations published by Sheeley and his co-workers leave unanswered one important question, namely, the extent to which the photospheric network is present in truly quiet regions. However, this point can be clarified by examining a spectroheliogram obtained by Frazier [1970] with the sensitive multi-channel magnetograph of the Kitt Peak Observatory. This shows very clearly the bright network in the core of Fe I λ5250 in a carefully chosen quiet region near the disk centre. Moreover, the brightness distributions in λ5250 and the K line of Ca II are closely correlated (compare Frazier's Figs. 3 and 4). This correlation is also illustrated in Fig. 3.3, which is taken from the paper of Tanenbaum *et al.* [1969].

According to Chapman and Sheeley [1968], observations in λ5250 and in the continuum near the limb show the same structure. Moreover, Frazier's observations in a quiet region near the disk centre demonstrate that the brightness variations in λ5250 are correlated with (very weak) variations in continuum brightness [Frazier, 1970, Fig. 8]. It seems likely, therefore, that the distribution of white-light faculae observed near the limb outlines the photospheric network present in an enhanced form in the vicinity of an active region. Indeed, examination of a good-quality photograph of a facular region [e.g., Bray and Loughhead, 1967, Plate 3.1] does give the impression of a foreshortened picture of a rather untidy network, whose average cell size appears to agree with the cell size of the K network, viz., ~33 000 km.

3.4.3 THE SUPERGRANULATION

As we shall see in Section 3.4.5, the chromospheric network is generally believed to owe its ultimate origin to the supergranulation. Basically a phenomenon of the upper photosphere, the supergranulation may be described as a regular, cellular pattern of mainly horizontal motions, uniformly present over the entire surface of the quiet Sun. The observed flow pattern of any given cell consists of a horizontal outflow from the vicinity of its centre, together with a slow upflow at its centre and a slow downflow at its boundary. The supergranulation bears a close relationship to the chromospheric and photospheric networks and to the distribution of weak magnetic fields outside active regions. Simon and Leighton [1964] have suggested that it owes its origin to convective instability arising from single and double ionization of helium at depths of the order of a few thousand kilometres below the photosphere [see discussion by Bray and Loughhead, 1967, pp. 101–103]. Its importance from the point of view of our understanding of the chromosphere is that the horizontal currents associated with each supergranule are believed to sweep magnetic fields from declining active regions into the boundaries of the cell, where they produce excess heating which, in turn, is the cause of the chromospheric network. A possible explanation of the excess heating in terms of enhanced shock wave dissipation in the presence of a magnetic field is discussed in Section 7.4.8.

Detailed descriptions of the basic properties of the supergranulation have been given by the present authors [Bray and Loughhead, loc. cit., Section 3.4 and pp. 101–103] and by Noyes [1967, pp. 309–312, 316–319] and need not be repeated here. These accounts include material published up to and including 1965. More recent work has added important new information on several aspects of the phenomenon: (*a*) the vertical velocities at the supergranule cell boundaries; (*b*) the correlation between the supergranulation, the photospheric magnetic field, and the chromospheric, photospheric, and very weak continuum networks; (*c*) the 'profiles' of various parameters across an average supergranule cell; and (*d*) the extension of the supergranulation into the low chromosphere. A description of this new work is given below, with emphasis on those aspects which enlarge our understanding of the relationship between the supergranulation and the chromospheric network.

Recent observations of the *vertical* velocities have proved particularly fruitful because they enable the correlation between the supergranulation and other phenomena to be studied at the *centre of the disk*. Earlier work was concentrated on the *horizontal* velocities, which are larger than the vertical velocities but of necessity have to be observed away from the disk centre: accordingly, these studies have been hampered by foreshortening. The modern approach is made possible by the greater sensitivity of the more recent equipment to the small line-of-sight velocities involved, which are only ~0.1 km sec^{-1} or less. A second crucial innovation has been the application of new methods of data collection, processing, and display: amongst other advantages, these have enabled one to average the data over a time long compared to the lifetime of the 5-min oscillations (Section 3.5.1) or of the photospheric granules (~10 min). In this way the small-amplitude velocities associated with the supergranulation can be separated from the generally larger velocities associated with the latter phenomena.

Outstanding among the modern equipment is the multi-channel magnetograph of the Kitt Peak Observatory, designed by Livingston [1968]. Using this equipment, Frazier [1970] obtained simultaneous raster scans across a carefully selected quiet region near the disk centre of no less than ten quantities: Doppler shift, brightness, and longitudinal magnetic field, measured in the wings of each of the two Fe lines λλ5250 and 5233; core brightnesses of λ5250, Hα, and the K line of Ca II (K_{232}); and the continuum brightness. With a scanning aperture of only 2″.4 x 2″.4 of arc square, these observations are well adapted to elucidating the relationship between the supergranulation and the network. Our account, accordingly, is based largely on Frazier's results.

(1) *Vertical velocities; correlation with the chromospheric network.* Figure 1 of Frazier's paper gives a photographic-type representation of the pattern of upward and downward velocities, the scaling being such that velocities between

+0.115 km sec^{-1} (upward) and -0.152 km sec^{-1} (downward) are recorded as patches varying from white to black. The dark patches (downflows) do not appear to form a well-defined network; however, they do correlate well with the longitudinal photospheric magnetic field and with the brightness fields in the cores of λ5250 and the K line (K_{232}) – i.e., with the photospheric and chromospheric networks (compare Frazier's Fig. 1 with his Figs. 2, 3, and 4). The correlation between the two networks and the downward velocity is also well demonstrated by the linear scans shown in Fig. 3.3, taken from Tanenbaum *et al.* [1969]. It is again shown – but with less conviction – by observations made by Musman and Rust [1970, cf. Fig. 4] at the Sacramento Peak Observatory: these observations, however, were made 17° from the disk centre, so interpretation of the line-of-sight velocities as downflows is complicated by the presence of horizontal velocities.

If we define a supergranule boundary as the locus of the downward velocities, then the above observations demonstrate the close correlation between the supergranulation and the chromospheric network. The same conclusion was reached by Simon and Leighton [1964] on the basis of a comparison of a Doppler spectroheliogram exhibiting the *horizontal* velocities in cells located away from the disk centre with K_2 and K_3 spectroheliograms.

The magnitude of the vertical downflows has been determined by several authors. Averaging the velocities measured at 39 points of maximum brightness in λ5250, Frazier [loc. cit.] found an average velocity of $\bar{V} = -0.09$ km sec^{-1} (see his Fig. 12). Frazier's Fig. 10 records individual values as high as $V = -0.22$ km sec^{-1}. Musman and Rust's smaller figure of $\bar{V} = -0.02$ to -0.04 km sec^{-1} for the average downflow probably reflects their larger scanning aperture, 6″ of arc. The presence of small vertical velocities had earlier been detected by Simon and Leighton [1964]; Simon estimated the average downflow to be -0.1 to -0.2 km sec^{-1} [see Bray and Loughhead, 1967, p. 59], in reasonable agreement with the more recent measurements. Note that the downward velocities at the cell boundaries are markedly smaller than the peak horizontal velocities within the individual cells, 0.3–0.5 km sec^{-1}.

(2) *Supergranule profiles: brightness, velocity, and magnetic field.* Figure 3.4, taken from Frazier [1970], shows the average distribution across a supergranule of the following parameters: longitudinal magnetic field, measured in Fe λ5233; brightness in K_{232}; brightness in the continuum; and vertical velocity, also measured in λ5233. Each profile represents the average for 20 cells along lines crossing the cells from one vertex to the most distant vertex, a vertex being defined as the meeting-point of several cells. The noteworthy feature of the profiles is the occurrence of sharp peaks or troughs at the cell boundaries; even the continuum intensity shows noticeable maxima, amounting to some 0.5 per cent of the background. Note also the small upflow at the centre of the cell,

whose magnitude is about one-half that of the downflow at its boundaries.

The profiles in Fig. 3.4 refer to cuts across the cells from vertex to vertex. When similar profiles are plotted for cuts across the cells from *wall to wall*, the

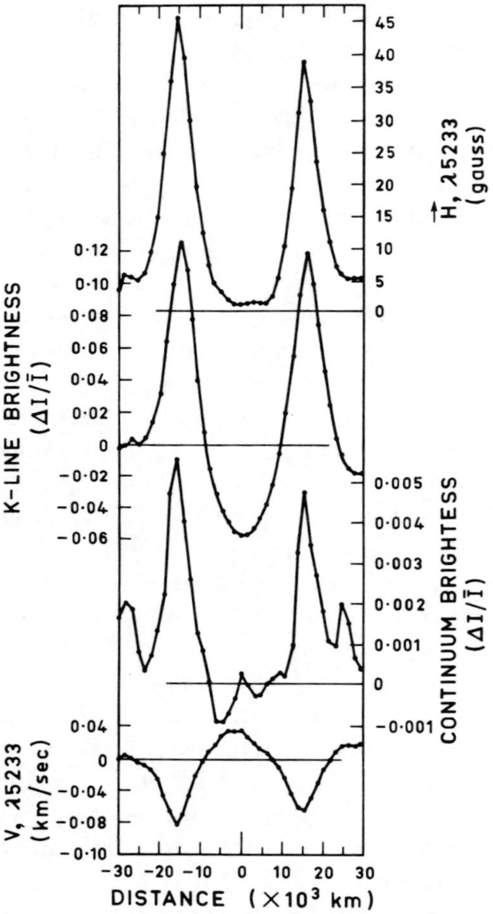

Fig. 3.4. Average 'vertex to vertex' supergranule profiles [Frazier, 1970]. Each profile represents the average for the same 20 cells, individual curves first being normalized to the same average cell diameter (30 000 km).

various maxima and minima, although still present, are much less prominent [see Frazier, loc. cit., Fig. 14]. This implies that the various parameters do not remain constant along the cell boundaries and, accordingly, some doubt is cast on the validity of the cell concept. To test this objectively, Frazier compared profiles of the various parameters obtained firstly along a cell wall and secondly, in the direction of the cell centre. Not surprisingly, the network showed up best in the

K brightness distribution, some 60 per cent of the profiles showing very clearly identifiable walls. On the basis of all the curves, Frazier concluded that: (*a*) the supergranule cells are real; (*b*) the cell walls are very weak in the photosphere but strengthen as they rise into the chromosphere; (*c*) the walls, although present, are always very weak compared to the vertices of the network; and (*d*) at any time roughly two-thirds of the total possible number of walls are present.

(3) *Cell size.* Figure 3.4 leads to a vertex to vertex dimension of some 30 000 km for the supergranulation, whereas Frazier's wall to wall profiles lead to a diameter of 24 000 km [Frazier, 1970, cf. Fig. 14] ; see also Tanenbaum *et al.* [1969]. Counts of the total number of supergranules on the visible hemisphere give a cell size of ~35 000 km [cf. Bray and Loughhead, 1967, pp. 58–59]. These measurements are in reasonable agreement with the cell size of the K network derived from autocorrelation analyses, namely, 33 000 km (Section 3.3.2).

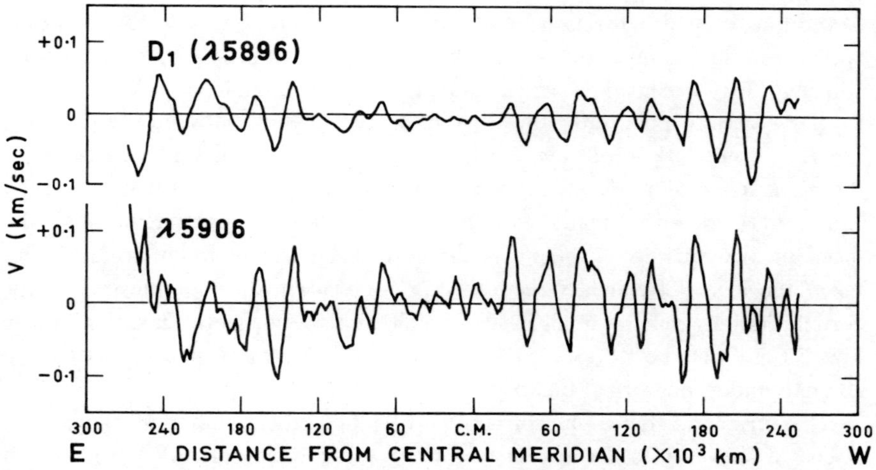

Fig. 3.5. Extension of supergranule velocities into the low chromosphere [Tanenbaum *et al.*, 1969]. The upper trace refers to the D_1 line of Na I, the lower to Fe λ5906. Each trace is the average of 300 scans crossing the disk centre. Away from the central meridian the velocities recorded are probably a mixture of both horizontal and vertical velocities.

(4) *Extension of the supergranulation into the low chromosphere.* It has been customary to regard the supergranulation as a purely photospheric phenomenon, despite its close relationship to the chromospheric network. However, recent observations by Tanenbaum *et al.* [loc. cit.] at the Mt. Wilson Observatory show unmistakably that the upward and downward velocities at the cell centres and boundaries respectively penetrate – albeit with reduced amplitude – at least into the low chromosphere. Tanenbaum *et al.'s* measurements are shown in Fig.

3.5; the upper trace refers to the D_1 line of Na I, the lower to the weak photospheric line Fe λ5906. Each trace is the average of 300 scans occupying some 3 h 20 min made along a line crossing the disk centre; a 10″ of arc scanning aperture was employed. Both upward and downward velocity peaks occur at the same locations in the two lines, which Tanenbaum *et al.* estimate to be separated in height by several hundred kilometres. Observations in the b_2 line of Mg I (λ5173) and the nearby Fe line λ5180 show a similar agreement. The presence of the supergranulation in the D_1 line had earlier been detected by Leighton *et al.* [1962].

3.4.4 PHOTOSPHERIC MAGNETIC FIELDS

Observations made by Frazier [1970] with the multi-channel magnetograph of the Kitt Peak Observatory have established on a very firm foundation the close correlation between the chromospheric network and the photospheric magnetic field distribution. Figure 2 of his paper gives a photographic-type representation of the longitudinal field, measured in the Fe λ5233 line with a 2″.4 x 2″.4 of arc aperture. Visual comparison of this picture with a simultaneous $K_{2\,32}$ picture (Frazier's Fig. 4) shows a virtually one-to-one correspondence between them. Frazier gives no quantitative measure of the correlation between the field and the K line network. However, he does compute the correlation coefficient for the *photospheric* line Fe λ5250, whose brightness distribution – like that of the K line network – also appears to show a very close correspondence with the magnetic field picture. In this case the normalized cross-correlation coefficient takes the relatively high value $C = 0.72$. A scatter diagram illustrating the correlation between the brightness in λ5250 and the magnetic field is shown in Fig. 3.6. It will be noticed that the brightness increases linearly with field strength, independently of polarity.

The magnetic field, in common with the supergranulation downflows discussed above, does not appear to form a well-defined network but instead is concentrated mainly in the form of patches. Both polarities are present and the patches correspond exactly with the downflows, independently of polarity; the measured value of the correlation coefficient is $C = 0.51$. Frazier's observations (see his Fig. 12) lead to an average diameter of a magnetic patch (full width at half maximum field strength) of ~7000 km. However, Livingston [1968] has resolved fields associated with the K network into elements having a diameter of 1000–2000 km, while Sheeley [1967a] has observed magnetic regions (closely correlated with $K_{2\,32}$ emission) ranging in size from several thousand kilometres down to only 500 km.[12]

[12] The close correlation between the longitudinal magnetic field and the chromospheric and photospheric networks has been demonstrated by numerous other workers, including Howard [1959], Simon and Leighton [1964], Leighton [1965], Tsap [1966], Sheeley [1967a], Bappu *et al.* [1968], Chapman and Sheeley [1968], and Sheeley and Engvold

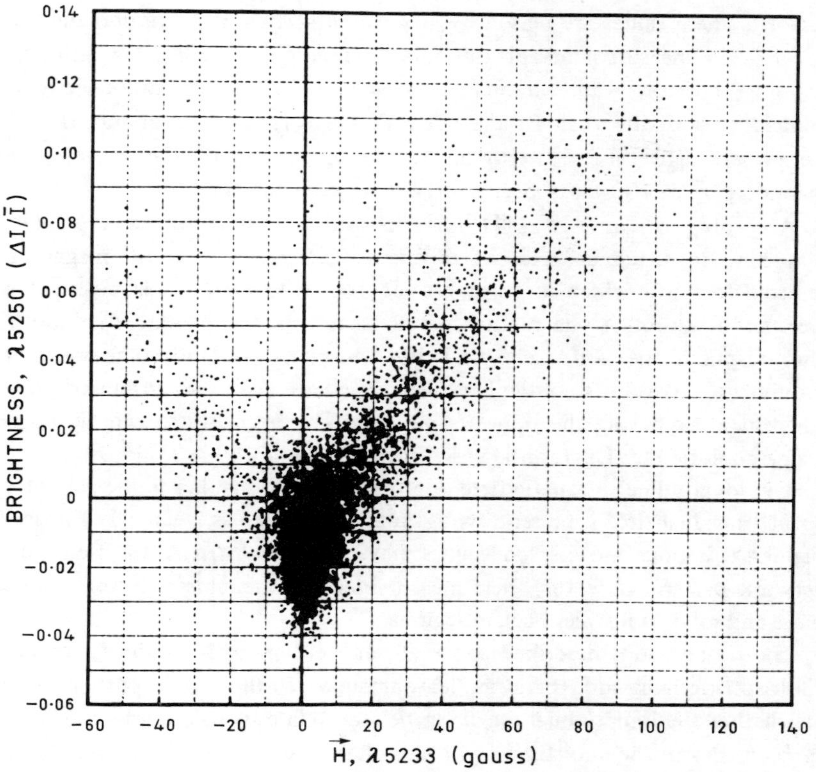

Fig. 3.6. Scatter diagram illustrating the close correlation between the photospheric network and the magnetic field distribution [Frazier, 1970]. Ordinate: brightness measured in the core of Fe λ5250. Abscissa: longitudinal magnetic field, measured in Fe λ5233. The occasional points representing high brightness but almost no field are actually small *bipolar* patches.

Various authors have given values for the magnitude of the longitudinal field associated with the network. Frazier finds a field strength of 50 gauss for the average region of downflow; the stronger ones attain field strengths of 100 gauss. Fields as weak as 5 gauss are accompanied by noticeable calcium emission. As one would expect, the fields in the network walls are strongest in the neighbourhood of the vertices and much weaker elsewhere (~10 gauss according to Frazier, loc. cit., Fig. 14). Values quoted for the network field by various other authors are as follows: < 40 gauss [Tsap, 1966, Fig. 3]; greater than 40 to 80 gauss [Noyes, 1967]; ~25 gauss [Bappu *et al.*, 1968]; > 25 gauss [Livingston, 1968]; ~40 gauss [Musman and Rust, 1970, Fig. 3] (a brighter than average

[1970]. In general, however, their observations are inferior to Frazier's in terms of either sensitivity, spatial resolution, or data processing techniques. Some of them, moreover, refer to regions fairly close to sunspots, whereas Frazier carefully chose a very quiet region with very weak calcium emission.

region). These values are reasonably concordant when one considers the sources of error in measurements of this kind including, especially, the influence of scanning aperture size. Summarizing, we conclude that photospheric fields coinciding with the walls of the network range from 10 to about 50 gauss or somewhat higher. Field strengths at the upper end of this range, possibly extending up to 100 gauss, occur at the vertices of the network.

According to Bappu *et al.* [1968], opposite sides of a given supergranule cell may have longitudinal fields of different polarity. In Frazier's magnetogram referred to above, which covers an area equivalent to several cells, both polarities are present although one predominates. Near an active region, where both the chromospheric network cells and the 'magnetic' cells become particularly prominent, large areas containing many cells have the same dominant polarity, adjacent areas having the opposite polarity. This is clearly shown on magnetograms published by Leighton [1965, Fig. 4] and Vrabec [1971, Fig. 4].

The longitudinal magnetic field does not vanish altogether in the cell interior: a small residual field is present, whose strength is given as 2 gauss by Frazier (see Fig. 3.4). Bappu and his colleagues have measured *transverse* fields in the network pattern; however, their measurements are reliable only for fields >80 gauss and refer to incipient active regions.

The observations described above all refer either to the K line or to various photospheric lines and, therefore, leave unanswered the very important question of whether the longitudinal magnetic field correlates with the *bright* or with the *dark* mottles of the undisturbed chromosphere observed in the Hα line. A partial answer to this question has been provided by some very high-quality magnetograms obtained recently by Vrabec [1971] at the San Fernando Observatory (California) of the Aerospace Corporation. Portion of one of these is shown in Plate 3.13b, together with a simultaneous photograph taken at the centre of Hα (Plate 3.13a). (It should be noted that the region illustrated is centred on a large active region, and therefore cannot be regarded as typical of the quiet chromosphere.) There is a well-marked correlation between the field and the bright Hα emission, both within the active region and on its outskirts. Furthermore, careful examination by the authors of a film prepared by Vrabec showing Hα wing photographs as well as Hα centre photographs and magnetograms demonstrates that the field coincides with *bright* rather than with *dark* material.

It seems likely that a similar correlation extends to the fainter bright material visible at the centre of Hα which outlines the chromospheric network in regions remote from spot groups (Section 3.3.3), but magnetic observations demonstrating this are so far lacking. One may also speculate that the correlation is still present on the scale of individual bright mottles. If this picture is correct, the rosette structure described in Section 3.2.5 possesses a longitudinal field coinciding with the bright mottle material concentrated towards its centre. Such a correlation between the longitudinal field and the bright Hα emission is of course to be expected in view of (*a*) the correlation between the field and the bright K network, and (*b*) the spatial coincidence between K flocculi and the *centres* of rosettes (Section 3.2.8).

3.4.5 CONCLUDING REMARKS

The correlation between the chromospheric network and the photospheric magnetic field distribution is so close that, as remarked by Leighton [1965], one may regard a photograph of the chromospheric network as being essentially a map of the magnitude of the longitudinal field component. Most workers believe that the field is responsible for localized excess heating which, in turn, manifests itself as a bright network in the K line of Ca II, the centre of Hα, and the photospheric lines discussed in Section 3.4.2. The discovery by Frazier [1970] of a small brightening of the continuum at the network boundaries is further evidence for a non-thermal source of heating: evidently, if the supergranulation were a purely convective phenomenon, not involving a magnetic field, then one would expect a small *darkening* at these locations. A discussion of the heating of the chromosphere in the neighbourhood of the network is given in Section 7.4.8.

As we have already remarked in Section 3.4.3, the magnetic fields are thought to be concentrated at the cell boundaries by the action of the supergranule flow pattern. The condition that hydrodynamic forces predominate over magnetic forces in an ionized or partially-ionized gas is, crudely speaking,

$$\frac{1}{2}\rho v^2 \gg \frac{H^2}{8\pi}$$

[see, for example, Bray and Loughhead, 1964, p. 266]. Here the term $\frac{1}{2}\rho v^2$ represents the kinetic energy of the gas per unit volume and $H^2/8\pi$ the magnetic energy density. Substituting $v = 0.4$ km sec^{-1} (Section 3.4.3), $\rho = 1.7 \times 10^{-7}$ gm cm^{-3} (Table 4.6) as values appropriate for the horizontal supergranule flow at the level $\tau = 0.1$, we obtain the result that the field will be swept to the cell boundaries provided that $H \ll 60$ gauss. Since the measured field in the cell interior at the photospheric level is only ~2 gauss (Section 3.4.4), this criterion is evidently satisfied.

The situation is, however, quite different at the chromospheric level in the cell *boundaries*. At a height of only 1000 km, for example, ρ has dropped to 8.1×10^{-12} gm cm^{-3} (Table 4.18), while we assume that the field H ranges from ~10 to ~100 gauss, the measured photospheric values. With this value of ρ we now find that in regions with fields of 10 and 100 gauss, for example, magnetic forces will dominate hydrodynamic forces unless v exceeds ~10 and

~100 km sec^{-1} respectively. On the other hand, it must be stressed that material can always move freely *along* the magnetic lines of force.

The supergranulation is a basic property of the Sun and, presumably, of stars of similar spectral class. Its significance goes far beyond its role in producing the chromospheric and photospheric networks. For example, it appears to play a part in the formation of new active regions. Moreover, when an active region breaks up, the supergranule motions are believed to be responsible for distributing its magnetic field throughout a large area of the 'quiet' Sun by a process which Leighton has likened to a 'random walk'. During all these events the chromospheric network, together with the enhanced network near active regions and the chromospheric faculae within them, behaves as a tracer for the more fundamental magnetic field. These topics, however, lie beyond the scope of this book and the interested reader is referred to discussions by Leighton [1964], Simon and Leighton [1964, pp. 1143–1144], Howard [1967a], and Bappu *et al.* [1968].

3.5 Chromospheric Velocities and Magnetic Fields

3.5.1 THE OSCILLATORY VELOCITY FIELD IN THE UPPER PHOTOSPHERE AND LOW CHROMOSPHERE

In addition to the regime of horizontal and vertical velocities associated with the supergranulation (Section 3.4.3), the photosphere possesses two further velocity fields which, apparently, are unrelated to the supergranulation. These are the convective field of upward and downward velocities associated with individual photospheric granules and the oscillatory velocity field of the upper photosphere. There are two reasons why the latter field is of importance to a study of the chromosphere. In the first place, it extends up at least into the low chromosphere: this is shown by its observation in lines such as Na D_1 and Mg b_2. Secondly, as discussed in Section 7.3.4, it is probably a manifestation of the existence, at the chromospheric level, of progressive waves of the gravity type.

Detailed descriptions of the oscillatory velocity field have been published by the authors [Bray and Loughhead, 1967, pp. 35–45] and Noyes [1967, pp. 297–309, 315–316]. In this section, accordingly, we shall confine ourselves to discussing more recent observations which, in one way or another, extend or modify these earlier accounts.

(1) *Size of oscillating elements.* Earlier rough measurements left the impression that the scale of the velocity pattern was somewhat larger than the scale of the photospheric granulation. This conclusion was reached by measuring the distance on the Sun over which the *amplitude* of a given oscillation had roughly the same value. However, if instead we define the size in terms of the distance over which the *phase* remains constant, then we arrive at much larger values. This is clearly shown in Fig. 3.7, due to Musman and Rust [1970], which

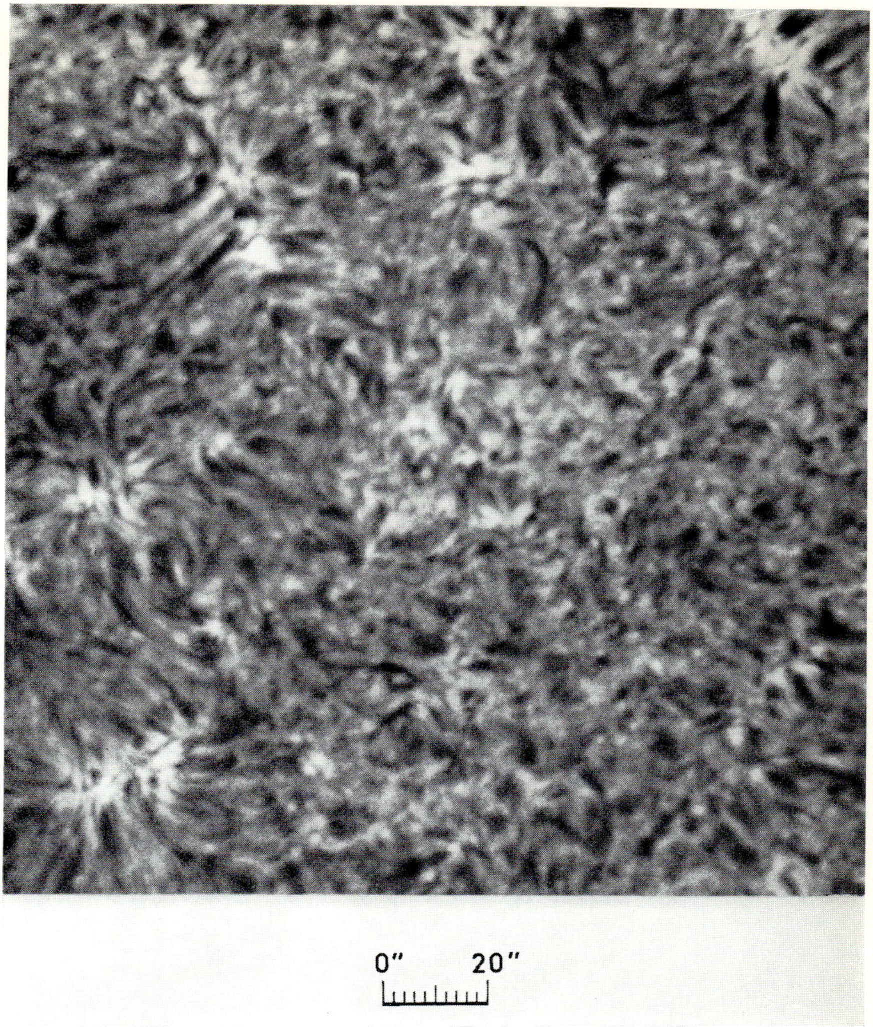

PLATE 3.1. Quiet chromosphere in central portion of disk photographed at the centre of Hα ($\frac{1}{8}$ Å filter + 1 Å pre-filter).

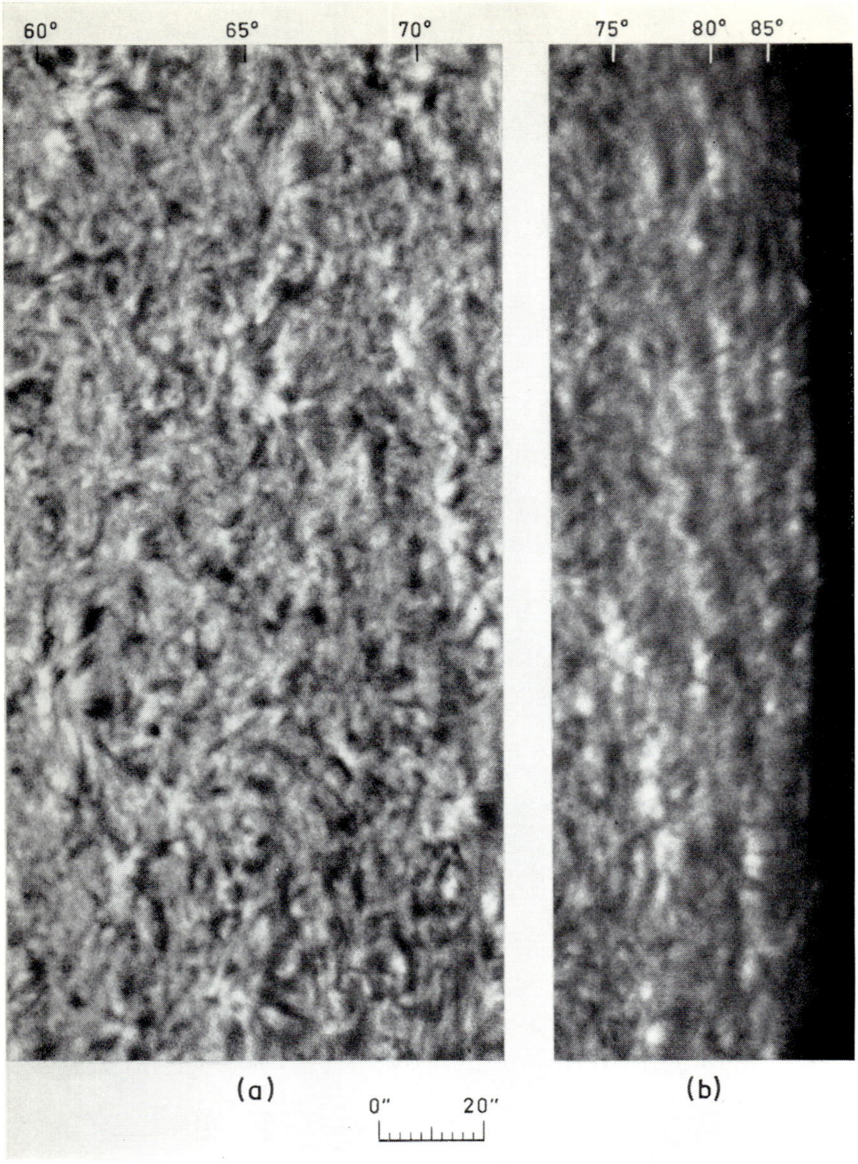

PLATE 3.2. Bright mottles at Hα line centre photographed on separate occasions (a) at intermediate heliocentric angles, and (b) near the extreme limb (0·5 Å Halle filter). The heliocentric angle is shown at the top. Note that near the limb the bright mottles stand out as distinct structures, whereas the dark mottles almost completely disappear.

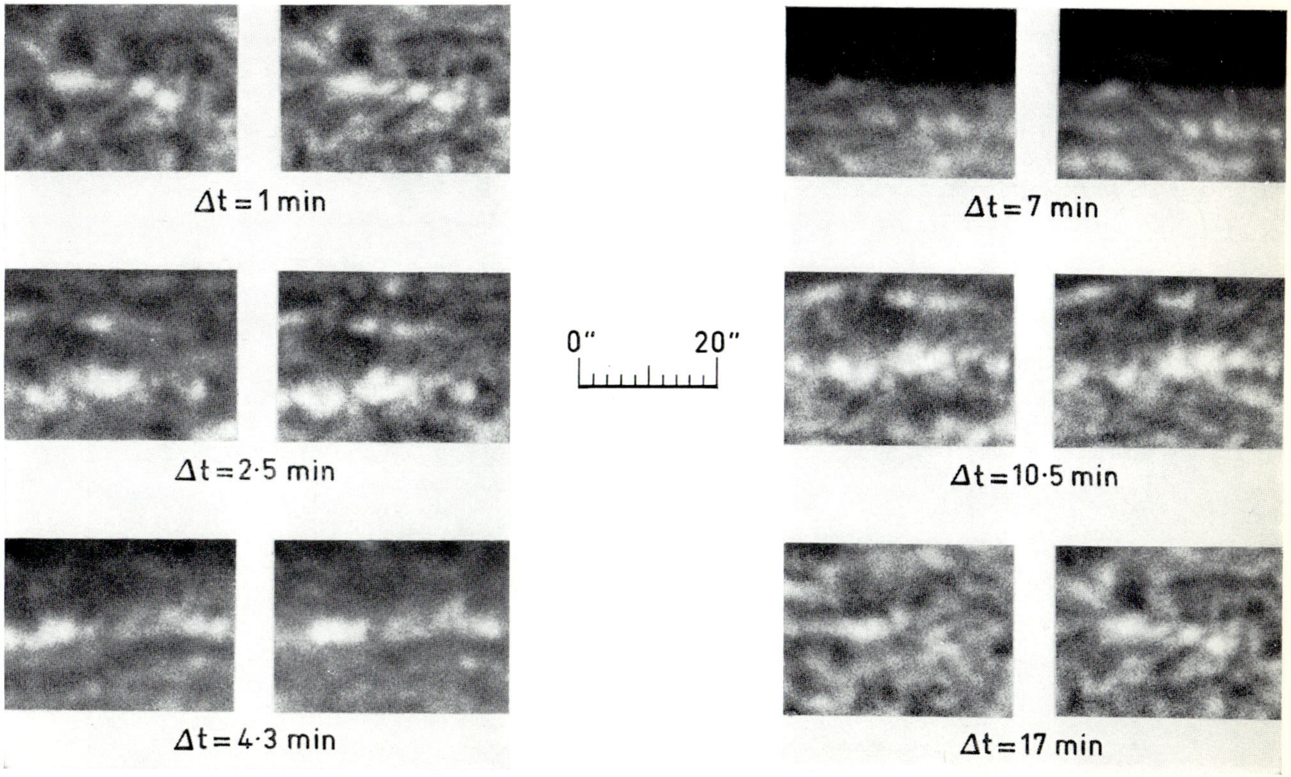

PLATE 3.3. Pairs of photographs at Hα line centre showing changes in individual bright mottles over intervals ranging from 1 min to 17 min (0·5 Å Halle filter). Each pair refers to the same region but different pairs correspond to different regions. All photographs were obtained in the general neighbourhood of the north limb.

PLATE 3.4. Dark mottles near disk centre photographed at $H\alpha + 0{\cdot}5$ Å ($\frac{1}{8}$ Å filter + 1 Å pre-filter). Note the spiky appearance of many mottles at this wavelength.

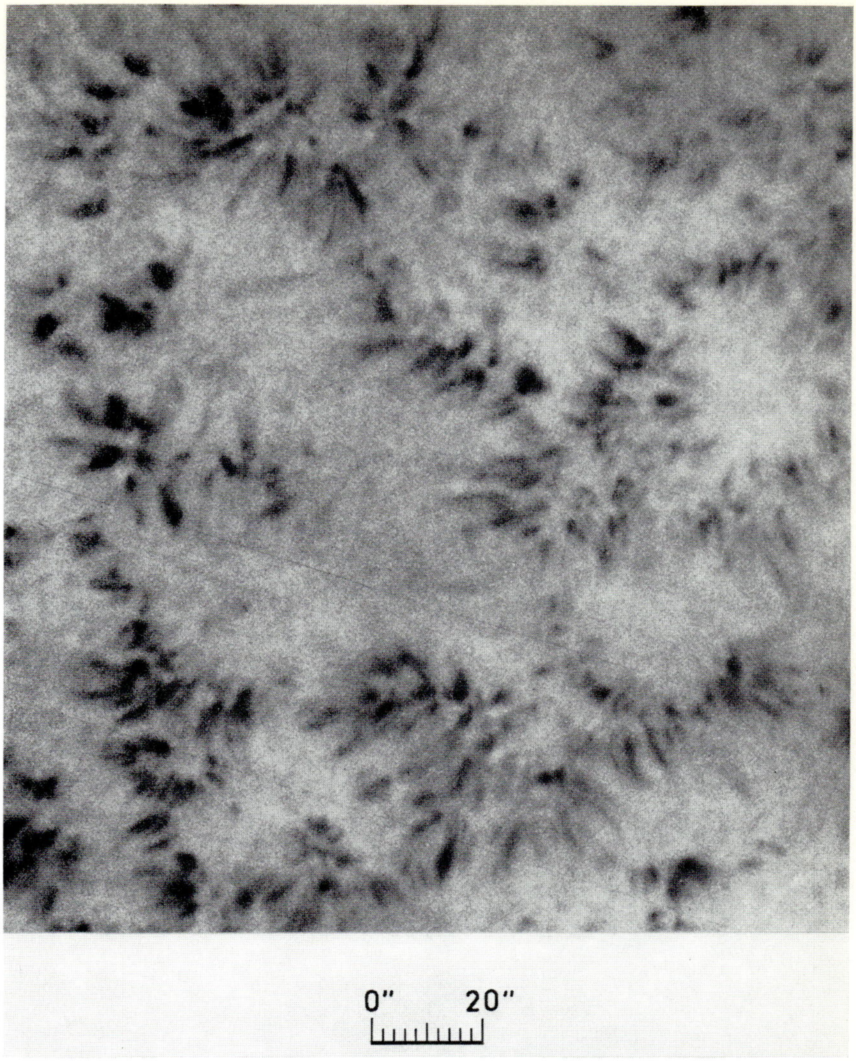

PLATE 3.5. Dark mottles near disk centre photographed at Hα + 0·75 Å ($\frac{1}{8}$ Å filter + 1 Å pre-filter).

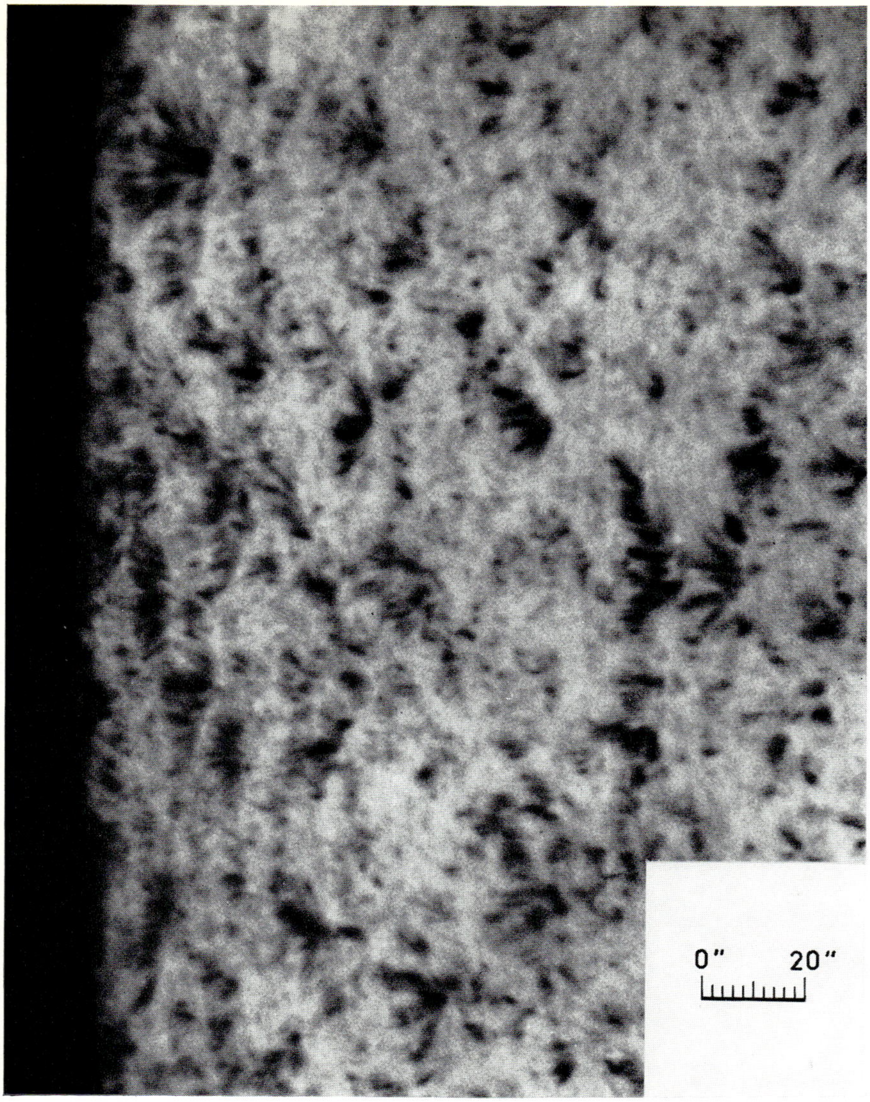

PLATE 3.6. Dark mottles near north limb photographed at Hα + 0·75 Å ($\frac{1}{8}$ Å filter + 1 Å pre-filter). Note the elongation of many mottles in the direction of the limb, implying vertically-standing structures.

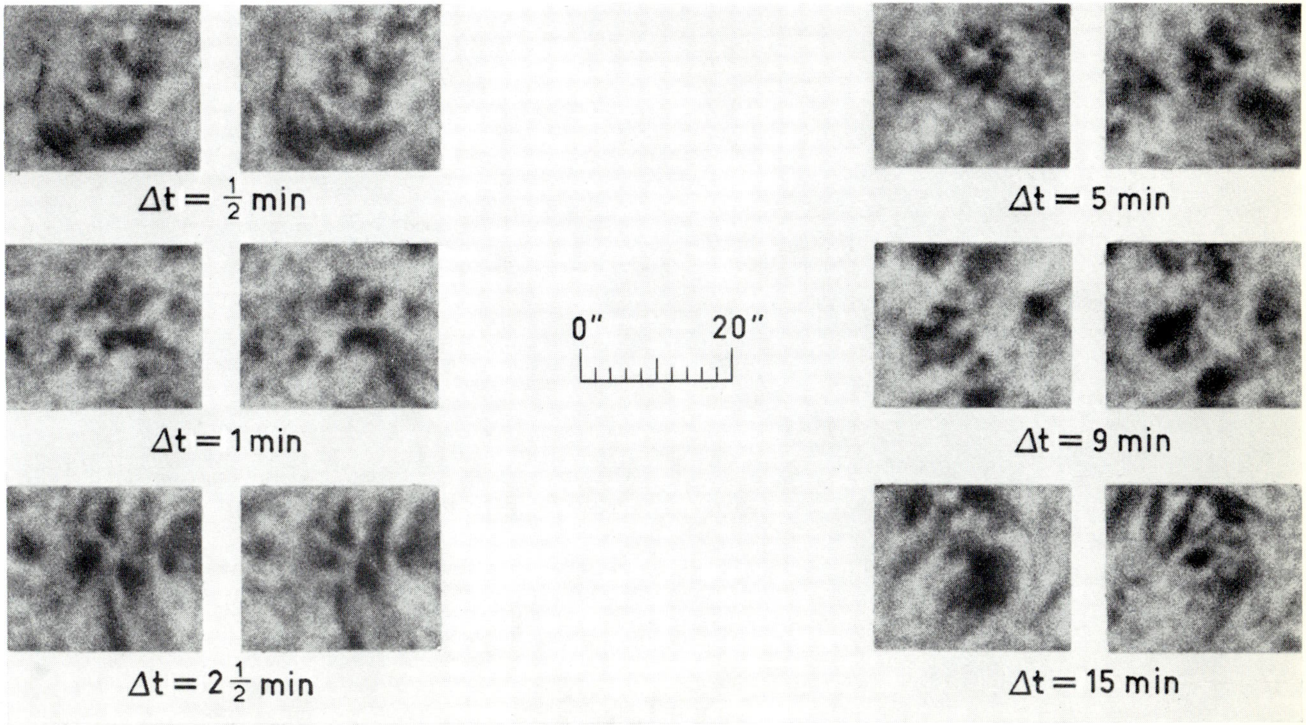

PLATE 3.7. Pairs of photographs showing changes in individual dark mottles over intervals ranging from 0·5 min to 15 min. Each pair refers to the same region but different pairs correspond to different regions. The photographs were obtained with a double-peaked profile with peaks at H$\alpha \pm 0{\cdot}3$ Å (0·5 Å Halle filter).

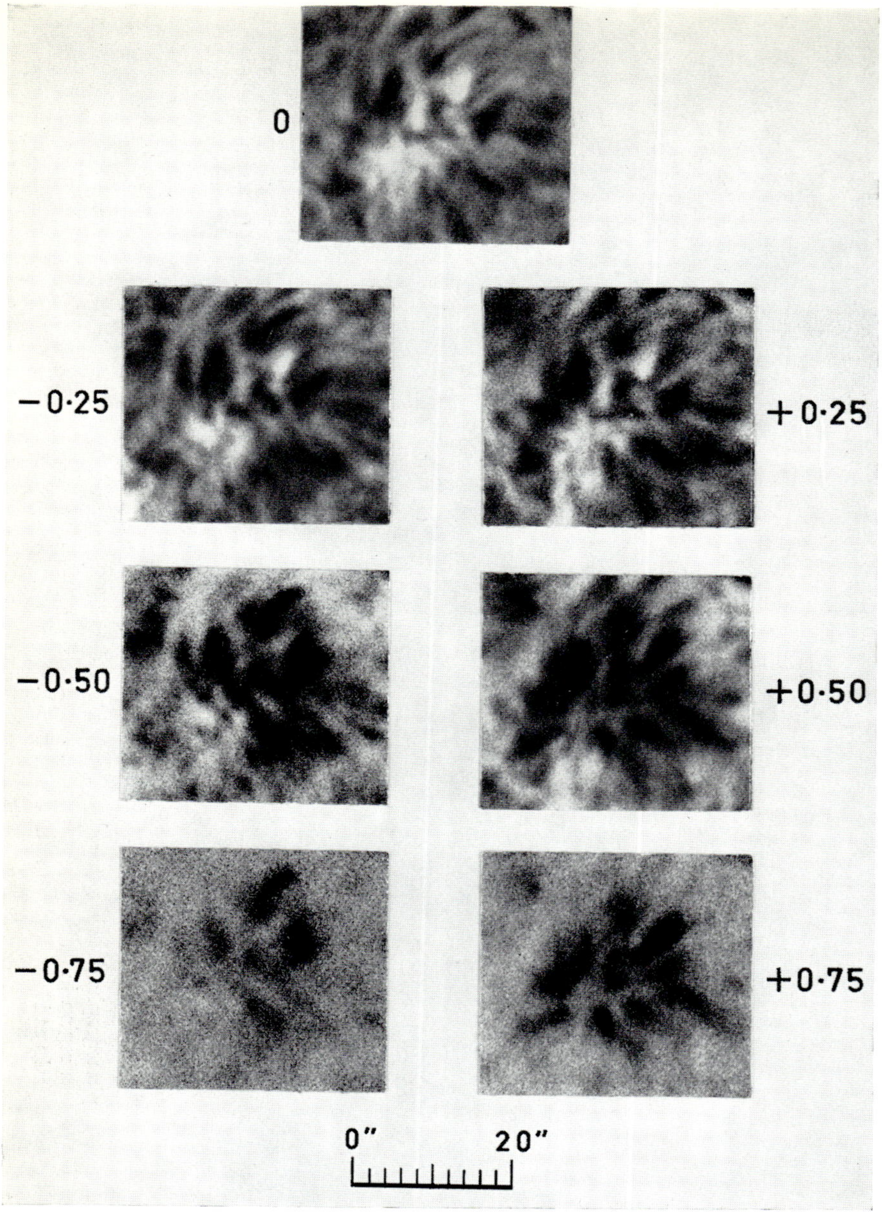

PLATE 3.8. 'Rosette' structure photographed nearly simultaneously at seven wavelengths ranging from Hα − 0·75 Å to Hα + 0·75 Å ($\frac{1}{8}$ Å filter + 1 Å pre-filter). The interval between the first and last photographs is 33 sec. For description, see text (Section 3.2.5).

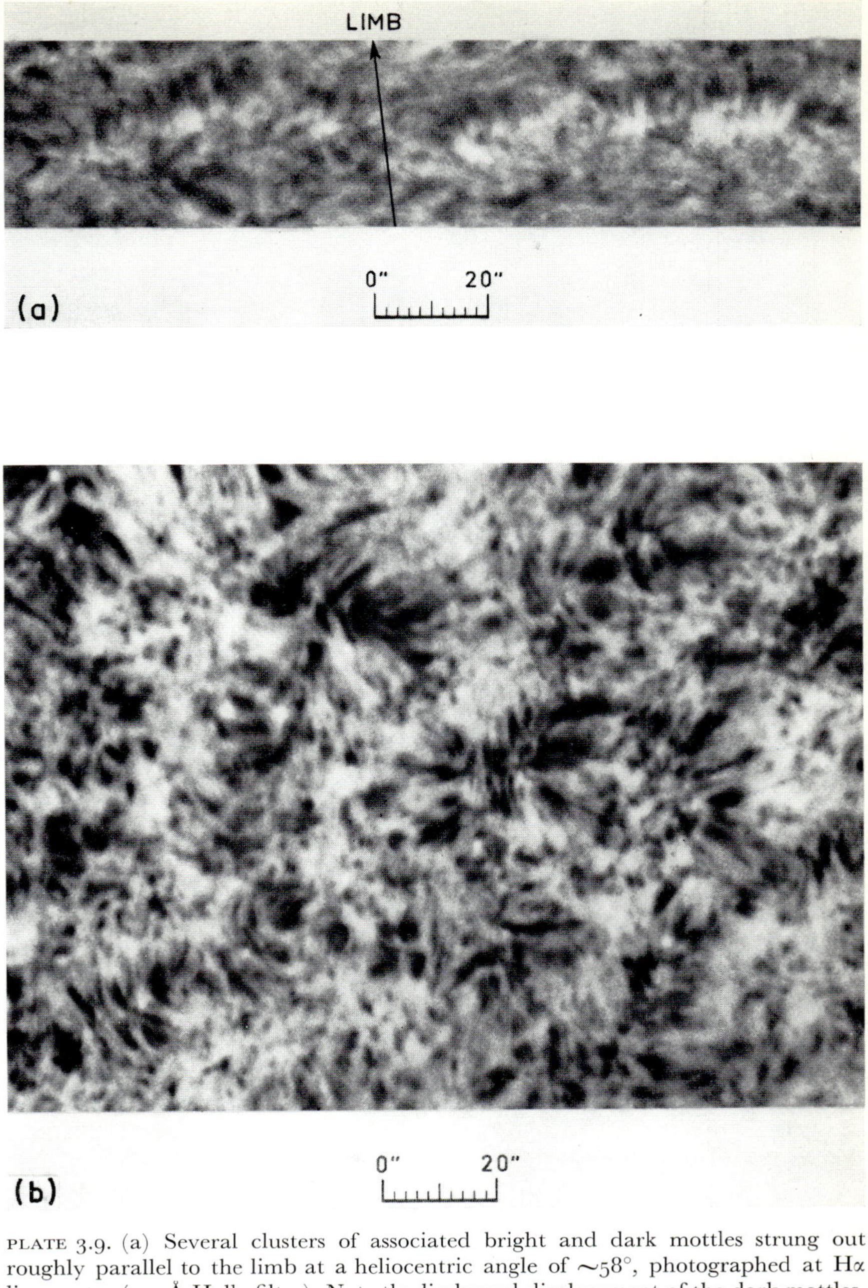

PLATE 3.9. (a) Several clusters of associated bright and dark mottles strung out roughly parallel to the limb at a heliocentric angle of $\sim 58°$, photographed at Hα line centre (0·5 Å Halle filter). Note the limbward displacement of the dark mottles, implying that they lie higher than the bright mottles. (b) Dark Hα 'grains' photographed at Hα − 0·5 Å ($\frac{1}{8}$ Å filter + 1 Å pre-filter); they are distinguishable from the dark mottles by their smaller size.

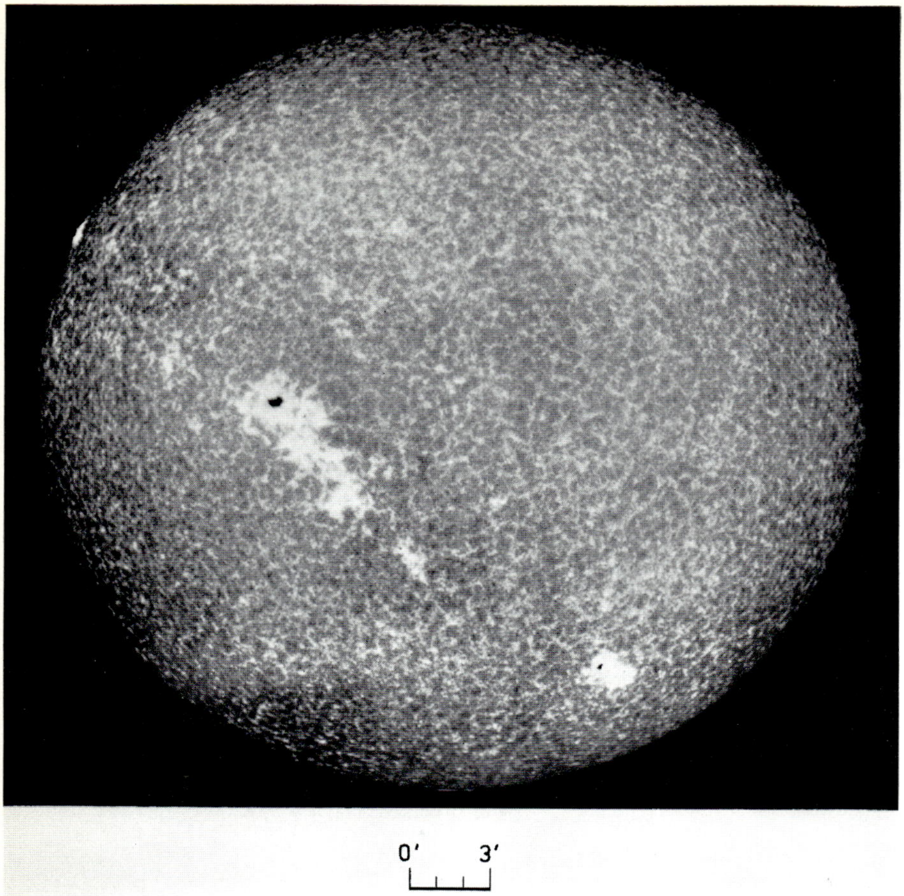

PLATE 3.10. Chromospheric network photographed in the K_{2R} emission peak of Ca II by R. B. Leighton, Mt. Wilson Observatory, 2 May 1962 [Title, 1966a, Plate 43].

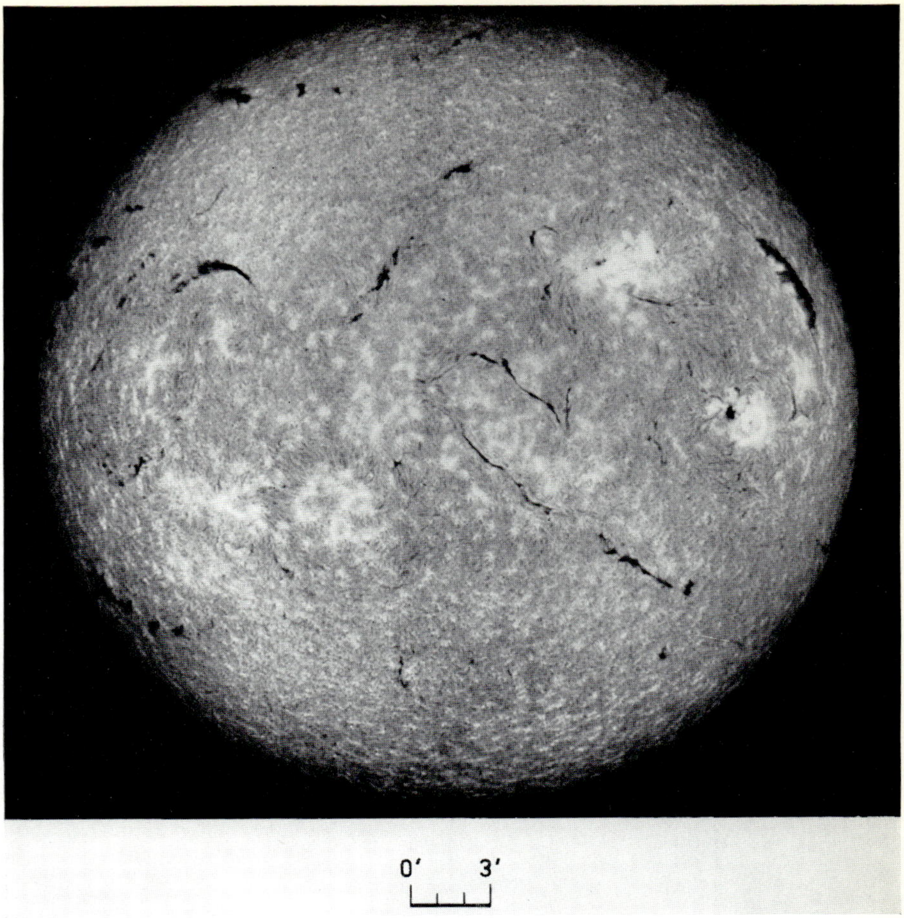

PLATE 3.11. Chromospheric network photographed at the centre of Hα by R. B. Leighton, Mt. Wilson Observatory, 12 Sept. 1961 [Title, 1966a, Plate 6].

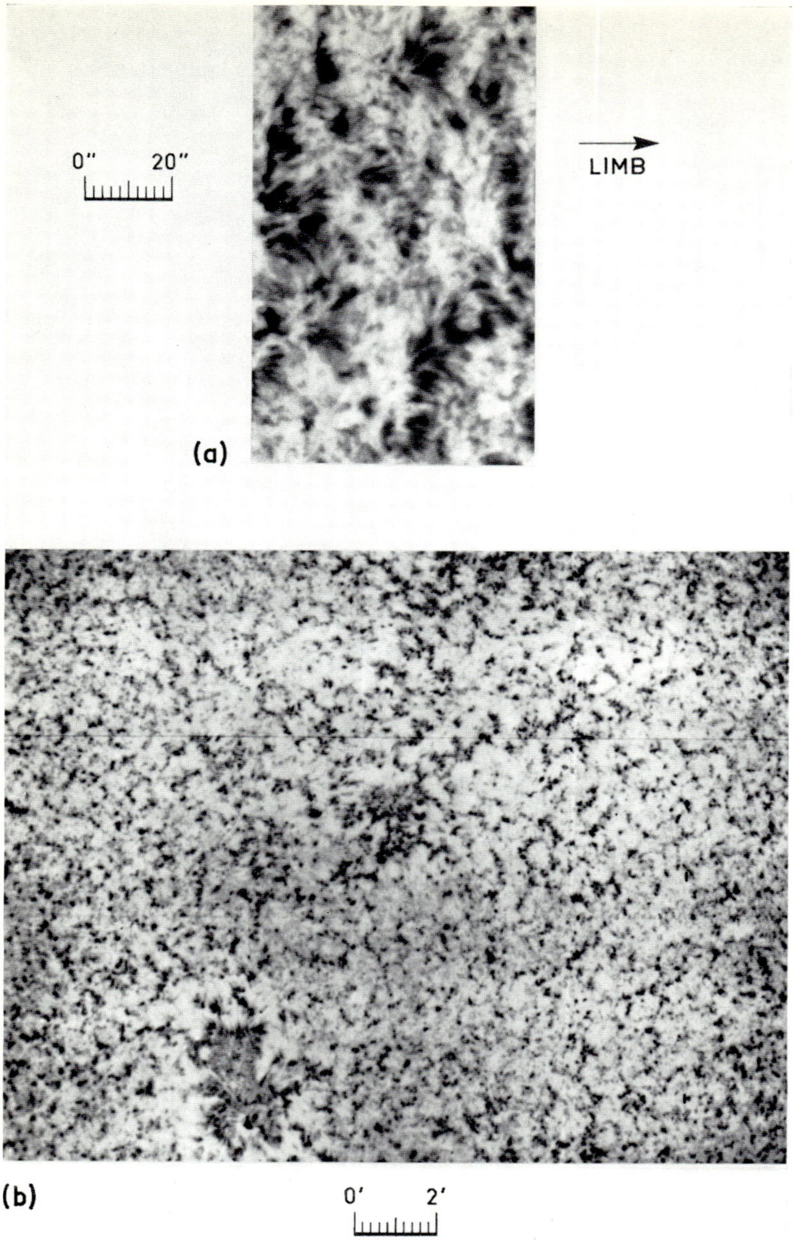

PLATE 3.12. (a) High-resolution photograph of a single network cell viewed obliquely at a heliocentric angle of 63° ($H\alpha - 0{\cdot}5$ Å; $\frac{1}{8}$ Å filter + 1 Å pre-filter); (b) medium-resolution photograph of dark $H\alpha$ network obtained with the Mt. Wilson spectroheliograph at $H\alpha + 0{\cdot}56$ Å [Title, 1966a, Plate 9a].

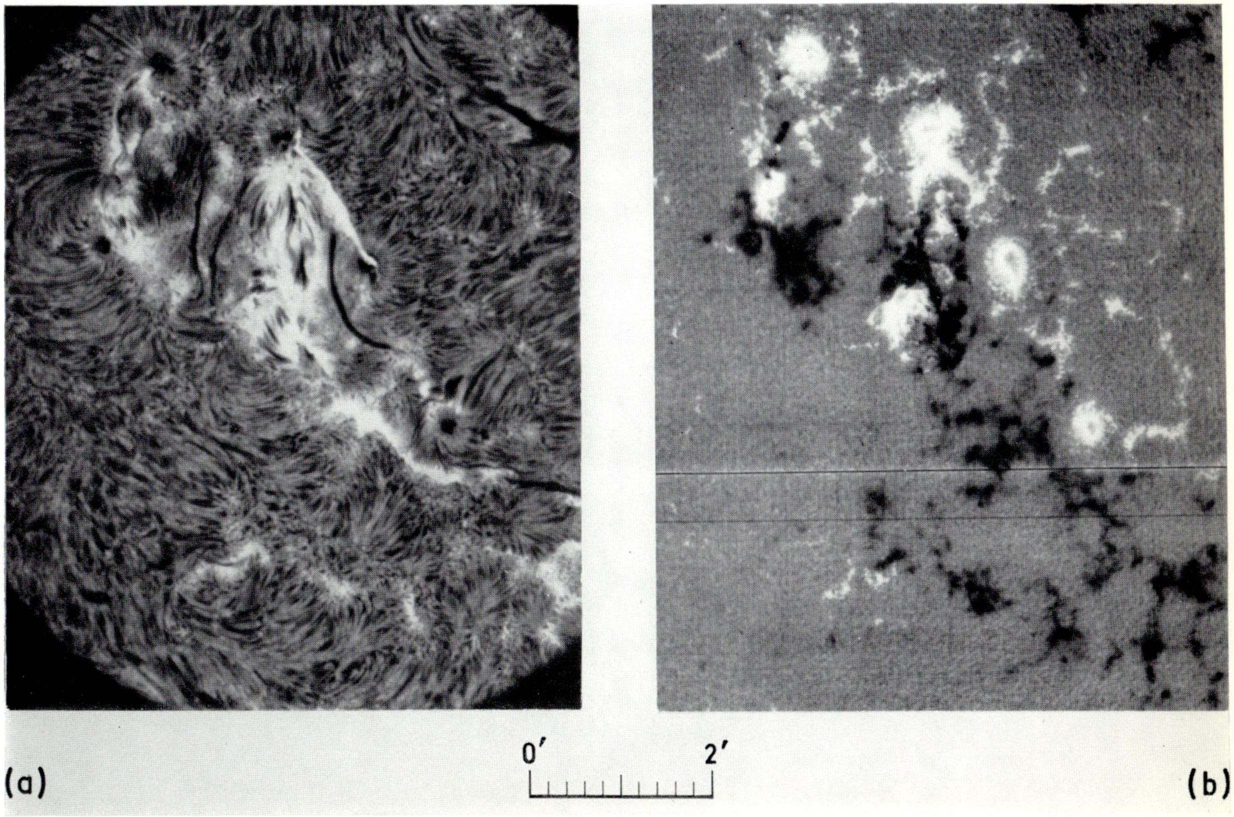

PLATE 3.13. Correlation between longitudinal magnetic field and bright Hα emission: (a) large active region and its surroundings photographed at the centre of Hα with a Zeiss $\frac{1}{4}$ Å filter on 26 Jan. 1970; (b) simultaneous magnetogram obtained by D. Vrabec in λ6103 of Ca I. (*By courtesy of Aerospace Corp.*)

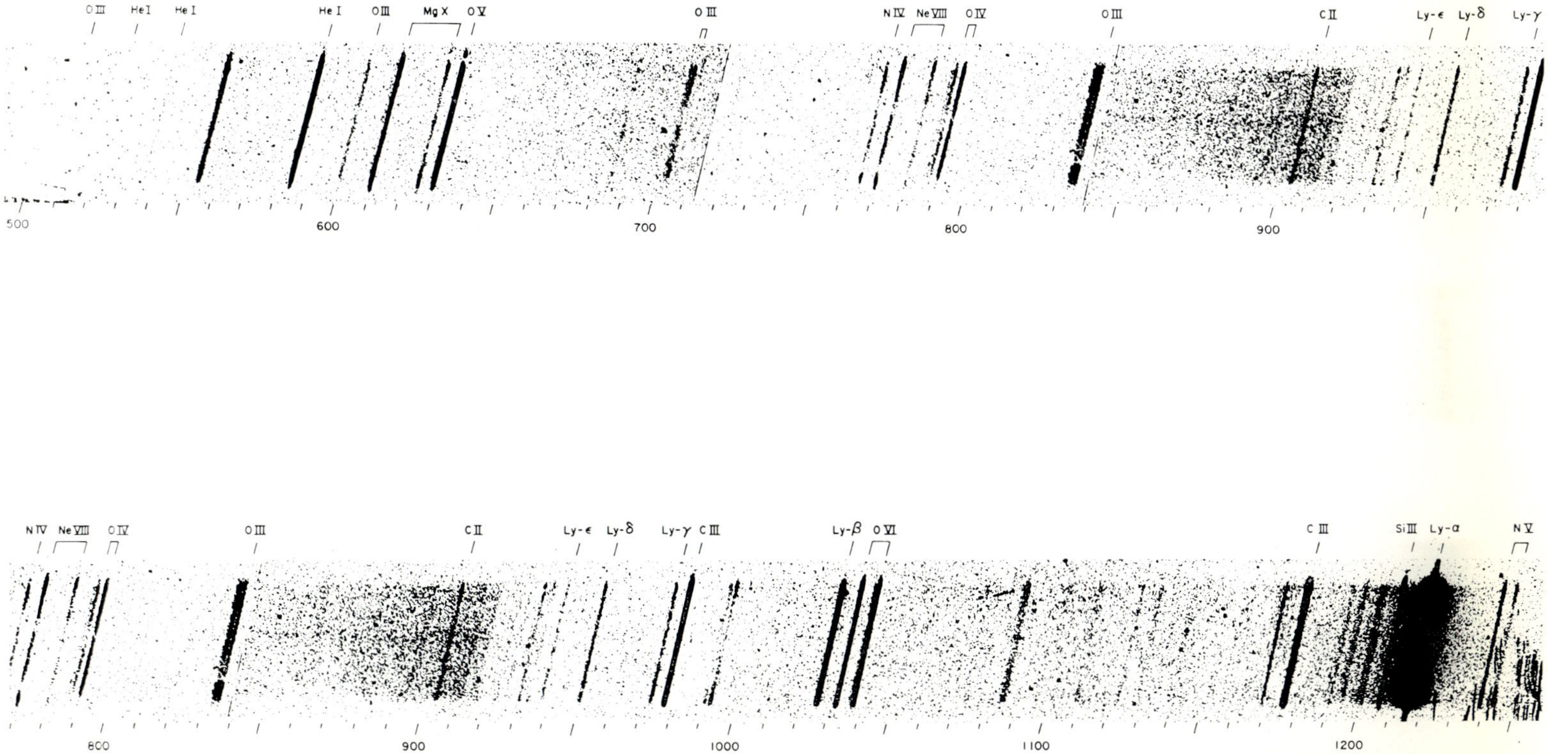

PLATE 4.1. Spectrum of the solar ultraviolet photographed from an Aerobee 150 rocket on 1 Feb. 1966. Note the Lyman continuum at $\lambda < 912$ Å and the strong chromospheric lines of H I (Lyman series), He I, O III–IV, and C II–III. (*By courtesy of Naval Research Laboratory.*)

reproduces plots of velocity against time for a number of adjacent points on the Sun separated by about 2200 km. It is easy to identify regions on this diagram, extending up to some tens of thousand kilometres in diameter, over which the phase – although not necessarily the amplitude – remains sensibly constant.

From a detailed analysis Musman and Rust have concluded that the phase is coherent over regions some 50 000 km in diameter, i.e. the phase changes more

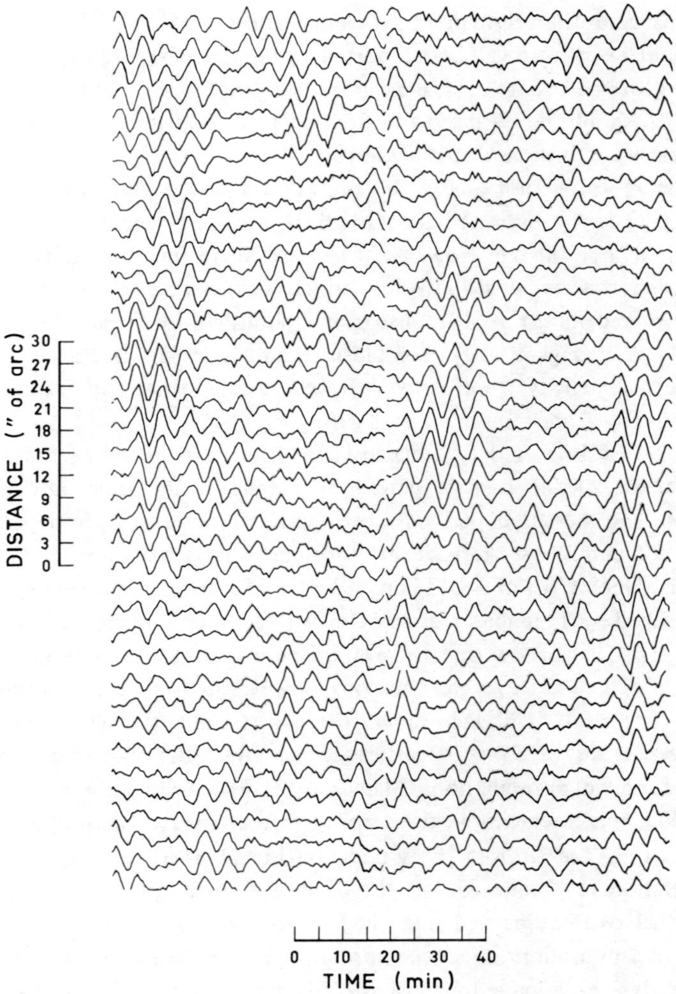

Fig. 3.7. Velocity as a function of time for a number of adjacent points separated by about 2200 km [Musman and Rust, 1970]. The measurements were made in the photospheric line Fe λ5250. The velocity scale is such that the distance between adjacent curves corresponds to 0.4 km sec^{-1}. Note the enormous distances over which the phase remains more or less constant.

or less uniformly over such a region. Moreover, the extent of a region over which the *amplitude* changes is at least five times smaller. Similar conclusions have been reached by Gonczi and Roddier [1969], who give a figure of 6″ of arc (4400 km) for the coherence scale of the amplitude and at least 20″ (14 500 km) for the coherence scale of the phase. Earlier, Roddier [1966] found that the measured amplitude of the oscillations was little affected by increasing the diameter of the measuring aperture from 5″ to 10″, or even to 15″; the oscillations were still well-marked with an aperture of 25″.

It should be mentioned that reported changes of the size of the oscillating elements with line strength [cf. Bray and Loughhead, 1967, p. 38], and therefore with height, should be regarded with caution in view of the difficulty of arriving at an objective definition of this parameter.

(2) *Period.* Inspection of Fig. 3.7 shows immediately that the period of the oscillations, i.e. the interval between adjacent velocity maxima or minima, is about 5 min, or 300 sec. However, when extensive measurements are made a large dispersion is revealed, individual values ranging from about 150 to 400 sec. This is clearly demonstrated by histograms published by Orrall [1965, Fig. 3], Howard [1967b, Figs. 8–10], and Musman and Rust [1970, Fig. 11], and also by a 'spectral density' plot derived from a two-dimensional Fourier analysis carried out by Tanenbaum *et al.* [1969, Fig. 5].

Table 2.6 of Bray and Loughhead [1967] summarizes the results of various determinations of the *average* period by means of the time autocorrelation technique. The measured periods range from ~295 sec for photospheric lines down to 285 and 286 sec respectively for the chromospheric lines λ5173 (b_2) of Mg I and λ5896 (D_1) of Na I. Concordant values for three of these lines have been published by Tanenbaum *et al.* [1969] (see Table 3.4 below). These results, therefore, appear to show that the average period undergoes a definite decrease with height. There are two good reasons, however, for exercising caution before accepting this result at its face value. In the first place, the average period for a given line has only a limited significance in view of the wide dispersion in the periods of individual oscillations. Many workers, accordingly, have preferred to present their measurements in the form of histograms (see above). Secondly, as pointed out by the authors [Bray and Loughhead, 1967, pp. 36, 41–42], the autocorrelation technique ignores the dispersion in the periods and amplitudes of individual oscillations and may, in fact, give average periods differing from those derived by a more direct measuring procedure [see also Howard, 1967b]. Consequently, the apparent precision of determinations made in this way may well be spurious.

According to Orrall [1965] and Howard [loc. cit.], the period is not significantly different in regions of weak, non-spot magnetic fields such as those present, for example, in the chromospheric network or in faint chromospheric faculae (cf. Section 3.4.4).

(3) *Velocity amplitude.* The velocity amplitude also displays a large dispersion, which is illustrated by histograms published by Orrall [1965, Fig. 3] and Howard [1967b, Figs. 8 and 11]. Orrall's histogram indicates a range of 0.08–0.28 km sec^{-1}, with a most probable value of ~0.17 km sec^{-1}. Howard's histograms yield a comparable most probable amplitude, namely 0.13 or 0.18 km sec^{-1}, depending on the size of aperture employed. However, Howard also finds a substantial number of oscillations having amplitudes as high as 0.3 to 0.4 km sec^{-1}. These measurements are in reasonable agreement with earlier results [cf. Bray and Loughhead, 1967, p. 41].

Orrall found no significant dependence of the velocity amplitude on the presence of weak, non-spot magnetic fields, but Howard found the amplitudes are reduced by 25 per cent in regions where the field exceeds 80 gauss. Howard also found that the amplitude depends on the period of the oscillations, albeit with a large scatter (see his Fig. 12); the shorter periods have the smaller amplitudes.

In agreement with earlier results, Howard's histograms demonstrate that the velocity amplitude declines markedly as the limb is approached, indicating that the oscillatory motions occur chiefly in the direction at right angles to the solar surface. Roddier [1966] finds that the oscillations virtually disappear beyond a heliocentric angle of about 70°.

The modern observations appear to have added little or no new information bearing on the question of any possible height variation of the velocity amplitude; earlier observations implied an increase with height [Bray and Loughhead, 1967, p. 41].

(4) *Lifetime.* The lifetime of an individual wave train is rather hard to determine with much accuracy, but inspection of the velocity traces in Fig. 3.7 shows that a well-defined, large-amplitude oscillation may persist for quite a number of cycles before decaying into a small-amplitude oscillation or background noise. Howard [1967b] estimates that a typical 'burst' of 5-min oscillations occupies 30–45 min, corresponding to 6–9 complete periods. Gonczi and Roddier [1969] attempted to derive the average lifetime by measuring the decay time of the time autocorrelation function for several clearly marked trains, but obtained a value of only ~10 min. However, the use of the autocorrelation function for such a purpose is questionable in view of the dispersion in amplitudes and periods described above.

(5) *Progressive or standing waves?* One question of considerable interest is whether the oscillatory motions in the upper photosphere and low chromosphere represent progressive waves moving vertically upwards or, alternatively, standing waves. In the first case one would expect to observe measurable phase

lags between the instantaneous velocities at different heights, whereas in the second case the phase would be the same at all heights. In principle, this question can be answered by simultaneously measuring the time variation of the Doppler shifts of two lines having widely different strengths and thus originating at different heights.

Early observations, using the strong Mg b_2 line and a weak Ti line, indicated a definite phase lag between the two lines without, however, establishing a precise value for its magnitude [cf. Bray and Loughhead, 1967, p. 40]. More recent observations have supplied this information. Tanenbaum *et al.* [1969] used two pairs of lines, the b_2 line (λ5173) and a weak Fe line nearby (λ5180), and the D_1 line of Na (λ5896) and a second weak Fe line (λ5906). In all cases the oscillations at the higher level lagged behind those at the lower level, the phase lag averaging 10 ± 3 sec for the first pair of lines and 14 ± 3 sec for the second pair. Blondel [1971] measured a phase lag of 9–15 sec between the chromospheric line λ4227 of Ca I and the photospheric line Fe I λ5049. It is not possible to derive the phase velocity from these results in the absence of contribution curves for the lines, which would enable heights of formation of those portions of the profiles used in the velocity measurements to be calculated. If the phase velocity were equal to the velocity of sound, ~7 km sec^{-1} throughout the region concerned (cf. Table 7.3), then a phase difference of 14 sec would imply a height difference of only ~100 km. Since this figure is rather small Tanenbaum *et al.* conclude that in the low chromosphere the vertical oscillations represent progressive waves propagating upwards with a phase velocity exceeding the local sound velocity. As we shall see in Section 6.3.2, this behaviour is characteristic of acoustic-type waves as opposed to gravity-type waves. However, this contradicts the conclusion of Section 7.3.4, based on a discussion of the critical frequencies involved, that the oscillations observed at this level probably represent gravity-type waves.

At lower levels, on the other hand, the situation appears to be quite different. Frazier [1968] has made simultaneous velocity measurements in three weak lines, formed at different heights, which reveal *no* measurable phase lags. Contribution curves for these lines (Frazier's Fig. 1) indicate that they cover a height of some 250 km, extending from the top of the convection zone into the temperature minimum. In this region of the atmosphere, accordingly, it is likely that we are observing standing oscillations (cf. Section 7.3.4).

(6) *Brightness–velocity phase relationship.* A second type of observation of potential importance in elucidating the physical nature of the oscillations is measurement of the phase relationship between the oscillations in velocity and in brightness observed in the core and wings of the various lines. The continuum brightness may also be measured. The brightness variations in both line and continuum are very weak and, accordingly, hard to measure. The best modern

results[13] are those of Tanenbaum *et al.* [1969], who measured velocity and wing brightness as functions of time in the weak carbon line λ5380, the Fe line λ5250, the Na D_1 line, and the Mg b_2 line. In order to smooth the curves, the method of 'superimposed epochs' was employed. The resulting velocity and brightness oscillations for D_1 are shown in Fig. 3.8.[14] A noteworthy feature is that the brightness maxima lead the corresponding upward velocity maxima by

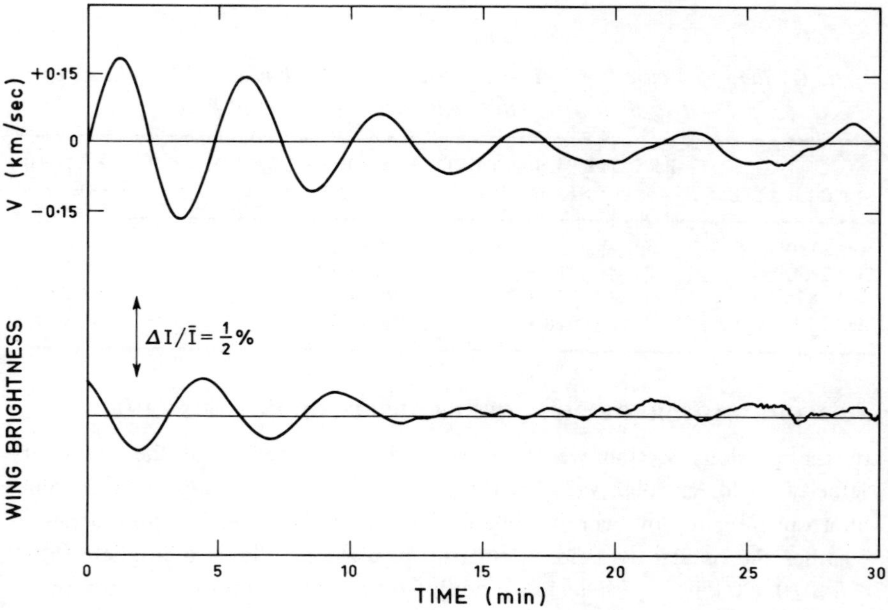

Fig. 3.8. Superimposed epoch curves for velocity and wing brightness in the Na D_1 line [Tanenbaum *et al.*, 1969]. Note that brightness leads upward velocity by more than a quarter of a period. These observations were made at Mt. Wilson with a 10″ aperture.

(for this line) approximately one-quarter of a period; the measured brightness–velocity phase lead for all four lines is given in Table 3.4. The last column gives Tanenbaum *et al.'s* values for the periods of the velocity oscillations in the various lines.

Referring to Table 3.4, we see that the wing brightness in the chromospheric lines Na D_1 and Mg b_2 leads the (upward) velocity by more than 90°, whereas in

[13] References to earlier attempts to investigate the relationship between velocity and brightness in lines and in the continuum are given by Bray and Loughhead [1967, p. 44].

[14] Figure 4 of Tanenbaum *et al.'s* paper gives similar curves for Fe λ5250 but includes, in addition, simultaneous measurements of brightness in the core of this line, in the continuum, and in K_{232}.

the photospheric lines Fe λ5250 and C λ5380 it leads the velocity by less than 90°; the phase lead for the continuum is also less than 90°. Frazier's [1968] measurements of the Fe λ6355 line confirm that the brightness leads the velocity by a relatively large phase angle. He measured the central intensity rather than the wing brightness and obtained a phase lead of 117°. This line is formed near the temperature minimum in the low chromosphere.

TABLE 3.4

The Oscillatory Velocity Field: Brightness–Velocity Phase Lead for Lines of Different Strength [Tanenbaum et al., *1969]*

SPECTRAL LINE	BRIGHTNESS–VELOCITY PHASE LEAD SECONDS	DEGREES	PERIOD (SEC)
C λ5380	34 ± 13	40° ± 15°	306 ± 6
Fe λ5250	52 ± 8	64° ± 10°	298 ± 2
Na λ5896 (D_1)	78 ± 7	98° ± 8°	288 ± 6
Mg λ5173 (b_2)	78 ± 5	100° ± 14°	280 ± 11

3.5.2 CHROMOSPHERIC VELOCITIES OBSERVED IN THE K LINE

In the previous section we have outlined the properties of the oscillatory velocity field as observed in Fraunhofer lines originating in the upper photosphere and low chromosphere. The important question now arises of whether the oscillatory field extends up into the middle chromosphere ($5000 \leqslant h \leqslant 10\,000$ km) – perhaps in a modified form – or alternatively, of whether it is replaced by an entirely different type of velocity regime. The answer to this question is locked up in somewhat fragmentary and hard to interpret data derived from measurements in the K line of Ca II and in Hα, discussed in this and the following section.

In general, two methods are widely used in measuring line-of-sight velocities on the Sun: (*a*) selecting two points of equal intensity in the profile of the line core (assumed symmetrical), and measuring the wavelength of the point midway between them; and (*b*) determining the wavelength of the Doppler-shifted line centre. Method (*a*) is not feasible in the case of the H and K lines of Ca II owing to their extreme broadness. Method (*b*) is very hard to apply owing to the complicated nature of the centres of these lines: this is illustrated in Fig. 3.2, which shows profiles at the centre of K for various solar features, observed with medium or low spatial resolution. This method becomes even more suspect when we recall (Section 3.2.8) that with *high* spatial resolution the K line loses its double peak K_{2V} and K_{2R}, leaving only a single peak (red or violet) at most locations [Zirin, 1966, p. 230; Linsky and Avrett, 1970; Pasachoff, 1970].

Despite these difficulties, velocity measurements in H and K have been attempted by Stepanov [1960], Jensen and Orrall [1963], and Orrall [1966]. Stepanov used a magnetograph in its velocity mode, the two entrance slits being centred on the H_3 portion of the H line; he claimed the detection of large-scale, long-lived, upward and downward motions of a *non*-oscillatory nature. Jensen and Orrall measured apparent Doppler shifts of K_3 and of the combined $K_{2\,V}/K_{2\,R}$ peak; they also observed intensity oscillations in $K_{1\,V}$, $K_{1\,R}$, $K_{2\,V}$, $K_{2\,R}$, and K_3. Orrall confined his measurements to K_3 and interpreted his data as demonstrating the existence of an oscillatory velocity field at the chromospheric levels involved. This differed from the field at lower levels (Section 3.5.1) in having a higher velocity amplitude (1.1–1.2 km sec^{-1}) and a much broader distribution of periods; moreover, the peak occurred at ~180 instead of ~300 sec. Diagrams illustrating this apparent change of power spectrum with height, ranging from the photosphere to the level of the K line, have been reproduced by Orrall [1966, Fig. 4] and Noyes [1967, Fig. 1].

Until recently these results have been widely accepted, but Pasachoff [1970], on the basis of observations of higher spatial resolution (~1″ of arc), has questioned whether Orrall's velocity measurements are physically meaningful. Pasachoff observed *single* K_2 peaks, which he attributes to separate spatial elements of the chromosphere and calls 'K_2 emission features'. These elements have a diameter of ~1000 km and assumed upward or downward velocities of ~10–20 km sec^{-1}, thus explaining the measured displacements of the K_{2V} and $K_{2\,R}$ peaks from the line centre. The half-widths of individual peaks can be accounted for by a temperature of 15 000 K and a non-thermal velocity (cf. Section 4.3.3) of 3 km sec^{-1}. Pasachoff raises the question of the possible identification of the emission features with *spicules*, pointing out the similarity in radial velocities.

Pasachoff is careful to list a number of objections to the hypothesis of the existence of discrete emission features, among which we may mention: (*a*) the absence of K_2 emission at wavelengths *close* to the line centre, the emission being confined, instead, to discrete wavelengths on the violet and red sides; and (*b*) the observed increase in the separation of the red and violet peaks as the limb is approached, implying that the emission features have horizontal velocity components exceeding those in the radial direction. An objection specifically to the identification of the emission features with spicules is that they presumably form part of the structure of bright K flocculi which, as we have seen in Section 3.2.8, coincide with the *bright* $H\alpha$ mottles; on the other hand, it is the *dark* $H\alpha$ mottles that we have identified with spicules (Section 3.2.6).

In summary, there have been several investigations which purport to determine the velocity field in the chromosphere at the levels where the K line is formed, but interpretation of the data is subject to great uncertainty. Pasachoff's results demonstrate that a spatial resolution sufficient to reveal the fine structure is mandatory. Moreover, the line transfer theory employed in analysing such measurements evidently needs to incorporate not only effects arising from inhomogeneous structure but also the influence of discrete moving elements. An attempt to take into account the effect of velocities on the line profile for the case of the Hα line is described below.

3.5.3 CHROMOSPHERIC VELOCITIES OBSERVED IN THE Hα LINE

Owing to the absence of complicated central emission peaks, Hα is better suited than the K line to elucidating the chromospheric velocity field. Moreover, any such measurements are easier to relate to known identifiable structures. On the other hand, the measurements are again subject to great difficulties of interpretation, as we shall see below.

(1) *Existence of velocities in Hα.* The best a priori evidence for the existence of velocities at the levels where Hα is formed comes from the so-called 'wiggly-line' spectrograms. Typical examples have been reproduced by numerous authors, including Prokofyeva [1957, Fig. 1]; Schröter [1958, Fig. 1]; Dubov [1960, Figs. 4, 8, and 9]; Beckers [1964a, Fig. 28]; Title [1966b, Fig. 10]; and Elliott [1969, Fig. 1]. The spatial scale of the streaks on these spectrograms is comparable with the sizes of the dark mottles, i.e. a few seconds of arc. Various authors have derived apparent r.m.s. velocities from measurement of such spectrograms, e.g. Goldberg *et al.* [1957], Schröter [1958], and Title [1966b]. These values are all less than 1 km sec^{-1} and are therefore much smaller than the measured velocities of *spicules* (cf. Section 2.4.2). However, we shall not dwell on the magnitude of these apparent velocities in view of the difficulty of interpreting such measurements, a topic which is discussed below.

(2) *Is there an oscillatory velocity field in Hα?* The available evidence bearing on the important question of the presence, or otherwise, of *oscillatory* velocities in Hα is contradictory. Leighton *et al.* [1962] found no evidence for oscillatory velocities in Hα, but both Michard [1967, Fig. 16] and Elliott [1969, Fig. 3] reproduce power spectra having fairly prominent peaks. Michard's curve has a rather broad maximum peaking at a period of ~230 sec; however, the broadness of the maximum may be merely a consequence of the rather short observation time involved, 540 sec. In Elliott's power spectrum the most prominent peak is located at 287 ± 3 sec, a value much closer than Michard's to the corresponding figures for lines such as Na D_1 and Mg b_2 (cf. Table 3.4). Elliott's spectrum shows, in addition, a high-frequency 'tail'. Finally, Title [1966b], using measurements of the time variation of the *areas* of upward- and downward-moving regions on Doppler spectroheliograms[15] derived a period much larger than Michard's or Elliott's, namely, 550 ± 25 sec.

[15] That is, a photograph produced by subtracting two spectroheliograms taken at selected wavelengths ±Δλ equally spaced on either side of the line centre.

According to Deubner [1969], the majority of the oscillating elements observed in Hα have a fairly uniform peak velocity of ~3 km sec^{-1}, whereas a curve published by Elliott [1969, Fig. 5] shows velocity amplitudes lying in general below 1.5 km sec^{-1}.

According to Noyes and Leighton [1963], there is no indication of *brightness* oscillations in the core of Hα, in contrast to the behaviour of other lines (Section 3.5.1).

(3) *Interpretation of Hα velocity measurements.* In interpreting apparent Doppler shifts measured in Hα in terms of the actual velocities of identifiable features, we must recognize the existence of a major uncertainty, namely, the effect on the profile of possible velocity gradients along the line-of-sight. When such gradients are present, the observed profile is built up over a number of layers having different relative velocities. The presence of such velocities modifies both the normalized atomic absorption profile ϕ_ν and, to a lesser extent, the source function S.

Athay [1970] has computed Hα profiles for a number of idealized velocity models consisting of either a finite moving layer having a certain height and thickness or of a semi-infinite moving layer having a certain lower boundary (see Table 3.5). Figure 3.9 gives the computed profiles for the case $y = 2$, where y is

TABLE 3.5

Idealized Velocity Models for Hα Profile Computations [Athay, 1970]

FINITE-LAYER MODELS h_1, h_2 (km)		SEMI-INFINITE MODELS h_1 (km)	
−240, −200	(A)	−200	(β)
−200, 0	(B)	0	(γ)
0, 500	(C)	500	(δ)
500, 1000	(D)	1000	(ϵ)
1000, 1500	(E)		
1500, 1900	(F)		
> 1900	(G)		

Note: h_1, h_2 represent the heights of the lower and upper boundaries of the moving layer; the computed profiles of Fig. 3.9 are keyed to the letters A – G, etc.

the dimensionless parameter

$$y = V \Big/ \left(\frac{2kT_e}{m_H} + \xi^2\right)^{1/2}.$$

In this formula, V is the assumed (downward) velocity of the moving layer and ξ is the non-thermal contribution to the Doppler half-width (cf. Section 4.3.3), taken from a solar model. The value $y = 2$ corresponds to a velocity V of ~20 km sec^{-1}; this in turn corresponds to a Doppler shift of the line absorption coefficient of just under 0.5 Å. The wavelength shift corresponding to this value

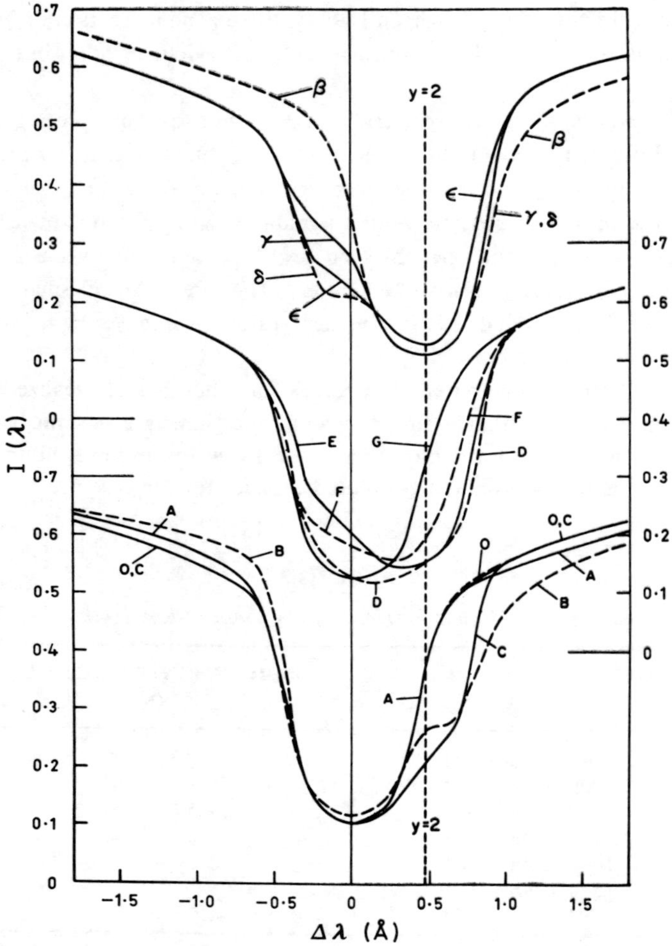

Fig. 3.9. Computed Hα profiles for the idealized velocity models given in Table 3.5 [Athay, 1970]. The calculations were performed for $y = 2$, corresponding to a (downward) velocity of ~20 km sec^{-1} (see text). The profile O refers to a stationary atmosphere.

of V is shown as a vertical dashed line in Fig. 3.9. The profile labelled O is computed for a stationary atmosphere, i.e. $V = 0$ and $y = 0$.

Referring to Fig. 3.9, we see that if we decide to make our velocity measurements by determining the wavelength of the *lowest point* of the profile,[16] the measured velocity would be correct only in the case of models β, γ, δ, and ϵ, the displacement corresponding to $y = 2$. Of these, γ, δ, and ϵ show prominent distortions on the violet side. By contrast, models A, B, and C give *zero* velocity, the effect of the moving layer being to produce distortions on the red side rather than to shift the line as a whole. Models D, E, and F are intermediate cases, producing shifts smaller than the true Doppler shift [cf. Beckers, 1968, p. 420], together with considerable distortions. The profile G appears unshifted and undistorted, presumably because the formation of the line has already taken place before the moving portion of the atmosphere is encountered. Note also that the cores of most of the profiles in Fig. 3.9 tend to be considerably broader than that for a stationary atmosphere, profile O.

In a somewhat analogous calculation, Avery and House [1969] have computed the contrast of a disk spicule rising with a velocity of 25 km sec^{-1} at several wavelengths in the K line, as a function of heliocentric angle and height. It would be desirable to carry out a detailed analysis of this kind for Hα where, of course, relevant observations are much more numerous.

(4) *Properties of the observed Hα profiles.* Athay's calculations are of considerable interest in that they clarify the effect of one important factor – velocity – on the profiles (and hence contrast) of Hα features such as the dark mottles. However, it would be too much to expect such simplified models to yield results in complete agreement with observations. For example, some of the profiles of Fig. 3.9 describe features that would appear dark on the red side of the line and bright on the violet side. On the other hand, this is *not* observed, as an examination of Plate 3.8 quickly demonstrates.

To gain a qualitative insight into the actual profiles of the fine structures of the quiet chromosphere the authors have carefully examined high-quality photographs taken nearly simultaneously at seven wavelengths in the Hα line. These were obtained with the 12-inch refractor of the CSIRO Solar Observatory, equipped with the tunable 1/8 Å filter plus a 1 Å pre-filter in tandem (Section 3.2.2), and the chosen wavelength shifts were $\Delta\lambda = 0, \pm 0.25, \pm 0.50$, and ± 0.75 Å. Intercomparison of pairs of photographs was carried out with the aid of a 'picture correlator' specially constructed for the purpose. (This device is similar in principle to the blink comparator familiar to stellar observers, but apart from other differences, is designed to handle conveniently 35-mm film.) Each photograph was carefully compared with the other six in the wavelength sequence, making 21 intercomparisons in all.

Considering firstly the dark mottles, we reached the following conclusions:

(*a*) when photographs taken on the *same* side of the line are compared the

[16] An alternative procedure is to measure the wavelength of the point midway between two selected points of equal intensity in the profile. Errors arising from this procedure for the various models are given in Table IV of Athay's paper.

agreement is very close, although the contrast varies; maximum contrast occurs at about Hα ± 0.5 Å [cf. Beckers, 1963, Fig. 11; 1964a, Fig. 29; Dubov, 1960];

(*b*) when photographs taken on opposite sides of the line are compared, e.g. at Hα + 0.5 Å and Hα – 0.5 Å, the general location of any group of dark mottles is the same; however, *there is no detailed agreement in the size, shape, and location of individual mottles;*

(*c*) there are many examples of dark mottles being prominent at Hα + 0.5 Å but not at Hα – 0.5 Å, and vice versa; the numbers present on the two sides of the line are comparable;

(*d*) at no wavelength do dark mottles turn into bright mottles: the latter are distinct entities in their own right and are located *between* the dark mottles (cf. Section 3.2.5).

It is evident that these results are consistent with profiles that – for some, at any rate, of the dark mottles – are not shifted but instead are broadened or distorted on either the red or violet side. Such profiles are qualitatively similar to the profiles B and C of Fig. 3.9 and to their mirror images, corresponding to downward and upward velocities respectively. However, at the present time the data are too qualitative to permit a precise identification, although a profile such as β is evidently ruled out. The fact that comparable numbers of dark mottles are found on the two sides of the line implies more or less equal numbers of upward- and downward-moving elements.

Spectroscopic observations by Prokofyeva [1957] and Schröter [1958] confirm the complexity of the profiles of individual dark features inferred above. Prokofyeva observed Hα profiles at adjacent positions on the disk and found considerable variation in width and markedly irregular shapes; broadening occurred on both sides of the line or on only one side. Schröter found shifted asymmetric profiles; he noticed, in addition, a correlation between the *direction* of the shift and the width of the Doppler core, in the sense that red shifts are associated with larger core widths than violet shifts.

Turning now to the bright mottles, we conclude from our observations that they have much simpler profiles than the dark mottles. They are brightest at $\Delta\lambda = 0$ and, although fainter at Hα ± 0.25 Å, are still well visible (see Plate 3.8). Thereafter their contrast rapidly decreases and very few remain visible at Hα ± 0.5 Å. This indicates an unshifted, narrow profile whose core lies somewhat higher than the 'stationary' profile O of Fig. 3.9. Such a profile implies an object whose temperature T_e exceeds that of its surroundings (cf. Section 4.3.6).

The only quantitative measurements of the profiles of the Hα fine structures which have so far been published are those of Grossmann-Doerth and von Uexküll [1971, Fig. 4]. These authors find that their results cannot be explained on the basis of the Athay model described above, but in most cases are in good agreement with the predictions of the earlier 'cloud' model of Beckers [1964a]. Profiles obtained by the present authors, as yet unpublished, confirm this conclusion for mottles *near the centre of the disk;* however, neither Athay's nor Beckers' model is capable of reproducing the observed profiles of mottles near the limb. In contrast to the observations of Grossmann-Doerth and von Uexküll, the authors' profiles for bright and dark mottles differ markedly, reflecting differences in their physical conditions.*

(5) *Doppler spectroheliograms in Hα.* Several workers have obtained Doppler spectroheliograms (see footnote 15) in Hα which purport to reveal the chromospheric velocity field; examples have been reproduced by Leighton *et al.* [1962, Figs. 10c, d; 13]; Leighton [1965, Fig. 8]; Title [1966a, Plates 8–11, 28, 44, 45]; and Zirin [1966, Figs. 9.14, 9.16] [see also Title, 1966b]. However, the validity of this technique depends on the assumption that the profiles of the structures under study remain symmetrical and are simply Doppler-shifted like, for example, the profile β in Fig. 3.9. On the other hand, in the case of the dark mottles there is strong evidence that this is far from the truth, as we have just seen. Accordingly, deductions derived from such Doppler spectroheliograms are highly questionable and for further details we shall merely refer the interested reader to the references cited above.

Evidently there remains an urgent need for more spectroscopic observations of Hα profiles having a spatial resolution adequate to enable them to be unambiguously related to identifiable chromospheric fine structures, i.e. the bright and dark mottles. Moreover, existing theory should be extended to take proper account of factors additional to velocity likely to influence the profiles, for example temperature and density inhomogeneities. A detailed comparison between the observed and theoretical profiles would then enable one to deduce the velocities of the mottles. In this way we may ultimately hope to unite in a single comprehensive theory our knowledge of the dynamics, morphology, and physical state of the chromospheric fine structure. However, for the time being our knowledge of the velocities of the chromospheric fine structures must continue to rest mainly on limb observations (Sections 2.4.2–2.4.3).

3.5.4 OBSERVATIONS OF CHROMOSPHERIC MAGNETIC FIELDS

The observation and interpretation of magnetic fields in the chromosphere, as opposed to the photosphere, are subject to a number of intrinsic difficulties. On the observational side, the great width – compared with photospheric lines – of lines such as Hα and Hβ inevitably reduces the sensitivity of measurements made, for example, with magnetographs of the photoelectric type [cf. Bray and Loughhead, 1964, p. 187; Vyal'shin and Krat, 1965]. Moreover, the theory of

**Note added in proof:* see Bray [1973] and Loughhead [1973] – *Solar Phys.* (in press).

formation of Fraunhofer lines in a magnetic field has so far been generally worked out only for relatively weak lines with simple triplet Zeeman patterns. Accordingly, there is no really adequate theoretical basis for the quantitative interpretation of records obtained in strong chromospheric lines such as Hα, Hβ, and the H and K lines of Ca II. Results obtained with these lines are therefore best regarded as qualitative indications of the presence of magnetic fields at the corresponding levels of the chromosphere rather than as reliable measurements of actual field strengths.

The suitability of the various chromospheric lines for magnetic field measurements has been discussed by Severny [1966]. He regards Hβ as quite unsuitable on account of its strong blending with lines of photospheric origin; the magnitudes of possible errors under various circumstances arising from the use of this line have been assessed by Ioshpa and Obridko [1966]. Severny also criticizes the use of the K_3 absorption core lying between the two K_2 emission peaks (Fig. 3.2), pointing out that fluctuations in the core from feature to feature will change the *gradient* of the profile, $\partial r_\lambda / \partial \lambda$, and hence the measured field strength, which is proportional to the gradient. The same criticism evidently applies to the use of the H_3 portion of the H line of Ca II. Severny, accordingly, regards Hα as the most suitable line to employ for measurement of chromospheric magnetic fields. Nevertheless, in spite of these difficulties various authors, including Stepanov [1961], Vyal'shin and Krat [1965], and Ioshpa and Obridko [1966], have made use of H_3 or the Hβ line.[17]

Existing chromospheric magnetic measurements are decidedly meagre; moreover, in some of the published accounts it is not always made clear whether the results refer to *quiet* or to *active* regions and their neighbourhood. Severny [1966], using Hα, found that in ~70 per cent of cases photospheric magnetic features, detectable in the λ5250 line, are also present at the chromospheric level, although sometimes displaced in position. In the remaining 30 per cent of cases the chromospheric field appears to be of greater strength than the underlying photospheric field or even to be of opposite sign. Stepanov [1961] attempted to determine the field gradient, using measurements in H_3 and λ5250.

Evidence bearing on the relationship between magnetic field and *velocity* in the chromosphere is contradictory. Using Hβ, Vyal'shin and Krat [1965] concluded that large-scale regions of chromospheric magnetic field and of radial velocity closely correspond with one another, maxima of velocity and field coinciding. On the other hand, Gopasyuk and Tsap [1970] found from observations made in Hα that the magnetic maxima in the neighbourhood of active regions generally coincide with locations of *zero* radial velocity.

[17] For references to earlier attempts to measure chromospheric magnetic fields using Hβ or the H or K line of Ca II, see Bray and Loughhead [1964, p. 195].

Our knowledge of any possible relationships between chromospheric magnetic fields and corresponding chromospheric structures is virtually nonexistent. On the other hand, we do have a fairly clear understanding of the various relationships between *photospheric* fields and chromospheric structures, both for quiet and active regions; these latter relationships are outlined in Section 3.4.4 of this chapter and in Sections 5.2.4 and 5.3.3 of Chapter 5. At the chromospheric level, magnetic forces will generally dominate hydrodynamic forces even for quite small field strengths (Section 3.4.5). Accordingly, it is possible that individual elements of the fine structure of both quiet and active chromosphere simply reflect changes in the magnitude and direction of the pervading field. Thus a 'rosette' structure, for example, may give us a detailed picture of the distribution of the field at a point on the boundary of a network cell, but testing this speculation would evidently require magnetic observations far superior to any currently available.

REFERENCES

ABETTI, G. [1929] 'Solar physics', *Handbuch der Astrophysik*, ed. G. Eberhard, A. Kohlschütter, and H. Ludendorff, vol. 4, p. 57; Springer, Berlin.

ALISSANDRAKIS, C. E., and MACRIS, C. J. [1971] 'A study of the fine structure of the solar chromosphere at the limb', *Solar Phys.* **20**, 47.

ALISSANDRAKIS, C. E., and MACRIS, C. J. [1972] 'Isophotometry of the chromospheric bright and dark mottles on the solar disk', *Astrophys. Lett.* **10**, 59.

ATHAY, R. G. [1964] 'The challenge of chromospheric physics', *Science* **143**, 1129.

ATHAY, R. G. [1970] 'Velocity effects on the profiles of Hα and two Fe I lines', *Solar Phys.* **12**, 175.

AUSTIN, W. E., PURCELL, J. D., SNIDER, C. B., TOUSEY, R., and WIDING, K. G. [1967] 'Recent extreme ultraviolet solar spectra and spectroheliograms', *Space Research VII*, p. 1252; North-Holland Publ. Co., Amsterdam.

AVERY, L. W., and HOUSE, L. L. [1969] 'A solar spicule model based upon calcium II K line radiative transfer studies', *Solar Phys.* **10**, 88.

D'AZAMBUJA, L. [1930] 'Recherches sur la structure de la chromosphère solaire', *Ann. Obs. Paris-Meudon* **8**, Part 2.

BANOS, G. J., and MACRIS, C. J. [1970] 'New observational results for the solar chromosphere', *Solar Phys.* **12**, 106.

BAPPU, M. K. V., GRIGORJEV, V. M., and STEPANOV, V. E. [1968] 'On the development of magnetic fields in active regions', *Solar Phys.* **4**, 409.

BAPPU, M. K. V., and SIVARAMAN, K. R. [1971] 'K emission-line widths and the solar chromosphere', *Solar Phys.* **17**, 316.

BECKERS, J. M. [1963] 'Study of the undisturbed chromosphere from Hα-disk filtergrams, with particular reference to the identification of spicules', *Astrophys. J.* **138**, 648.

BECKERS, J. M. [1964a] 'A study of the fine structures in the solar chromosphere', Thesis, Utrecht; AFCRL–Environmental Research Paper No. 49.

BECKERS, J. M. [1964b] 'On the relation between solar granules and spicules', *Astrophys. J.* **140**, 1339.

BECKERS, J. M. [1966] 'Observations of chromospheric spicules', *The Fine Structure of the Solar Atmosphere*, ed. K. O. Kiepenheuer, p. 80; Franz Steiner Verlag, Wiesbaden.

BECKERS, J. M. [1968] 'Solar spicules', *Solar Phys.* **3**, 367.

BHAVILAI, R. [1965] 'The structure of the solar chromosphere. I. Identification of spicules on the disk', *Mon. Not. R.A.S.* **130**, 411.

BHAVILAI, R. [1966] 'Time changes in the Hα chromospheric features', *The Fine Structure of the Solar Atmosphere*, ed. K. O. Kiepenheuer, p. 96; Franz Steiner Verlag, Wiesbaden.

BLONDEL. M. [1971] 'Étude statistique comparée du champ de vitesse dans une région active et deux régions calmes de la photosphère du Soleil', *Astron. Astrophys.* **10**, 342.

BRAY, R. J. [1964] 'The helium D_3 absorption line in the spectra of chromospheric flares', *Z. Astrophys.* **60**, 207.

BRAY, R. J. [1968] 'High-resolution photography of the solar chromosphere. IV. Size, shape, and evolution of the dark mottles', *Solar Phys.* **5**, 323.

BRAY, R. J. [1969] 'High-resolution photography of the solar chromosphere. VI. Properties of the bright mottles', *Solar Phys.* **10**, 63.

BRAY, R. J., and LOUGHHEAD, R. E. [1964] *Sunspots*, Chapman and Hall, London.

BRAY, R. J., and LOUGHHEAD, R. E. [1967] *The Solar Granulation*, Chapman and Hall, London.

BRAY, R. J., and WINTER, J. G. [1970] 'High-resolution photography of the solar chromosphere. VIII. Computer control of a tunable 1/8 Å filter', *Solar Phys.* **15**, 309.

BRUZEK, A. [1959] 'Über die Spiculen auf der Sonnenscheibe', *Z. Astrophys.* **47**, 191.

CHAPMAN, G. A., and SHEELEY, N. R. [1968] 'The photospheric network', *Solar Phys.* **5**, 442.

CRAGG, T., HOWARD, R., and ZIRIN, H. [1963] 'Vertical structures in the chromosphere', *Astrophys. J.* **138**, 303.

DEUBNER, F.-L. [1969] 'Some properties of velocity fields in the solar photosphere. II. The spatial distribution of the oscillatory field', *Solar Phys.* **9**, 343.

DOYLE, R. O. (ed.) [1969] *A Long-Range Program in Space Astronomy, NASA SP-213*, NASA, Washington, D.C.

DUBOV, E. E. [1960] 'A possible interpretation of spectroheliograms by the hypothesis of non-homogeneous structure of the chromosphere', *Izv. Crim. Astrophys. Obs.* **24**, 301.

DUBOV, E. E. [1969] 'Study of spectroheliograms in metal lines and the chromospheric structure', *Izv. Crim. Astrophys. Obs.* **39**, 279.

DUNN, R. B. [1969] 'Sacramento Peak's new solar telescope', *Sky and Telescope* **38**, No. 6, p. 368.

ELLIOTT, I. [1969] 'Power spectra of Hα Doppler shifts', *Solar Phys.* **6**, 28.

FRAZIER, E. N. [1968] 'An observational study of the hydrodynamics of the lower solar photosphere', *Astrophys. J.* **152**, 557.

FRAZIER, E. N. [1970] 'Multi-channel magnetograph observations. II. Supergranulation', *Solar Phys.* **14**, 89.

FREDGA, K. [1969] 'Spectroheliograms in the Mg II line at 2795.5 Å', *Solar Phys.* **9**, 358.

GIOVANELLI, R. G., HALL, D. N. B., and HARVEY, J. W. [1972] 'A comparison between the helium 10830 Å and the hydrogen Hα chromospheres', *Solar Phys.* **22**, 53.

GOLDBERG, L. [1967] 'Ultraviolet and X rays from the Sun', *Ann. Rev. Astron. Astrophys.* **5**, 279.

GOLDBERG, L., MOHLER, O. C., and BROWN, J. D. [1957] 'A connexion between the granulation and the structure of the low chromosphere', *Nature* **179**, 369.

GONCZI, G., and RODDIER, F. [1969] 'A Fourier spectrum analysis of long samples of solar line oscillations', *Solar Phys.* **8**, 255.

GOPASYUK, S. I., and TSAP, T. T. [1970] 'Relationship between velocity fields and magnetic fields in the solar atmosphere', *Astron. J. U.S.S.R.* **47**, 1012 (*Sov. Astron. AJ* **14**, 813).

GRIETHUYSEN, I. G. VAN, and HOUTGAST, J. [1959] 'Quantitative photometric measurements with a Lyot filter', *Bull. Astron. Inst. Neth.* **14**, 279.

GROSSMANN-DOERTH, U., and VON UEXKÜLL, M. [1971] 'Spectral investigation of chromospheric fine structures', *Solar Phys.* **20**, 31.

HOWARD, R. [1959] 'Observations of solar magnetic fields', *Astrophys. J.* **130**, 193.

HOWARD, R. [1967a] 'Magnetic field of the Sun (observational)', *Ann. Rev. Astron. Astrophys.* **5**, 1.

HOWARD, R. [1967b] 'Velocity fields in the solar atmosphere', *Solar Phys.* **2**, 3.

HOWARD, R., and HARVEY, J. W. [1964] 'Photospheric magnetic fields and chromospheric features', *Astrophys. J.* **139**, 1328.

IOSHPA, B. A., and OBRIDKO, V. N. [1966] 'On the problem of measuring magnetic fields in the chromosphere in the Hβ line', *Soln. Dann.*, 1966, No. 5, p. 68.

JAGER, C. DE [1957] 'The interpretation of hydrogen spectroheliograms', *Bull. Astron. Inst. Neth.* **13**, 133.

JAGER, C. DE [1959] 'Structure and dynamics of the solar atmosphere', *Handbuch der Physik*, ed. S. Flügge, vol. 52, p. 80; Springer, Berlin.

JANSSENS, T. J. [1970] 'Long term observations of the Hα chromospheric network', *Solar Phys.* **11**, 222.

JEFFERIES, J. T., and LOUGHHEAD, R. E. [1956] 'Correlation patterns in the chromospheric granulation', *Nature* **177**, 180.

JENSEN, E., and ORRALL, F. Q. [1963] 'Observational study of macroscopic inhomogeneities in the solar atmosphere. IV. Velocity and intensity fluctuations observed in the K line', *Astrophys. J.* **138**, 252.

KIEPENHEUER, K. O. [1957] 'Erfahrungen mit einem Lyotschen Hα-Filter sowie vorläufige Ergebnisse über die Struktur der Chromosphäre', *Z. Astrophys.* **42**, 209.

KIEPENHEUER, K. O. [1960a] 'Über die Struktur der gestörten Sonnenatmosphäre', *Z. Astrophys.* **49**, 73.

KIEPENHEUER, K. O. [1960b] 'The optical phenomena forming a solar center of activity', Soc. Ital. Fisica, Rendiconti Scuola Internazionale di Fisica 'Enrico Fermi' (Director, G. Righini), p. 39.

KIEPENHEUER, K. O. [1966] Introduction to *The Fine Structure of the Solar Atmosphere,* ed. K. O. Kiepenheuer, p. 11; Franz Steiner Verlag, Wiesbaden.

KIEPENHEUER, K. O. [1967] 'Morphology of an active region on the Sun', *Solar Physics*, Proc. NATO Conf., Athens, ed. J. N. Xanthakis, p. 383; Interscience, London.

KRAFFT, M. [1968] 'Spectroscopic investigation of a birefringent Lyot-filter for Hα', *Solar Phys.* **5**, 462.

LEIGHTON, R. B. [1964] 'Transport of magnetic fields on the Sun', *Astrophys. J.* **140**, 1547.

LEIGHTON, R. B. [1965] 'Introductory report – small scale solar magnetic fields, plages, filaments: observations', *Stellar and Solar Magnetic Fields*, I.A.U. Symp. No. 22, ed. R. Lüst, p. 158; North-Holland Publ. Co., Amsterdam.

LEIGHTON, R. B., NOYES, R. W., and SIMON, G. W. [1962] 'Velocity fields in the solar atmosphere. I. Preliminary report', *Astrophys. J.* **135**, 474.

LINSKY, J. L., and AVRETT, E. H. [1970] 'The solar H and K lines', *Publ. Astron. Soc. Pac.* **82**, 169.

LIVINGSTON, W. C. [1968] 'Magnetograph observations of the quiet Sun. I. Spatial description of the background fields', *Astrophys. J.* **153**, 929.

LOUGHHEAD, R. E. [1969] 'High-resolution photography of the solar chromosphere. VII. Structure of the low chromosphere', *Solar Phys.* **10**, 71.

LOUGHHEAD, R. E., BRAY, R. J., TAPPERE, E. J., and WINTER, J. G. [1968] 'High-resolution photography of the solar chromosphere. I. The 30-cm refractor of the C.S.I.R.O. Solar Observatory', *Solar Phys.* **4**, 185.

LOUGHHEAD, R. E., and TAPPERE, E. J. [1971] 'High-resolution photography of the solar chromosphere. IX. Limb observations of high spectral purity', *Solar Phys.* **19**, 44.

MACRIS, C. J. [1957] 'Recherches sur les spicules en projection sur le disque solaire', *Ann. Astrophys.* **20**, 179.

MACRIS, C. J. [1962] 'Studies on the flocculi of the solar chromosphere. Part I. Life time of the flocculi', *Mem. Soc. Astron. Ital.* **33**, 3.

MACRIS, C. J., and ALISSANDRAKIS, C. E. [1970] 'Lifetime of the dark and bright mottles of the solar chromosphere', *Solar Phys.* **11**, 59.

McMATH, R. R., MOHLER, O. C., PIERCE, A. K., and GOLDBERG, L. [1956] 'Preliminary results with a vacuum solar spectrograph', *Astrophys. J.* **124**, 1.

MICHARD, R. [1967] 'Nonhomogeneous structure of the solar atmosphere', *Solar Physics*, Proc. NATO Conf., Athens, ed. J. N. Xanthakis, p. 95; Interscience, London.

MOORE, C. E., and MERRILL, P. W. [1968] 'Partial Grotrian Diagrams of Astrophysical Interest', *National Standards Reference Data Series, NSRDS-NBS 23*; National Bureau of Standards, Washington.

MUSMAN, S., and RUST, D. M. [1970] 'Vertical velocities and horizontal wave propagation in the solar photosphere', *Solar Phys.* **13**, 261.

NOYES, R. W. [1967] 'Observational studies of velocity fields in the solar photosphere and chromosphere', *Aerodynamic Phenomena in Stellar Atmospheres*, ed. R. N. Thomas, p. 293; Academic Press, London.

NOYES, R. W., and LEIGHTON, R. B. [1963] 'Velocity fields in the solar atmosphere. II. The oscillatory field', *Astrophys. J.* **138**, 631.

ORRALL, F. Q. [1965] 'Observational study of macroscopic inhomogeneities in the solar atmosphere. VI. Photospheric oscillations and chromospheric structure', *Astrophys. J.* **141**, 1131.

ORRALL, F. Q. [1966] 'Observational study of macroscopic inhomogeneities in the solar atmosphere. VIII. Vertical chromospheric oscillations measured in K_3', *Astrophys. J.* **143**, 917.

PAGEL, B. E. J. [1964] 'The structure of the solar chromosphere', *Ann. Rev. Astron. Astrophys.* **2**, 267.

PASACHOFF, J. M. [1970] 'Fine structure in Ca II on the solar disc', *Solar Phys.* **12**, 202.

PROKOFYEVA, I. A. [1957] 'The structure of the chromosphere as shown by the Fraunhofer lines Hα and Hβ', *Izv. Pulkovo Astron. Obs.* **21**, No. 160, p. 156.

RAMSAY, J. V., NORTON, D. G., and MUGRIDGE, E. G. V. [1968] 'The testing and adjustment of Lyoten–Öhman filters', *Solar Phys.* **4**, 476.

REEVES, E. M., and PARKINSON, W. H. [1970] 'An atlas of extreme-ultraviolet spectroheliograms from OSO-IV', *Astrophys. J. Supp.* **21**, No. 181.

REICHEL, M. [1953] 'Zur Theorie der Spektroheliogramme', *Z. Astrophys.* **33**, 79.

RODDIER, F. [1966] 'Étude à haute résolution de quelques raies de Fraunhofer par observation de la résonance optique d'un jet atomique. III. Les oscillations périodiques des raies', *Ann. Astrophys.* **29**, 639.

ROGERS, E. H. [1970] 'Lifetime of the Hα chromospheric network', *Solar Phys.* **13**, 57.

ROGERSON, J. B. [1955] 'A statistical analysis of spectroheliograms', *Astrophys. J.* **121**, 204.

SCHRÖTER, E. H. [1958] 'Chromosphärische Strukturen in den Balmerlinien', *Z. Astrophys.* **45**, 68.

SEVERNY, A. B. [1966] 'Magnetic fields at various depths in the solar atmosphere', *Astron. J. U.S.S.R.* **43**, 465 (*Sov. Astron. AJ* **10**, 367).

SHEELEY, N. R. [1967a] 'Observations of small-scale solar magnetic fields', *Solar Phys.* **1**, 171.

SHEELEY, N. R. [1967b] 'The average profile of the solar K-line during the sunspot cycle', *Astrophys. J.* **147**, 1106.

SHEELEY, N. R. [1969] 'The evolution of the photospheric network', *Solar Phys.* **9**, 347.

SHEELEY, N. R., and ENGVOLD, O. [1970] 'Simultaneous measurements of magnetic fields and brightness fields using a 4-image spectroheliograph', *Solar Phys.* **12**, 69.

SIMON, G. W., and LEIGHTON, R. B. [1964] 'Velocity fields in the solar atmosphere. III. Large-scale motions, the chromospheric network, and magnetic fields', *Astrophys. J.* **140**, 1120.

STEEL, W. H., SMARTT, R. N., and GIOVANELLI, R. G. [1961] 'A 1/8 Å birefringent filter for solar research', *Aust. J. Phys.* **14**, 201.

STENFLO, J. O. [1966] 'Solar photometry in Hα', *Arkiv för Astron.* **3**, 481.

STEPANOV, V. E. [1960] 'The motion of Ca^+ in the chromosphere and the connection of the motion with magnetic fields', *Izv. Crim. Astrophys. Obs.* **23**, 184.

STEPANOV, V. E. [1961] 'The determination of the mean gradient of the chromospheric magnetic field', *Izv. Crim. Astrophys. Obs.* **25**, 174.

SYKORA, J. [1970] 'Time and shape changes of the supergranular network', *Solar Phys.* **13**, 292.

TANENBAUM, A. S., WILCOX, J. M., FRAZIER, E. N., and HOWARD, R. [1969] 'Solar velocity fields: 5-min oscillations and supergranulation', *Solar Phys.* **9**, 328.

TITLE, A. [1966a] *Selected Spectroheliograms*, Calif. Inst. Tech., Pasadena.

TITLE, A. M. [1966b] 'A study of velocity fields in the Hα chromosphere by means of time-lapse Doppler movies', Thesis, Calif. Inst. Tech., Pasadena.

TOUSEY, R. [1967] 'Some results of twenty years of extreme ultraviolet solar research', *Astrophys. J.* **149**, 239.

TOUSEY, R. [1971] 'Observations of the extreme ultraviolet solar spectrum', *Phil. Trans. Roy. Soc.* A **270**, 59.

TSAP, T. T. [1966] 'Photospheric magnetic fields and the structure of the undisturbed chromosphere', *Izv. Crim. Astrophys. Obs.* **35**, 161.

UNSÖLD, A. [1955] *Physik der Sternatmosphären*, Springer, Berlin.

VRABEC, D. [1971] 'Magnetic field spectroheliograms from the San Fernando Observatory', *Solar Magnetic Fields*, I.A.U. Symposium No. 43, ed. R. Howard, p. 329; Reidel, Dordrecht-Holland.

VYAL'SHIN, G. F., and KRAT, V. A. [1965] 'The magnetic fields in the solar chromosphere', *Izv. Pulkovo Astron. Obs.* **24**, No. 2, p. 26.

WATTS, R. N. [1970] 'The Apollo Telescope Mount', *Sky and Telescope* **40**, No. 4, p. 202.

ZIRIN, H. [1966] *The Solar Atmosphere*, Blaisdell Publ. Co., Waltham, Mass.

CHAPTER 4

Physical Conditions in the Quiet Chromosphere

4.1 Introduction

The total vertical extent of the solar chromosphere is only about 20 000 km, yet over this short distance the temperature increases by more than two orders of magnitude – from $T \simeq 5000$ K at the top of the relatively cool photosphere to $T \simeq 1\,000\,000$ K at the base of the corona. Equally spectacular changes are exhibited by other physical parameters such as density and degree of ionization, and the same is true of certain physical processes, for example, the mode of energy transport. It is no wonder, therefore, that the investigation of the physical conditions in the chromosphere has attracted the attention of a large number of workers, who have employed many different observational techniques. In fact, our present knowledge of the physical conditions in the chromosphere and their variation with height has been derived from an almost bewildering array of spectral data, pertaining to both the continuum and line radiation of the chromosphere. The various types of data available are summarized in Tables 4.2 and 4.7, in which the last columns indicate, for each kind of data, the section containing a detailed discussion. In wavelength the relevant observations span the range from centimetric radio wavelengths to the extreme ultraviolet. The observational techniques are equally diverse and both eclipse and non-eclipse observations have contributed to the overall picture.

Our aim in the present chapter is threefold: (*a*) to summarize the relevant continuum and spectral line data; (*b*) to outline the astrophysical techniques employed to extract the desired physical information; and (*c*) to compare chromospheric models derived by various workers which incorporate the results of such investigations. Extended treatment of one important topic has been omitted, namely the difficult theoretical problem of excitation and ionization in the absence of local thermodynamic equilibrium (LTE). This omission is justified by the very full account of non-LTE methodology and its application to the chromosphere to be found in Thomas and Athay's [1961] well-known monograph; however, a brief description of the basic principles involved is included in Section 4.3.4. Other astrophysical techniques are outlined to an extent governed by their success in producing solid results from the particular data involved, their future potential, or their intrinsic interest.

A substantial part of our existing knowledge of the physical conditions in the chromosphere has been derived from continuum observations. Accordingly, we start with a discussion of the continuum data (Section 4.2), dealing firstly with radio, infrared, and ultraviolet observations of the disk and a model of the low chromosphere based on this data (Sections 4.2.2–4.2.5). Then we turn to eclipse observations of the Balmer continuum and other free-bound continua of hydrogen and helium and the electron scattering continuum at λ4700 (Sections 4.2.6–4.2.9); these observations extend our knowledge of the physical conditions to greater heights.

Chromospheric spectral line data provides valuable information not derivable from continuum data, although up to the present time it has not been exploited to the same extent. The main reason for this has been the difficulties encountered in applying non-LTE theory with the necessary rigour to realistic situations. Nevertheless, there have been some notable successes such as, for example, the determination of the physical conditions in spicules; moreover, certain types of line data have a promising potential for future analyses following further improvements in the application of non-LTE and line formation theories. Therefore, in Section 4.3 we have included fairly complete discussions of individual topics even when the information derived so far is somewhat meagre, dealing successively with the integrated emission and logarithmic gradients of flash spectrum lines (Section 4.3.2); profiles, half-widths, and total intensities of spicule emission lines (Sections 4.3.3–4.3.4); disk profiles of chromospheric lines (Section 4.3.5); Hα central intensities of bright and dark mottles (Section 4.3.6); and integrated fluxes of ultraviolet emission lines (Section 4.3.7).

A *homogeneous* model of the chromosphere is one in which the physical conditions are assumed to be uniform over any horizontal plane, i.e. parallel to the solar surface. (Some authors prefer to use the phrase 'spherically symmetric'.) At the photospheric level, the presence of granulation close to the limb demonstrates the existence of inhomogeneities at least up to the level corresponding to $\tau = 0.1$ [see Bray and Loughhead, 1967, Section 2.3], i.e. up to about 100–200 km below the base of the chromosphere. If we couple this to inhomogeneities at the chromospheric levels indicated by the presence of spicules (Section 2.3.2) and, at lower heights, by the presence of bright mottles (Section 2.6), we arrive at the conclusion that it is doubtful whether *any* region of the photosphere or chromosphere can be regarded as homogeneous. This conclusion is reinforced by the existence of oscillatory velocities in the photosphere–chromosphere transition region (Section 3.5.1). To some extent, therefore, the concept of a homogeneous model is an artificial one since there may be no portion of the real chromosphere to which it strictly applies. Accordingly, the principal functions of such a model are to serve as a zero-order

TABLE 4.1

*Principal Symbols Used in Chapter 4**

A	abundance of element
A_{kl}	spontaneous transition probability
B_{kl}	transition probability for absorption
$B(\nu,T_e)$	Planckian source function
b_k	LTE departure coefficient
C_{kl}	collisional transition rate
c	velocity of light (2.998×10^{10} cm sec^{-1})
$E(h)$	integrated emission
F_ν	mean intensity; surface intensity
g	acceleration due to gravity (2.74×10^4 cm sec^{-2} at solar surface)
g_{fb}	free-bound Gaunt factor
h	geometrical height
h'	Planck constant (6.625×10^{-27} erg sec)
k	Boltzmann constant (1.380×10^{-16} erg deg^{-1})
L	mean Earth–Sun distance (1.496×10^{13} cm)
m_H	mass of hydrogen atom (1.673×10^{-24} gm)
N_{HI}	number density of neutral hydrogen
N_{HII}	number density of ionized hydrogen
N_e	electron number density
n	principal quantum number
p	gas pressure
p_e	electron pressure
Q	fractional area occupied by spicules
R_{kl}	radiative transition rate
$R_\odot$	solar radius (6.960×10^{10} cm at photosphere)
T_B	brightness temperature (radio, infrared, or ultraviolet)
T_e	electron temperature
T_{ex}	excitation temperature
T_{ion}	ionization temperature
T_{kin}	kinetic temperature
x_H	degree of hydrogen ionization
α	most probable speed
β	logarithmic emission gradient
ϵ	volume emission coefficient
θ	heliocentric angle
κ_λ	(linear) absorption coefficient
λ	wavelength
$\Delta\lambda_{1/2}$	half-width of line profile
μ	mean molecular weight; molecular weight of element
ν	frequency
ξ	line-of-sight component of non-thermal velocity
ρ	gas density
σ_T	Thomson scattering coefficient (6.65×10^{-25} cm^2)
τ_λ	optical depth
ϕ	flux at top of Earth's atmosphere
χ_i	ionization potential

*C.g.s. units are used throughout.

approximation to the actual chromosphere and to provide a basis for reconciling observational data from various sources. Existing homogeneous models are described and compared in Section 4.4.

Inhomogeneous models, on the other hand, attempt to take account of the actual inhomogeneity of the chromosphere and thus purport to give physical conditions separately for spicules and the interspicular regions. Existing inhomogeneous models are described and compared in Section 4.5. Our adopted model is given in Table 4.23; it is characterized by dense spicules embedded in a much more tenuous medium. The spicules have a constant temperature of 16 000 K over most of their vertical extent, whereas the interspicular medium is cooler than the spicules at the lower levels ($h < 11\,000$ km) of the chromosphere but rapidly approaches coronal temperatures at greater heights.

How reliable are these homogeneous and inhomogeneous models? In some cases we can obtain an objective assessment by comparing values for a given parameter derived by different workers, and this procedure has been attempted in the account that follows. Often, however, the number of truly independent models covering the same range of heights is inadequate for this purpose. In this account we have concentrated on the more recent models, but it is quite possible that some of the earlier ones, e.g. those compared by de Jager [1959, p. 142], could turn out to be equally worthy of consideration. It would certainly be naïve to assume that the very latest model is necessarily the most reliable! In general, the reliability of any particular model is likely to decrease with height as we proceed from the low ($h < 5000$ km) to the middle ($5000 \leqslant h \leqslant 10\,000$ km) and high ($10\,000$ km $< h < 20\,000$ km) chromosphere.

The important question of evaluating the net radiative energy loss from the chromosphere as a function of height – a quantity that must equal the input of mechanical and conductive energy at each height – is postponed to Chapter 7, Section 7.2.

4.2 Chromospheric Continuum Data

4.2.1 INTRODUCTION

Many regions of the continuous spectrum of the chromosphere yield information about the physical conditions prevailing there. In wavelength, the useful portions range from centimetric radio wavelengths through the infrared and visible continua out to the far ultraviolet. The observational methods employed are equally diverse: they include, for example, measurements made with precision radio dishes, slitless spectra obtained during total eclipses, and photon counts made with rocket-borne instruments. Even so, there are still gaps in our data such as, for example, the region between $\lambda \simeq 0.3$ mm (infrared) and $\lambda = 1$ mm (radio). Similarly, more extensive measurements of the He I and He II continua in the far ultraviolet are desirable.

Our aim in the following eight sections (Sections 4.2.2–4.2.9) is to outline the large mass of continuum data available, laying stress on the astrophysical basis of each method employed to extract physical information from each particular kind of raw data. Little emphasis is given to purely observational or instrumental topics, but the interested reader will have no difficulty in tracking down such information from the references cited.

The topics discussed are so diverse that, as an aid to the reader, we present in Table 4.2 a summary of the various types of observations involved and the information which – in practice or in principle – can be extracted from each. The last column indicates the section or sections wherein each topic is discussed in detail.

4.2.2 RADIO OBSERVATIONS AT CM AND MM WAVELENGTHS

Radio emission from the Sun at wavelengths $\lambda < 3$ cm ($\nu > 10$ GHz) originates in the chromosphere and is therefore a potential source of information about the physical conditions prevailing there. At greater wavelengths the chromospheric contribution becomes hard to separate from emission from the corona and, in particular, coronal active regions. The radio emission originates from free-free transitions of electrons in the field of positive ions (mainly protons). Accordingly, complicated non-LTE effects are absent. Observations have been made of both (*a*) the radio brightness temperature $T_B(\lambda)$ at the centre of the disk as a function of λ, and (*b*) the centre-limb variation of $T_B(\lambda)$ at fixed wavelengths. The available observational material and the methods of extracting the information may be conveniently summarized under the following headings:

(1) *Radio brightness temperatures.* The radio brightness temperature at a given wavelength λ and heliocentric angle θ is defined as the temperature of a black body having the same brightness at that wavelength as the given point on the Sun. Using the Rayleigh–Jeans approximation to the radiation law and the usual integral equation of limb darkening, we can immediately write down a relation between $T_B(\lambda,\theta)$ and $T_e(h)$, the electron temperature as a function of height, namely

$$T_B(\lambda,\theta) = \int_0^\infty T_e(h) e^{-\sec\theta.\tau_\lambda} \sec\theta \, . \, d\tau_\lambda, \tag{4.1}$$

where

$$\tau_\lambda = \int_h^\infty \kappa_\lambda \, dh \tag{4.2}$$

is the optical depth and κ_λ is the (linear) absorption coefficient.

Measurements of T_B at the centre of the disk ($\theta = 0$) have been made by numerous workers at a variety of wavelengths. A convenient compilation of these measurements is given by Shimabukuro and Stacey [1968], an extract

TABLE 4.2

Chromospheric Continuum Data: Summary

NATURE OF OBSERVATIONS	WAVELENGTH	ECLIPSE OR NON-ECLIPSE	DISK OR LIMB	HIGH OR LOW CHROMOSPHERE	INFORMATION YIELDED	SECTION
Radio brightness temperature T_B at $\theta = 0$ as a function of λ	1 mm–3.2 cm	Non-eclipse	Disk	High + low	Check of homogeneous or inhomogeneous models derived from other data	4.2.2 4.2.5
Centre-limb variation of radio brightness temperature T_B	1.2–8.6 mm	Non-eclipse	Disk	Low	Approx. location of temperature minimum; qualitative evidence for inhomogeneity of chromosphere	4.2.2
Infrared brightness temperature T_B at $\theta = 0$ as a function of λ	0.004–0.312 mm (4–312 μm)	Non-eclipse	Disk	Low	Check of homogeneous model of low chromosphere derived from other data	4.2.3 4.2.5
Centre-limb variation of infrared continuum (relative measurements)	0.020–0.024 mm (20–24 μm)	Non-eclipse	Disk	Low	Location and extent of temperature minimum	4.2.3 4.2.5

Ultraviolet brightness temperature T_B at $\theta = 0$ as a function of λ	1525–2900 Å	Non-eclipse	Disk	Low	Value of $(T_e)_{min}$; temperatures in the region of the temperature minimum	4.2.4 4.2.5
Centre-limb variation of ultraviolet brightness temperature T_B	1525–2900 Å	Non-eclipse	Disk	Low	As above	4.2.4 4.2.5
Balmer continuum of H I: integrated emission $E(h)$ as a function of h	< 3646 Å	Eclipse	Limb	Middle + low	$N_e^2 T_e^{-3/2}$ in spicules as a function of height	4.2.7
Electron scattering continuum: integrated emission $E(h)$ as a function of h	~4700 Å	Eclipse	Limb	High + low	N_e in the inter-spicular medium as a function of height	4.2.8
Lyman continuum of H I: total flux. and its variation with λ	< 912 Å	Non-eclipse	Disk (integrated)	Low	T_e in layer assumed responsible; in conjunction with a model, values of N_{HII}	4.2.9
He I and He II continua: total fluxes and their variation with λ	< 504 Å < 228 Å	Non-eclipse	Disk (integrated)	High + middle	T_e in layer assumed responsible; in conjunction with a model, values of N_{HeII} and N_{HeIII}	4.2.9

from which appears in Table 4.3. The decline in T_B as λ decreases towards mm wavelengths is very apparent and illustrates the height discrimination made possible by observing at different wavelengths; nevertheless, even at $\lambda = 1.0$ mm the chromospheric temperature minimum has yet to be reached.

TABLE 4.3

Central Radio Brightness Temperature as a Function of Wavelength [Shimabukuro and Stacey, 1968]

λ (mm)	ν (GHz)	$T_B(\lambda)$ (K)	λ (mm)	ν (GHz)	$T_B(\lambda)$ (K)
32	9.4	19 300 ± 1351	5.62	53.4	6900 ± 600
32	9.4	17 000	5.62	53.4	6100 ± 600
32	9.4	16 000	5.62	53.4	6500 ± 600
31.6	9.5	11 800 ± 2500	5.61	53.5	6750 ± 600
20	15.0	15 100 ± 1057	5.50	54.5	6900 ± 600
17.6	17.0	9100 ± 200	4.3	70.0	8000 ± 700
16	18.8	8000	4.3	70.0	7100 ± 200
15.8	19.0	10 800 ± 700	4.3	70.0	9600 ± 500
14.3	21.0	10 800 ± 700	4.3	70.0	7000 ± 700
13.5	22.0	11 000 ± 700	3.3	91.0	6375 ± 574
12.8	23.4	10 700 ± 700	3.2	94.0	7860 ± 500
11.8	25.4	9800 ± 700	3.2	94.0	6402 ± 215
11.8	25.4	8870 ± 980	3.0	100.0	5870 ± 950
8.7	34.5	5280	2.8	107.0	6800 ± 500
8.6	35.0	10 420 ± 730	2.73	110.0	5500 ± 715
8.5	35.3	6740 ± 674	2.4	125.0	6500 ± 400
8.5	35.3	6500	2.2	136.0	6800 ± 400
8.0	37.5	7500 ± 900	2.15	140.0	5433 ± 500
8.0	37.5	6400 ± 800	2.0	150.0	5670 ± 230
7.5	40.0	5700	1.8	167.0	5300 ± 700
7.5	40.0	6000 ± 500	1.3	231.0	6000 ± 500
6.0	50.0	4500	1.3	231.0	5900 ± 400
6.0	50.0	4500	1.2	250.0	5600 ± 400
5.74	52.29	6900 ± 500	1.0	300.0	5900 ± 500

Note: for references to the original investigations, see the last column of Shimabukuro and Stacey's Table 1.

Mathematical expressions for the absorption coefficient κ_λ in (4.2) can be derived from plasma theory. In particular, Thomas and Athay [1961, p. 262] quote the formula

$$\kappa_\lambda = 2.5 \times 10^{-23}\, \lambda^2 N_e^2\, T_e^{-3/2}\ [3.23 + \log (T_e N_e^{-1/3})]\ \text{cm}^{-1}, \quad (4.3)$$

where λ is expressed in cm. Expressions showing a similar functional dependence on λ, N_e, and T_e, but differing somewhat in the numerical factors involved have been given by other authors [see, for example, Shklovskii and Kononovich, 1958; Zirin and Dietz, 1963; Pagel, 1964; Beckers, 1968].

Equations (4.1)–(4.3) form the mathematical basis for extracting model data from observations of T_B as a function of λ and θ, but it is evident that the quantities $N_e(h)$ and $T_e(h)$ are too inextricably mixed to be separately derivable from radio observations alone. Accordingly, the radio observations find their greatest use either in checking models (homogeneous or inhomogeneous) derived from entirely different data or in extracting a single parameter when one or more of the other parameters is known from other sources (see below).

(2) *Checking a chromospheric model.* A typical example of the first application is provided by Noyes *et al.'s* [1968] check of the Bilderberg Continuum Atmosphere (BCA), a homogeneous model of the low chromosphere (Section 4.2.5). From the model these authors calculated the variation of T_B with λ and compared the resulting curve with a number of actual measurements. Their curve is shown in Fig. 4.1, but for the purpose of the comparison we have made use of the more extensive measurements listed in Table 4.3. Figure 4.1 shows that the BCA successfully reproduces the observed values of T_B within the fairly large errors of measurement. Of course, a check of this kind is only partial, because other distributions of N_e and T_e could be devised which would produce equally good agreement with observation.

(3) *Extraction of model information.* For an inhomogeneous model chromosphere we have the task of determining no less than five parameters, each a function of h. These parameters are N_e, T_e (spicules); N_e^*, T_e^* (interspicular regions); and Q, the fractional area at any given height occupied by spicules. Clearly radio data alone is insufficient; however, in conjunction with other data, observations of $T_B(\lambda)$ can be used to derive one or more of the parameters. Beckers' [1968] determination of T_e^* over the range 0–20 000 km (Table 4.21) provides an example.

Beckers starts with an equation for the average electron density

$$\bar{N}_e = N_e Q + N_e^* (1 - Q), \tag{4.4}$$

in which $\bar{N}_e$ is determined from eclipse observations of the continuum at λ4700 (Section 4.2.8), N_e from the total intensities of spicule emission lines (Section 4.3.4), and Q from spicule counts (Section 2.3.6). Equation (4.4) then yields N_e^*. For an inhomogeneous atmosphere Beckers replaces (4.1) by

$$T_B(\lambda) = Q \int_0^\infty T_e e^{-\tau_\lambda} d\tau_\lambda + (1 - Q) \int_0^\infty T_e^* e^{-\tau_\lambda^*} d\tau_\lambda^*, \tag{4.5}$$

valid at the disk centre. The first integral is put equal to 16 000 K, the derived value of the spicule temperature (Section 4.3.4), which is found to be largely independent of h. A run of T_e^* with h is then *assumed* and values of $T_B(\lambda)$ are calculated with the aid of (4.2), (4.3), and (4.5), using the known values of N_e^*. Comparison of the computed values of $T_B(\lambda)$ with those observed finally yields a distribution of $T_e^*(h)$ and thus a model consistent with the radio observations.

A basically similar procedure for determining T_e^* has been followed by Kanno [1966]. However, Kanno assumes a temperature of 10^6 K for the lower

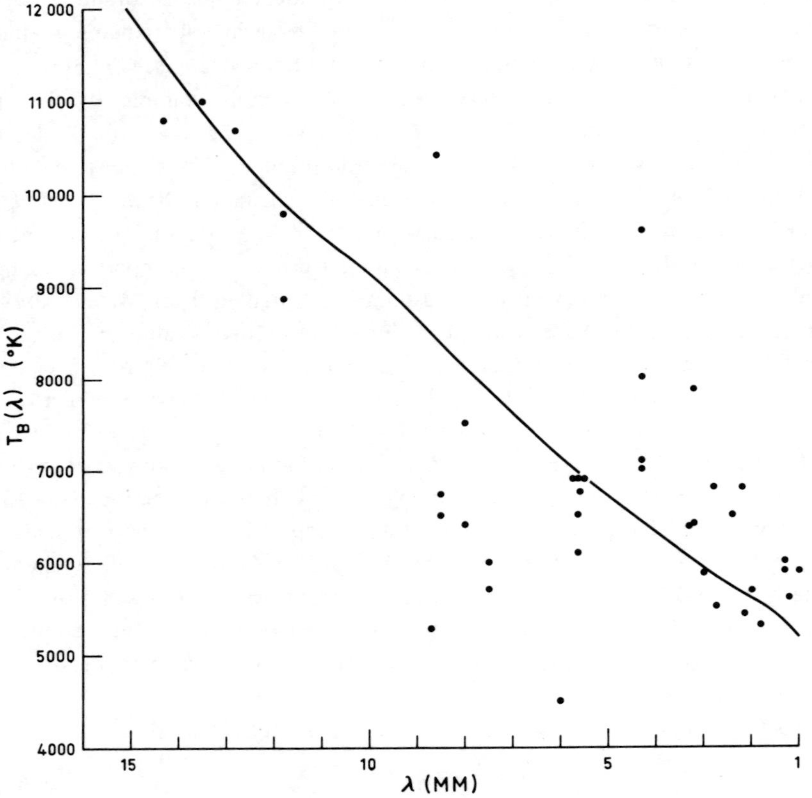

Fig. 4.1. Central radio brightness temperature as a function of wavelength. The curve represents the run predicted by the Bilderberg Continuum Atmosphere, according to the calculations of Noyes *et al.* [1968]. The filled circles give measured values (Table 4.3).

'boundary' of the corona and gives T_e^* only over the height range 6000–16 000 km (Table 4.21).

Finally, Tominaga [1967] has used cm and dm radio data, together with ultraviolet emission line fluxes (Section 4.3.7), to derive a homogeneous model of the upper chromosphere. Tominaga starts with an empirical expression for T_B

as a function of frequency, viz.

$$T_B(\nu) = a + b\nu^{-1} + c\nu^{-2}. \tag{4.6}$$

Then, following an analysis of Piddington [1950, 1954], he derives from (4.6) a relation between N_e and T_e of the form

$$N_e^2 \frac{dh}{dT_e} = \frac{2b^2}{\pi\zeta} \cdot \frac{T_e^{3/2}}{(T_e - a)^3}, \tag{4.7}$$

valid for $10^4 \lesssim T_e \lesssim 10^5$, where $a = 7000$, $b = 6.8 \times 10^{13}$, and $\zeta = 0.14$. A further relation between N_e and T_e is obtained by assuming the upper chromosphere to be in *hydrostatic equilibrium*, so that

$$\frac{d}{dh}(N_e T_e) = -\frac{\mu m_H g}{k} N_e, \tag{4.8}$$

where μ is the mean molecular weight (= 0.69) and, in view of the almost complete ionization of hydrogen for $T_e \gtrsim 20\,000$ K, the gas density is taken to be proportional to the electron density. It must be stressed, however, that the assumption of hydrostatic equilibrium is of questionable validity, and the derived values of N_e and T_e obtained from (4.7) and (4.8) should accordingly be treated with caution. Tominaga's distributions of electron, hydrogen, and helium densities are given in Fig. 4.14.

(4) *Centre-limb variation of* T_B. Measurements of the centre-limb variation of T_B are potentially capable of yielding useful information, but existing data are scanty and suffer from inadequate spatial resolution. For a homogeneous atmosphere whose temperature increases outwards one would evidently expect to see a measurable limb brightening at wavelengths λ such that the level $\tau_\lambda = 1$ lies above the chromospheric temperature minimum. Curves obtained by Coates *et al.* [1958] at $\lambda = 8.6$ mm during an eclipse show, on the contrary, a more or less constant brightness out to $\sin\theta = 0.98$ and beyond this, a sharp spike. This behaviour can be taken as indirect evidence for the inhomogeneous nature of the chromosphere at the levels involved [Thomas and Athay, 1961, p. 264; Pagel, 1964] and, in fact, the spike at the extreme limb is attributed by Zirin and Dietz [1963] and Pagel [1964] to emission from spicules.

Drift curves showing a small degree of limb brightening have been obtained by Noyes *et al.* [1968] and Newstead [1969] at $\lambda = 1.2$ mm. These curves have not been analysed in detail, although Noyes *et al.* have concluded that the presence of limb brightening at $\lambda \simeq 1$ mm implies that the temperature minimum must lie below the level $\tau_{1\,\mathrm{mm}} = 1$, corresponding to $\tau_{5000} \simeq 0.000015$. However, more recent observations at $\lambda = 1.2$ mm [Kundu, 1971] and 1.4 mm [Shimabukuro, 1971] cast doubt on the reality of the limb brightening observed earlier. Similar uncertainty attaches to the reality or otherwise of limb brightening observed in the range 3.2–3.4 mm [see Takahashi, 1967; Shimabukuro and Stacey, 1968; Shimabukuro, 1970; Simon *et al.*, 1970].

4.2.3 OBSERVATIONS OF THE INFRARED CONTINUUM

Figure 4.1 demonstrates that even at a radio wavelength as low as $\lambda = 1$ mm the chromospheric temperature minimum has yet to be reached. Infrared measurements, on the other hand, yield information about the physical conditions in this region and, in particular, about the magnitude and location of the temperature minimum.

Let $(T_e)_{min}$ be the electron temperature at the minimum; then for those wavelengths originating largely in the immediate neighbourhood of the minimum, (4.1) becomes, to a first approximation,

$$T_B(\lambda, \theta) \simeq (T_e)_{min} \int_0^\infty e^{-\sec\theta . \tau_\lambda} \sec\theta \, . \, d\tau_\lambda,$$

and hence

$$T_B(\lambda, \theta) \simeq (T_e)_{min}. \tag{4.9}$$

Two consequences follow from (4.9): firstly, for these wavelengths there is neither limb darkening nor limb brightening and, secondly, the electron temperature is approximately equal to the measured brightness temperature.

Gingerich and de Jager [1968] have concluded that the centre-limb observations of Léna [1968] and Noyes *et al.* [1968] indicate a variation going from a definite limb darkening at 20 μm to an essentially flat curve at 24 μm. It follows from the argument given above that, at or near the latter wavelength (and beyond), the bulk of the observed radiation originates in the immediate neighbourhood of the temperature minimum.

The centre-limb measurements referred to above are only relative and therefore do not give values of T_B and hence T_e. Other workers have made absolute determinations of T_B at the disk centre, or averaged over the disk, for values of λ ranging from 312 μm down to 4.0 μm; these are collected in Table 4.4. The first three entries give values for T_B which, within the experimental error, lie close to the value of $(T_e)_{min}$ given by the Bilderberg Continuum Atmosphere, described in Section 4.2.5, namely 4600 K. The other temperatures in Table 4.4, i.e. for $\lambda = 12$ to 4 μm, demonstrate the rise which occurs as one proceeds downwards past the minimum and penetrates into the upper photosphere.

In summary, both absolute values of T_B at infrared wavelengths and its centre-limb variation provide useful tools for checking models of the low chromosphere. As in the case of radio observations, the sources of continuous absorption must be known and evaluated before detailed calculations are possible. Existing observations, however, are somewhat scanty. In particular, the relative paucity of accurate observations in the wavelength range lying between the shortest available radio wavelength ($\lambda \simeq 1.0$ mm) and the easily accessible portions of the infrared ($\lambda < 0.02$ mm) leads to uncertainty in our knowledge of the physical conditions in the region immediately above the temperature minimum. However, this gap is rapidly closing as a result of improved

TABLE 4.4

Central Infrared Brightness Temperature as a Function of Wavelength

λ (mm)	λ (μm)	$T_B(\lambda)$ (K)	REFERENCE
0.238–0.312	238–312	4370 ± 260*	Eddy *et al.* [1969]
0.100–0.200	100–200	4600 ± 400*	Gay [1970]
0.02015	20.15	4590 ± 110	Koutchmy and Peyturaux [1968]
0.01202	12.02	5050 ± 80	Saiedy [1960]
0.0111	11.1	5036 ± 30	Saiedy and Goody [1959]
0.00863	8.63	5160 ± 40	Saiedy [1960]
0.005	5.0	5270 ± 150*	Murcray *et al.* [1964]
0.004	4.0	5626 ± 100*	Murcray *et al.* [1964]

*Averaged over solar disk.

observations from both balloons and high-altitude aircraft. At both radio and infrared wavelengths further exploitation of the eclipse technique should yield more extensive data of high quality on the centre-limb variation of T_B.

4.2.4 OBSERVATIONS OF THE ULTRAVIOLET CONTINUUM

The third portion of the electromagnetic spectrum which yields important information about the low chromosphere and, in particular, the region of the temperature minimum, is the ultraviolet. Since the wavelengths of interest do not penetrate the Earth's atmosphere, the relevant observations must be made from high-altitude rockets or satellites. As before, the significant data comprise measurements as a function of λ of both absolute intensity (or total flux) and the centre-limb variation. For the latter some form of biaxial solar guider is required.

Just as in the case of the infrared observations, the vanishing of the centre-limb variation at a certain wavelength implies that the bulk of the corresponding radiation originates in the immediate neighbourhood of the temperature minimum. The observations show that the variation disappears in the region 1525–1683 Å [Gingerich and de Jager, 1968] ; at greater and lesser wavelengths there is limb darkening and limb brightening, respectively [Goldberg, 1967; Bonnet and Blamont, 1968; Gingerich and Rich, 1968] .

That this region corresponds to the temperature minimum can be shown more directly by plotting $T_B(\lambda)$ as a function of λ. This is illustrated in Fig. 4.2, which is based on measurements of rocket spectra made by Detwiler, Garrett, Purcell, and Tousey [Goldberg, 1967, Fig. 7]. The broad temperature minimum roughly centred on $\lambda = 1600$ Å is well shown. The value of T_B at this wavelength is ~4700 K; in accordance with (4.9) this should also be the value of $(T_e)_{min}$,

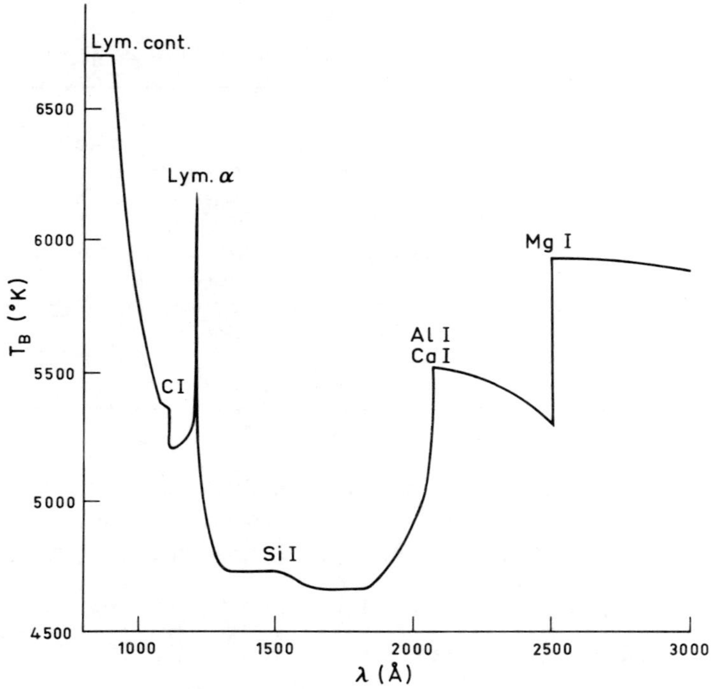

Fig. 4.2. Ultraviolet brightness temperature as a function of wavelength [Goldberg, 1967]. Note the discontinuities due to the ionization continua of various neutral metals. The curve refers to the whole disk, not to the disk centre.

and the figure quoted is close to that given by the Bilderberg Continuum Atmosphere (Section 4.2.5), namely 4600 K. More recent rocket observations have yielded values of 4400 K [Parkinson and Reeves, 1969] and 4670 ± 100 K [Widing *et al.*, 1970].

In the region of the temperature minimum, the photoionization continua of various neutral metals provide the principal contribution to the ultraviolet opacity. Several of the corresponding continua limits reveal their presence in Fig. 4.2 by causing well-marked discontinuities in the curve for $T_B(\lambda)$. On the long wavelength side of the minimum, each continuum limit is characterized by a sudden drop in T_B as λ decreases, reflecting an increase in absorption and therefore a level of observation closer to the minimum. On the short wavelength side, T_B increases as each limit is passed. Figure 4.2 also shows a discontinuity due to ionization of hydrogen from the ground state (the Lyman continuum limit) and a large apparent increase in T_B at $\lambda \simeq 1200$ Å due to the Lyman α line.

Provided the sources of opacity can be evaluated, the ultraviolet data can be used to check a proposed chromospheric model using procedures similar to those already described in connection with radio and infrared observations. Thus Gingerich and Rich [1968] explicitly considered absorption cross-sections associated with the ionization transitions listed in Table 4.5. As additional

TABLE 4.5

Principal Sources of Ultraviolet Opacity in the Region of the Temperature Minimum [Gingerich and Rich, 1968]

IDENTIFICATION	CONTINUUM LIMIT (Å)
Si I (^{3}P)	1525
(^{1}D)	1680
(^{1}S)	1980
Mg I (^{2}S)	1620
(^{3}P^0)	2510
Al I (^{2}P)	2071

Note: the table lists the identifications of the ionization continua which for reasons of abundance, etc., are the most significant contributors to the absorption; some of the continuum limits are identifiable in Fig. 4.2.

sources of opacity they included H^-, H, H_2^+, and Rayleigh scattering, and on this basis they compared ultraviolet fluxes and limb darkening calculated from the Bilderberg Continuum Atmosphere with the observational data. A more recent discussion of possible sources of ultraviolet opacity has been published by Noyes [1971]; this paper also includes an up-to-date review of modern rocket and satellite observations in the ultraviolet.

4.2.5 A HOMOGENEOUS MODEL OF THE LOW CHROMOSPHERE BASED ON CONTINUUM DATA: THE BILDERBERG CONTINUUM ATMOSPHERE

The Bilderberg Continuum Atmosphere (BCA) is the product of an international study week held in 1967 near Arnhem, Holland, with the aim of producing an internationally acceptable model of the photosphere and low chromosphere [Gingerich and de Jager, 1968]. The chromospheric portion of the model is based on the types of observations we have discussed in Sections 4.2.2–4.2.4, namely absolute intensities and centre-limb measurements of the radio, infrared, and ultraviolet continua. Model data derived from these widely separated regions are found to be in close agreement, thereby establishing a certain degree of confidence in the adopted model. Nevertheless, there remain some discrepancies and, moreover, certain portions of the model are interpolated rather than directly determined; at this stage, therefore, the BCA is best regarded as a working tool rather than as a definitive model.

In order to supplement the continuum data, an attempt was made to use certain line profile and equivalent width measurements in constructing the model. However, further consideration showed this attempt to be premature, although future advances in spectral line theory should ultimately enable such observations to make their proper contribution (cf. Section 4.3.5).

The observational basis for the various portions of the BCA may be summarized as follows:

(1) *The photosphere:* $\tau_{5000} > 0.1$. Temperatures and other physical conditions in this region are derived by well-established methods from measurements of absolute intensities and limb darkening in the visible and near infrared regions of the spectrum.

(2) *The transition region, photosphere to temperature minimum:* $\tau_{5000} = 0.1$ *to* 0.01. This portion of the model is poorly determined and the adopted temperature distribution is simply a smooth curve connecting the portions on either side. In principle, as discussed in Section 4.2.4, temperatures in this range could have been established from observations in a suitable ultraviolet region of the spectrum, e.g. 1900–2800 Å, as suggested by Gingerich and de Jager [1968]. However, according to these authors, it was not possible to obtain a satisfactory interpretation of Bonnet and Blamont's [1968] limb darkening data in this region, although the BCA satisfactorily gives their observed intensities at the disk centre.

(3) *The temperature minimum:* $\tau_{5000} = 0.01$ *to* 0.0001. The value of 4600 K adopted for $(T_e)_{min}$ rests mainly on ultraviolet intensities in the region of 1600 Å (Section 4.2.4); in addition, adoption of this figure makes possible a smooth transition from the photospheric to the chromospheric layers. The *extent* of the temperature minimum, i.e. the approximate locations of its upper and lower boundaries, is derived from the infrared limb darkening measurements of Léna and of Noyes *et al.* referred to in Section 4.2.3.

(4) *The temperature rise:* $\tau_{5000} = 0.0001$ *to* 0.00001. This portion of the model is poorly determined and once again recourse to interpolation is necessary. Radio observations of shorter wavelength than are currently available would fill this gap.

(5) *The temperature rise:* $\tau_{5000} = 0.00001$ *to* 0.000001. This is derived from

brightness temperatures in the millimetre and far ultraviolet regions of the spectrum. The results from the two widely separated regions show excellent agreement.

(6) *The lower part of the low chromosphere:* τ_{5000} = *0.000001 to 0.0000001.* Temperatures in this range are derived from millimetre observations but are poorly determined.

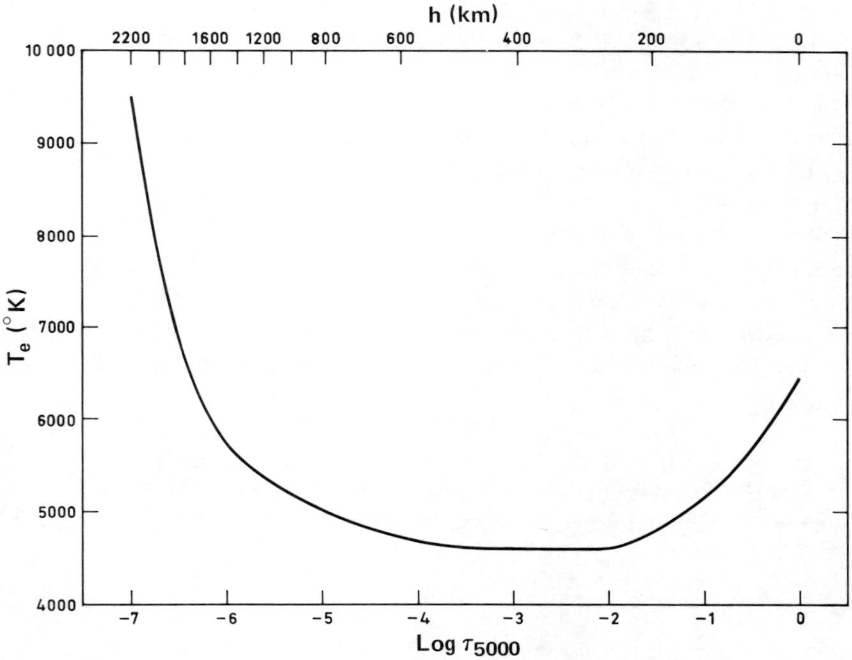

Fig. 4.3. The chromospheric temperature minimum, as described by the Bilderberg Continuum Atmosphere [Gingerich and de Jager, 1968]. The upper abscissa scale gives geometrical height, the zero level corresponding to unit optical depth.

The top of the BCA corresponds to a height of about 2200 km above the level τ_{5000} = 1; this represents the maximum height to which physical conditions have currently been derived exclusively from *disk* observations of the continuum. The chromospheric and upper photospheric portions of the model are given in Table 4.6, while in Fig. 4.3 temperature is plotted against optical depth and geometrical height. Note that from τ_{5000} = 0.02 to 0.00008, T_e remains within the range 4600–4700 K. The corresponding geometrical thickness of this quasi-isothermal layer is some 350 km.

Although the portion of the model given in Table 4.6 covers a height range of only slightly more than 2000 km, we notice that the gas density and pressure decrease by factors of 10^7 and 3×10^6 respectively. Moreover, over the same range there occurs a transition from a degree of hydrogen ionization of only $\sim 10^{-5}$ to virtually complete ionization.[1]

The following assumptions were made in computing the model:

(1) the model was assumed to be homogeneous and in hydrostatic equilibrium, with an effective temperature of 5780 K;

(2) except for a hydrogen:helium ratio of 10:1 (in terms of number of atoms per unit volume), the element abundances were taken from Goldberg *et al.* [1960];

(3) the electron contributions of the six most productive elements besides hydrogen were considered explicitly, i.e. Si, Mg, Fe, C, Na, and Al;

(4) the sources of opacity included were H, H^-, H_2^+, metals, and Rayleigh and electron scattering;

(5) the abundance of H^- was calculated on the assumption of local thermodynamic equilibrium; a non-LTE version of the model has also been derived, but it differs little from the BCA [Gingerich and de Jager, 1968].

The BCA represents a distinct advance over earlier models of the photosphere and low chromosphere, e.g. the Utrecht Reference Model of 1964.

TABLE 4.6

A Model of the Low Chromosphere Based on Continuum Data: the Bilderberg Continuum Atmosphere [Gingerich and de Jager, 1968]

τ_{5000}	h (km)	T_e (K)	p (dyne cm^{-2})	p_e (dyne cm^{-2})	ρ (gm cm^{-3})	x_H
0.1000	124	5140	5.518×10^4	3.234×10^0	1.66×10^{-7}	9.12×10^{-6}
0.0794	139	5060	4.879	2.681	1.49	6.51
0.0631	153	4990	4.307	2.254	1.34	4.83
0.0501	168	4920	3.797	1.897	1.20	3.53
0.0398	183	4860	3.346	1.614	1.07	2.71
0.0316	197	4800	2.946	1.373	9.51×10^{-8}	2.06
0.0251	211	4750	2.593	1.179	8.46	1.65
0.0200	225	4700	2.282	1.012	7.52	1.32
0.0158	239	4660	2.007	8.772×10^{-1}	6.67	1.11
0.0126	253	4630	1.764	7.673	5.90	1.01
0.01000	267	4610	1.551	6.777	5.21	9.72×10^{-7}
0.00794	281	4605	1.363	6.074	4.59	1.04×10^{-6}
0.00631	295	4600	1.199	5.441	4.04	1.12
0.00501	309	4600	1.054	4.896	3.55	1.24
0.00398	323	4600	9.264×10^3	4.402	3.12	1.38
0.00316	337	4600	8.143	3.954	2.74	1.54
0.00251	351	4600	7.156	3.550	2.41	1.71
0.00200	365	4600	6.286	3.184	2.12	1.91

[1] However, according to our adopted homogeneous model given in Table 4.18, complete ionization is delayed until h = 8000 km.

TABLE 4.6 *(continued)*

τ_{5000}	h (km)	T_e (K)	p (dyne cm^{-2})	p_e (dyne cm^{-2})	ρ (gm cm^{-3})	x_H
0.00158	379	4600	5.520×10^{3}	2.853×10^{-1}	1.86×10^{-8}	2.13×10^{-6}
0.00126	393	4600	4.844	2.554	1.63	2.38
0.001000	407	4600	4.247	2.283	1.43	2.66
0.000794	421	4600	3.720	2.039	1.25	2.98
0.000631	436	4600	3.254	1.818	1.10	3.35
0.000501	451	4605	2.841	1.625	9.55×10^{-9}	3.90
0.000398	465	4610	2.476	1.450	8.32	4.54
0.000316	481	4620	2.152	1.298	7.21	5.50
0.000251	496	4630	1.864	1.160	6.23	6.66
0.000200	512	4640	1.608	1.035	5.36	8.07
0.000159	529	4650	1.381	9.215×10^{-2}	4.60	9.81
0.000126	546	4665	1.179	8.249	3.91	1.23×10^{-5}
0.0001000	564	4680	9.999×10^{2}	7.382	3.31	1.55
0.0000794	583	4700	8.419	6.660	2.77	2.00
0.0000631	603	4720	7.029	6.022	2.31	2.58
0.0000501	624	4750	5.817	5.572	1.90	3.49
0.0000398	646	4780	4.773	5.191	1.55	4.69
0.0000316	669	4810	3.880	4.865	1.25	6.25
0.0000251	694	4840	3.121	4.578	9.98×10^{-10}	8.26
0.0000200	720	4870	2.479	4.316	7.88	1.09×10^{-4}
0.0000158	748	4910	1.943	4.190	6.12	1.49
0.0000126	778	4950	1.500	4.066	4.69	2.03
0.0000100	810	5000	1.138	4.055	3.52	2.87
0.0000079	845	5055	8.476×10^{1}	4.087	2.59	4.13
0.0000063	883	5115	6.177	4.148	1.87	6.04
0.0000050	924	5175	4.376	4.154	1.31	8.88
0.0000040	971	5240	2.987	4.142	8.81×10^{-11}	1.34×10^{-3}
0.0000032	1020	5305	1.940	4.018	5.65	2.06
0.0000025	1090	5370	1.176	3.754	3.38	3.25
0.0000020	1160	5435	6.485×10^{0}	3.332	1.84	5.35
0.0000016	1260	5500	3.153	2.763	8.80×10^{-12}	9.34
0.0000013	1360	5610	1.387	2.444	3.76	1.93×10^{-2}
0.0000010	1480	5700	5.807×10^{-1}	1.968	1.52	3.81
0.0000008	1610	5830	2.437	1.705	6.02×10^{-13}	8.22
0.0000006	1720	6010	1.167	1.662	2.58	1.82×10^{-1}
0.0000005	1810	6250	6.960×10^{-2}	1.801	1.28	3.84
0.0000004	1880	6550	5.044	1.920	7.38×10^{-14}	6.76
0.0000003	1940	6900	4.055	1.819	5.02	8.95
0.0000003	2000	7300	3.392	1.594	3.81	9.75
0.0000002	2050	7740	2.887	1.371	3.03	9.94
0.0000002	2100	8220	2.489	1.184	2.46	9.99
0.0000001	2160	8780	2.173	1.035	2.01	1.00×10^{0}
0.0000001	2210	9500	1.922	9.204×10^{-3}	1.63	1.00

Note: the height h is measured upwards, the zero level corresponding to unit optical depth; the last column gives the degree of ionization of hydrogen.

Nevertheless, various inadequacies may be briefly noted: (*a*) there are marked discrepancies between the predictions of the BCA and the observed intensities in that portion of the visible spectrum which lies to the violet of 4500 Å [Gingerich and de Jager, 1968]; (*b*) the BCA fails to interpret satisfactorily the limb darkening observations of Bonnet and Blamont [1968] in the region 1800–2900 Å; (*c*) there is still some uncertainty over how well it accounts for the observed ultraviolet fluxes in the same region, although the position is satisfactory at shorter wavelengths [see Gingerich and Rich, 1968, Fig. 1]; (*d*) the method used to average various limb darkening measurements in the visible and near infrared in deriving the photospheric portion of the model has been criticized by Elste [1968].

On the credit side, the BCA gives a basically correct account of a number of important observed features, including (*a*) the observed intensities and centre-limb darkening in the visible and near infrared; (*b*) the observed centre-limb darkening in the 20–24 μm region (Section 4.2.3); (*c*) the observed radio brightness temperatures in the mm region (see Fig. 4.1); and (*d*) the virtual absence of any centre-limb variation in the region 1525–1683 Å [see Gingerich and de Jager, 1968, Fig. 3].

A revision of the BCA, called the 'Harvard-Smithsonian Reference Atmosphere' (HSRA), has recently been proposed by Gingerich *et al.* [1971]. It differs from the BCA mainly in the following respects: (*a*) $(T_e)_{min}$ is set at 4170 K instead of 4600 K, although this figure is admittedly a somewhat arbitrary choice; (*b*) the extent of the region of temperature minimum is narrower than in the BCA; and (*c*) above this region the temperatures are higher than in the BCA (cf. Gingerich *et al.*'s Fig. 1). At the time of writing, insufficient time has elapsed for a considered judgement as to whether or not it will eventually supersede the BCA. Most calculations in succeeding chapters are based on the latter model.

4.2.6 CIRCUMSTANCES OF ECLIPSE OBSERVATIONS OF THE CHROMOSPHERIC CONTINUUM AND FLASH SPECTRUM LINES

Hitherto in this chapter we have been concerned with portions of the continuum which yield information about the chromosphere from *disk* observations, namely the radio, infrared, and ultraviolet continua. In the next two sections we shall deal with portions of the chromospheric continuum which are observable only at the *limb* during a total solar eclipse, namely the Balmer continuum, $\lambda < 3646$ Å, and the electron scattering continuum at $\lambda \simeq 4700$ Å. Outside of an eclipse both these continua are swamped by photospheric radiation. They have proved to be of great importance as a means of deriving physical conditions in the middle and low chromosphere but, before proceeding to discuss the continua themselves, it is convenient to outline the circumstances of such eclipse observations and to present a few relevant definitions.

Suppose the eclipsed Sun is observed with a telescope equipped with a slitless spectrograph; let $R_\odot$ be the solar radius and h the projected height in the chromosphere of the Moon's limb at any instant (see Fig. 4.4). The spectrum obtained will consist of a number of arcs orientated in the x-direction, each corresponding to an emission line, superimposed on a faint continuum [see, e.g., Dunn *et al.*, 1968, Plates 1 and 3]. Let us first consider the energy $E(h)$ in a

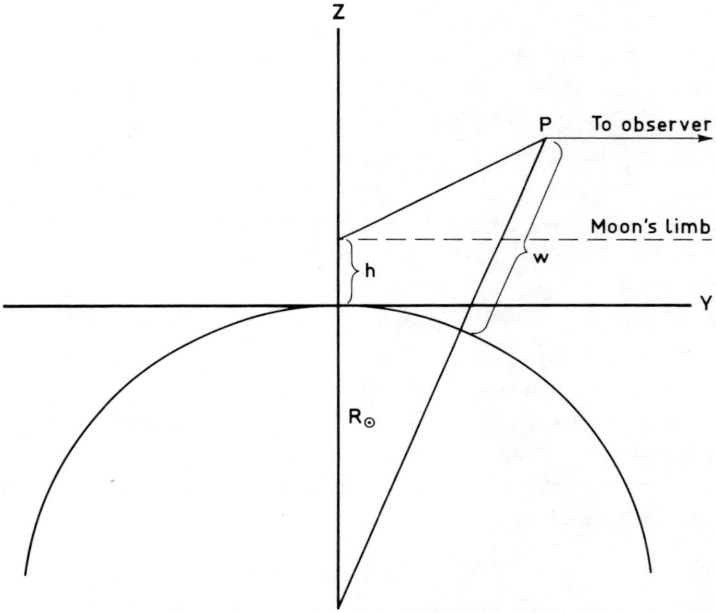

Fig. 4.4. Coordinate system used in discussing eclipse observations of the chromospheric continuum and flash spectrum lines.

portion of one of these 'flash spectrum' arcs. By convention, the energy evaluated is that which originates in a thin slice of the chromosphere 1 cm thick in the x-direction. Then

$$E(h) = \int_{z=h}^{\infty} \int_{y=-\infty}^{\infty} \int_{x=-½}^{½} \int_{\lambda=0}^{\infty} \epsilon(\lambda)\, e^{-\tau_\lambda}\, d\lambda\, dx\, dy\, dz, \quad (4.10)$$

where $\epsilon(\lambda)$ is the volume emission coefficient at wavelength λ at any point in the chromosphere and τ_λ is the optical thickness along the line-of-sight from that point to the observer.

Owing to the absence of a slit, only total line intensities, not profiles, can be derived from such observations; therefore it is convenient to put

$$\epsilon = \int_0^{\infty} \epsilon(\lambda)\, d\lambda.$$

Furthermore, if it can be assumed that the chromosphere is optically thin for the line or continuum wavelength in question, i.e. that $\tau_\lambda \simeq 0$, then the observed emission per 1 cm slice reduces to

$$E(h) = \int_{z=h}^{\infty} \int_{y=-\infty}^{\infty} \epsilon \, \mathrm{d}y \, \mathrm{d}z. \tag{4.11}$$

The quantity $E(h)$ is called the *integrated emission* and is usually given in units of erg sec^{-1} or erg sec^{-1} ster^{-1}, per 1 cm slice of the chromosphere.

It is clear from (4.11) that, in effect, a double differentiation is required in order to derive the desired volume emission ϵ from the observed run of $E(h)$. For this reason considerable precision is required in the measurement of E and, in fact, by taking closely-spaced spectrograms between second and third contacts during the course of a total eclipse it has proved possible to achieve a height discrimination of a mere 100 km [Zirker, 1958; Dunn *et al.*, 1968]. (The accuracy of the determination of the zero point of the height scale is, of course, much lower.) This figure represents an effective resolution some five times superior to the best attained under ordinary circumstances with a modern high-resolution solar telescope (cf. Section 3.2.2)! This important advantage of slitless spectrograms goes some way towards offsetting the well-known disadvantages of eclipse observations, such as the short time available and the often primitive equipment and inconvenient site.

If the observed integrated emission $E(h)$ is represented as the sum of a number of exponential terms, i.e.

$$E(h) = \sum_i B_i \exp(-\beta_i h), \tag{4.12}$$

then it can be shown that the volume emission coefficient ϵ at a height w above the limb is given by the formula

$$\epsilon(w) = \left(\frac{1}{2\pi R_\odot}\right)^{1/2} \sum_i B_i \beta_i^{3/2} \exp(-\beta_i w) \tag{4.13}$$

[Thomas and Athay, 1961, p. 33]. The consistency of (4.13) and (4.12) can readily be verified by substituting (4.13) into (4.11), bearing in mind that since $w \ll R_\odot$, $w \simeq z + y^2/2R_\odot$ (cf. Fig. 4.4), an approximation which at moderate heights leads to only small errors [Pecker and Pottasch, 1969].

The above analysis is valid for both line and continuum data provided that, in the case of the latter, due correction is made for the superimposition on the spectrogram of emission from both the east and west limbs [see Thomas and Athay, 1961, Appendix I]. Equation (4.13) then provides a basis for determining $T_e(h)$ and $N_e(h)$ from eclipse measurements of the Balmer and λ4700 continua, as we shall see in the next two sections. Whereas each flash spectrum line is integrated over its entire apparent width, the continua must, of course, be measured over finite wavelength intervals. The slight lack of 'spectral purity' in the continua spectra due to the absence of a slit is of no consequence because of the small height of the chromosphere and the relatively slow variation of the continua with λ. Once again it should be emphasized that the validity of (4.13) rests on the correctness of the assumption that the optical depth is small.

The *logarithmic emission gradient* is defined as the quantity

$$\beta = -\frac{\mathrm{d}}{\mathrm{d}h} \ln E(h) \ \ \mathrm{cm}^{-1}. \tag{4.14}$$

The reciprocal of β is termed the *emission scale height*. If the expansion for $E(h)$ given in (4.12) contains only one term then evidently $\beta = \beta_1$ and the logarithmic gradient and scale height are independent of h. One term does indeed suffice for many flash spectrum lines; for others and for the Balmer and λ4700 continua, additional terms are required and, accordingly, β changes with h. We shall return to the integrated emissions and logarithmic gradients of flash spectrum lines in Section 4.3.2.

4.2.7 THE BALMER CONTINUUM

Emission in the Balmer continuum arises from the recombination of electrons and protons at the quantum level $n = 2$ (see Fig. 4.5). The head of the continuum (corresponding to electrons of zero velocity) occurs at a wavelength of 3646 Å. A theoretical expression for the emission in the various hydrogen continua, including the Balmer continuum, can be derived by computing the capture cross-section for an electron of velocity v and evaluating the corresponding recombination rate [Aller, 1963, pp. 187–188]. In the units of Thomas and Athay [1961, p. 195] and Hiei [1963] the volume emission per unit frequency interval in all directions is given by the formula

$$\epsilon(\nu) = 2.15 \times 10^{-32}\, g_{\mathrm{fb}}\, n^{-3}\, N_{\mathrm{e}} N_{\mathrm{H\,II}} T_{\mathrm{e}}^{-3/2} \ \exp \frac{h'}{kT_{\mathrm{e}}}(\nu_{\mathrm{n}} - \nu) \ \ \mathrm{erg\,cm^{-3}\,sec^{-1}}, \tag{4.15}$$

where h' is Planck's constant.

For the Balmer continuum, $n = 2$ and the free-bound Gaunt factor g_{fb} = 0.876. Also, in the immediate neighbourhood of the continuum head $\nu_{\mathrm{n}} \simeq \nu$ and hence the exponential term may be put equal to unity. Finally, in all but the lowest regions of the chromosphere the free electrons originate overwhelmingly from the ionization of hydrogen, so that $N_{\mathrm{e}} \simeq N_{\mathrm{H\,II}}$, the proton density. Accordingly, (4.15) simplifies to

$$\epsilon = 2.35 \times 10^{-33}\, N_{\mathrm{e}}^2\, T_{\mathrm{e}}^{-3/2} \ \ \mathrm{erg\,cm^{-3}\,sec^{-1}}. \tag{4.16}$$

Equation (4.16), together with equations (4.12) and (4.13) above, forms the mathematical basis for deriving the value of $N_{\mathrm{e}}^2\, T_{\mathrm{e}}^{-3/2}$ as a function of height from eclipse observations of the Balmer continuum. The details of the procedure are well illustrated by Kanno's [1966] derivation of $N_e^2 \, T_e^{-3/2}$ in *spicules.*

Since the Balmer continuum varies as N_e^2 and, moreover, N_e *between* the spicules is nearly an order of magnitude less than within them (anticipating a later result), Kanno starts with the assumption that the observed Balmer continuum originates predominantly in the spicules, provided that $Q(h)$, the

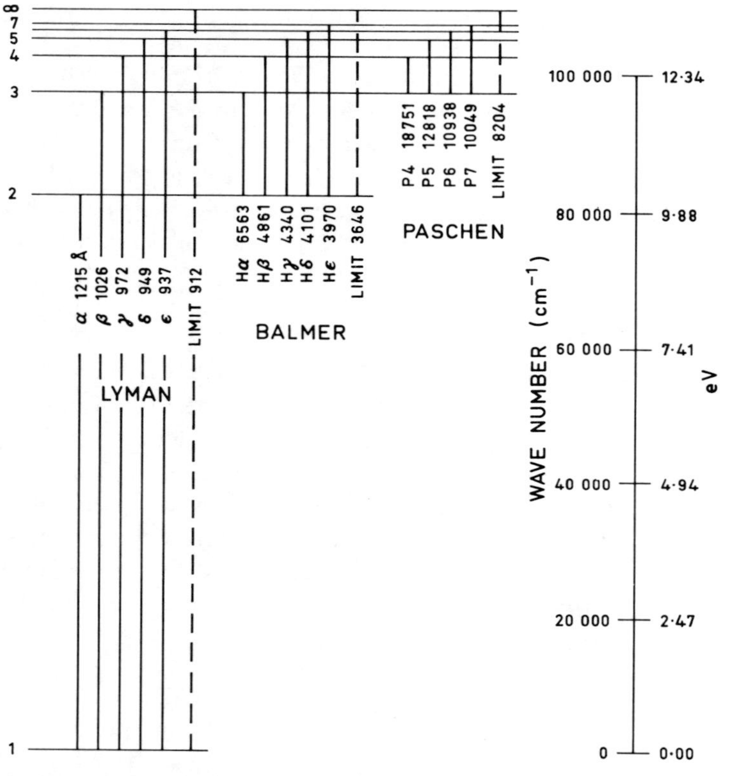

Fig. 4.5. Partial Grotrian diagram of hydrogen [Moore and Merrill, 1968]. The wavelengths of the various continua heads are shown to the right of the associated series, the corresponding free-bound transitions being shown dashed. Principal quantum numbers are shown to the left and an energy scale in cm^{-1} and electron volts to the right.

fractional area at any height occupied by spicules, $\gtrsim 0.01$. Furthermore, he assumes that $Q(h)$ and $\epsilon(h)$ can be represented by expressions of the form

$$Q(h) = C \exp(-\gamma h); \; \epsilon(h) = \sum_i A_i \exp(-\alpha_i h), \tag{4.17}$$

so that the effective emission per unit volume is given by

$$Q(h) \cdot \epsilon(h) = \sum_i CA_i \exp[-(\alpha_i + \gamma)h]. \tag{4.18}$$

Noting that the right-hand side of (4.18) is of the same functional form as that of (4.13), we see immediately on comparing (4.12), (4.13), and (4.18) that

$$E(h) = (2\pi R_\odot)^{1/2} \sum_i CA_i(\alpha_i + \gamma)^{-3/2} \exp[-(\alpha_i + \gamma)h] . \qquad (4.19)$$

Kanno's expression for the observed values of $E(h)$ is based on a formula given by Thomas and Athay [1961, equation 2-12] to represent results obtained at the 1952 Khartoum eclipse.[2] However, Kanno divides the observed emission by a factor of 3 in an attempt to correct the intensities for photometric errors which, on the basis of comparison with other eclipses, are thought to be present in the Khartoum data. This yields the formula, applicable at $\lambda = 3640$ Å,

$$[E(h)]_{\text{obs}} = 1000 \exp(-5 \times 10^{-8} h) + 120 \exp(-1.8 \times 10^{-8} h). \qquad (4.20)$$

Finally, introduction of a formula for $Q(h)$ which fits the spicule counts of Hiei [1963], namely

$$Q(h) = \exp[-1.1 \times 10^{-8}(h - 0.7 \times 10^8)] , \qquad (4.21)$$

gives immediately C and γ in (4.17). (Note that (4.21) implicitly assumes that the spicules start at $h = 700$ km, and that at this height they occupy the whole volume; see, however, Sections 2.3.2 and 2.3.6.)

By equating coefficients and exponents in (4.19) and (4.20) one can now evaluate the remaining constants, i.e. A_1, A_2, α_1, and α_2, and thus obtain an expression for $\epsilon(h)$ which, on substitution in (4.16), gives finally a formula for $N_e^2 T_e^{-3/2}$ in spicules as a function of height, viz.,

$$N_e^2 T_e^{-3/2} = 3.33 \times 10^{18} \exp(-3.9 \times 10^{-8} h) + 8.63 \times 10^{16} \exp(-0.7 \times 10^{-8} h), \qquad (4.22)$$

where h is in cm.

If the spicule temperature is known from other sources (4.22) yields directly $N_e(h)$. Kanno in fact takes T_e to be 10 000 K, independently of height; however, other workers, as we shall see in Section 4.5.2, have derived somewhat different values. Kanno's values of N_e over the height range 1000–6000 km are given in Table 4.20.

In conclusion, it should be pointed out that formulae such as (4.20) and (4.21) must not be taken too literally: they are merely useful expressions for representing the observations which allow solutions such as (4.22) to be obtained in a closed form.

[2] This formula and other expressions given by Thomas and Athay give the true emission in the Balmer continuum, i.e. separated from the background continuum due to other sources. The latter can be determined by measuring the emission at $\lambda > 3700$ Å and extrapolating backwards.

4.2.8 THE ELECTRON SCATTERING CONTINUUM AT λ4700

In the previous section we evaluated just a single contribution to the continuous emission of the chromosphere, namely the Balmer continuum at $\lambda < 3646$ Å due to free-bound transitions of neutral hydrogen involving the quantum state $n = 2$. Moreover, we pointed out that when measurements are made on both sides of the continuum head, the true Balmer emission can easily be separated from the continuum due to other sources, yielding ultimately $N_e^2 T_e^{-3/2}$ in spicules as a function of height.

In this section we shall consider these other contributions to the continuous emission and in particular, the electron scattering continuum, at wavelengths well removed from the various hydrogen continuum heads. We shall find that when the height dependence of the various continua is taken into account, the electron scattering continuum can be separated from the other sources, thus providing a convenient means of determining N_e in the interspicular medium in the height range concerned.

Again following Thomas and Athay [1961, p. 195] and Hiei [1963], we may express the emission per unit frequency interval in all directions due to the various sources, in units of erg cm^{-3} sec^{-1}, by the following formulae:

(1) *Free-bound transitions of neutral hydrogen.* This is given by (4.15), i.e.

$$\epsilon_{fb}(\nu) = 2.15 \times 10^{-32}\, g_{fb} n^{-3} N_e N_{HII} T_e^{-3/2} \exp \frac{h'}{kT_e}(\nu_n - \nu). \tag{4.15}$$

In the wavelength region with which we are concerned, the Brackett, Pfund, etc., continua produce only a negligible contribution owing to the influence of the exponential factor in (4.15). Accordingly, the only contribution to ϵ_{fb} is the Paschen continuum (see Fig. 4.5), for which $n = 3$ and $g_{fb} = 0.975$ at λ4700 and 0.987 at λ4117;

(2) *Electron scattering.* We have

$$\epsilon_{es}(\nu) = 2\pi F_\nu \sigma_T N_e, \tag{4.23}$$

where F_ν is the mean monochromatic intensity at the top of the photosphere as tabulated, for example, by Minnaert [1953, Table 1] and Allen [1963, p. 172], and $\sigma_T = 8\pi e^4/3m^2c^4 = 6.65 \times 10^{-25}$ cm^2 is the Thomson scattering coefficient;

(3) *Rayleigh scattering by neutral hydrogen.* Here we have

$$\epsilon_{Rs}(\nu) = 2\pi F_\nu \left(\frac{1026}{\lambda}\right)^4 \sigma_T N_{HI}, \tag{4.24}$$

where λ is in Å;

(4) *Emission by the negative hydrogen ion.* This is given by an approximate formula due to Zanstra [see Hiei, 1963], viz.,

$$\epsilon_{H^-}(\nu) = 4\pi B(\nu, T_e) a(\nu, T_e) kT_e N_e N_{HI}, \quad (4.25)$$

where $a(\nu, T_e)$ is the absorption coefficient of H^-.

The total emission at any wavelength λ and height h is the sum of the four contributions enumerated above. At first sight it might seem to be a rather daunting prospect to attempt to derive physical conditions from the corresponding observations since no less than four unknowns are involved, namely T_e, N_e, N_{HI}, and N_{HII}. However, the outlook appears brighter when we examine the height dependence of the various sources of emission. If we ignore the temperature dependence of terms in (4.15) and (4.25), then we see from the above formulae that ϵ_{fb}, ϵ_{es}, ϵ_{Rs}, ϵ_{H^-} are proportional to $N_e N_{HII}$, N_e, N_{HI}, and $N_e N_{HI}$ respectively. Now as h increases N_{HI} will quickly decline to very small values owing to the rapidly increasing ionization of hydrogen (cf. last column of Table 4.6). Hence the two contributions proportional to N_{HI}, namely ϵ_{Rs} and ϵ_{H^-}, will decline relative to ϵ_{fb} and ϵ_{es}. Moreover, at sufficient heights $N_{HII} \simeq N_e$, so that ϵ_{fb} becomes proportional to N_e^2 whereas ϵ_{es} remains proportional to N_e; since N_e also decreases with h, we therefore conclude that ultimately a height is reached at which the electron scattering continuum dominates the other sources of continuous emission. This is very convenient because knowledge of ϵ_{es} then yields N_e directly, as (4.23) shows.

A quantitative evaluation of the four sources of emission carried out by Hiei [1963, cf. Fig. 13-iii] over the height range 200–1000 km confirms the qualitative arguments given above. Hiei's calculations are based on his own homogeneous model of the low chromosphere and are valid for a wavelength of 4117 Å; the results are shown in Fig. 4.6. Note that for $h < 850$ km, emission from negative hydrogen provides the greatest contribution, but that for $h > 850$ km electron scattering is dominant. Thereafter, the relative contribution of the H^- continuum and Rayleigh scattering continue to decline while the Paschen continuum rises at a slower rate than the electron scattering continuum. Of course, the exact shapes of the curves must inevitably depend on the model adopted; nevertheless, the important conclusion that electron scattering dominates the visible continuum at heights greater than a few thousand kilometres is likely to remain valid for any reasonable model.

The above considerations have been used by Kanno [1966] to determine N_e in the *interspicular regions* over a height range $4000 \leqslant h \leqslant 16\,000$ km from eclipse observations of the continuum at $\lambda 4700$. At this wavelength $F_\nu = 2.28 \times 10^{-5}$ erg cm^{-2} sec^{-1} per unit frequency interval and solid angle so that, in the notation of Sections 4.2.6 and 4.2.7, equation (4.23) becomes

$$\epsilon = 9.52 \times 10^{-29} N_e \quad \text{erg cm}^{-3} \text{ sec}^{-1} \quad (4.26)$$

per unit frequency interval. As in the case of the Balmer continuum, consideration must be given to contributions from both spicules and the interspicular medium. However, Kanno concludes that while for $h < 3000$ km the spicules dominate, for $h > 4000$ km the predominant contribution comes from interspicular regions: at this height the latter – in accordance with (4.21) – occupy 97 per cent of the volume, a factor which outweighs the smaller value of N_e there.

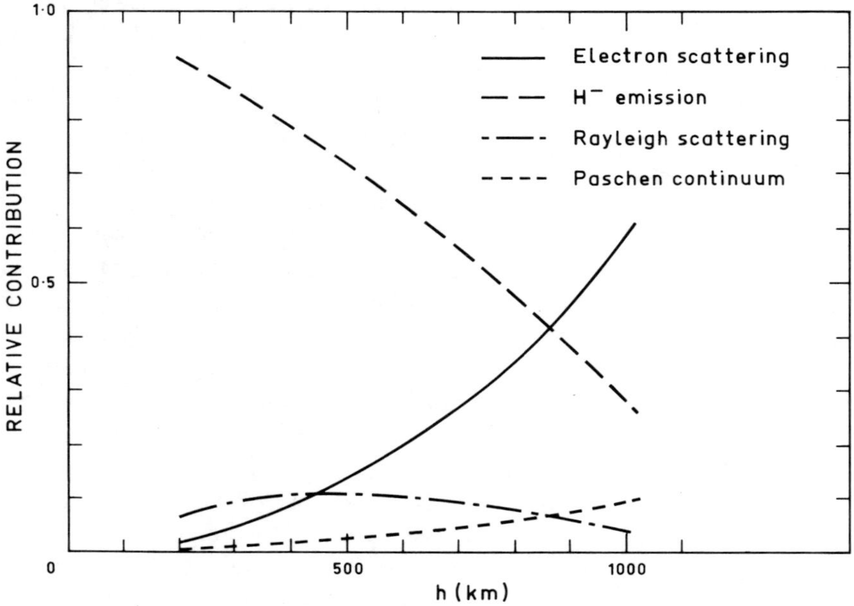

Fig. 4.6. Relative contributions of various sources to the continuous emission of the chromosphere [Hiei, 1963]. Note the dominance of electron scattering as h increases beyond ~ 1000 km. The curves refer to a wavelength of 4117 Å.

The derivation of N_e as a function of h proceeds similarly to the derivation of $N_e^2 T_e^{-3/2}$ from measurements of the Balmer continuum described in the previous section. Again use is made of the 1952 Khartoum eclipse data, but this time the observed integrated emission at λ4700, $[E(h)]_{obs}$, is represented as the sum of *six* rather than only two exponential terms. Kanno's final values for N_e in the interspicular medium are given in the third column of Table 4.22.

One obvious criticism which can be directed against the analyses outlined in this and the previous section is that the results obtained depend intimately on the reliability of the assumed $Q(h)$ which, in turn, depends on the accuracy of the corresponding spicule counts. Indeed, the same criticism applies to virtually all quantitative discussions of inhomogeneous chromospheric models. The observed spicule abundances are probably fairly reliable above $h = 8000$ km. However, in the low chromosphere large and uncertain corrections have to be applied to the raw data (see Section 2.3.6), and therefore the reliability of $Q(h)$ in this region depends to a large extent on the validity of the correction procedure.

4.2.9 FURTHER FREE-BOUND CONTINUA OF HYDROGEN AND HELIUM

In this section we shall briefly consider certain free-bound continua of hydrogen and helium which lie in the ultraviolet region of the spectrum, namely the Lyman continuum of H I at $\lambda < 912$ Å (see Fig. 4.5), the analogous He II continuum at $\lambda < 228$ Å, and the He I continuum at $\lambda < 504$ Å. All of these arise from re-combination at the ground level and all are observable only from outside the Earth's atmosphere. Existing rocket and satellite observations of these continua do not have the precision of the eclipse observations of the Balmer continuum described in Section 4.2.7. In particular, observations of the emission as a function of h are not yet available and for the most part we must be content with measurements of the total flux from the whole solar disk reaching the Earth and its variation with wavelength.

We have already given an expression for the volume emission in the free-bound continua of hydrogen, namely equation (4.15); if we convert the units of this formula to units more appropriate to the instrumentation often employed in extra-terrestrial observations, we arrive at the expression, following Zirin and Dietz [1963],

$$\epsilon(\lambda) = 3.25 \times 10^{-14}\, g_{\text{fb}} n^{-3} N_{\text{e}} N_{\text{H II}} T_{\text{e}}^{-3/2} \lambda^{-1} \exp \frac{h'c}{kT_{\text{e}}}\left(\frac{1}{\lambda_{\text{n}}} - \frac{1}{\lambda}\right) \tag{4.27}$$

photons cm^{-3} sec^{-1} Å^{-1}, where λ is in cm. This formula is valid for the Lyman continuum.

An expression for the corresponding flux received by 1 square centimetre at the Earth's distance is easily derived, namely

$$\phi(\lambda) = \frac{1}{2} \times \frac{1}{4\pi L^2} \int_V \epsilon(\lambda)\, \mathrm{d}V \text{ photons cm}^{-2} \text{ sec}^{-1} \text{ Å}^{-1}, \tag{4.28}$$

where L is the Earth–Sun distance and $\mathrm{d}V$ is an element of volume in the chromospheric shell responsible for the emission, assumed optically thin. The factor 1/2 accounts for photons emitted towards the photosphere and there absorbed.

Since ϕ depends on the volume or thickness of the chromospheric shell, which in the absence of further data is unknown, it is clear that flux measurements by themselves do not form an adequate basis for a detailed discussion of physical conditions. On the other hand, Zirin and Dietz [1963] have developed an ingenious method of determining the 'effective' temperature $\overline{T}_e$ in the shell (assumed uniform) from the *wavelength variation* of the observed flux: since N_e, N_{HII}, and $T_e = \overline{T}_e$ are assumed constant we have from (4.27) and (4.28)

$$\phi(\lambda) = \text{const.} \times \lambda^{-1} \exp\left(-\frac{h'c}{k\overline{T}_e} \cdot \frac{1}{\lambda}\right). \qquad (4.29)$$

Then differentiation with respect to λ leads, after a little algebra, to the following formula for $\overline{T}_e$:

$$\overline{T}_e = \frac{h'c}{k\lambda}\left[\frac{\lambda}{\phi(\lambda)}\frac{\mathrm{d}\phi(\lambda)}{\mathrm{d}\lambda} + 1\right]^{-1} \quad (\lambda \text{ in cm}). \qquad (4.30)$$

Substitution in (4.30) of values of ϕ and $\mathrm{d}\phi/\mathrm{d}\lambda$ derived from rocket observations of the Lyman continuum yields $\overline{T}_e = 7000$ K, with a probable error not exceeding a few thousand degrees. The Lyman continuum thus originates in the low chromosphere, a region where the physical conditions are better determined from the more reliable and extensive data discussed in the previous sections. (Nevertheless, Noyes and Kalkofen [1970] have applied OSO IV measurements of the *centre-limb variation* of the Lyman continuum and other data to deriving a model of the low chromosphere.)

On the other hand, the application of a similar analysis to the He I continuum at $\lambda < 504$ Å by Zirin and Dietz leads to a much higher temperature, $\overline{T}_e = 17\,000$ K. This continuum therefore originates in the middle chromosphere.

Inspection of (4.27) shows that the volume emission in the Lyman continuum $\epsilon(\lambda)$, and hence the observed flux $\phi(\lambda)$, depends on the proton density N_{HII}. Similarly, the fluxes in the λ504 and λ228 continua depend on N_{HeII} and N_{HeIII} respectively. It follows that if we use as a starting point a chromospheric model expressing T_e and N_e as functions of height, the observed fluxes in the various continua can be used to derive information about N_{HII}, N_{HeII}, and N_{HeIII}, and their variation with height. Of course, no great precision can be expected in the height variation since the particle densities are located under the integral sign in (4.28): the most one can do is to derive values consistent with the measured fluxes. This procedure has been followed, in effect, by Tominaga [1967], using the rocket observations of Hinteregger and his group. The particle densities derived by Tominaga partially with the aid of this data are given in Fig. 4.13.

The He I and He II continua originate in the higher levels of the chromosphere where reliable information about the physical conditions is hard to come by. However, fuller exploitation of these continua must await observations from accurately-guided satellites with a resolution comparable to that of eclipse observations of the Balmer continuum. Instead of flux, a more appropriate quantity to measure would be the *surface intensity* of the chromosphere beyond the limb as a function of height, $F(h)$; this is related to the integrated emission normally measured in eclipse observations by the equation

$$F(h) = -\frac{\mathrm{d}E(h)}{\mathrm{d}h}. \tag{4.31}$$

Subsequent analysis of the data would follow procedures similar to those outlined in previous sections.

Disk measurements at various wavelengths in the λ912 and λ504 continua derived from OSO IV observations have been tabulated by Reeves and Parkinson [1970] and Dupree and Reeves [1971]. A spectrometer trace which depicts both continua is reproduced in Fig. 4.12.

4.3 Chromospheric Spectral Line Data

4.3.1 INTRODUCTION

In the following six sections we turn from the continuum data to consider the equally large mass of spectral line data pertaining to the chromosphere currently available. All of the material discussed, which is summarized in Table 4.7, has yielded or is potentially capable of yielding information about physical conditions in the chromosphere; thus it supplements the information derived from the various regions of the continuous spectrum.

Our aim in the sections that follow is once again to outline the types of data available and the astrophysical methods employed in extracting the information. Particular stress is laid on types of data which have already yielded solid information or which by virtue of likely improvements in observational techniques or in methods of analysis seem to have a promising future. As before, little emphasis is given to purely instrumental topics.

4.3.2 INTEGRATED EMISSION AND LOGARITHMIC GRADIENTS OF FLASH SPECTRUM LINES

The emission spectrum of the chromosphere observed at a total eclipse of the Sun is called the *flash spectrum*. It comprises a large number of emission lines and a continuum, present both in the visual region of the spectrum and beyond the Balmer limit ($\lambda < 3646$ Å). The Balmer and visual continua have already been extensively discussed in Sections 4.2.7 and 4.2.8. Moreover, in Section 4.2.6 we have outlined the circumstances of eclipse observations of the flash spectrum lines and continua by the slitless spectrograph technique, defined the terms

TABLE 4.7

Chromospheric Spectral Line Data: Summary

NATURE OF OBSERVATIONS	WAVELENGTH (Å)	ECLIPSE OR NON-ECLIPSE	DISK OR LIMB	HIGH OR LOW CHROMOSPHERE	INFORMATION YIELDED	SECTION
Integrated emission $E(h)$ and logarithmic gradients β of flash spectrum lines	~3200–9100	Eclipse	Limb	High + low	Number density of hydrogen N_H and gas density ρ, assuming knowledge of element abundance A and departures from LTE (homogeneous model)	4.3.2
Half-widths of spicule emission lines	Visible–infrared	Eclipse + non-eclipse	Limb	High + low	In principle, T_e and ξ (non-thermal velocity) in spicules as functions of height; in practice, only crude upper limits are derivable	4.3.3

Central and total intensities of spicule emission lines	Visible – infrared	Non-eclipse	Limb	High + low	T_e and N_e in spicules as functions of height	4.3.4
Disk profiles of Hα and Lyman α lines of H I and of H and K lines of Ca II and Mg II	6563, 1215, 3968, 3934, 2803, 2795	Non-eclipse	Disk	Low	Qualitative information on variation of source function with height	4.3.5
Hα central intensities of bright and dark mottles	6563	Non-eclipse	Disk	Middle + low	T_e and N_e in bright and dark mottles	4.3.6
Integrated fluxes of ultraviolet emission lines	~300–1800	Non-eclipse	Disk (integrated)	High	$A\int N_e^2\, dh$, where A = abundance of element and the integration is performed over the region of line formation; in conjunction with other data, a model of the upper chromosphere	4.3.7

TABLE 4.8

The Strongest Lines in the Flash Spectrum (3200–9100 Å)

WAVELENGTH (Å)	ELEMENT	NAME OF LINE	$E(h) \times 10^{-12}$ (erg sec^{-1} ster^{-1} per 1 cm slice) SPECTR. NO. 1	SPECTR. NO. 2
6562.82	H I	Hα	241.81	278.78
4861.33	H I	Hβ	85.55	101.87
3933.66	Ca II	K	67.28	70.31
3968.47	Ca II	H	54.05	61.25
5875.65	He I	D_3	40.91	44.85
4340.47	H I	Hγ	39.09	40.48
8542.09	Ca II		–	37.74
4101.74	H I	Hδ	18.85	23.25
8662.14	Ca II		–	19.79
3888.65*	He I		10.28	17.36
3889.05*	H I	H_8	10.28	17.36
3970.07	H I	Hε	10.12	14.50
4471.69	He I		4.09	5.28
3835.39	H I		3.71	7.02
7065.19*	He I		–	5.64
7065.72*	He I		–	5.64
8498.02	Ca II		–	4.66
3797.90	H I		2.49	4.60
3349.04*	Ti II		–	3.44
3770.63	H I		1.80	3.33
3685.19	Ti II		1.58	2.49
3750.15	H I		1.43	2.62
3759.29	Ti II		1.43	2.24
3761.32*	Ti II		1.30	2.21
3761.87*	Ti II		–	2.21
3372.80	Ti II		–	2.16
3734.37	H I		1.19	2.07
7771.96	O I		–	1.84
3838.29	Mg I		0.87	1.55

Note: asterisks indicate possible blends; ‘Spectr. No. 1’ and ‘Spectr. No. 2’ refer to heights of the Moon’s limb of 1988 and 1981 km respectively, the corresponding values of $E(h)$ being taken from Tables 3C and 4B of Dunn *et al.* [1968]; wavelengths in the first column 'integrated emission' $E(h)$ and 'logarithmic gradient' β, and obtained an expression for the volume emission coefficient $\epsilon(h)$. In this section we shall be concerned solely with the lines.

The general characteristics of the flash spectrum have been summarized by van de Hulst [1953] and Pagel [1964]. It bears a rough resemblance to the Fraunhofer absorption line spectrum of the solar disk but is not an exact reversal; this is not surprising in view of the differences in physical conditions and conditions of excitation generally between the chromosphere and the photosphere. The flash spectrum contains more high-excitation lines and lines of ionized metals than the Fraunhofer spectrum. For example, certain lines of He I and He II are prominent members whereas on the quiet disk the helium lines are absent, apart from λ10830 of He I, which is faintly present in the chromospheric network and near the limb. The lines which extend to the highest levels, e.g. $h > 10\,000$ km, are the lower members of the Balmer and Paschen series of hydrogen, the H and K lines of Ca II, λ4472 and other lines of He I, and λ4686 of He II. Emission bands due to CH and CN are present at low heights. Different groups of lines behave differently with respect to height, some extending to great heights but others dropping off sharply with height.

Tables 2 and 4 of van de Hulst's review list published compilations and photometric investigations of flash spectrum lines carried out by various workers during the period 1900–1952, while Pagel's Table I updates the list to 1958. The most extensive modern compilation is that of Dunn *et al.* [1968, Table 2], which covers the wavelength region 3200–9100 Å and includes some 3500 lines; Tables 3 and 4 of this publication give values of $E(h)$ for h ranging up to 4500 km for the lines. This huge amount of data is the product of a well-equipped expedition to the 1962 eclipse at Lae, New Guinea, organized jointly by the High Altitude and Sacramento Peak Observatories and the U.S. National Bureau of Standards. The same eclipse was also successfully observed by a Japanese expedition and Kurokawa *et al.* [1969] have compared the Japanese and American measurements of the intensities and gradients of a number of lines. A catalogue of approximately 11 500 emission lines in the region 3040–9266 Å observed *outside* of an eclipse at the Kitt Peak Observatory has been published by Pierce [1968].

Table 4.8 lists the twenty-nine strongest lines in the flash spectrum in order of strength, culled from the data given in Tables 3C and 4B of Dunn *et al.*'s paper; the values of $E(h)$ refer to a height of about 2000 km. Note that the eighteen strongest lines belong to just three elements, namely H I, He I, and Ca II; only towards the end of the list do lines of other elements appear.

Table 4.9 gives values of the logarithmic gradient, i.e. the quantity

$$\beta = -\frac{\mathrm{d}}{\mathrm{d}h} \ln E(h), \qquad (4.14)$$

TABLE 4.9

Logarithmic Gradients and Scale Heights of Flash Spectrum Lines

		$\beta \times 10^8$ (cm^{-1})			
LINE	h (km)	HOUTGAST	THOMAS & ATHAY	ZIRKER	SCALE HT. (km)
Hα	2000	0.70	0.35	–	1400, 2900
	5000	1.40	–	–	710
	'Max. ht.'	–	1.30	–	770
Hβ	2000	0.75	0.87	–	1300, 1100
	5000	1.05	–	–	950
	'Max. ht.'	–	1.60	–	620
Hγ	2000	1.00	1.20	–	1000, 830
	5000	1.00	–	–	1000
	'Max. ht.'	–	1.80	–	560
Hδ	2000	1.10	1.36	–	910, 740
	5000	1.25	–	–	800
	'Max. ht.'	–	1.80	–	560

D_3 (He I)	2000	0.75	–	–	1300
	5000	0.80	–	–	1250
λ4472 (He I)	2000	0.50	–	–	2000
	4000	0.90	–	–	1100
H (Ca II)	2000	0.30	–	–	3300
	12 000	1.00	–	–	1000
	All hts.	–	–	0.68 ± 0.02	1500
K (Ca II)	2000	0.25	–	–	4000
	12 000	1.00	–	–	1000
λ7772 (O I)	All hts.	–	–	1.57 ± 0.07	640
D_2 (Na I)	All hts.	–	–	2.12 ± 0.11	470
D_1 (Na I)	All hts.	–	–	2.26 ± 0.07	440
λ3838 (Mg I)	All hts.	–	–	1.74 ± 0.22	570
λ3685 (Ti II)	All hts.	–	–	1.85	540
Other neutral and singly-ionized metals	All hts.	–	–	0.7–4.6	1400–220

for several of the strong lines of Table 4.8 and also for a few selected weaker lines according to measurements of Zirker [1958], Thomas and Athay [1961, Table A-3], and Houtgast [1962]. The last column gives the *emission scale height*, $1/\beta$, expressed in kilometres: if $E(h)$ can be approximated by a single exponential term, then this quantity represents a constant height over which the volume emission $\epsilon(h)$ changes by a factor of e (cf. equation 4.13). This is indeed the case for most of the moderately strong lines of neutral and singly-ionized metals and oxygen [Zirker, 1958]. On the other hand, β does not remain constant with height for some of the stronger lines, as Table 4.9 shows: for example, for the Hα line β doubles over the range 2000–5000 km.

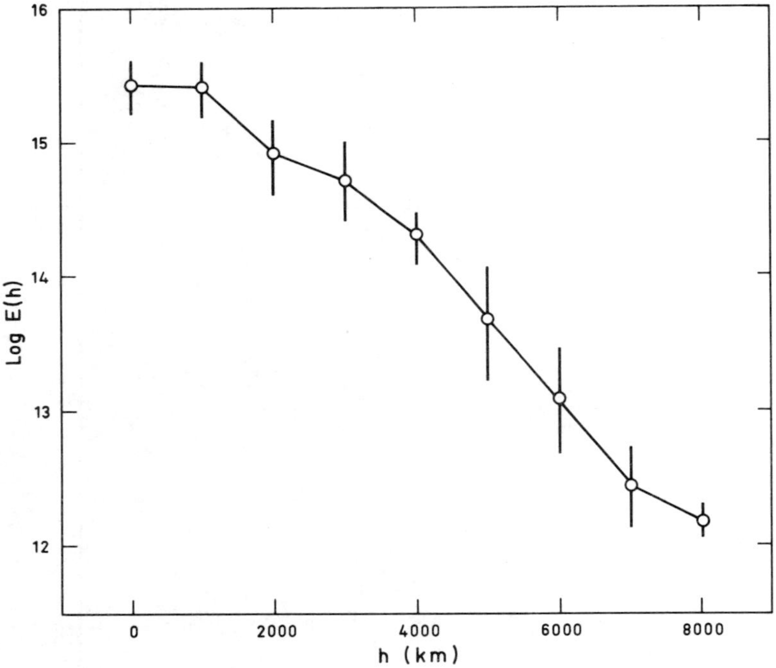

Fig. 4.7. Integrated emission of the Hα line measured at the 1954 Gotland eclipse [Houtgast, 1962]. Ordinates: $E(h)$ in erg sec^{-1} ster^{-1} per 1 cm slice of chromosphere; abscissae: height of Moon's limb in kilometres. Vertical lines indicate observational scatter.

Another way of demonstrating the variation in β for certain lines is by plotting log $E(h)$ against h. This is shown for Hα in Fig. 4.7, which is taken from Houtgast [1962, Fig. 14]. The steepening of the curve as h increases beyond 3000 km is quite noticeable. Figs. 14–16 of Houtgast's paper show additional log $E(h)$ curves for Hβ–H$_8$, λ5876 and λ4472 of He I, and the H and K lines of

Ca II. Similar plots are given by Zirin and Dietz [1963, Fig. 1] for Hβ, D_3, λ4686 of He II, λ3827 of Fe I, and the Balmer and λ4700 continua, and by Kurokawa *et al.* [1969, Figs. 3–6] for Hα and D_3. The variation in β for many of these lines is almost certainly due partly to the influence of *self-absorption*: in such cases the simple theory given in Section 4.2.6 for deriving the volume emission $\epsilon(h)$ from the observed integrated emission $E(h)$ is no longer valid – except possibly at great heights.

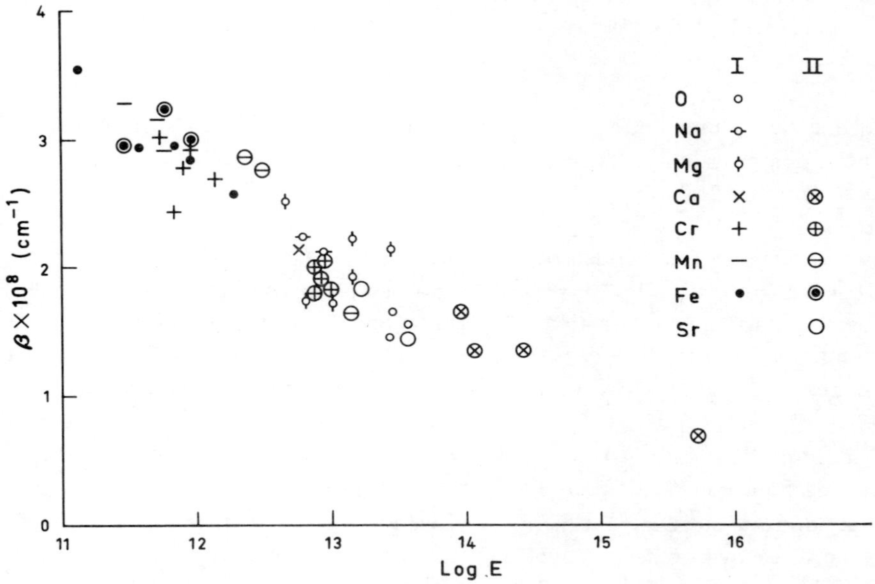

Fig. 4.8. Logarithmic emission gradients β of neutral (I) and singly-ionized (II) metals as a function of integrated emission at h = 2000 km [Zirker, 1958]. The units of E are erg sec^{-1} per 1 cm slice of chromosphere.

What determines the magnitude of β? When β is plotted against ionization potential for a number of neutral and singly-ionized metals there appears to be no correlation whatsoever [Zirker, 1958, Fig. 1]. On the other hand, when β is plotted against the value of log $E(h)$ at some fixed height a strong correlation appears in the sense that β decreases with increasing line strength. This is shown in Fig. 4.8, which is taken from Zirker. This correlation suggests, as Zirker remarks, that the primary factor controlling the magnitude of β is self-absorption rather than the conditions of excitation, although the latter largely determine the line strength. Zirker accordingly takes the 'true' value of β for a given line to be the value measured at the 'plate limit', i.e. at the height where the chromospheric arc fades into invisibility.

The quantity of data that has been collected for the various flash spectrum lines is enormous but it must be admitted that so far, in contrast to the corresponding eclipse data for the continuum, it has yielded little solid information about physical conditions in the chromosphere. This is understandable if we consider exactly what can be derived from measurement of a given line.

Suppose the line in question involves a transition between an upper state k and a lower state l and assume (*a*) that self-absorption is absent, and (*b*) that the integrated emission can be represented by a single exponential term. Then from (4.12) and (4.13) we have

$$\epsilon_{\mathrm{kl}}(h) = [\beta^{3/2}/(2\pi R_{\odot})^{1/2}]\ E_{\mathrm{kl}}(h), \qquad (4.32)$$

giving the volume emission as a known function of h. Moreover, if A_{kl} is the transition probability, which we may assume to be known, then

$$\epsilon_{\mathrm{kl}} = h'\nu_{\mathrm{kl}}A_{\mathrm{kl}}N_{\mathrm{k}} \quad \mathrm{erg\ cm^{-3}\ sec^{-1}}, \qquad (4.33)^{3}$$

where N_{k} is the number density of atoms in the state k.

Combining (4.32) and (4.33) we see immediately that, without further assumptions, the information obtainable from a single flash spectrum line is limited to the number density of the upper level N_{k} as a function of h. Of course, if in addition we knew the excitation and ionization temperatures T_{ex} and T_{ion} from other sources, then we could establish the total populations of the ionization stage of which k is a member and of the other ionization stages (using Boltzmann's and Saha's equations) and hence the total number density of the atom in question. Furthermore, if we knew the abundance of the element relative to hydrogen, we could also calculate the number density of hydrogen and hence the gas density ρ. Alternatively, knowing the hydrogen density from some other source, we could calculate the relative abundance of the element.

The main difficulty in following such a procedure is that, under chromospheric conditions, local thermodynamic equilibrium does not obtain and hence

$$T_{\mathrm{ex}} \neq T_{\mathrm{ion}} \neq T_{\mathrm{e}}.$$

Consequently, T_{ex} and T_{ion} or, alternatively, the b_{k} factors used by Thomas and Athay [1961, p. 80] to describe departures from LTE, would ideally have to be calculated for each atomic level by detailed consideration of its mechanism of excitation and de-excitation to and from other levels and the continuum (cf. Section 4.3.4). Unfortunately, such a programme can be followed with the necessary rigour only with great difficulty. Accordingly, existing attempts to derive useful information from flash spectrum lines often involve a complex set

[3] The consistency of the units in (4.32) and (4.33) is ensured by expressing the measured values of E_{kl} in erg sec^{-1} per 1 cm slice.

of assumptions which inevitably affect the reliability of the information derived. The following references, which have been selected specifically to illustrate some of the principles considered above, provide a useful cross-section of such attempts: de Jager and de Groot [1957] and Athay and Johnson [1960] (non-LTE calculations of helium line strengths); Zirker [1958] (self-absorption and evaluation of the density gradient); Thomas and Athay [1961, Chapter 7] (use of helium line data to improve a chromospheric model); Gulyaev *et al.* [1963] (determination of number density of hydrogen); and Pecker and Pottasch [1969] (determination of certain metal abundances).

4.3.3 PROFILES AND HALF-WIDTHS OF SPICULE EMISSION LINES

Profiles of spicule emission lines, photographed with a slit spectrograph either inside or outside of an eclipse, provide a potential means of determining two parameters: (*a*) the kinetic temperature of the spicules, T_{kin}, and (*b*) the magnitude of any non-thermal velocities which may be present along the line-of-sight, i.e. parallel to the solar surface. So far the application of this method has not proved very rewarding, but its principle is simple. In the absence of self-absorption or other sources of broadening the observed profile mimics the Gaussian distribution of the line-of-sight component of the atomic velocities and is given by

$$I(\lambda)\, d\lambda = I_0 \exp\left[-\frac{c^2(\lambda-\lambda_0)^2}{\alpha^2\lambda_0^2}\right] d\lambda, \qquad (4.34)$$

where I_0 is the intensity at the centre of the line, λ_0 is the wavelength of the line centre, and α is the most probable speed [Aller, 1963, pp. 108–111].

If non-thermal motions are absent, then

$$\alpha = \sqrt{\frac{2kT_{\text{kin}}}{\mu m_{\text{H}}}} = \sqrt{\frac{2kT_{\text{e}}}{\mu m_{\text{H}}}}, \qquad (4.35)$$

the last step being justified by the close equality between the kinetic and electron temperatures under chromospheric conditions [Pagel, 1964, p. 268]. Non-thermal motions are customarily included by assuming the presence of small clouds of gas whose velocity distribution in the line-of-sight is also Gaussian, i.e. $N(v) \propto \exp(-v^2/\xi^2)$. With this assumption, the profile remains Gaussian but α^2 in (4.34) becomes

$$\alpha^2 = \frac{2kT_{\text{e}}}{\mu m_{\text{H}}} + \xi^2 \qquad (4.36)$$

[cf. Weatherburn, 1947, pp. 61–62; Unsöld, 1955, p. 286; and Aller, 1963, pp. 387–388, 503–504].

In passing, it should be recognized that although (4.36) has been much used by stellar and solar astronomers in discussing 'turbulence' in the atmospheres of the Sun and stars, there is actually no observational or theoretical foundation for the assumption of a Gaussian distribution for the material motions. The quantity ξ is therefore best regarded as a parameter indicating the *presence* of non-thermal motions of some kind along the line-of-sight but not, in the absence of a knowledge of the nature of the velocity field, specifying the *magnitude* of the velocities.

A convenient property of the profile to measure is the half-width, $\Delta\lambda_{1/2}$, i.e. the wavelength difference between the points whose intensity is $\frac{1}{2} I_0$. From (4.34) and (4.36) we obtain

$$\Delta\lambda_{1/2} = 2\sqrt{\ln 2} \cdot \frac{\lambda_0}{c} \cdot \sqrt{\frac{2kT_e}{\mu m_H} + \xi^2}, \tag{4.37}$$

which reduces to

$$\Delta\lambda_{1/2} = 7.16 \times 10^{-7}\, \lambda_0 \sqrt{\frac{T_e}{\mu}}$$

when $\xi = 0$. In principle, therefore, measurement of $\Delta\lambda_{1/2}$ for a number of different elements (i.e. different μ) provides a simple means of determining both T_e and ξ with the aid of (4.37).

However, certain difficulties arise when an attempt is made to apply this procedure to the spicule emission lines. Firstly, in the case of non-eclipse observations of lines other than helium, the observed profile must be corrected for scattering of the corresponding Fraunhofer line from the disk spectrum. Secondly, the later members of the Balmer series may be expected to show some broadening due to Stark effect [Clube, 1958]. Thirdly, as we have seen in the previous section, self-absorption is likely to be present in many of the lines, particularly in the strong lines at low heights. Finally, in carrying out the observations it is difficult, owing to atmospheric seeing, to avoid imaging clumps of unresolved spicules on the spectrograph slit [Zirker, 1962; Livshits, 1966]; in fact, regardless of the resolution, it is impossible to avoid such superimposition if the spicules are partially transparent over the Doppler core of the line under study. Since each spicule is likely to have its own outward or inward velocity (Section 2.4.2) and its own inclination to the line-of-sight (Section 2.3.7) it is evident that the net effect, like the other effects listed above, is to broaden the line profile. Accordingly, values of T_e and ξ determined from spicule line profiles represent *upper limits* to the true values.

Observations of emission line profiles and measurements of half-widths (in some cases as a function of h) have been made by numerous authors, including Smith [1957, Hα]; Treanor [1957, D_3]; Clube [1958, Hα and D_3]; Michard

[1958, Hα]; Zirker [1962, Hα, Hβ, D_3, H and K of Ca II]; Athay and Bessey [1964, Hα, D_3, H and K]; Nikol'skii [1965, Hα and D_3]; and Livshits [1966, Balmer lines, D_3- and λ10830 of He I, H of Ca II]. Some of these results purport to refer to individual spicules but in other cases the resolution employed was manifestly inadequate. Half-widths measured by these and other workers have been carefully collected and compared by Beckers [1968, Tables VII–X]. Table 4.10 summarizes these measurements, but before considering the application of (4.37) to them it is instructive to discuss certain qualitative aspects of the profiles obtained.

TABLE 4.10

Half-Widths of Spicule Emission Lines [Beckers, 1968]

LINE	h (km)	$\Delta\lambda_{1/2}/\lambda_0$	T_e (K)
Hα	6000	1.8×10^{-4}	63 000
Hβ	6000	1.6	50 000
Hγ	4500	1.65	53 000
Hδ	4500	1.4	38 000
Hε	3000	1.4	38 000
H_8	2000	1.2	28 000
D_3 (He I)	5500	0.95	70 000
λ10830 (He I)	5000	1.0	78 000
λ3889 (He I)	2000	0.9	63 000
λλ7772/4 (O I)	2500	0.44	61 000
H (Ca II)	5000	1.4	1 500 000
K (Ca II)	5000	1.5	1 800 000
λ8498 (Ca II)	5000	1.1	950 000
λ8542 (Ca II)	5000	1.5	1 800 000

Note: the first three columns are taken from Beckers [1968, Table XI]; values of T_e are calculated from (4.37) assuming $\xi = 0$.

Many workers have asserted that with the exception of the strong lines at low heights, the profiles are Gaussian. However, this conclusion has been contradicted by Livshits [1966]. He finds that the profiles show great variety, some being flat-topped or notched by several peaks. Moreover, the line widths vary markedly from one spicule to another; the largest values are found for *groups* of spicules. Therefore, it seems likely that the discrepancies between different observers can be largely attributed to differences in spatial resolution and, accordingly, once again we have to conclude that any derived values of T_e or ξ represent upper limits.

The strongest lines show conspicuous evidence of self-absorption, particularly at low heights. For example, Hα shows a well-marked self-reversal for $h \leqslant 5000$ km; this is illustrated in Fig. 4.9, which is due to Smith [1957]. Beyond this level the profiles become narrower and somewhat flat-topped,

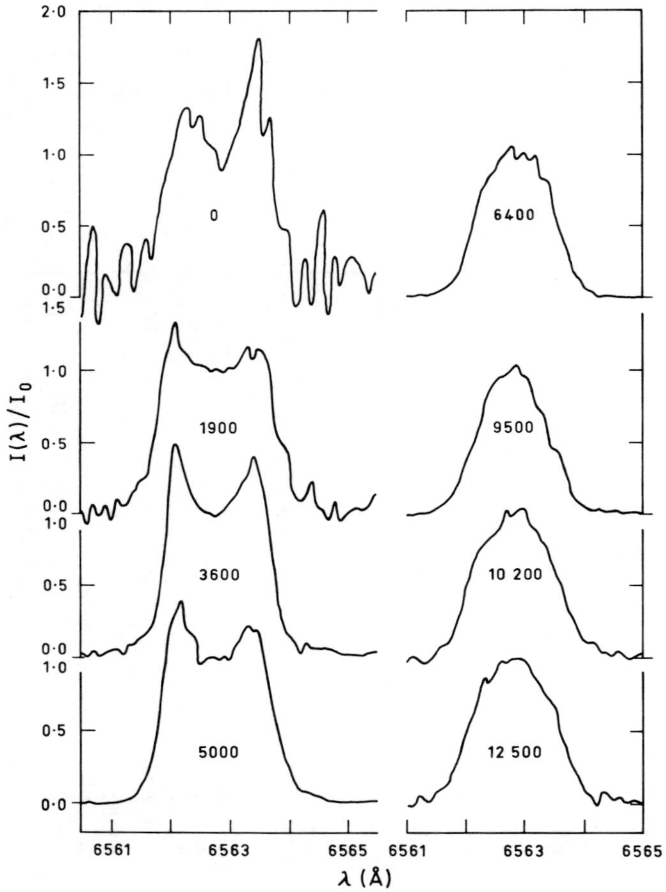

Fig. 4.9. Chromospheric Hα emission profiles as a function of height [Smith, 1957]. The numbers within each profile give the height in km. The spatial resolution of these observations is estimated as 2–4″ of arc (1450–2900 km). The zero point of the height scale is uncertain and could be in error by up to 1500 km.

becoming more and more Gaussian as h increases but never quite attaining a Gaussian shape. A similar behaviour is also characteristic of Hβ and the H and K lines of Ca II [Zirker, 1962]; on the other hand, D_3 remains Gaussian throughout.

The fourth column of Table 4.10 gives the value of T_e for lines of the various elements, calculated from (4.37) under the assumption that non-thermal velocities are absent. On examining these figures it is immediately apparent that the observed half-widths are inconsistent with this equation. In the first place, the derived temperatures for the hydrogen lines extend over the uncomfortably large range of 28 000–63 000 K, so that no possible pair of values (T_e, ξ) can possibly reproduce them all. Secondly, to account for the widths of the Ca II lines we evidently have to admit the existence of a large non-thermal velocity ξ; however, this is inconsistent with the hydrogen, helium, and oxygen data, which is fitted – albeit very roughly – by $T_e = 60\,000–70\,000$ K and $\xi = 0$. Thomas and Athay [1961, pp. 73–77] have sought to explain the discrepancy by postulating a non-thermal mechanism special to the Ca II ions and not shared by the neutral hydrogen and helium atoms; however, this approach seems artificial in view of the probable presence of other broadening mechanisms, e.g. self-absorption.

We have to conclude, therefore, that existing observations of spicule line profiles fail to provide a reliable guide to ξ and give only a rough upper limit to T_e. Any future observations should evidently concentrate on *weak* lines of *isolated* spicules at heights where self-absorption can be neglected, preferably during a total eclipse.

4.3.4 CENTRAL INTENSITIES AND TOTAL INTENSITIES OF SPICULE EMISSION LINES

In Section 4.3.2 we pointed out that theoretical complications arising from departures from local thermodynamic equilibrium (LTE) represent a major obstacle facing attempts to derive useful information from lines in the flash spectrum. Exactly the same difficulty applies, of course, to the interpretation of the emission lines of individual spicules. Nevertheless, in certain cases the formidable theoretical problems involved have been largely overcome, and in this section we shall outline derivations by Giovanelli [1967a,b] and Beckers [1968] of T_e and N_e in spicules from measured central or total intensities of Hα, Hβ, and H_8; $\lambda\lambda$3889, 10830, and D_3 of He I; and the H and K lines of Ca II.

The primary quantity given by the observational data, as in the case of the flash spectrum lines, is the population (number density) of the upper state of the transition integrated over the volume of the emitting region, i.e. the quantity $\int (N_k)_{obs}\, dV$; the volume V can be evaluated from the known width of a spicule (Section 2.3.3). The task facing a non-LTE analysis is therefore to calculate a *theoretical* population, $(N_k)_{theor}$, on the basis of assumed values of T_e and N_e – without applying the Boltzmann–Saha equation – and then by comparing the result with $(N_k)_{obs}$ to obtain refined values of T_e and N_e. Not surprisingly, data from several emission lines are required to uniquely determine T_e and N_e.

The starting point of the analysis is the assumption of a *statistically steady state*, i.e. the assumption that the rate of departure from a given state k equals the rate of arrival at that state [Thomas and Athay, 1961, Chapter 4; Jefferies, 1968, Chapter 6].[4] Then if $P_{kl}N_k$ represents the transition rate from state k to state l, we have

$$N_k \sum_{l=1}^{\infty} P_{kl} = \sum_{l=1}^{\infty} N_l P_{lk}, \quad (l \neq k; k = 1 \ldots \infty) \tag{4.38}$$

the summation being symbolically extended over all states, including the continuum. The transition coefficient P_{kl} can be expressed as the sum of radiative and collisional terms, viz.,

$$P_{kl} = R_{kl} + C_{kl} \cdot N_e,$$

so that (4.38) becomes

$$N_k \sum_{l=1}^{\infty} (R_{kl} + C_{kl} \cdot N_e) = \sum_{l=1}^{\infty} N_l(R_{lk} + C_{lk} \cdot N_e). \quad (l \neq k; k = 1 \ldots \infty) \tag{4.39}$$

The coefficients C_{kl} in (4.39) are functions only of T_e and in general are known from calculation or from laboratory data, although not always with great precision. When k represents a discrete upper level, R_{kl} is simply the Einstein coefficient A_{kl} (spontaneous transition probability), if we ignore stimulated emission. When k represents a lower level, R_{kl} depends both on the Einstein coefficient B_{kl} (transition probability for absorption) and on the *radiation field surrounding the absorbing atom.* As in the case of the C_{kl}, the coefficients A_{kl} and B_{kl} are usually known. The system of equations (4.39), together with an auxiliary relation expressing the total number density of the element in question,

$$N = \sum_{l=1}^{\infty} N_l,$$

suffices to determine the population densities N_k of all the states in terms of N_e, T_e, certain known constants, and the radiation field.[5]

[4] This assumption is less restrictive than the assumption of LTE, where each individual process is balanced by its corresponding inverse process. Nevertheless, its validity in the case of spicules has been questioned by Beckers [1968] on the ground that the time required for some of the transitions to take place is comparable to the time during which the spicule may change significantly.

[5] Thomas and Athay [1961, p. 80] introduce a parameter b_k into (4.39) to represent departures from LTE. This quantity is equal to the ratio N_k/N_k^*, where N_k is the actual population and N_k^* is the population that would obtain in the *absence* of departures from LTE with the same values of N_e, T_e, and N^+, the (actual) population of the next higher stage of ionization. The use of the parameters b_k rather than the population densities themselves has the advantage of pinpointing the various terms in the equations which depart significantly from LTE.

Several difficulties arise when an attempt is made to apply this procedure to a practical problem [Giovanelli, 1967b]. In the first place, the number of energy levels is infinite, so as a first approximation one has to consider only a finite number of levels (e.g., three) plus the continuum. This is done by lumping the highly excited states, which are populated approximately as in LTE, with the continuum. (This procedure is satisfactory for hydrogen and helium but not for the H and K lines.) A second difficulty is the specification of the radiation field inside the spicule: if the spicule is optically thin in the spectral line in question no problem arises since the radiation field is simply the photospheric radiation diluted by the appropriate factor (0.5). On the other hand, if it is optically thick the field can be established only by solving the equation of radiative transfer within the spicule. Since this itself presupposes a knowledge of the populations of the various excited states, it is clear that some of the coefficients in the system of equations (4.39) involve the very parameters that we wish to determine. Moreover, in advance it may be difficult to decide whether the spicule is optically thick or thin. Evidently, therefore, an iterative process generally has to be employed.

The details of the procedure have been described by Giovanelli [1967a,b] and Beckers [1968]. Giovanelli considers first the Hα line of spicules for $h > 6000$ km. The measurements of Dunn [1960] and Zirker [1962] give respectively the total intensity (i.e. integrated over the profile) and the central intensity, both at various heights; from these data $(N_3)_{obs}$ may be derived, assuming a spicule thickness of 1000 km. The values of $(N_3)_{obs}$ are given as a function of h in Giovanelli's [1967b] Table 2. The next step is to calculate N_3 for various combinations of N_e and T_e. These values, which are shown in Giovanelli's Table 3, were calculated on the basis of a structure thickness of 2000 km; however, according to Giovanelli, the calculations are not sensitive to this parameter. Next, pairs of values (T_e, N_e) are plotted for which

$$(N_3)_{calc} = (N_3)_{obs}.$$

The resulting curve (Giovanelli's Fig. 2) accordingly represents physical conditions in the spicule consistent with the observed Hα intensities.

The final step is to plot similar curves for D_3 and the H and K lines. Where all the curves intersect, the values of T_e and N_e represent the unique combination which, at the height in question, correctly reproduces the observed spectral line intensities. Giovanelli concludes that, giving the greatest weight to the Hα analysis, the observations at $h = 6000$ km are reasonably well satisfied by $N_e = 3 \times 10^{10}$ cm^{-3} and $T_e = 20\,000$–$30\,000$ K. These figures may be compared with values derived by Beckers [1968] for the same height, namely $N_e = 1.15 \times 10^{11}$ cm^{-3} and $T_e = 14\,000$ K (see below).

Beckers follows a basically similar procedure, although his method for the simultaneous solution of (4.39) and the equation of radiative transfer differs from Giovanelli's. Moreover, he assumes a spicule thickness of 815 km, this being the mean value, corrected for instrumental profile, determined by Dunn [1960] from high-quality spicule photographs taken at Sacramento Peak

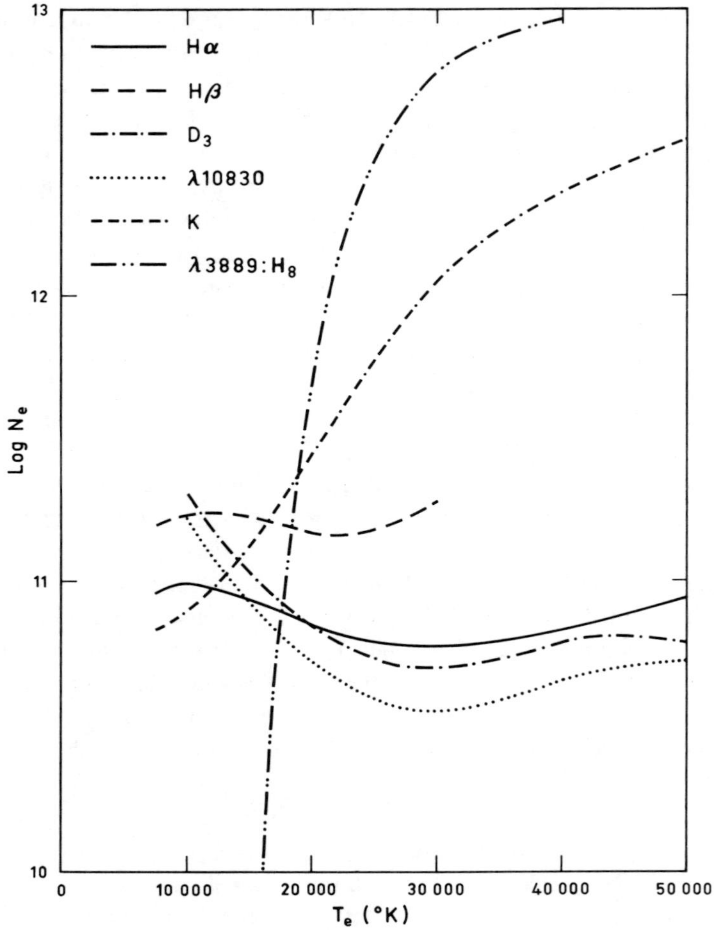

Fig. 4.10. (log N_e, T_e) combinations that reproduce the observed total intensities of various spicule emission lines [Beckers, 1968]. This diagram refers to a height h = 6000 km. The values of N_e and T_e derived from this set of curves are 1.15×10^{11} cm^{-3} and 14 000 K respectively.

Observatory (cf. Section 2.3.3). Beckers carries out calculations for Hα, Hβ, and H_8 of H I; λλ3889, 10830, and D_3 of He I; and λλ8498, 8542 and K of Ca II, using the measured total intensities of the various lines; his data cover a height range of 2000–10 000 km. Figure 4.10, which is taken from his paper, shows plots of pairs of values (log N_e, T_e) that reproduce the observed intensities of Hα, Hβ, D_3, λ10830, and K, and also the intensity ratio λ3889 (He I): H_8, all at $h = 6000$ km. Note that Hα and D_3 serve to establish N_e almost independently of T_e, whereas the K line depends strongly on both N_e and T_e. Taken together the six curves enclose a relatively small area that effectively establishes the values of N_e and T_e at the height in question. From similar sets of curves referring to other heights, Beckers is finally led to the values of N_e and T_e set out in Table 4.11.

TABLE 4.11

T_e and N_e in Spicules Derived from their Emission Line Intensities [Beckers, 1968]

h (km)	T_e (K)	N_e (cm^{-3})
2000	(17 000)	(2.2×10^{11})
4000	17 000	2.0
6000	14 000	1.15
8000	15 000	6.5×10^{10}
10 000	(15 000)	(3.4)

Note: figures in parentheses are uncertain.

Table 4.11 forms the basis of Beckers' spicule model, which is presented in full in Tables 4.19–4.20. In view of the near-constancy of the figures in the second column of Table 4.11, T_e is taken in the model to have a constant value of 16 000 K above $h = 2000$ km. Below this level it is assumed to decline linearly with height to the value 4700 K at $h = 0$. Log N_e is assumed to follow a linear relation over the whole range 0–20 000 km, the line being fitted to the figures in Table 4.11 over the range 2000–10 000 km.

4.3.5 DISK PROFILES OF CHROMOSPHERIC LINES: Hα AND LYMAN α OF H I AND THE H AND K LINES OF Ca II AND Mg II

Turning now to *disk* observations, we shall apply the term 'chromospheric line' to any strong spectral line which reveals substantial evidence of chromospheric structure on spectroheliograms made at or near the line centre (cf. Section 3.3). The principal chromospheric lines and the ones most frequently subjected to theoretical analysis are Hα and Lyman α (λ1215), the H and K lines of Ca II, and the analogous H and K lines of Mg II ($\lambda\lambda$2803, 2795); the Mg II pair are nearly as strong as the corresponding Ca II pair, the strongest absorption lines in the Fraunhofer spectrum of the Sun. It is evident that the physical conditions in the chromosphere must influence the observed profiles of these lines; conversely, measurements of the profiles, if adequately analysed, might be expected to yield information about the prevailing physical conditions.[6]

Observed with *low* spatial resolution, profiles of the lines referred to above fall into three categories: (*a*) a bell-shaped absorption profile characteristic of Hα and the other Balmer lines [White, 1962, Figs. 6–8] ; (*b*) a broad absorption line with a self-reversed emission peak at its centre characteristic of the H and K lines of Ca II and Mg II [Athay and Skumanich, 1968, Figs. 1 and 2]; and (*c*) a self-reversed emission profile standing well above the weak continuum characteristic of Lyman α and Lyman β [Pottasch, 1964, Fig. 11] .

Current attempts to analyse the profiles of these lines have mostly been directed at explaining their broad characteristics in terms of certain rough properties of assumed atomic and chromospheric models [see, for example, Thomas and Athay, 1961, pp. 120–133 and Chapter 9; Obridko, 1963; Engvold, 1966; Athay and Skumanich, 1968; Cuny, 1968; Jefferies, 1968, pp. 149–162] . The complexity of the subject forces the use of drastic and sometimes unrealistic simplifications, e.g. the assumption of a two-level atom. Moreover, the non-LTE theory of line formation is not yet fully established as a standard working technique in spite of rapid progress in recent years. Finally, in trying to bring about agreement between theory and observation it is sometimes necessary to make arbitrary judgements concerning which parameters to vary.

Despite these difficulties, some success has been achieved in explaining qualitative features of the profiles: for example, the self-reversed emission cores of the H and K lines have been attributed to the influence of a source function increasing with height in the chromosphere. On the other hand, such analyses have so far yielded little or no solid information about the prevailing physical conditions. However, further progress in the non-LTE theory of line formation may make the analysis of chromospheric line profiles an important means of determining the physical conditions in the years to come.

In conclusion, it should be pointed out that the importance of the chromospheric H and K lines extends well beyond the field of solar physics. The observation of similar central emission peaks in these lines in the spectra of late-type stars supplies direct evidence of the existence of chromospheres in such stars. Moreover, it has been established that the widths of the central peaks bear a close relationship to absolute magnitude, independently of spectral type (Wilson–Bappu effect). References to discussions by O. C. Wilson and his

[6] In passing, we should mention that it is now widely believed that the central cores of many Fraunhofer lines much weaker than those referred to above – even down to Rowland Intensity <10 in some instances [Thomas and Athay, 1961, p. 353] – are formed in the low chromosphere. Zirker [1958] has tabulated the chromospheric height corresponding to radial optical depth unity at the line centre for the lines of a number of neutral and singly-ionized metals.

co-workers of the stellar H and K lines have been given by Jefferies [1968, p. 154].

4.3.6 Hα CENTRAL INTENSITIES OF BRIGHT AND DARK MOTTLES

Continuing our discussion of disk observations of the chromosphere we may ask ourselves whether it is possible to determine physical conditions in the bright and dark mottles by measuring their emission at the centre of Hα or,

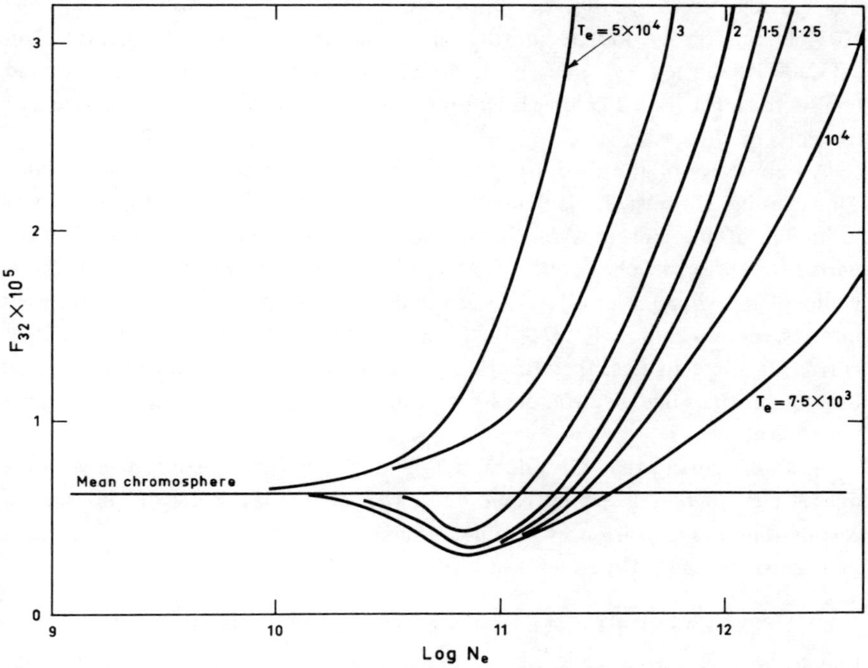

Fig. 4.11. Calculated surface intensity F_{32} at Hα line centre of a model structure illuminated by the surrounding chromosphere and having uniform T_e and N_e and a thickness of 2000 km [Giovanelli, 1967b]. Portions of the curves above and below the horizontal line represent values of T_e and N_e consistent with bright and dark mottles respectively. The units of F_{32} are erg cm^{-2} sec^{-1} ster^{-1} per unit frequency interval.

equivalently, their contrast against the background. Giovanelli [1967b] has provided an affirmative answer to this question, using non-LTE methods of calculation similar to those he employed in his analysis of spicule intensities (Section 4.3.4).

Figure 4.11 shows the results of the calculations: the ordinate F_{32} represents the surface intensity at the centre of Hα of a model structure illuminated by

Balmer and Lyman line and continua radiation from the surrounding chromosphere and having a thickness of 2000 km and a given N_e and T_e, assumed uniform. Proper allowance has been made for the transmission and (non-coherent) scattering of Hα radiation by the structure. The horizontal line represents the brightness of the 'mean' chromosphere.

Even without making use of contrast measurements, we can make several useful deductions from Fig. 4.11. Firstly, for a dark mottle T_e must be less than ~20 000 K, regardless of the value of N_e; the lower the value of T_e, the darker the mottle. Secondly, $\log N_e$ for a dark mottle lies roughly in the range 10.5–11.5. Thirdly, bright mottles are characterized by a T_e greater than ~25 000 K, the $\log N_e$ values being about the same as for dark mottles. And finally, for $\log N_e < 10$ neither bright nor dark mottles are possible, regardless of the value of T_e.

A knowledge of the contrast enables one to refine these estimates somewhat. Thus, using contrast measurements of Beckers and Bhavilai together with estimates of the opacities of the mottles, Giovanelli has concluded that the darkest mottles are characterized by $T_e < 10\,000$ K and $N_e \simeq 2 \times 10^{11}$ cm^{-3}, while for less opaque mottles $T_e < 20\,000$ K and $N_e \simeq 10^{11}$ cm^{-3}. For bright mottles, he concludes that T_e must be $\gtrsim 25\,000$ K and $N_e \gtrsim 5 \times 10^{10}–10^{11}$ cm^{-3}. On these figures, therefore, it appears that bright and dark mottles could have much the same electron density but that the bright mottles have the higher temperature.

It is noteworthy that the values of T_e and N_e for dark mottles agree with the figures for *spicules* given in Table 4.11; therefore, the results of Giovanelli's calculations are consistent with the conclusion (Section 3.2.6) that the dark mottles are to be identified with spicules.

4.3.7 INTEGRATED FLUXES OF ULTRAVIOLET EMISSION LINES

Radiation of wavelengths shorter than about 3000 Å is absorbed by the Earth's atmosphere and our current knowledge of the ultraviolet spectrum of the Sun is accordingly derived entirely from observations made from rockets and satellites. Rockets have made a large contribution to this field ever since they first became available for scientific purposes shortly after World War II (cf. Plate 4.1). In recent years spectroheliograms of the solar disk have been obtained in various exotic ultraviolet lines from both rockets and satellites, which have led to new qualitative information about the structure of the middle and upper chromosphere (Section 3.3.5). However, quantitative information about the physical conditions in the chromospheric layers responsible has come mainly from rocket measurements of the integrated fluxes of certain ultraviolet resonance lines. These lines originate from various moderately-ionized elements and are our concern in this section.

TABLE 4.12

Some Observed Ultraviolet Resonance Lines and Their Measured Fluxes [Pottasch, 1963]

ION	IONIZ. POT. (χ_i) (volts)	λ(Å)	FLUX (ϕ) (erg cm^{-2} sec^{-1})	$\overline{T}_e$ (K)
C II	24.4	1335.7	0.050	31 000
		1334.5	0.050	
C III	47.9	977.0	0.090	55 000
C IV	64.5	1550.7	0.060	90 000
		1548.2	0.110	
N II	29.6	1085.7	0.009	34 000
		1085.5		
		1084.6		
		1084.0		
N III	47.4	991.5	0.012	60 000
		989.8		
N IV	77.5	765.1	0.008	100 000
N V	97.9	1238.8	0.004	145 000
		1242.8	0.003	
O II	35.1	832.8	0.016	43 000
		833.3		
		834.5		
O III	54.9	833–835	0.016	80 000
O IV	77.4	790.1	0.013	125 000
		787.7	0.012	
O V	113.9	629.7	0.045	160 000
O VI	138.1	1037.6	0.026	220 000
		1031.9	0.039	
Si II	16.3	1817.2	0.450	17 000
		1808.0	0.150	
Si III	33.5	1206.5	0.048	38 000
Si IV	45.1	1402.7	0.013	49 000
		1393.7	0.030	
Si XII	523.2	521.1	0.010	1 600 000
		499.3	0.021	

TABLE 4.12 *(continued)*

ION	IONIZ. POT. (χ_i) (volts)	λ(Å)	FLUX (ϕ) (erg cm^{-2} sec^{-1})	$\overline{T}_e$ (K)
S II	23.3	1250.5 1253.8 1259.5	0.004	26 000
S III	34.7	1190–1202	0.008	42 000
S IV	47.3	1073.3 1062.7	0.003	66 000
S VI	88.0	944.5 933.4	0.004	140 000
Ne VII	207.3	465.1	0.013	350 000
Ne VIII	239.0	780.3 770.4	0.006^5 0.013	480 000
Mg IX	327.9	368.1	0.034	700 000
Mg X	367.4	625.3 609.8	0.013 0.025	820 000
Fe XV	457	284.3	0.029	1 300 000
Fe XVI	489.4	361.7 336.6	0.018 0.027	1 600 000
He I	24.6	584.3	0.054	32 000
He II	54.4	303.8	0.240	85 000

Note: ϕ represents the flux measured at the top of the Earth's atmosphere from the whole solar disk, integrated over the line profile; $\overline{T}_e$ represents the temperature at which the function $G(T_e)$ (defined in the text) reaches its maximum value – roughly speaking, it represents the temperature of the region emitting the line in question.

The general nature of the solar ultraviolet spectrum has been well described by Tousey [1963], Zirin and Dietz [1963], Pottasch [1964], and Goldberg [1967]. Between ~3000 Å and ~2085 Å the spectrum is simply an extension of the ordinary Fraunhofer spectrum, showing dark absorption lines and limb darkening in the continuum. The region 2085–1200 Å is characterized by a transition from limb darkening to limb brightening (cf. Section 4.2.4) and from a dark line to a bright line spectrum; the bright and dark lines become equally numerous in the region 1725–1750 Å. Most of the chromospheric and coronal emission lines show limb brightening, which is generally more pronounced the higher the ionization potential [Goldberg]. The most prominent emission lines are resonance lines of H I,[7] He I, and He II, together with resonance lines of intermediate stages of ionization of C, N, O, Si, and S, and of highly-ionized stages of Si, Ne, Mg, and Fe (cf. Fig. 4.12 and Table 4.12). Another prominent feature is the Lyman continuum at $\lambda < 912$ Å; also detectable are the He I and He II continua at $\lambda < 504$ Å and $\lambda < 228$ Å respectively. (These continua have been discussed in Section 4.2.9.) Lines from intermediate stages of ionization yield information about the upper chromosphere. Lines from highly-ionized elements, on the other hand, originate in the corona or in active regions, as do X-rays. Table 4.12, which is taken from Pottasch [1963], lists the properties of a number of emission lines of interest observed from rockets over the wavelength region 284–1817 Å, together with adopted values of their measured fluxes. More extensive lists derived from both rocket and satellite data have been published by Burton *et al.* [1967], Burton and Ridgeley [1970], Reeves and Parkinson [1970], and Dupree and Reeves [1971].

Since the measured fluxes are integrated over the solar disk and over the line profile, it is evident that we cannot expect the ultraviolet observations, taken alone, to yield detailed information about physical conditions in the relevant layers. And in fact, existing analyses of such observations incorporate data from other sources in deriving chromospheric models consistent with the ultraviolet data or, alternatively, focus attention on determining the abundances of the elements involved [Ivanov-Kholodnyi and Nikol'skii, 1961, 1962; Koyama, 1963; Pottasch, 1963, 1964; Zirin and Dietz, 1963; Pagel, 1964; Dupree and Goldberg, 1967; Goldberg, 1967; Tominaga, 1967]. In order to illustrate just what information the ultraviolet observations yield, we shall outline the analysis of Pottasch [1964].

Suppose that the emission in a given resonance line originates in an optically-thin spherical shell of volume V. Then if ϵ is the emission per unit volume, the flux received at the Earth's surface is (cf. Section 4.2.9)

$$\phi = \frac{1}{2} \times \frac{1}{4\pi L^2} \int_V \epsilon \, dV \quad \text{erg cm}^{-2}\ \text{sec}^{-1}, \tag{4.28}$$

where L is the Earth–Sun distance and the factor 1/2 accounts for the fraction of the emitted radiation absorbed by the photosphere. It is now assumed that the dominant process in raising an ion from the ground state to its lowest excited state is *collisional excitation* and, furthermore, that each such excitation is followed by the emission of a resonance line quantum. Then if C_{lu} represents

[7] The Lyman series.

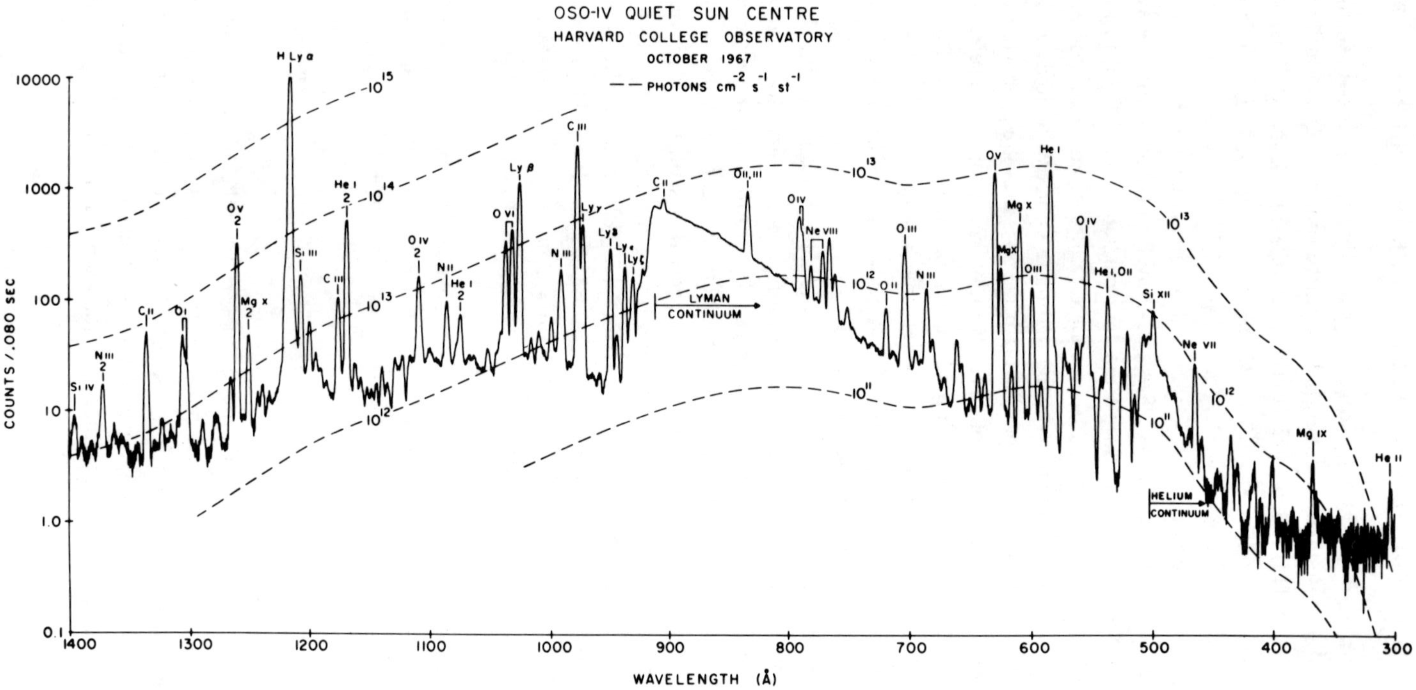

Fig. 4.12. Spectral scan at the centre of the solar disk obtained with the OSO IV spectrometer. Note the Lyman continuum of H I at $\lambda < 912$ Å and the He I continuum at $\lambda < 504$ Å, in addition to many prominent emission lines. (*By courtesy of Harvard College Observatory.*)

the collisional rate (per electron) as defined in Section 4.3.4,

$$\epsilon = h'\nu \cdot N_i \cdot C_{\mathrm{lu}} N_{\mathrm{e}}, \tag{4.40}$$

where N_i is the number density of the ion in its ground state. Apart from atomic constants, the quantity C_{lu} depends only on T_{e}, i.e. $C_{\mathrm{lu}} = F(T_{\mathrm{e}})$. Pottasch [loc. cit., equation 1] gives an explicit formula for $F(T_{\mathrm{e}})$.

If we let ΣN_i represent the total number density of the element (i.e., summed over all stages of ionization), then the abundance relative to hydrogen is $A = (\Sigma N_i)/N_{\mathrm{H}}$. Let us assume that the electrons come only from hydrogen and helium so that, approximately, $N_{\mathrm{H}} = 0.8\, N_{\mathrm{e}}$. Combining (4.28) and (4.40) we have

$$\phi = \frac{1}{8\pi L^2} \int_V h'\nu \cdot N_i \cdot C_{\mathrm{lu}} N_{\mathrm{e}} \, \mathrm{d}V$$

$$= \frac{h'\nu}{8\pi L^2} \int_V N_i N_{\mathrm{e}} F(T_{\mathrm{e}}) \, \mathrm{d}V.$$

But

$$N_i \equiv \frac{N_i}{\Sigma N_i} \cdot \frac{\Sigma N_i}{N_{\mathrm{H}}} \cdot N_{\mathrm{H}} = \frac{N_i}{\Sigma N_i} \cdot A \cdot 0.8\, N_{\mathrm{e}},$$

and hence

$$\phi = \frac{0.1\, h'\nu}{\pi L^2} \cdot A \int_V N_{\mathrm{e}}^2 F(T_{\mathrm{e}}) \frac{N_i}{\Sigma N_i} \, \mathrm{d}V. \tag{4.41}$$

To evaluate the term $N_i/\Sigma N_i$ in (4.41) certain assumptions are necessary concerning the mode by which ionization equilibrium is maintained. Pottasch assumes that since the solar ultraviolet radiation is relatively weak, ions whose ionization potential exceeds 15 eV are produced solely by collisional ionization. The rate of production of ions in the $i + 1$ state is then $N_{\mathrm{e}} N_i C_{i,i+1}$, where $C_{i,i+1}$ depends only on T_{e}. In addition, Pottasch assumes that for $N_{\mathrm{e}} < 10^{10}$ cm^{-3} the principal means of recombination is radiative recapture of an electron: the rate for this process is $N_{\mathrm{e}} N_{i+1} \alpha$, where the radiative recombination rate α also depends only on T_{e}. In equilibrium,

$$N_{\mathrm{e}} N_i C_{i,i+1} = N_{\mathrm{e}} N_{i+1} \alpha,$$

so that

$$\frac{N_{i+1}}{N_i} = \frac{C_{i,i+1}}{\alpha}. \tag{4.42}$$

Equations (3) and (4) of Pottasch's paper give explicit formulae for $C_{i,i+1}$ and α.

With the aid of (4.42) the quantity $G(T_e) = F(T_e)N_i/\Sigma N_i$ appearing in (4.41) can now be evaluated as a function of T_e for any particular ion. It turns out that $G(T_e)$ is a highly-peaked function, having a sharp maximum over a small range of temperatures and falling off rapidly at lower and higher values. To a good approximation, $G(T_e)$ (in the neighbourhood of its maximum) may be replaced by 0.7 times its maximum value, G_{max}, so that (4.41) finally becomes

$$\phi = \text{const.} \times AG_{max} \int_R N_e^2 \, dh, \qquad (4.43)$$

the integral now being confined to the region where $G(T_e)$ is non-zero – i.e. the region where the line is formed – and the element of volume dV of the spherical shell being replaced by an element of height $dh = dV/4\pi R_\odot^2$.

It is evident from (4.43) that since G_{max} can be calculated, measurement of ϕ for a given line of a given ion yields essentially the abundance of the element A multiplied by the integral of N_e^2 over the height range responsible for the origin of the line. Moreover, since $G(T_e)$ peaks at widely different values of T_e for different ions and elements, as shown in the fifth column of Table 4.12, the data as a whole open the possibility of checking a model of the upper chromosphere over a wide range of heights.

It remains to enquire to what extent the ultraviolet data have so far been applied in actual practice. The most thorough discussion of the data is due to Pottasch [1963, 1964], who considers several topics in detail, including (*a*) the derivation of abundances *without* assuming a specific model; (*b*) the consistency of the ultraviolet with the radio data; and (*c*) in conjunction with N_e derived from other sources, the derivation of T_e as a function of h. Pottasch finds that the radio and ultraviolet data are mutually consistent for a *homogeneous model*, although neither type of observation alone is adequate to fully define the model. However, to describe this model as 'homogeneous' may be inappropriate since there are good reasons for identifying it with a model of the interspicular medium:[8]

(1) many of the temperatures $\overline{T}_e$ in Table 4.12 fall within the range of temperatures T_e for interspicular regions given in our inhomogeneous model for heights running from 10 000–20 000 km (see Table 4.23);

(2) the electron densities assumed by Pottasch [1964, Table VIII] in accounting for the ultraviolet data agree well with the interspicular densities given in Table 4.23;

[8] Zirin and Dietz [1963] disagree with this conclusion. They contend that the ultraviolet lines of chromospheric origin are anomalously weak compared with the coronal lines, bearing in mind that $(N_e)_{chrom} \gg (N_e)_{corona}$ and that, ignoring temperature factors, the flux depends on N_e^2 (equation 4.43). Accordingly, they believe that the chromospheric lines originate in scattered hot elements having temperatures in the range 30 000–80 000 K, which they identify with spicules. On the other hand, Zirin and Dietz's identification of spicules with hot elements has been severely criticized by Nikol'skii [1965].

(3) our adopted spicule temperature of 16 000 K (Section 4.5.2) is much too low to produce many of the ions of Table 4.12;

(4) the spicules occupy a very small fraction of the total volume of the chromosphere at the heights in question (Table 4.23).

It should be pointed out, however, that Pottasch's temperature distribution shows a much steeper rise to near-coronal values than either our interspicular model (Table 4.23) or our homogeneous model (Table 4.18), T_e attaining a value of 10^6 K at a height of only 9000 km. In Pottasch's model, as in our homogeneous model, the pressure p shows a slight increase with height.

Other homogeneous models which are based partially on measurements of ultraviolet line fluxes are those of Ivanov-Kholodnyi and Nikol'skii [1961, 1962], Koyama [1963], and Tominaga [1967]. The various authors differ somewhat from Pottasch and from one another in the procedures employed and assumptions made, although their assumptions concerning the modes of excitation and ionization which are operative are substantially the same; for details the reader should consult the original papers. Our adopted homogeneous model (Table 4.18) incorporates the results of Koyama, Tominaga, and Gulyaev *et al.* [1963]; the latters' model is closely tied to that of Ivanov-Kholodnyi and Nikol'skii.

Observations of ultraviolet emission lines are likely to play an even more important role in establishing the physical conditions in the upper chromosphere when improved techniques lead to more refined data. For example, one difficulty in using existing total flux measurements is that active regions can exert a large effect on the observed values. As a result large and uncertain corrections may be necessary in order to extrapolate to the 'quiet' chromosphere. To eliminate this difficulty, observations capable of resolving the disk are required. Such observations would also yield centre-limb data, thus allowing a further refinement in the theoretical analysis. A final refinement might be a more detailed evaluation of the possible modes of excitation and ionization of the atomic states giving rise to the lines under study.

4.4 A Homogeneous Model of the Quiet Chromosphere

4.4.1 INTRODUCTION

In previous sections of the chapter we have outlined the various types of continuous and line spectrum data pertaining to the chromosphere currently available and the astrophysical techniques which have been employed to extract knowledge of the physical conditions there. In this and the following section (Section 4.5) we shall tabulate a selection of homogeneous and inhomogeneous chromospheric models which in recent years have been derived to represent this data.

While an inhomogeneous model (it is to be hoped) correctly describes the physical conditions in the spicule and interspicular regions, a homogeneous model represents, to a certain extent, an academic exercise. For we know from direct observations that the chromosphere *is* inhomogeneous, from heights corresponding to the tops of the spicules (~15 000 km) right down to the photosphere–chromosphere transition region, as we have already explained in Section 4.1. A homogeneous model therefore represents a mythical mixture of material in which the physical conditions *may* be intermediate to those of the spicules and interspicular regions but whose principal function is to correctly reproduce the observed data in the simplest possible way. In effect, it represents a zero-order approximation to the actual chromosphere.

How can the reliability of any particular homogeneous model be assessed? One method is to compare it with other models derived independently, preferably from different data. This method has been attempted in the discussion that follows, but with only partial success. One difficulty is the scarcity of models covering the full range of chromospheric heights. Secondly, there is often heavy reliance on the same set of primary data, e.g. slitless spectrograms from the 1952 Khartoum eclipse. Finally, the true independence of different model derivations by different authors cannot always be assumed.

In the discussion that follows we employ homogeneous models of Gulyaev *et al.* [1963], Koyama [1963], Tominaga [1967], and – at the lowest levels – the Bilderberg Continuum Atmosphere (Section 4.2.5) to derive an 'adopted' homogeneous model of the quiet chromosphere running from $h = 0$ to $h = 20\,000$ km (Table 4.18). If we judge purely on the basis of internal consistency, the electron temperature T_e is the most reliable parameter, differences in log T_e derived by different authors ranging from 0.01 to 0.27, apart from one difference of 0.44. However, errors in N_e in the adopted model could amount to one-half an order of magnitude and errors in p and ρ to over 1½ orders of magnitude. It is apparent, therefore, that even in the absence of systematic error the model must be regarded as rather tentative; nevertheless, it serves as a useful tool in cases where only rough values of the various parameters are required, as we shall see in subsequent chapters.

4.4.2 TEMPERATURE DISTRIBUTION

Table 4.13 gives the distributions of log T_e with h according to Koyama [1963, Table 4 – 'sunspot minimum'] and Gulyaev *et al.* [1963, Table 4 – 'undisturbed']; also tabulated are the values for $h = 1000$ and 2000 km taken from the Bilderberg Continuum Atmosphere (Table 4.6). The height scale of the BCA has been made consistent with that of the other models by shifting its zero point from $\tau_{5000} = 1$ to $\tau_{5000} = 0.01$. This latter value, according to Pagel [1964], is the optical depth of the limb as seen in the continuum. The shift of the zero point involves subtracting 267 km from the h values in the second column of Table 4.6.[9]

At $h = 1000$ km the BCA value for log T_e agrees exactly with that of Gulyaev *et al.*, but at $h = 2000$ km there is a fairly large discrepancy. However, the BCA temperatures are thought to be unreliable at this level so the BCA value at $h = 2000$ km has been discarded. From $h = 6000$ to 20 000 km the agreement

TABLE 4.13

Homogeneous Models of the Quiet Chromosphere: log T_e Distribution

	log T_e			
h (km)	KOYAMA	GULY. *et al.*	BCA	ADOPTED
1000	–	3.74	3.74	3.74
2000	–	3.74	[4.04]	3.74
3000	–	3.75	–	3.75
4000	–	3.76	–	3.76
5000	–	3.78	–	3.78
6000	4.257	3.82	–	4.04
7000	4.429	4.20	–	4.31[5]
8000	4.654	4.55	–	4.60
9000	4.928	4.83	–	4.88
10 000	5.273	5.06	–	5.16[5]
11 000	5.542	(5.27)	–	5.40[5]
12 000	5.708	5.45	–	5.58
13 000	5.804	(5.61)	–	5.70[5]
14 000	5.863	5.74	–	5.80
15 000	5.903	(5.85)	–	5.87[5]
16 000	5.931	5.94	–	5.93[5]
17 000	5.952	(6.01)	–	5.98
18 000	5.969	6.07	–	6.02
19 000	5.985	(6.13)	–	6.06
20 000	6.000	6.20	–	6.10

Note: parentheses indicate interpolated values; the value in square brackets has been discarded.

between Koyama and Gulyaev *et al*. is mostly very good, and our adopted values are simply the mean of the two; from 1000 to 5000 km we adopt Gulyaev *et al.'s* values. The adopted model fits smoothly onto the BCA at $h = 1000$ km. It may be noted that there is a 'plateau' in Gulyaev *et al.'s* model (and hence in our adopted model) between 1000 and 5000 km, the steep temperature rise

[9] Allen [1963, p. 164] takes the base of the chromosphere and the solar limb to correspond to the level $\tau_{5000} = 0.004$; the height difference between this level and $\tau_{5000} = 0.01$ is only 56 km, which is quite negligible in the context of chromospheric heights.

commencing several thousand kilometres higher than in the BCA. The adopted temperature distribution is plotted in Fig. 4.13.

Gulyaev *et al.'s* model is closely tied to that of Ivanov-Kholodnyi and Nikol'skii [1962]. At its lower levels it is based on (*a*) the Balmer continuum observed at the 1952 Khartoum eclipse; (*b*) the surface brightness and logarithmic gradient β of a Sr II resonance line observed at the same eclipse; and

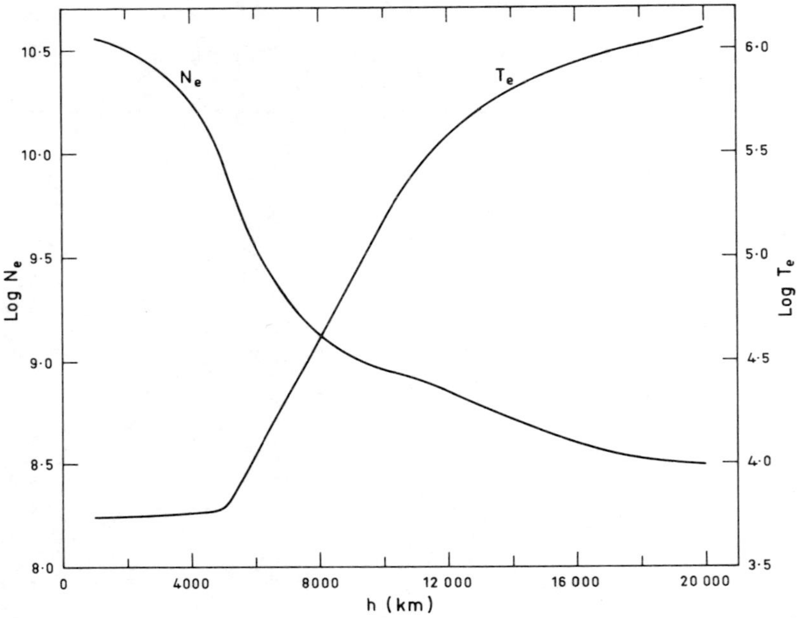

Fig. 4.13. Electron temperature and electron density distributions according to the adopted homogeneous model of the quiet chromosphere (Table 4.18).

(*c*) the known abundance ratio of strontium to hydrogen. The method of analysis unfortunately involves several somewhat risky assumptions whose justifications are not adequately discussed by the authors. In effect (*a*) yields $N_e^2 T_e^{-3/2}$ as a function of h (cf. Section 4.2.7), while (*b*) and (*c*) taken together give N_{HI} as a function of h (cf. Section 4.3.2). Finally, the assumption of local thermodynamic equilibrium permits the use of Saha's law (with $N_{\mathrm{HII}} = N_e$),

$$N_{\mathrm{HI}} = 4.13 \times 10^{-16}\, N_e^2 T_e^{-3/2}\, \exp \chi_i/kT_e, \qquad (4.44)$$

from which T_e and N_e can now be determined.

At its upper levels [see Ivanov-Kholodnyi and Nikol'skii, 1962] Gulyaev *et al.'s* model is based on (*a*) measurements of the absolute intensities of a number of ultraviolet emission lines in the region 60–1300 Å (cf. Section 4.3.7); (*b*) extrapolation of the Balmer continuum measurements from 6000 to 20 000 km;

and (*c*) observations at radio wavelengths (cf. Section 4.2.2). Such a wide extrapolation of the Balmer continuum data is a particularly unsatisfactory feature of the model's derivation, as its architects have pointed out.

Koyama [1963] likewise makes use of the absolute intensities of ultraviolet lines, together with measurements of the Balmer continuum at the 1952 eclipse; he also extrapolates the latter to 20 000 km. In addition, he assumes that $\log T_e = 6.00$ and $\log N_e = 8.30$ at the level $h = 20\,000$ km. Koyama compares his model with the earlier homogeneous models of Piddington [1954] and Athay *et al.* [1955] and the inhomogeneous models of Hagen [1957], Coates [1958], and de Jager [1959, p. 141].

TABLE 4.14

Homogeneous Models of the Quiet Chromosphere: log N_e Distribution

h (km)	log N_e			
	KOYAMA	GULY. *et al.*	BCA	ADOPTED
1000	–	11.41	10.56	10.56
2000	–	11.00	[9.75]	(10.50)
3000	–	10.73	–	(10.40)
4000	–	10.25	–	10.25
5000	–	9.94	–	9.94
6000	9.42	9.68	–	9.55
7000	9.10	9.51	–	9.30^{5}
8000	8.87	9.39	–	9.13
9000	8.76	9.29	–	9.02^{5}
10 000	8.72	9.20	–	8.96
11 000	8.70	(9.12)	–	8.91
12 000	8.64	9.05	–	8.84^{5}
13 000	8.56	(8.99)	–	8.77^{5}
14 000	8.49	8.93	–	8.71
15 000	8.42	(8.88)	–	8.65
16 000	8.37	8.83	–	8.60
17 000	8.32	(8.79)	–	8.55^{5}
18 000	8.30	8.75	–	8.52^{5}
19 000	8.30	(8.71)	–	8.50^{5}
20 000	8.30	8.68	–	8.49

Note: parentheses indicate interpolated values; the value in square brackets has been discarded.

4.4.3 ELECTRON DENSITY DISTRIBUTION

The log N_e distributions of Koyama [1963] and Gulyaev *et al.* [1963] are given in Table 4.14. Also given are the BCA values for $h = 1000$ and 2000 km, calculated from the electron pressures in the fifth column of Table 4.6 with the aid of the formula $p_e = kN_eT_e$. The agreement between Koyama and Gulyaev *et al.* is reasonably good, the differences amounting at most to about one-half an order of magnitude; over the range 6000–20 000 km our adopted value of $\log N_e$ at any height is the mean of the two values. At h = 1000 km we adopt the BCA value, which is nearly an order of magnitude smaller than Gulyaev *et al.'s*, but once again we discard the BCA value for h = 2000 km. The adopted values for h = 2000 km and 3000 km are interpolated from a curve drawn to connect Gulyaev *et al.'s* curve smoothly to the BCA.

The adopted $\log N_e$ distribution is shown in Fig. 4.13. It should be noted that the flattening of the curve in the region 1000–4000 km probably has no real existence but merely reflects the result of smoothly connecting Gulyaev *et al.'s* curve to the BCA.

4.4.4 PRESSURE DISTRIBUTION

Knowing the particle densities of hydrogen and helium in their various stages of ionization and ignoring all other elements, we may calculate the gas pressure from the equation of state of a perfect gas, viz.:

$$p = k(N_e + N_{\mathrm{HI}} + N_{\mathrm{HII}} + N_{\mathrm{HeI}} + N_{\mathrm{HeII}} + N_{\mathrm{HeIII}})T_e. \qquad (4.45)$$

Table 4.15 gives the pressure distribution calculated in this way from the particle densities determined by Gulyaev *et al.* [1963, Table 4 – 'undisturbed'] and Tominaga [1967, Fig. 5]. Also given are the BCA values for h = 1000 and 2000 km, taken directly from Table 4.6.

There is a large discrepancy between the BCA value and Gulyaev *et al.'s* at 1000 km; similar discrepancies are present between Gulyaev *et al.* and Tominaga over the temperature range for which both give figures. Nevertheless, in the absence of other data, we adopt the mean of their $\log p$ values over the height range 6000–10 000 km. At h = 1000 km we accept the BCA value, while over the range 14 000–20 000 km we adopt Gulyaev *et al.'s* figures; all other values (figures in parentheses) are interpolated.

A disturbing feature of Gulyaev *et al.'s* pressure distribution is a *rise* in pressure over the range 6000–20 000 km amounting to an order of magnitude. Numerically, this result is due to the circumstance that as h increases, the temperature in their model rises more rapidly than the particle density falls. However, such a pressure rise implies a downward force on the portion of the chromosphere involved; this force would give rise to mass motions which are not only apparently unobserved but which, if present, would destroy the assumed homogeneity of the model (see also Section 7.5). Two possible explanations may be offered for the apparent paradox: (*a*) errors in the derived particle densities, or (*b*) a basic failure of the initial assumption of homogeneity leading, in effect, to an internal contradiction.

TABLE 4.15

Homogeneous Models of the Quiet Chromosphere: log p Distribution

$\log T_e$	h (km) GULY. *et al.*	$\log p$ GULY. *et al.*	TOMINAGA	BCA	ADOPTED
3.74	1000	2.12	–	0.48	0.48
3.74	2000	1.47	–	[−1.79]	(0.11)
3.75	3000	0.69	–	–	(−0.23)
3.76	4000	−0.24	–	–	(−0.54)
3.78	5000	−1.21	–	–	(−0.82)
3.82	6000	−1.90	−0.24	–	−1.07
4.20	7000	−1.82	−0.60	–	−1.21
4.55	8000	−1.60	−0.69	–	-1.14^5
4.83	9000	−1.42	−0.72	–	−1.07
5.06	10 000	−1.28	−0.73	–	-1.00^5
5.45	12 000	−1.04	–	–	(−0.93)
5.74	14 000	−0.87	–	–	−0.87
5.94	16 000	−0.77	–	–	−0.77
6.07	18 000	−0.72	–	–	−0.72
6.20	20 000	−0.66	–	–	−0.66

Note: Tominaga's and Gulyaev *et al.'s* values of $\log p$ have been averaged at congruent values of T_e, not h; parentheses indicate interpolated values; the value in square brackets has been discarded.

The sources of Gulyaev *et al.*'s values of N_{HI} and N_e have already been mentioned in Section 4.4.2; in addition, they assumed that $N_{\mathrm{HII}} = N_e$. The particle densities of helium in its various stages of ionization were derived from a detailed discussion of the statistical steady state equations governing the ionization of He I and He II. The particle densities obtained in this way account

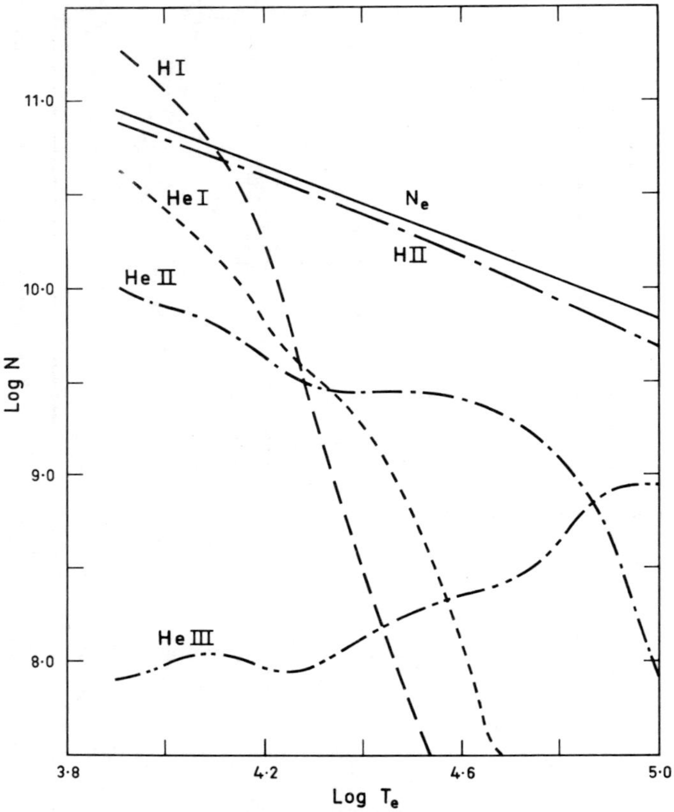

Fig. 4.14. Particle densities of hydrogen and helium in their various ionization states according to Tominaga [1967]. Hydrogen becomes approximately 90 per cent ionized at a level corresponding to T_e = 20 000 K, the equivalent chromospheric height being $h \simeq$ 7000 km (Fig. 4.13). The second ionization of helium is 90 per cent complete at T_e = 100 000 K ($h \simeq$ 9500 km).

reasonably well for the observed fluxes in the He I and He II continua at $\lambda < 504$ Å and $\lambda < 228$ Å respectively (cf. Section 4.2.9).

Tominaga's model is based essentially on (*a*) intensities of ultraviolet emission lines (cf. Section 4.3.7); (*b*) observations at radio wavelengths (cf. Section 4.2.2); and (*c*) the assumption of *hydrostatic equilibrium* (cf. Section 4.2.2). In

Figs. 2 and 3 of his paper he compares his model with the earlier models of Woolley and Allen [1950], Piddington [1954], Oster [1956], Pottasch [1960], and Ivanov-Kholodnyi and Nikol'skii [1962]. He obtains the hydrogen and helium particle densities in their various ionization states by calculation and compares the calculated fluxes in the Lyman continuum of hydrogen ($\lambda < 912$ Å) and, like Gulyaev *et al.*, in the two helium continua, with the observed values. Tominaga's particle densities are shown in Fig. 4.14: this diagram strikingly illustrates the changes in the populations of the various ionized states of hydrogen and helium as T_e increases from about 8000 K to 100 000 K.

4.4.5 GAS DENSITY AND HYDROGEN IONIZATION DISTRIBUTIONS

The gas density and the degree of hydrogen ionization can readily be calculated from the particle densities determined by Gulyaev *et al.* [1963] and Tominaga [1967] with the aid of the formulae

$$\rho = m_H(N_{HI} + N_{HII}) + m_{He}(N_{HeI} + N_{HeII} + N_{HeIII}) \qquad (4.46)$$

and

$$x_H = N_{HII}/(N_{HI} + N_{HII}), \qquad (4.47)$$

where m_H and m_{He} are the masses of the hydrogen and helium atoms respectively; all other elements are ignored. The results are given in Tables 4.16 and 4.17, together with the BCA values for log ρ and log x_H at $h = 1000$ and 2000 km taken directly from Table 4.6.

As in the case of the gas pressure, the gas densities reveal serious discrepancies amongst the different authors. Nevertheless, between 6000 and 10 000 km we take for our adopted values of log ρ the mean of Gulyaev *et al.'s* and Tominaga's figures. Between 14 000 and 20 000 km we accept Gulyaev *et al.'s* values, and at $h = 1000$ km the BCA value. All other figures are interpolated.

The log x_H values show less serious discrepancies than log p and log ρ. Gulyaev *et al.* and Tominaga agree in placing the level at which the ionization of hydrogen becomes virtually complete at about $h = 8000$ km. According to the BCA, on the other hand, total ionization is reached at only $h = 2000$ km, but this is ignored in our adopted distribution. At $h = 1000$ km, Gulyaev *et al.* and the BCA are in reasonable agreement.

4.4.6 THE ADOPTED HOMOGENEOUS MODEL

The values of T_e, N_e, p, ρ, and x_H adopted in Sections 4.4.2–4.4.5 are collected together in Table 4.18. Once again, we should emphasize the tentative nature of the adopted model and indeed of any current chromospheric model; the magnitudes of the possible errors in the various physical quantities have already been roughly estimated in Section 4.4.1.

TABLE 4.16

Homogeneous Models of the Quiet Chromosphere:
log ρ Distribution

$\log T_e$	h (km) GULY. *et al.*	$\log \rho$ GULY. *et al.*	TOMINAGA	BCA	ADOPTED
3.74	1000	−9.43	–	−11.09	−11.09
3.74	2000	−10.08	–	[−13.91]	(−11.19)
3.75	3000	−10.88	–	–	(−11.36)
3.76	4000	−11.82	–	–	(−11.62)
3.78	5000	−12.85	–	–	(−12.10)
3.82	6000	−13.71	−11.90	–	-12.80^{5}
4.20	7000	−14.11	−12.76	–	-13.43^{5}
4.55	8000	−14.25	−13.28	–	-13.76^{5}
4.83	9000	−14.36	−13.61	–	-13.98^{5}
5.06	10 000	−14.45	−13.89	–	−14.17
5.45	12 000	−14.60	–	–	(−14.49)
5.74	14 000	−14.72	–	–	−14.72
5.94	16 000	−14.82	–	–	−14.82
6.07	18 000	−14.90	–	–	−14.90
6.20	20 000	−14.97	–	–	−14.97

Note: Tominaga's and Gulyaev *et al.'s* values of $\log \rho$ have been averaged at congruent values of T_e, not h; parentheses indicate interpolated values; the value in square brackets has been discarded.

TABLE 4.17

Homogeneous Models of the Quiet Chromosphere:
log x_H Distribution

log T_e	h (km)	log x_H			
	GULY. *et al.*	GULY. *et al.*	TOMINAGA	BCA	ADOPTED
3.74	1000	−2.79	–	−2.01	−2.40
3.74	2000	−2.55	–	[0.00]	(−2.21)
3.75	3000	−2.02	–	–	(−1.95)
3.76	4000	−1.56	–	–	−1.56
3.78	5000	−0.83	–	–	(−1.00)
3.82	6000	−0.24	−0.60	–	−0.42
4.20	7000	−0.02	−0.16	–	−0.09
4.55	8000	0	0	–	0
4.83	9000	0	0	–	0
5.06	10 000	0	0	–	0
5.45	12 000	0	–	–	0
5.74	14 000	0	–	–	0
5.94	16 000	0	–	–	0
6.07	18 000	0	–	–	0
6.20	20 000	0	–	–	0

Note: Tominaga's and Gulyaev *et al.'s* values of log x_H have been averaged at congruent values of T_e, not h; parentheses indicate interpolated values; the value in square brackets has been discarded.

The adopted model given in Table 4.18 is (fortuitously) in excellent agreement with a homogeneous model proposed by Allen [1963, p. 174]; his model was also derived by averaging a number of independent models. It does not show the anomalous rise in gas pressure discussed in Section 4.4.4, but p, T_e, and N_e are otherwise in good agreement with the values presented in Table 4.18. On the other hand, the adopted model differs from models proposed by Dupree

TABLE 4.18

Adopted Homogeneous Model of the Quiet Chromosphere

h (km)	T_e (K)	N_e (cm^{-3})	p (dyne cm^{-2})	ρ (gm cm^{-3})	x_H
1000	5500	3.63×10^{10}	3.02×10^{0}	8.13×10^{-12}	3.98×10^{-3}
2000	5500	(3.16)	(1.29)	(6.46)	(6.17)
3000	5620	(2.51)	(5.89×10^{-1})	(4.37)	(1.12×10^{-2})
4000	5750	1.78	(2.88)	(2.40)	2.75
5000	6030	8.71×10^{9}	(1.51)	(7.94×10^{-13})	(1.00×10^{-1})
6000	11 000	3.55	8.51×10^{-2}	1.57	3.80
7000	20 700	2.02	6.17	3.67×10^{-14}	8.13
8000	39 800	1.35	7.16	1.72	1.00×10^{0}
9000	75 900	1.06	8.51	1.04	1.00
10 000	146 000	9.12×10^{8}	9.89	6.76×10^{-15}	1.00
11 000	254 000	8.13	–	–	–
12 000	380 000	7.00	(1.18×10^{-1})	(3.24×10^{-15})	1.00
13 000	507 000	5.96	–	–	–
14 000	631 000	5.13	1.35×10^{-1}	1.91×10^{-15}	1.00
15 000	750 000	4.47	–	–	–
16 000	861 000	3.98	1.70×10^{-1}	1.51×10^{-15}	1.00
17 000	955 000	3.59	–	–	–
18 000	1 050 000	3.35	1.91×10^{-1}	1.26×10^{-15}	1.00
19 000	1 150 000	3.20	–	–	–
20 000	1 260 000	3.09	2.19×10^{-1}	1.07×10^{-15}	1.00

Note: parentheses indicate interpolated values; for $h < 1000$ km, use the BCA (Table 4.6) with the h values reduced by 267 km.

and Goldberg [1967], Burton *et al.* [1971], and Reimers [1971] in that the latter all show *much steeper temperature gradients.* In fact, the gradients are so steep that T_e can almost be said to be discontinuous! For example, Reimers finds that T_e increases from 25 000 K to 250 000 K over a height of only 90 km, while Burton *et al.* obtain an order of magnitude increase in only 25 km. But there is no general agreement as to the *height* at which such a discontinuity is supposed to occur: Dupree and Goldberg assume a height of 2000 km, a figure subsequently accepted by Withbroe [1970] and Noyes [1971], while Burton *et al.* and Reimers give values of 6500 and 3000 ± 1000 km respectively. For comparison, in our adopted model the more gradual rise of T_e to coronal values begins at $h \simeq 5000$ km (cf. Fig. 4.15).

While it may be natural to regard such discontinuities in temperature with a certain degree of suspicion, the steep-gradient models – like the earlier models discussed in preceding pages – are nevertheless based on sound observational data. These include fluxes of ultraviolet emission lines, radio brightness temperatures, etc. Moreover, according to Withbroe [loc. cit.] a modified version of the Dupree–Goldberg model satisfactorily accounts for the limb brightening of some (but not all) of the ultraviolet lines observed from OSO IV.

However, an alternative explanation of these observations suggests itself: if the emission were confined mainly to the network *boundaries*, then evidently geometrical effects alone could give rise to the brightening observed near the limb. If so, there is no necessity to invoke a steep temperature gradient. To test this explanation would require the application of non-uniform radiative transfer theory to a two-component model of the chromosphere over the height range involved, the two components being the network boundaries and the network cell interiors. Such a model has recently been proposed by Dubov [1971]. In the cell interiors his model is characterized by a gradual rise of T_e to coronal values, resembling in this respect both our adopted homogeneous and interspicular models. In the network boundaries, on the other hand, there is a sharp discontinuity in T_e located at a height of $h = 9000$ km.

The arguments favouring a sharp temperature rise at low heights have recently been set out by Noyes [1971], but perhaps final judgement should be reserved until further relevant observations have become available. The most convincing evidence – one way or the other – would come from rocket or satellite spectrograms of ultraviolet lines taken beyond the limb with better spatial resolution and a well-determined height scale.

4.5 An Inhomogeneous Model of the Quiet Chromosphere

4.5.1 INTRODUCTION

In the sections that follow we discuss the temperature and electron density distributions in spicules and interspicular regions derived by Thomas and Athay [1961], Hiei [1963], Kanno [1966], and Beckers [1968]. Together with $Q(h)$, the fractional area at any height occupied by spicules, these parameters constitute an inhomogeneous chromospheric model. However, in contrast to the homogeneous model given in Table 4.18, gas pressures and densities and the degree of hydrogen ionization are not available for existing inhomogeneous models.[10] The adopted model, which is virtually identical to Beckers' [1968]

[10] In regions where the hydrogen ionization can be assumed to be virtually complete, $N_{HI} \simeq 0$, $N_{HII} \simeq N_e$, and $\Sigma N_{He} \simeq 0.1\, N_e$. Rough values of p and ρ can then be obtained via equations (4.45) and (4.46).

model, is tabulated in Table 4.23; the adopted temperatures and electron densities are compared with those of the homogeneous model of Table 4.18 in Figs. 4.15 and 4.16 respectively. The more recent, purely spicule models of Avery and House [1969] and Beckers [1972] differ only in relatively minor respects from our adopted model and from Beckers' original model.

4.5.2 TEMPERATURE DISTRIBUTION IN SPICULES

Table 4.19 gives the temperature distribution in spicules according to Thomas and Athay [1961, Table 7-6], Kanno [1963, Table 1], and Beckers [1968, Table XX]. Both Kanno and Beckers take the spicule temperature to be constant with height over most of the chromosphere. Kanno's value of 10 000 K is not directly based on observation but is more in the nature of a tentative assumption. Beckers' figure of 16 000 K, on the other hand, is based on the absolute

TABLE 4.19

Inhomogeneous Models of the Quiet Chromosphere:
T_e Distribution in Spicules

h (km)	T_e (K) BECKERS	KANNO	THOMAS & ATHAY	ADOPTED
0	4700	–	–	4700
1000	10 350	10 000	7500	10 350
2000	16 000	10 000	7500–10 000	16 000
3000	16 000	10 000	8000–10 000	16 000
4000–6000	16 000	10 000	50 000	16 000
7000–16 000	16 000	10 000	–	16 000
17 000–20 000	16 000	–	–	16 000

intensities of a number of spicule emission lines, including lines of hydrogen, helium, and calcium; the details of Beckers' derivation of his spicule model are outlined in Section 4.3.4.

Finally, Thomas and Athay quote a figure of 50 000 K for the range h = 4000–6000 km but do not evaluate the temperature beyond this point. This surprisingly high value is based on the intensities of certain helium emission lines coupled with a non-LTE analysis which follows the procedures of Chapter 4 of their monograph and includes both triplet and singlet states and their mutual coupling via collisions to and from the ground state and the continuum. However, we shall adopt Beckers' temperatures, which are founded on more extensive observational data. Avery and House's [1969] temperature shows a more gradual rise with height, attaining a value of 15 000 K only at h = 11 500 km; however, their values are based on more limited data, namely profiles of the Ca K line observed beyond the limb.

We should point out that there is a certain degree of artificiality in allowing the spicule model – as opposed to the model of the interspicular regions – to extend as high as 20 000 km (cf. Tables 4.19 and 4.20), since very few spicules are observed beyond about h = 15 000 km (see Table 2.2).

4.5.3 ELECTRON DENSITY DISTRIBUTION IN SPICULES

The log N_e distributions of Beckers, Kanno, and Thomas and Athay are given in Table 4.20, together with two values for the low chromosphere derived by Hiei [1963, Table 4]. In general, the various independent values show reasonable

TABLE 4.20

Inhomogeneous Models of the Quiet Chromosphere: log N_e Distribution in Spicules

	log N_e				
h (km)	BECKERS	KANNO	THOMAS & ATHAY	HIEI	ADOPTED
0	11.79	–	–	–	11.79
1000	11.66	11.52	11.5	11.11	11.45
2000	11.53	11.18	11.4–12.5	10.94	11.40
3000	11.41^5	11.01	11.6–11.9	–	11.39
4000	11.30	10.86	10.6–11.3	–	(11.30)
5000	11.18	10.71	11.3–12.1	–	11.20
6000	11.08	10.56	11.0–11.8	–	(11.07)
7000	10.93	–	–	–	10.93
8000	10.81	–	–	–	10.81
9000	10.68	–	–	–	10.68
10 000	10.56	–	–	–	10.56
11 000	10.41^5	–	–	–	10.41^5
12 000	10.30	–	–	–	10.30
13 000	10.20	–	–	–	10.20
14 000	10.08	–	–	–	10.08
15 000	9.94	–	–	–	9.94
16 000	9.82	–	–	–	9.82
17 000	9.70	–	–	–	9.70
18 000	9.58	–	–	–	9.58
19 000	9.46	–	–	–	9.46
20 000	9.34	–	–	–	9.34

Note: parentheses indicate interpolated values.

agreement but there are occasional differences exceeding one-half an order of magnitude. The adopted distribution is given in the last column: from h = 7000 to 20 000 km we adopt Beckers' values but where there are several determinations we take the mean, excepting the two values in parentheses, which are interpolated from a smooth curve connecting adjacent points.

For the origin of Beckers' spicule model we refer the reader once again to the discussion in Section 4.3.4, and for the origin of Thomas and Athay's values to Chapter 7 of their monograph. Kanno's values for N_e are derived from his determination of $N_e^2 T_e^{-3/2}$ as a function of height from Balmer continuum observations (described in Section 4.2.7), together with his assumption that T_e in spicules equals 10 000 K, independently of height. Since our adopted figure for T_e is 16 000 K (Beckers' value) we could have 'corrected' Kanno's log N_e values by adding ¾ log (16 000/10 000) = 0.15 to them; however, Kanno's figures in the third column of Table 4.20 have been allowed to stand. Finally, Hiei's log N_e values for the low chromosphere are based essentially on eclipse observations of the continuum at λ4117: one contribution to this continuum, namely the electron scattering continuum, is directly proportional to the electron density, as we have seen in Section 4.2.8.

4.5.4 TEMPERATURE DISTRIBUTION IN INTERSPICULAR REGIONS

Table 4.21 gives the temperature distributions in the interspicular regions derived by Beckers and Kanno over the height ranges 0–20 000 km and 5000–16 000 km respectively, together with a few rough values derived by Thomas and Athay. Comparison of the figures shows that there is no possibility of reconciliation between Beckers and Kanno: according to the latter, T_e has already reached 710 000 K at a height of only 6000 km whereas according to Beckers, T_e is still only 100 000 K at h = 14 000 km. In fact, in Beckers' model the steep rise which ultimately takes T_e up to near-coronal values is delayed until $h \simeq$ 10 000 km, while in Kanno's model it begins at $h <$ 5000 km.

Beckers' method of deriving T_e in the interspicular regions involves prior knowledge of T_e in spicules, N_e in both spicules and interspicular regions, and $Q(h)$; it relies basically on observations of the radio brightness temperature $T_B(\lambda)$ and its wavelength variation, and is fully described in Section 4.2.2.

One piece of evidence in favour of Beckers' interspicular temperature distribution and, indeed, of his inhomogeneous model generally is that values of $T_B(\lambda)$ calculated from it agree closely with the observed values.[11] This is shown in Beckers' Table XXI, which covers a wavelength range running from λ = 1.5 mm to 60 cm (at the longer wavelengths the emission is presumably largely coronal). On the other hand, repeating his calculations on the basis of Kanno's temperature distribution Beckers finds a marked *disagreement* with observation, even though Kanno's model is also based partly on radio data. In fact, Beckers remarks that disagreement with the observed radio brightness temperatures is characteristic of all models which, like Kanno's, predict a

[11] Of course, in view of the number of parameters involved, this agreement alone cannot uniquely define the model.

temperature increasing rapidly at low heights to near-coronal values [for references, see Beckers, 1968, pp. 410–411].

However, there is one piece of evidence in *favour* of coronal temperatures at low heights which has been pointed out by Kanno and other authors: this is the observation at relatively low heights of a coronal line due to highly-ionized iron (λ7892 of Fe XI) by Athay and Roberts [1955] at the 1952 eclipse. These

TABLE 4.21

Inhomogeneous Models of the Quiet Chromosphere:
T_e Distribution in Interspicular Regions

h (km)	T_e (K) BECKERS	KANNO	THOMAS & ATHAY	ADOPTED
0	4600	–	–	Beckers'
1000	4600	–	–	values
2000	5010	–	50 000	throughout
3000	5460	–	50 000	
4000	5950	–	>50 000	
5000	6490	<100 000	>50 000	
6000	7080	710 000	>50 000	
7000	7710	860 000	–	
8000	8410	920 000	–	
9000	9170	950 000	–	
10 000	10 000	970 000	–	
11 000	17 800	–	–	
12 000	31 600	980 000	–	
13 000	56 200	–	–	
14 000	100 000	990 000	–	
15 000	145 000	–	–	
16 000	211 000	1 000 000	–	
17 000	307 000	–	–	
18 000	447 000	–	–	
19 000	650 000	–	–	
20 000	650 000	–	–	

authors concluded that the maximum emission from the line occurred at $h < 10\,000$ km where, according to Beckers' model, the spicule and interspicular temperatures do not exceed 16 000 K and 10 000 K respectively.

Finally, it should be mentioned that Beckers' interspicular model exhibits a pressure inversion, an unsatisfactory feature it shares with our adopted homogeneous model given in Table 4.18. However, at the present time the balance of the evidence seems to favour Beckers' model.

4.5.5 ELECTRON DENSITY DISTRIBUTION IN INTERSPICULAR REGIONS

The $\log N_e$ distributions of Beckers, Kanno, and Thomas and Athay are presented in Table 4.22. This time the agreement between Beckers and Kanno is remarkably close, the two log values differing throughout the common range of heights by only 0.06–0.21. From $h = 4000$ to 16 000 km we therefore adopt the mean of their values; from $h = 17\,000$ to 19 000 km we take the figures from a smooth curve drawn to meet Beckers' value at $h = 20\,000$ km. At

TABLE 4.22

Inhomogeneous Models of the Quiet Chromosphere: $\log N_e$ Distribution in Interspicular Regions

			$\log N_e$	
h (km)	BECKERS	KANNO	THOMAS & ATHAY	ADOPTED
0	11.54	–	–	11.54
1000	11.11	–	–	11.11
2000	10.56	–	11.0–11.3	10.85^{5}
3000	10.08	–	10.8	10.44
4000	9.83	9.64	< 11.0	9.73^{5}
5000	9.63	9.56	9.7–11.3	9.59^{5}
6000	9.51	9.42	9.6–10.2	9.46^{5}
7000	9.34	9.28	–	9.31
8000	9.26	9.14	–	9.20
9000	9.15	9.01	–	9.08
10 000	9.08	8.90	–	8.99
11 000	8.99	(8.81)	–	8.90
12 000	8.93	8.73	–	8.83
13 000	8.87	(8.67)	–	8.77
14 000	8.82	8.61	–	8.71^{5}
15 000	8.77	(8.56)	–	8.66^{5}
16 000	8.73	8.52	–	8.62^{5}
17 000	8.69	–	–	(8.60)
18 000	8.64	–	–	(8.58)
19 000	8.60	–	–	(8.57)
20 000	8.56	–	–	8.56

Note: parentheses indicate interpolated values.

$h = 2000$ and 3000 km we take the average of Beckers' and Thomas and Athay's figures and at $h = 0$ and 1000 km we adopt Beckers' values.

Beckers' values for the electron density are based on the reasonable assumption that eclipse observations of the electron scattering continuum at λ4700 yield (via equation 4.26 of Section 4.2.8) the *average* electron density $\bar{N}_e$, which is related to the electron densities at the same height in the spicules

TABLE 4.23

Adopted Inhomogeneous Model of the Quiet Chromosphere

	SPICULES			INTERSPICULAR REGIONS		
h (km)	T_e (K)	N_e (cm^{-3})	Q	T_e (K)	N_e (cm^{-3})	$1 - Q$
0	4700	6.2×10^{11}	3.8×10^{-2}	4600	3.5×10^{11}	9.62×10^{-1}
1000	10 350	2.8	3.2	4600	1.3	9.68
2000	16 000	2.5	2.7	5010	7.2×10^{10}	9.73
3000	16 000	2.5	2.2	5460	2.8	9.78
4000	16 000	(2.0)	1.7	5950	5.4×10^{9}	9.83
5000	16 000	1.6	1.2	6490	3.9	9.88
6000	16 000	(1.2)	8.6×10^{-3}	7080	2.9	9.91
7000	16 000	8.5×10^{10}	5.8	7710	2.0	9.94
8000	16 000	6.4	3.7	8410	1.6	9.96
9000	16 000	4.8	2.2	9170	1.2	9.98
10 000	16 000	3.6	1.2	10 000	9.8×10^{8}	9.99
11 000	16 000	2.6	6.5×10^{-4}	17 800	7.9	9.99
12 000	16 000	2.0	3.2	31 600	6.8	1.00×10^{0}
13 000	16 000	1.6	1.7	56 200	5.9	1.00
14 000	16 000	1.2	8.6×10^{-5}	100 000	5.2	1.00
15 000	16 000	8.8×10^{9}	4.9	145 000	4.6	1.00
16 000	16 000	6.6	2.8	211 000	4.2	1.00
17 000	16 000	5.0	1.5	307 000	(4.0)	1.00
18 000	16 000	3.8	8.0×10^{-6}	447 000	(3.8)	1.00
19 000	16 000	2.9	5.0	650 000	(3.7)	1.00
20 000	16 000	2.2	2.0	650 000	3.6	1.00

Note: parentheses indicate interpolated values.

and interspicular regions, say N_e and N_e^*, by the equation

$$\overline{N}_e = N_e Q + N_e^* (1 - Q), \tag{4.4}$$

Q being the fractional area occupied by spicules. Beckers takes $\overline{N}_e$ from discussions by Athay and his colleagues of the λ4700 data obtained at the 1952 eclipse and Q from his own analysis of spicule counts (Section 2.3.6). Moreover,

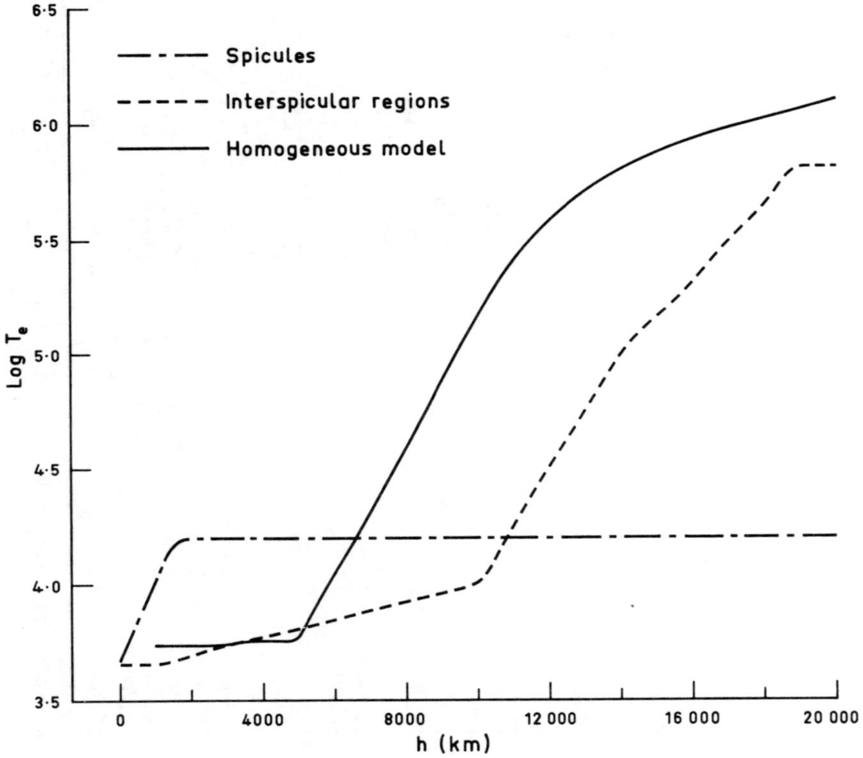

Fig. 4.15. Comparison between homogeneous and inhomogeneous model temperature distributions; the spicule and interspicular temperatures are tabulated in Table 4.23, the homogeneous model temperatures in Table 4.18. For description, see text.

N_e is known from his spicule model (Table 4.20), so that (4.4) finally yields N_e^* in the interspicular regions.

Kanno used exactly the same λ4700 data in his determination but assumed an exponential formula for $Q(h)$ (equation 4.21 of Section 4.2.7). Moreover, instead of employing (4.4) he concluded that for $h > 4000$ km, the spicule contribution to the λ4700 emission – represented by the first term on the right-hand side of (4.4) – could be neglected. A detailed description of Kanno's method of deriving the interspicular electron density from the raw data is to be found in Section 4.2.8. For the origin of Thomas and Athay's values we refer the reader once again to Chapter 7 of their monograph.

4.5.6 THE ADOPTED INHOMOGENEOUS MODEL

The adopted values of T_e and N_e in spicules and interspicular regions given in Tables 4.19–4.22 are collected together in Table 4.23; also given are Beckers' values of $Q(h)$. The inhomogeneous model of Table 4.23 is virtually the same as Beckers' model, although the adopted values of N_e are influenced to some extent by results obtained by other workers, as explained in previous sections.

The log T_e distributions corresponding to spicules and interspicular regions and to the homogeneous model given in Table 4.18 are compared in Fig. 4.15. It will be noticed that for $h < 11\,000$ km, the interspicular regions are *cooler* than the spicules; for $h > 11\,000$ km, they are hotter, the interspicular temperature rapidly rising to near-coronal values. The homogeneous model temperature disagrees with the temperatures of both components of the inhomogeneous model except over the range $h = 1000$–5000 km, where it agrees with the interspicular temperature. In the homogeneous model the temperature begins its rapid rise to high values at $h \simeq 5000$ km, whereas in the interspicular model the rise starts at $h \simeq 10\,000$ km. As a consequence, over a considerable range of heights the temperatures differ by an order of magnitude although the slopes of the two curves, $d(\log T_e)/dh$, are similar. Logically one could reasonably expect a better agreement between the two models since the chromosphere consists largely of interspicular material at the relevant heights; the discrepancy is probably due to differences in the raw data and in the methods of analysis employed in deriving the models in question.

The electron densities for the three models are shown in Fig. 4.16. The values for the homogeneous and interspicular models are here in very close agreement over the height range 6000–20 000 km; at lower heights the maximum difference is about one-half an order of magnitude.

It will be noticed that at all heights the electron density in the spicules exceeds that in the interspicular regions, the maximum difference amounting to slightly more than one and one-half orders of magnitude. The difference in electron density must, of course, reflect a similar difference in gas density: for if we neglect helium and all elements other than hydrogen and assume that above a certain height the hydrogen is largely ionized (cf. Table 4.18), then $N_{\mathrm{HII}} \simeq N_e$ and therefore, from equation (4.46), $\rho \simeq m_{\mathrm{H}} N_e$.

In summary, the inhomogeneous model of Table 4.23 is characterized by dense spicules which, over most of their vertical extent, have a constant temperature of 16 000 K. They are embedded in a much more tenuous medium which is cooler than the spicules at the lower levels ($h < 11\,000$ km) but which at greater heights rapidly approaches coronal temperatures. Undoubtedly further observations and improved methods of analysis will lead to revision – pernaps even drastic revision – of this model and also of our adopted homogeneous model of Section 4.4.6. However, in Chapter 7 we shall find the inhomogeneous

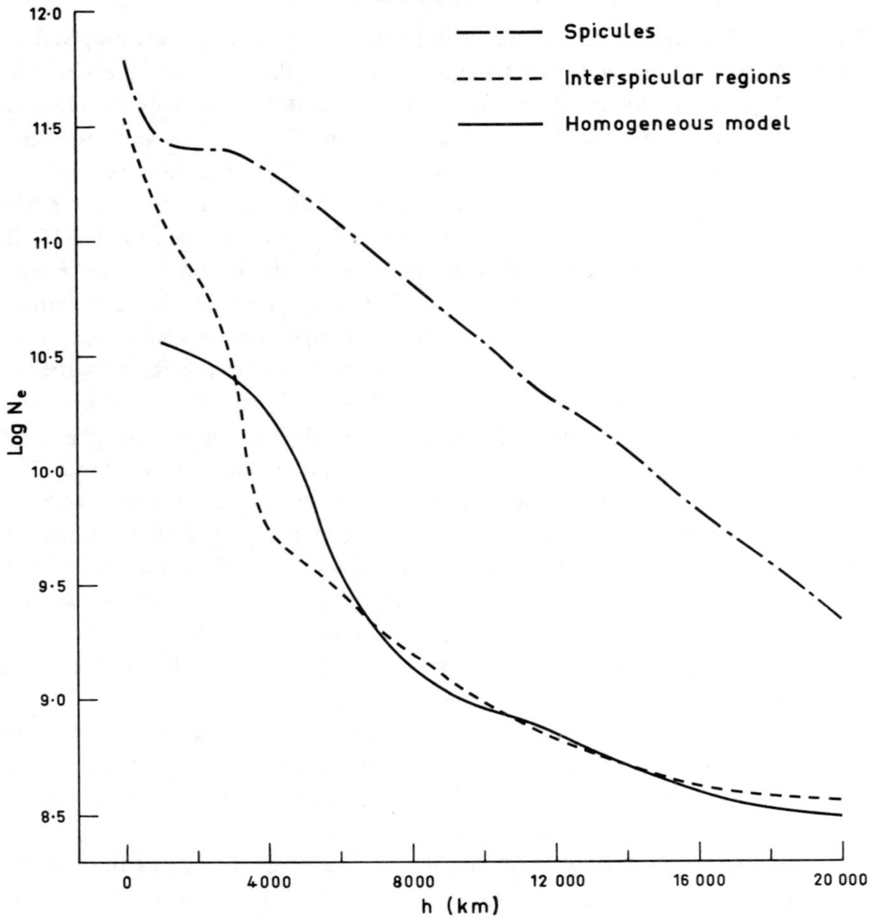

Fig. 4.16. Comparison between homogeneous and inhomogeneous model electron density distributions; the spicule and interspicular electron densities are tabulated in Table 4.23, the homogeneous model electron densities in Table 4.18. For description, see text.

model a useful guide in considering possible mechanisms for the origin of spicules. In addition, the homogeneous model will be used as a zero-order approximation to the actual physical conditions in elucidating possible modes of wave propagation in the low chromosphere.

REFERENCES

ALLEN, C. W. [1963] *Astrophysical Quantities*, Athlone Press, London.

ALLER, L. H. [1963] *Astrophysics – the Atmospheres of the Sun and Stars*, Ronald Press, New York.

ATHAY, R. G., and BESSEY, R. J. [1964] 'Doppler shifts and line broadening in spicules', *Astrophys. J.* **140**, 1174.

ATHAY, R. G., and JOHNSON, H. R. [1960] 'The excitation of He I in the solar spectrum', *Astrophys. J.* **131**, 413.

ATHAY, R. G., MENZEL, D. H., PECKER, J.-C., and THOMAS, R. N. [1955] 'The thermodynamic state of the outer solar atmosphere. V. A model of the chromosphere from the continuum emission', *Astrophys. J. Suppl.* **1**, 505.

ATHAY, R. G., and ROBERTS, W. O. [1955] 'Coronal line intensities at the Khartoum eclipse', *Astrophys. J.* **121**, 231.

ATHAY, R. G., and SKUMANICH, A. [1968] 'Emission cores in H and K lines. I. The optically thick chromosphere', *Solar Phys.* **3**, 181.

AVERY, L. W., and HOUSE, L. L. [1969] 'A solar spicule model based upon calcium II K line radiative transfer studies', *Solar Phys.* **10**, 88.

BECKERS, J. M. [1968] 'Solar spicules', *Solar Phys.* **3**, 367.

BECKERS, J. M. [1972] 'Solar spicules', *Ann. Rev. Astron. Astrophys.* **10**, 73.

BONNET, R. M., and BLAMONT, J. E. [1968] 'Limb-darkening observations between 1800 and 2900 Å', *Solar Phys.* **3**, 64.

BRAY, R. J., and LOUGHHEAD, R. E. [1967] *The Solar Granulation*, Chapman and Hall, London.

BURTON, W. M., JORDAN, C., RIDGELEY, A., and WILSON, R. [1971] 'The structure of the chromosphere–corona transition region from limb and disk intensities', *Phil. Trans. Roy. Soc.* **270A**, 81.

BURTON, W. M., and RIDGELEY, A. [1970] 'The solar limb emission spectrum between 300 Å and 2803 Å', *Solar Phys.* **14**, 3.

BURTON, W. M., RIDGELEY, A., and WILSON, R. [1967] 'The ultraviolet emission spectrum of the solar chromosphere and corona', *Mon. Not. R.A.S.* **135**, 207.

CLUBE, S. V. M. [1958] 'The profiles of chromospheric Hα and D_3 from interferometric observations', *Mon. Not. R.A.S.* **118**, 18.

COATES, R. J. [1958] 'A model of the chromosphere from millimeter wave-length observations', *Astrophys. J.* **128**, 83.

COATES, R. J., GIBSON, J. E., and HAGEN, J. P. [1958] 'The 1954 eclipse measurement of the 8.6-mm solar brightness distribution', *Astrophys. J.* **128**, 406.

CUNY, Y. [1968] 'Contribution à l'étude du spectre solaire de l'hydrogène', *Solar Phys.* **3**, 204.

DUBOV, E. E. [1971] 'A model of the chromosphere and transition zone. Radio and UV emission of these layers', *Solar Phys.* **18**, 43.

DUNN, R. B. [1960] 'Photometry of the solar chromosphere', Thesis, Harvard University.

DUNN, R. B., EVANS, J. W., JEFFERIES, J. T., ORRALL, F. Q., WHITE, O. R., and ZIRKER, J. B. [1968] 'The chromospheric spectrum at the 1962 eclipse', *Astrophys. J. Suppl.* **15**, 275.

DUPREE, A. K., and GOLDBERG, L. [1967] 'Solar abundance determination from ultraviolet emission lines', *Solar Phys.* **1**, 229.

DUPREE, A. K., and REEVES, E. M. [1971] 'The extreme-ultraviolet spectrum of the quiet Sun', *Astrophys. J.* **165**, 599.

EDDY, J. A., LÉNA, P. J., and MACQUEEN, R. M. [1969] 'Far infrared measurement of the solar minimum temperature', *Solar Phys.* **10**, 330.

ELSTE, G. [1968] 'Comments on the Bilderberg Continuum Atmosphere', *Solar Phys.* **3**, 106.

ENGVOLD, O. [1966] 'On the center-limb variation of the H and K lines of Ca II', *Astrophys. Norv.* **10**, 101.

GAY, J. [1970] 'Balloon observations in the submillimeter region: an absolute measurement of the brightness of the Sun and the transparency of the stratosphere above 25 km', *Astron. Astrophys.* **6**, 327.

GINGERICH, O., and JAGER, C. DE [1968] 'The Bilderberg model of the photosphere and low chromosphere', *Solar Phys.* **3**, 5.

GINGERICH, O., NOYES, R. W., KALKOFEN, W., and CUNY, Y. [1971] 'The Harvard-Smithsonian Reference Atmosphere'. *Solar Phys.* **18**, 347.

GINGERICH, O., and RICH, J. C. [1968] 'The far ultraviolet spectrum of the Sun', *Solar Phys.* **3**, 82.

GIOVANELLI, R. G. [1967a] 'Excitation of hydrogen and Ca II under chromospheric conditions', *Aust. J. Phys.* **20**, 81.

GIOVANELLI, R. G. [1967b] 'Structure of the normal chromosphere', *Solar Physics*, ed. J. N. Xanthakis, p. 353; Interscience, London.

GOLDBERG, L. [1967] 'Ultraviolet and X rays from the Sun', *Ann. Rev. Astron. Astrophys.* **5**, 279.

GOLDBERG, L., MÜLLER, E. A., and ALLER, L. H. [1960] 'The abundances of the elements in the solar atmosphere', *Astrophys. J. Suppl.* **5**, 1.

GULYAEV, R. A., NIKOL'SKAYA, K. I., and NIKOL'SKII, G. M. [1963] 'The structure of the solar atmosphere in active and undisturbed regions. Ionization of hydrogen and helium', *Astron. J. USSR* **40**, 433 (*Sov. Astron. AJ* **7**, 332).

HAGEN, J. P. [1957] 'The structure of the solar chromosphere from centimetre-wave radio observations', *Radio Astronomy*, ed. H. C. van de Hulst, p. 263; Cambridge Univ. Press.

HIEI, E. [1963] 'Continuous spectrum in the chromosphere', *Publ. Astron. Soc. Jap.* **15**, 277.

HOUTGAST, J. [1962] 'Absolute intensities of strong chromospheric emission lines of H, He and Ca II at the eclipse of 30 June 1954', *Bull. Astron. Inst. Neth.* **16**, 115.

HULST, H. C. VAN DE [1953] 'The chromosphere and the corona', *The Sun,* ed. G. P. Kuiper, p. 207; Univ. Chicago Press.

IVANOV-KHOLODNYI, G. S., and NIKOL'SKII, G. M. [1961] 'Ultraviolet solar radiation and the transition layer between the chromosphere and the corona', *Astron. J. USSR* **38**, 45 (*Sov. Astron. AJ* **5**, 31)

IVANOV-KHOLODNYI, G. S., and NIKOL'SKII, G. M. [1962] 'Extreme ultraviolet solar radiation and the structure of the solar atmosphere in active and undisturbed regions', *Astron. J. USSR* **39**, 777 (*Sov. Astron. AJ* **6**, 609).

JAGER, C. DE [1959] 'Structure and dynamics of the solar atmosphere', *Handbuch der Physik*, ed. S. Flügge, vol. 52, p. 80; Springer, Berlin.

JAGER, C. DE, and GROOT, B. H. M. DE [1957] 'The helium spectrum of the chromosphere', *Bull. Astron. Inst. Neth.* **14**, 21.

JEFFERIES, J. T. [1968] *Spectral Line Formation,* Blaisdell Publ. Co., Waltham, Mass.

KANNO, M. [1966] 'A model of the upper chromosphere with the spicule structure', *Publ. Astron. Soc. Jap.* **18**, 103.

KOUTCHMY, S., and PEYTURAUX, R. [1968] 'Sur l'intensité du continuum solaire à $\lambda = 20.15\ \mu$. Mesure absolue de la température de brillance du centre du disque', *C. R. Acad. Sci.* **267**, 905.

KOYAMA, S. [1963] 'Extreme ultraviolet solar emission lines and the transition layer between the chromosphere and the corona', *Publ. Astron. Soc. Jap.* **15**, 15.

KUNDU, M. R. [1971] 'Solar radio emission at 1.2 mm wavelength', *Solar Phys.* **21**, 130.

KUROKAWA, H., TOMINAGA, S., KUBOTA, J., and KAWAGUCHI, I. [1969] 'The flash spectrum observed at the total eclipse of February 5, 1962', *Publ. Astron. Soc. Jap.* **21**, 141.

LÉNA, P. J. [1968] 'Observations of the center-to-limb variation of the solar brightness in the far infrared (10 to 25 microns)', *Solar Phys.* **3**, 28.

LIVSHITS, M. A. [1966] 'Line widths in spicules', *Astron. J. USSR* **43**, 718 (*Sov. Astron. AJ* **10**, 570).

MICHARD, R. [1958] 'Etude spectroscopique des structures fines de la chromosphère', *C. R. Acad. Sci.* **246**, 3413.

MINNAERT, M. [1953] 'The photosphere', *The Sun,* ed. G. P. Kuiper, p. 88; Univ. Chicago Press.

MOORE, C. E., and MERRILL, P. W. [1968] *Partial Grotrian Diagrams of Astrophysical Interest,* National Standards Reference Data Series, NSRDS–NBS 23; National Bureau of Standards, Washington.

MURCRAY, F. H., MURCRAY, D. G., and WILLIAMS, W. J. [1964] 'The spectral radiance of the Sun from 4 μ to 5 μ', *Appl. Opt.* **3**, 1373.

NEWSTEAD, R. A. [1969] 'Solar limb brightening and enhancement measurements at 1.2 mm', *Solar Phys.* **6**, 56.

NIKOL'SKII, G. M. [1965] 'Observations of the spectrum of spicules and the fine structure of the solar chromosphere', *Astron. J. USSR* **42**, 86 (*Sov. Astron. AJ* **9**, 64).

NOYES, R. W. [1971] 'Ultraviolet studies of the solar atmosphere', *Ann. Rev. Astron. Astrophys.* **9**, 209.

NOYES, R. W., BECKERS, J. M., and LOW, F. J. [1968] 'Observational studies of the solar intensity profile in the far infrared and millimeter regions', *Solar Phys.* **3**, 36.

NOYES, R. W., and KALKOFEN, W. [1970] 'The solar Lyman continuum and the structure of the solar chromosphere', *Solar Phys.* **15**, 120.

OBRIDKO, V. N. [1963] 'The K_{232} line profile and the structure of the solar atmosphere', *Astron. J. USSR* **40**, 446 (*Sov. Astron. AJ* **7**, 342).

OSTER, L. [1956] 'Das Strahlungsfeld der Übergangsschicht Chromosphäre–Korona', *Z. Astrophys.* **40**, 28.

PAGEL, B. E. J. [1964] 'The structure of the solar chromosphere', *Ann. Rev. Astron. Astrophys.* **2**, 267.

PARKINSON, W. H., and REEVES, E. M. [1969] 'Measurements in the solar spectrum between 1400 and 1875 Å with a rocket-borne spectrometer', *Solar Phys.* **10**, 342.

PECKER, J.-C., and POTTASCH, S. R. [1969] 'On the abundance determination in the solar chromosphere', *Astron. Astrophys.* **2**, 81.

PIDDINGTON, J. H. [1950] 'The derivation of a model solar chromosphere from radio data', *Proc. Roy. Soc.* A **203**, 417.

PIDDINGTON, J. H. [1954] 'Model solar chromospheres', *Astrophys. J.* **119**, 531.

PIERCE, A. K. [1968] 'The chromospheric spectrum outside of eclipse, λλ3040 to 9266 Å', *Astrophys. J. Suppl.* **17**, 1.

POTTASCH, S. R. [1960] 'Use of the equation of hydrostatic equilibrium in determining the temperature distribution in the outer solar atmosphere', *Astrophys. J.* **131**, 68.

POTTASCH, S. R. [1963] 'The lower solar corona: interpretation of the ultraviolet spectrum', *Astrophys. J.* **137**, 945.

POTTASCH, S. R. [1964] 'On the interpretation of the solar ultraviolet emission line spectrum', *Space Sci. Rev.* **3**, 816.

REEVES, E. M., and PARKINSON, W. H. [1970] 'An atlas of extreme-ultraviolet spectroheliograms from OSO–IV', *Astrophys. J. Suppl.* **21**, No. 181.

REIMERS, D. [1971] 'On the solar transition layer and solar active regions', *Astron. Astrophys.* **14**, 198.

SAIEDY, F. [1960] 'Solar intensity and limb darkening between 8.6 and 13 μ', *Mon. Not. R.A.S.* **121**, 483.

SAIEDY, F., and GOODY, R. M. [1959] 'The solar emission intensity at 11 μ', *Mon. Not. R.A.S.* **119**, 213.

SHIMABUKURO, F. I. [1970] 'The 3.3-mm brightness distribution of the quiet Sun', *Solar Phys.* **12**, 438.

SHIMABUKURO, F. I. [1971] 'Observations of the 1.4 mm brightness distribution of the Sun', *Solar Phys.* **18**, 247.

SHIMABUKURO, F. I., and STACEY, J. M. [1968] 'Brightness temperature of the quiet Sun at centimeter and millimeter wavelengths', *Astrophys. J.* **152**, 777.

SHKLOVSKII, I. S., and KONONOVICH, E. V. [1958] 'Models of the solar chromosphere', *Astron. J. USSR* **35**, 37 (*Sov. Astron. AJ* **2**, 32).

SIMON, M., BUHL, D., COGDELL, J. R., SHIMABUKURO, F. I., and ZAPATA, C. [1970] 'Radio observations of the distribution of chromospheric brightness temperature', *Nature* **226**, 1154.

SMITH, E.v.P. [1957] 'Hα profiles in the chromosphere', *Astrophys. J.* **126**, 529.

TAKAHASHI, K. [1967] 'An investigation of solar emission at frequencies of 35, 70, and 94 Gc with a 4.88-meter diameter telescope', *Astrophys. J.* **148**, 497.

THOMAS, R. N., and ATHAY, R. G. [1961] *Physics of the Solar Chromosphere,* Interscience, New York.

TOMINAGA, S. [1967] 'Models of the transition region between chromosphere and corona', *Mem. Fac. Sci. Kyoto Univ.* **32**, 49.

TOUSEY, R. [1963] 'The extreme ultraviolet spectrum of the Sun', *Space Sci. Rev.* **2**, 3.

TREANOR, P. J. [1957] 'Chromospheric interferometry', *Mon. Not. R.A.S.* **117**, 22.

UNSÖLD, A. [1955] *Physik der Sternatmosphären,* Springer, Berlin.

WEATHERBURN, C. E. [1947] *Mathematical Statistics*, Cambridge Univ. Press.

WHITE, O. R. [1962] 'Limb-darkening observations of Hα, Hβ, and Hγ', *Astrophys. J. Suppl.* **7**, 333.

WIDING, K. G., PURCELL, J. D., and SANDLIN, G. D. [1970] 'The UV continuum 1450–2100 Å and the problem of the solar temperature minimum', *Solar Phys.* **12**, 52.

WITHBROE, G. L. [1970] 'Solar XUV limb brightening observations. II. Lines formed in the chromospheric-coronal transition region', *Solar Phys.* **11**, 208.

WOOLLEY, R. v. d. R., and ALLEN, C. W. [1950] 'Ultra-violet emission from the chromosphere', *Mon. Not. R.A.S.* **110**, 358.

ZIRIN, H., and DIETZ, R. D. [1963] 'The structure of the solar chromosphere. I. A picture based on extreme ultraviolet, millimeter, and λ10830 data', *Astrophys. J.* **138**, 664.

ZIRKER, J. B. [1958] 'The metallic chromosphere at the Khartoum eclipse', *Astrophys. J.* **127**, 680.

ZIRKER, J. B. [1962] 'Spectral observations of solar chromospheric spicules', *Astrophys. J.* **136**, 250.

CHAPTER 5

The Fine Structure of the Active Chromosphere

5.1 Introduction

Like the quiet chromosphere the active chromosphere shows a variety of fine structures, on a scale roughly comparable to those observed in quiet regions and described in Chapter 3. To what extent can a knowledge of the fine structures of *active* regions contribute to our understanding of the mechanisms responsible for the fine structures of the *quiet* chromosphere? Our main aim in the present chapter is to provide a partial answer to this question by summarizing the available observational knowledge of some of the basic active region fine structures. Our discussion is limited to fibrils (Section 5.2), chromospheric facular granules (Section 5.3), 'moustaches' (Section 5.4.2), and the fine structure of flares (Section 5.4.3). Moreover, we shall deliberately confine our attention to those aspects of their behaviour which appear to offer most hope of throwing additional light on the *quiet* chromosphere. Thus it is not our intention to give an extended account of the development and properties of active regions, nor of the many diverse and spectacular phenomena associated with them, such as flares, surges, prominences, etc. For fuller discussions of these topics the interested reader is referred to the references cited by Bray and Loughhead [1964, p. 247]; to the symposium proceedings edited by Hess [1964], Xanthakis [1967], Kiepenheuer [1968], and Öhman [1968]; and to the monographs by Smith and Smith [1963], Zirin [1966], and Tandberg-Hanssen [1967].

At the chromospheric level the first sign of the birth of a new active region is the appearance of a localized disturbance in the fine structure of the quiet chromosphere. On good-quality Hα photographs the structure of the disturbed area is characterized by an apparently well-ordered pattern of elongated dark elements, generally referred to as *fibrils*. At the same time, or shortly afterwards, the first facular material appears; it takes the form of a bright compact patch which is particularly prominent in the K line of Ca II. The newly-born facular region rapidly elongates in the east-west direction and, by the second day, the first sunspot appears at its western end. Meanwhile, the facular region continues to increase in brightness and area, while the fibril pattern surrounding it intrudes further and further into the quiet chromosphere.

On the third day another spot (or spots) of opposite magnetic polarity to the existing spot appears at the eastern end of the facular region; the spot group as a whole attains its maximum area between the 6th and 13th days. Throughout this development the size and brightness of the faculae continue to increase, as do the extent and complexity of the surrounding fibril structure. High-resolution Hα photographs of well-developed active regions show that the facular areas are composed of individual *facular granules* having diameters of about 1″ of arc (cf. Plate 5.8); their properties are described in Section 5.3.

Between the 14th and 30th days the Hα faculae begin to shrink. At the same time all the spots except the main (western) one disappear and a stable filament (i.e. prominence), some 50 000 km in length, points towards the surviving spot. The alignment of the fibril structure now appears to be governed by this filament. From the thirtieth to the sixtieth day the last spot slowly dissolves and the brightness of the Hα and K faculae decreases. On the other hand, the large filament grows steadily in length and divides the facular region into two halves. Subsequently, the Hα faculae disappear, while the K faculae dissolve into the bright K network characteristic of the quiet chromosphere, described in Section 3.3.2. Between the sixtieth and hundredth days the filament reaches its greatest length and is then almost parallel to the solar equator. During this period the fibril structure is still present and maintains its arrangement with respect to the filament. Subsequently, although the filament itself may last for 100–250 days, the fibril structure disappears and, after 135 days, the fine structure in the area of the former active region is indistinguishable from that of the quiet chromosphere.[1]

Examination of high-quality Hα photographs of active regions, such as those reproduced in Plates 5.4 and 5.5, reveals an enormous diversity and complexity of structure, which at first sight appears to bear no relationship to that of the quiet chromosphere. However, if one looks closely at the nature of the *fine* structures in disturbed and quiet regions, one detects several points of resemblance. In each case the fine structure is composed of just two basic elements, one of which is dark compared to its surroundings and the other bright. Furthermore, in both instances the dark and bright elements tend to occur in close association with one another. The basic elements of the fine structure of active regions are the fibrils and the chromospheric facular granules, while in the quiet chromosphere they are the dark and bright mottles discussed in Chapter 3.

[1] This description largely follows the account given by Kiepenheuer [1953, p. 432 et seq.] of the development of a *long-lived* active region. A table conveniently summarizing the various stages in chronological order is given by Bray and Loughhead [1964, Table 7.1]. In contrast to such long-lived active regions there are some which last for only a few days, or even hours [Kiepenheuer, 1967, p. 383]. The reader is referred to Bumba and Howard [1965] for a more recent description of the development of active regions which pays particular attention to the roles played by magnetic fields and the supergranulation.

The most distinctive of the various mottle groupings observed in quiet regions is the *rosette*, in which a small central cluster of bright mottles is surrounded by a number of dark mottles aligned more or less radially outwards (Section 3.2.5). Let us compare this typical rosette structure with that of a small active region at an early stage of its development, such as the one illustrated in Plates 5.1 and 5.2. There are two small facular areas in this particular active region, each one presumably a conglomerate of unresolved facular granules (cf. Section 5.3.1). These are surrounded by a large number of elongated fibrils extending outwards in some apparently ordered but complex pattern, in a way reminiscent of the clustering of dark mottles around the bright mottles at the centre of a rosette. Accordingly, one is prompted to ask whether a rosette should be regarded as an active region seen in microcosm, differing from the latter in scale and complexity but not in its essential physical nature.[2] Admittedly, this analogy is more difficult to sustain when one turns to large, well-developed active regions, but once again each bright area is more or less surrounded by an ordered pattern of fibrils (cf. Plates 5.4 and 5.5).

The similarity in the nature of the fine structures in active and quiet regions suggests that they owe their ultimate origin to the same physical agency which, in each case, is responsible for the existence of just two basic elements, one dark compared to its surroundings and the other bright. If this hypothesis is correct, then the explanation of the differences between quiet and active regions – and, for that matter, among different active regions themselves – must be sought in terms of alterations in the underlying physical mechanism.

But what is the actual mechanism? It is likely that the answer to this question will be found in improved observations of magnetic fields in the photosphere and chromosphere and in their correlation with the chromospheric fine structure. Existing magnetic observations of quiet regions have already been described in Sections 3.4.4 and 3.5.4. Although conclusive proof is lacking, these observations suggest that a rosette structure probably possesses a weak longitudinal field coinciding with the bright mottles concentrated at its centre. By contrast, the fields in active regions are much stronger and the distribution of lines of force is likely to be much more complicated. Throughout most of the chromosphere the electrical conductivity is very high and one may therefore expect any moving material to be guided along lines of force. In fact, as we shall see in Section 5.2.4, there is strong observational evidence – albeit of an indirect nature – to indicate that the individual fibrils in active regions trace out actual magnetic field lines.

On this picture the differences between the fine structures of quiet and active regions of the chromosphere must reflect differences in the magnitudes and

[2] A somewhat similar suggestion has been made by Howard and Harvey [1964].

directions of the associated magnetic fields, and in the changes in these parameters with time. If this view is correct, an understanding of the mode of origin of the dark and bright mottles may aid our understanding of the process of formation of active regions. And, conversely, a study of the properties of the fine structures of active regions and their changes with time may help to throw additional light on the nature of the physical processes operating in the quiet chromosphere.

The available observational knowledge of fibrils and chromospheric facular granules and of their relationship to photospheric and chromospheric magnetic fields is summarized in Sections 5.2 and 5.3. We then turn our attention to the fine structure exhibited by two of the transient phenomena associated with active regions, namely moustaches (Section 5.4.2) and flares (Section 5.4.3). These two features have been singled out for discussion partly on account of their intrinsic interest and partly because both probably owe their origin to some form of magnetic field activity.

5.2 Fibril Structure of Active Regions

5.2.1 FIBRIL STRUCTURE OF GROWING ACTIVE REGIONS

Plate 5.1 illustrates the rapid growth over a period of just under four hours of a new active region which was born on the eastern hemisphere of the Sun on the afternoon of the previous day.[3] The inner part of the active region appears to show a typically bipolar magnetic configuration, being composed of two small bright facular areas linked by a number of small dark filaments. These latter features are prominent in the early stages of the development of active regions and are referred to by Bruzek [1967, 1968] as 'arch' filaments [see also Weart and Zirin, 1969; Weart, 1970; Martres and Soru-Escaut, 1971; Zirin, 1971a; Frazier, 1972]. They appear to follow lines of force connecting regions of different magnetic polarity, although recent high-quality magnetograms obtained by Smith [1971] suggest that the fields at the ends of the filaments may actually be a complex mixture of both polarities. Observations show that arch filaments are associated with a flow of matter along their length and that their occurrence is closely related to that of the small transient features known as moustaches (see Section 5.4.2). According to Frazier [1972] the typical lifetime of an arch filament is of the order of 1/2 h.

The two small facular areas within the active region shown in Plate 5.1 are separated from the surrounding undisturbed chromosphere by a number of dark,

[3] Plates 5.1–5.12 were taken by the authors with the 12-inch refractor of the CSIRO Solar Observatory, Culgoora, equipped with a 0.5 Å Halle filter. In the case of Plates 5.1–5.3, 5.6–5.9, 5.11, and 5.12 the filter had a double-peaked profile with red and blue peaks at approximately ± 0.3 Å from the centre of Hα, while in the case of Plates 5.4, 5.5, and 5.10 it had a normal single-peaked profile [cf. Bray, 1968].

elongated features known as *fibrils.* On the whole, the fibrils tend to run more or less radially outwards, but there are places where they bend round from one facular area to the other, as if following curved lines of force. Apart from their greater elongation and radial alignment they bear some resemblance to the dark mottles of the quiet chromosphere (Section 3.2.4). The outer tips of the fibrils define a roughly elliptical area, whose size continually increases as the active region develops. In fact, according to Bumba and Howard [1965], the growth of the outer boundary proceeds with an outward velocity of approximately 0.2 km sec^{-1}. During this development the fibril structure intrudes further and further into the surrounding quiet chromosphere, while the facular areas grow rapidly in size. Moreover, the fibril pattern shows a continual increase in regularity and contrast. These three effects are well illustrated in Plate 5.2, which shows the active region of Plate 5.1 some 24 hours later. A further example of the fibril structure around a young active region is given in Plate 5.3, where the presence of arch filaments is again very noticeable. Other illustrations of the structure of young active regions, this time at a number of wavelengths in the Hα line, have been published by Bruzek [1967, Figs. 1 and 2; 1968, Fig. 2].

Bumba and Howard [1965] have pointed out that a structure resembling the fibril pattern described above is also observed on spectroheliograms taken in the K line, each Hα fibril having its K line counterpart at the same position in the form of a broad dark feature. On low-resolution K spectroheliograms, the resulting appearance is that of a very faint, dark ellipse, which is most visible at the time of maximum brightness of the associated faculae.[4] The sizes of these ellipses have been measured by Bumba and Howard [loc. cit., Fig. 5] in the case of nine young regions observed in Hα and four observed in K.

5.2.2 FIBRIL STRUCTURE OF WELL-DEVELOPED ACTIVE REGIONS

As an active region grows the size and complexity of the associated fibril structure increases until, at the peak of development, it may cover an area of up to about one-tenth of the visible hemisphere. Moreover, the fibril structures around adjacent active centres sometimes merge to form an ordered pattern connecting the two regions [Smith and Smith, 1963, p. 23]. Good-quality photographs illustrating the extent and complexity of the fibril patterns in active regions have been published by a number of workers including Kiepenheuer [1966, Figs. 1–3, 5], Smith [1968, Figs. 1–4], Foukal [1971a, Figs. 1, 4, 6, 10; 1971b, Figs. 2, 3], and Prata [1971, Figs. 1, 2, 4].

Perhaps the most remarkable property of the fibril structure of an active region is its overall stability. In fact, although the individual fibrils are relatively

[4] The dark ellipses around faculae on K_3 spectroheliograms were first noticed by H. Deslandres, who named them 'circumfacules' [see d'Azambuja, 1930, p. 25].

short-lived (see below), the general pattern of their arrangement may show little significant change over a period of an hour or more [Lippincott, 1955; Kiepenheuer, 1957; Loughhead, 1968a; Foukal, 1971a]. The sunspots, faculae, and filaments within an active region exert a strong influence on the form of the fibril pattern in their immediate proximity. This is well illustrated in the case of spots and faculae by Plate 5.4, which shows the characteristic 'iron filings' pattern around a bipolar spot group. An even more complicated pattern is visible in the neighbourhood of the bipolar group in Plate 5.5, where the arrangement of the individual fibrils is affected both by the spots and by the faculae near the top of the photograph. According to Smith [1968] the fibrils around the facular areas within an active region are generally shorter than those near the outer boundary of the region.

Large filaments (prominences) also influence the arrangement of the fibrils in their vicinity. Those immediately adjacent to a filament tend to run parallel to it, while those a little further removed may appear to bend away from it. Sometimes the fibrils at the ends of a filament may lie perpendicular to its axis, whereas in other cases the fibrils may continue parallel to the axis. A good-quality photograph of the fibril pattern in the neighbourhood of a large filament, taken at the Lockheed Solar Observatory, has been published by Smith [1968, Fig. 4]. A more recent study of the alignment of fibrils in the neighbourhood of filaments has been made by Foukal [1971a].

Despite all the effort that has been expended on the study of active regions, little attention has so far been paid to the investigation of the properties of the individual fibrils or to the way these properties may change from place to place in an active region. The meagre information available may be summarized as follows:

(1) *Shape and size*. Individual fibrils normally take the form of long thin dark streaks whose apparent widths are of the order of 1–3″ of arc (725–2200 km). Their lengths vary considerably and may change by a large amount during the course of their lives [Lippincott, 1955]; however, a figure of about 15″ (11 000 km) is usually quoted as a representative value [see, for example, Lippincott, 1955; Howard and Harvey, 1964; Smith, 1964]. In essential agreement with these estimates, Foukal [1971b, Table II] quotes representative values of 2000 and 15 000 km for the widths and lengths, respectively. It is not uncommon for adjacent fibrils to merge at their ends to produce a brush- or tuft-like appearance [Smith and Smith, 1963, p. 23].

(2) *Lifetime and evolution.* Smith [1964] states that individual fibrils have a lifetime of 10 to 20 min, an estimate which is in good agreement with the value of 17 min found by Loughhead [1968a] for the lifetime of fibrils in the immediate vicinity of isolated sunspots (see Section 5.2.3). Foukal [1971b] finds that the lifetime increases with the size of the feature, ranging from only

about one minute for the smallest fibrils up to 20 min for the largest. According to Lippincott [1955], Smith [1964], and Gimse [1967] fibrils show a distinct tendency to recur in approximately the same location.[5]

Apart from a single investigation of the fibrils in the immediate vicinity of isolated sunspots (Section 5.2.3), no systematic attempt to study the evolution of individual fibrils has so far been made.

(3) *Motion.* Using a sequence of photographs of an active region obtained by Lyot in 1942, Lippincott [1955] has investigated the apparent movements of the two ends of individual fibrils. She finds that these involve random fluctuations in position, the motion at one end bearing no relationship to that at the other. Sometimes there is a real or apparent merging with other fibrils, but there is no translation of the fibril as a whole. According to Lippincott the motion in any particular direction lasts, on the average, for 8 min; the apparent velocity during this period is 7 km sec^{-1}. On the other hand, from an examination of films taken at the Lockheed Solar Observatory, Smith [1968] claims to have detected continuous motion in fibrils in the direction of their lengths. This is supported by the recent work of Foukal [1971b], who measured changes in the lengths of a few individual fibrils and derived apparent velocities of 20–30 km sec^{-1}.

The question of the physical conditions obtaining in individual fibrils has recently been investigated theoretically by Pikelner [1971], who concludes, amongst other things, that they are maintained in pressure equilibrium with their surroundings by the presence of weak magnetic fields with strengths of a few gauss directed along their axes. On the other hand, Foukal [1971b] claims that the fields inside individual fibrils are greater than 100 gauss. He suggests that fibrils are produced by shock waves propagating along magnetic tubes of force, in a manner similar to that advanced by E. N. Parker to explain the origin of spicules (see Section 7.6.2).

5.2.3 SUPERPENUMBRAE OF ISOLATED SUNSPOTS

When an active region enters its declining phase the sunspots associated with it gradually dissolve until only a single spot remains. The fibril pattern in the immediate vicinity of such an isolated spot often takes a very distinctive and simplified form. As illustrated in Plate 5.6, the individual fibrils run more or less radially outwards from the spot to form a pattern very much resembling a greatly enlarged version of the white-light penumbra [see also Howard and

[5] This fact, coupled with somewhat inadequate resolution, may explain why Kiepenheuer [1957] and Gimse [1967] came to the conclusion that fibrils last for several hours.

Harvey, 1964, Fig. 2]. This resemblance is again well shown by Fig. 5.1, which gives a schematic drawing of the fibril pattern visible around the isolated spot of Plate 5.6a. (The white-light penumbra is discernible in Plate 5.6a as an area of reduced brightness surrounding the spot umbra; the approximate location of its outer boundary is shown by the dashed line in Fig. 5.1.) This is not to suggest, however, that there is necessarily any physical relationship between the chromospheric fibrils and the underlying photospheric penumbral filaments.

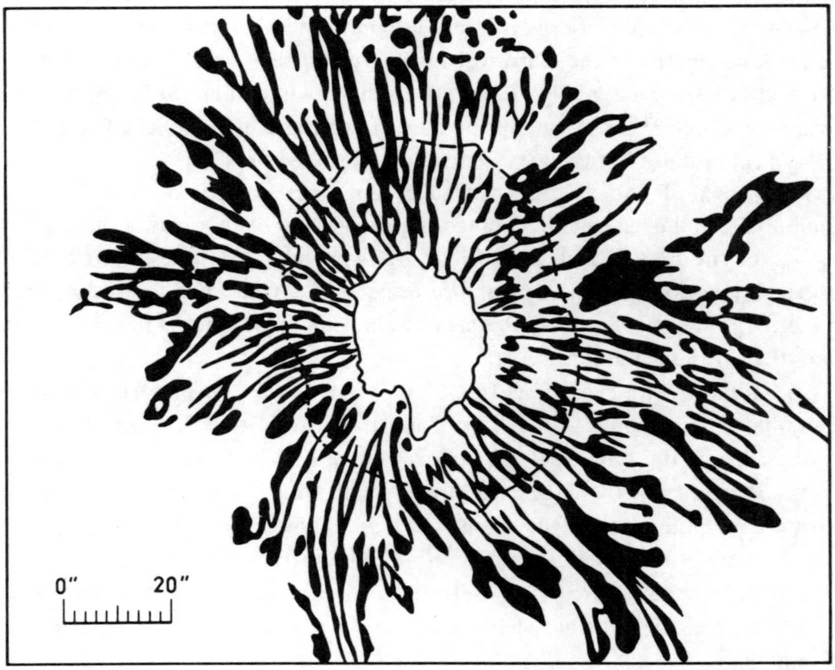

Fig. 5.1. Schematic drawing of the fibril pattern (superpenumbra) visible around the isolated sunspot of Plate 5.6a, demonstrating its resemblance to a greatly enlarged version of the white-light penumbra. Outlines of the umbra and penumbra are shown by solid and dashed lines respectively.

Nevertheless, the fibril pattern gives the unmistakable impression of being intimately related to the spot and may accordingly be described as a chromospheric 'superpenumbra'.

The only detailed investigation of the properties of superpenumbra fibrils so far undertaken is that of Loughhead [1968a], using films obtained with the 12-inch Culgoora refractor. The information obtained may be conveniently summarized under the following headings:

(1) *Shape and size*. As a general rule individual fibrils take the form of long thin dark streaks which begin within the penumbra and run outwards in a more

or less radial direction for upwards of a spot diameter (cf. Plate 5.6 and Fig. 5.1). Some fibrils begin at the border of the umbra – a fact also pointed out by Howard and Harvey [1964] – while a few lie entirely outside the penumbra. Despite their general tendency to run radially outwards, the fibrils may in places show a definite curvature. Moreover, many appear to branch or coalesce.

The fibrils are packed most closely together near the umbral border, where the distance between the centres of adjacent fibrils may be as little as 1″ of arc; further out their separation is typically 2–3″ but may be considerably greater. Outside the spot some fibrils increase in width and assume rather complicated shapes. The lengths of the individual fibrils are also very variable and, in the case of the spot illustrated in Fig. 5.1, range from a few sec of arc to 50 sec or more. A value of about 25″ may perhaps be quoted as a representative length and 2–3″ as a typical distance between the centres of adjacent fibrils.

Some idea of the size of the superpenumbra compared to that of the penumbra can be gauged by comparing the values of the ratios A_s/A_u and A_p/A_u for the spot of Plate 5.6, where A_s, A_p, and A_u denote the areas of the superpenumbra, penumbra, and umbra respectively. One finds that $A_s/A_u \simeq 18$ and $A_p/A_u \simeq 4$, showing that the area of the superpenumbra is more than four times that of the penumbra.

(2) *Lifetime and evolution.* Considered as a large-scale pattern, the superpenumbra is remarkably stable. This is illustrated in Plate 5.6, which shows two photographs of the same spot taken 1.5 hours apart. Not only does the general region occupied by the superpenumbra remain much the same, but even the curvature of individual fibrils at any particular place is largely unaltered. This accords with the earlier observation of Lippincott [1955], mentioned in Section 5.2.2, that there is no detectable change in the average *orientation* of the fibrils in the general neighbourhood of a sunspot group over a period of 45 min.

This is not to say, however, that the individual fibrils remain identifiable over a period of 1.5 hours. On the contrary, an inspection of Plate 5.6 shows that most of the individual fibrils can no longer be recognized as the same structures after this time, although the general pattern of their arrangement and, in particular, their curvature retain a well-marked similarity. A 42-min sequence of high-quality photographs of another isolated and fairly regular spot was used for the investigation of changes in the individual fibrils. Plate 5.7 shows selected pairs of photographs of the same region of the superpenumbra of this spot, separated by intervals ranging from 0.5 to 42 min. The changes in the individual fibrils occurring during the various intervals may be described as follows:

(i) after only 0.5 min any change is almost imperceptible;

(ii) after 3 min some fibrils show minor changes in shape and brightness. The majority, however, are still readily recognizable;

(iii) after 8.5 min major changes are apparent and only about half the fibrils retain sufficient recognizable features to be regarded as the same identifiable structures;

(iv) after 21 min the changes are so great that only a few fibrils can still be recognized as the same structures; and

(v) after 42 min possibly just one fibril remains identifiable, although the general pattern of their arrangement and, in particular, their curvature retain a well-marked similarity.

Summarizing, we may conclude that individual fibrils show minor changes over 3 min, major changes over 8.5 min, and are rarely identifiable after 21 min, but that the general pattern of their arrangement remains stable for at least 42 min and possibly much longer. It seems reasonable, therefore, to estimate the lifetime of the individual superpenumbra fibrils as about 17 min. Some are much shorter-lived, while an occasional one may last for twice this time. The value of 17 min is in good agreement with the estimate of 10–20 min given by Smith [1964, p. 6] for the lifetime of fibrils in general (see Section 5.2.2).

The lifetime determination described above is clearly rather subjective since the observer has to use his personal judgement in deciding how much change to allow a fibril before it is deemed to have become unrecognizable. A careful examination of all the photographs in the 42-min sequence illustrated in Plate 5.7 gives the impression that individual fibrils do not appear and disappear suddenly. On the contrary, they seem to undergo continual small changes in brightness, size, and shape, and as a result gradually become unrecognizable after a period of 20 min or so. In this respect they behave similarly to the ordinary dark mottles of the undisturbed chromosphere (Section 3.2.4).

Finally, it is of interest to enquire how the lifetime of the superpenumbra fibrils compares with that of (*a*) the dark chromospheric mottles and (*b*) the penumbral filaments observed in white light. As we have seen in Section 3.2.4, the lifetime of the dark mottles is about 5 min, so that they are considerably shorter-lived than the fibrils [see also Howard and Harvey, 1964]. On the other hand, the penumbral filaments are known to remain identifiable for periods of the order of several hours [Bray and Loughhead, 1964, p. 77] and hence, presumably, are different structures.

(3) *Change of contrast with wavelength.* Photographs taken with a carefully adjusted, single-peaked filter (see footnote 3) show that the superpenumbra as a whole is more prominent at the centre of Hα than at Hα – 0.5 Å or Hα + 0.5 Å.[6] On the other hand, the white-light penumbra becomes progressively

[6] Howard and Harvey [1964] claim that the fibrils in active regions show their greatest contrast in the blue wing of the line (Hα – 0.45 Å). However, Foukal [1971b] concludes that this is true for what he terms the 'roots' of the fibrils but not for their much longer 'horizontal sections'. The enhancement in the contrast of the 'roots' is approximately the same in the blue and red wings. Illustrations showing the appearance of the fibrils on the two sides of the line have been published by Kiepenheuer [1967, Fig. 3] and Foukal [1971b, Figs. 2b,c].

more distinct as one tunes outwards from the line centre. The umbra–penumbra boundary is diffuse at the line centre but sharp in the wings.

5.2.4 RELATIONSHIP TO MAGNETIC FIELD DISTRIBUTION

Within an active region the large dark filaments (prominences) demarcate the boundary between areas of opposite magnetic polarity [cf. Babcock and Babcock, 1955; Howard, 1959; Martres *et al.*, 1966; von Klüber, 1968; Bhatnagar, 1971; Prata, 1971; Smith, 1971; Bruzek, 1972; Frazier, 1972], while both sunspots and faculae are pervaded by magnetic fields of appreciable strength. Accordingly, the fact that all three features exert a strong influence on the form of the fibril pattern in their neighbourhood suggests that the arrangement of the fibrils in an active region is governed by the configuration of the magnetic field. This has led to the speculation that the individual fibrils may trace out the actual field lines themselves [see, for example, Kiepenheuer, 1953, p. 436; Loughhead, 1968a; Smith, 1968; Bhatnagar, 1971; Foukal, 1971a].

Several workers have attempted to compare the configuration of the fibrils with that of the longitudinal component of the underlying *photospheric* magnetic field, using magnetograms obtained at the Mt. Wilson Observatory [Howard and Harvey, 1964; Smith and Ramsey, 1967; Veeder and Zirin, 1970; Bhatnagar, 1971]. However the low spatial resolution of the magnetograms (5″ of arc at best), combined with a lack of knowledge of the *transverse* field component, makes it impossible to specify the exact direction of the field within any individual fibril. Moreover, it must be stressed that the field distribution in the chromosphere may differ appreciably from that observed in the photosphere [see, for example, Ioshpa *et al.*, 1964; Severny, 1966; Brückner, 1968; Brückner and Waldmeier, 1969; Tsap, 1970]. Nevertheless, Smith and Ramsey found that almost every area with a magnetic field strength of 3 gauss or greater shows an organized pattern of fibrils. Except for those running between areas of opposite polarity, the fibrils were found to be longer in areas of 'weak' fields (3–6 gauss) than in areas of 'strong' fields (6–12 gauss).

Recently, Vrabec [1971] has compared magnetograms obtained with the new 61-cm vacuum solar telescope and spectroheliograph at the San Fernando Observatory with simultaneous photographs of the same region taken at a number of selected wavelengths in the Hα line. The magnetograms are produced by means of the ingenious photographic subtraction technique devised by R. B. Leighton and show the distribution of the longitudinal component of the photospheric magnetic field with a spatial resolution of about 2.″5 of arc. Although their resolution is still too low to reveal the field associated with individual fibrils, Vrabec's observations do show that fibrils extending out from regions of the same magnetic polarity avoid connecting with one another [see also Foukal, 1971b]. This provides additional support – but not definite proof

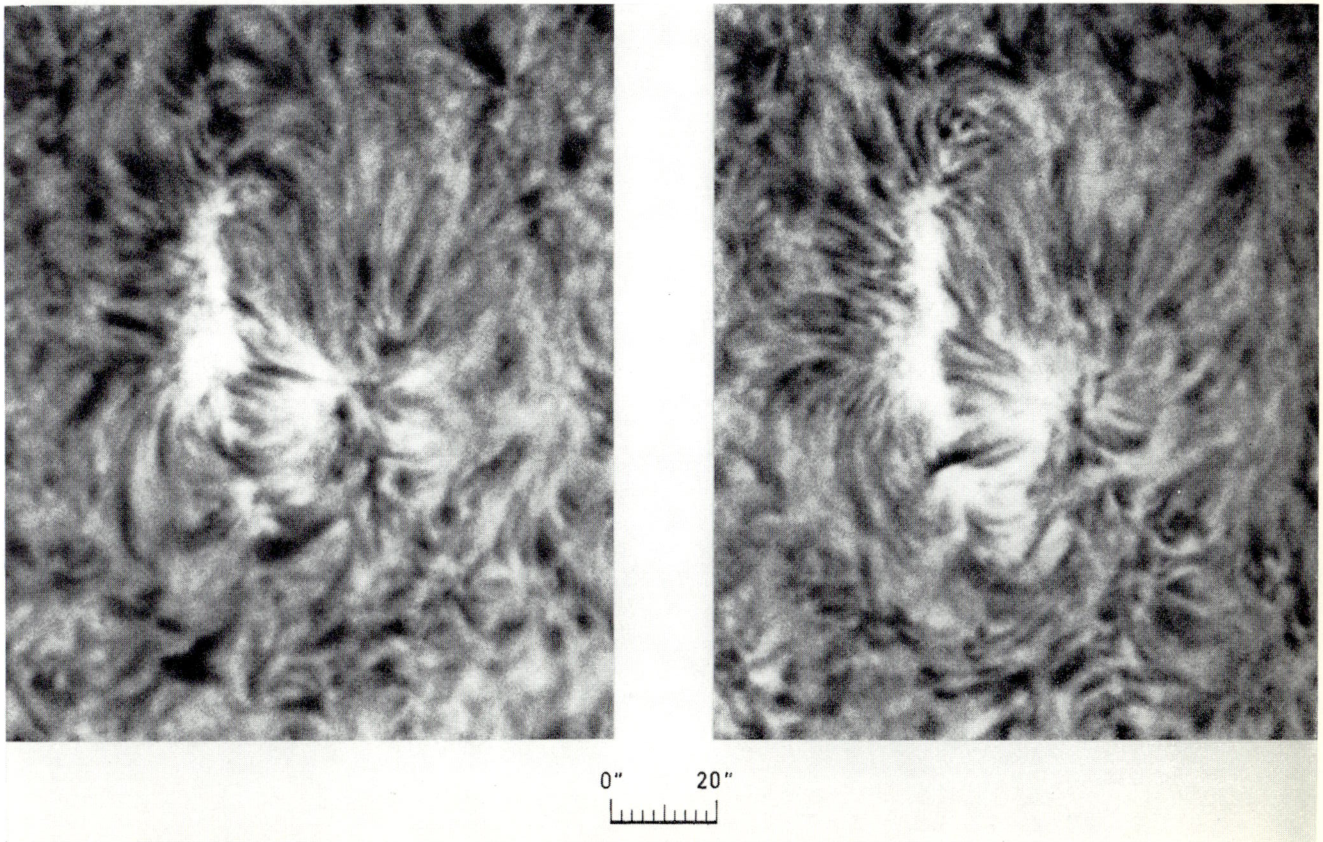

PLATE 5.1. Rapid growth of a new active region over a period of just under four hours. It is composed of two small bright facular areas separated from the surrounding undisturbed chromosphere by a number of elongated dark fibrils. The region was born on the eastern hemisphere on the afternoon of the previous day.

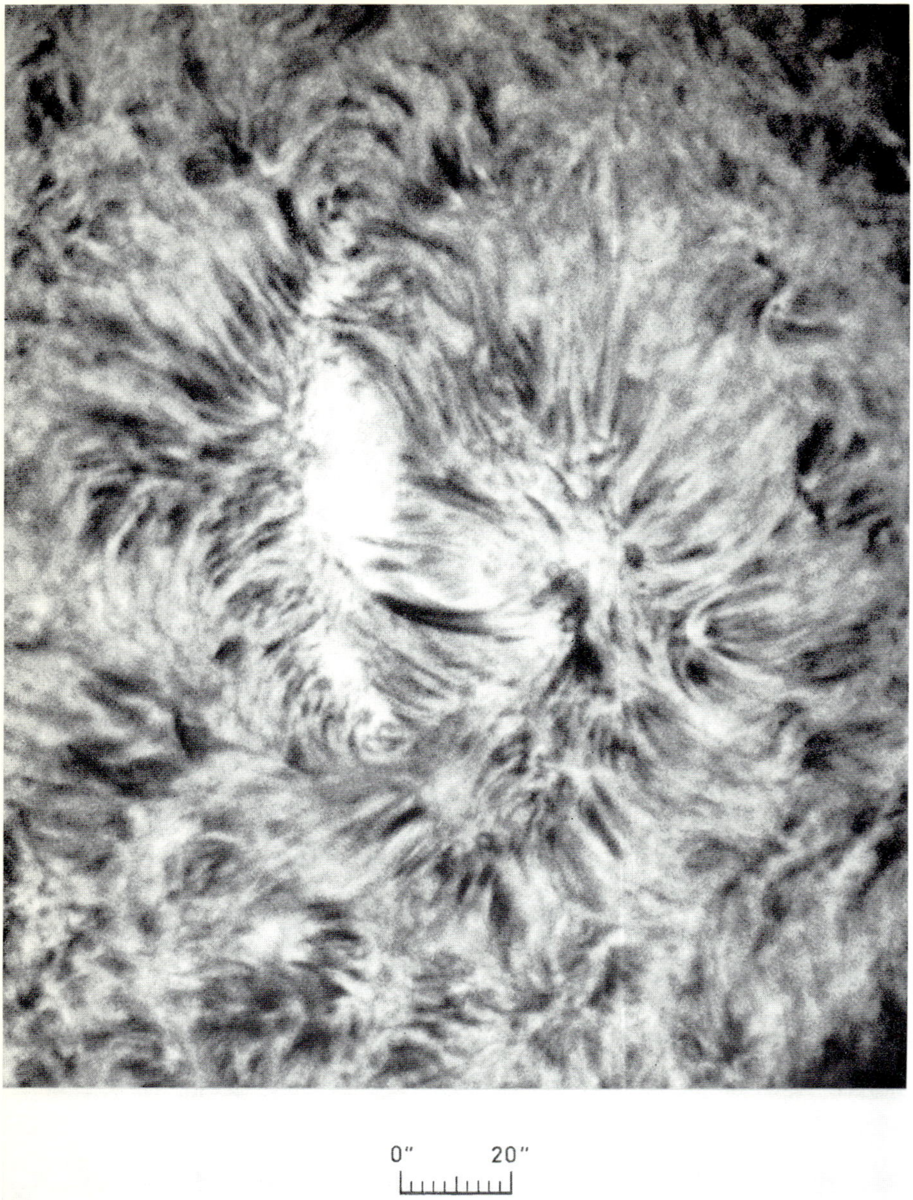

PLATE 5.2. Growth of a new active region (continued): the region of Plate 5.1 some 24 h later. The central facular areas are now considerably larger, while the associated fibril structure intrudes further into the surrounding chromosphere and shows an increase in regularity and contrast.

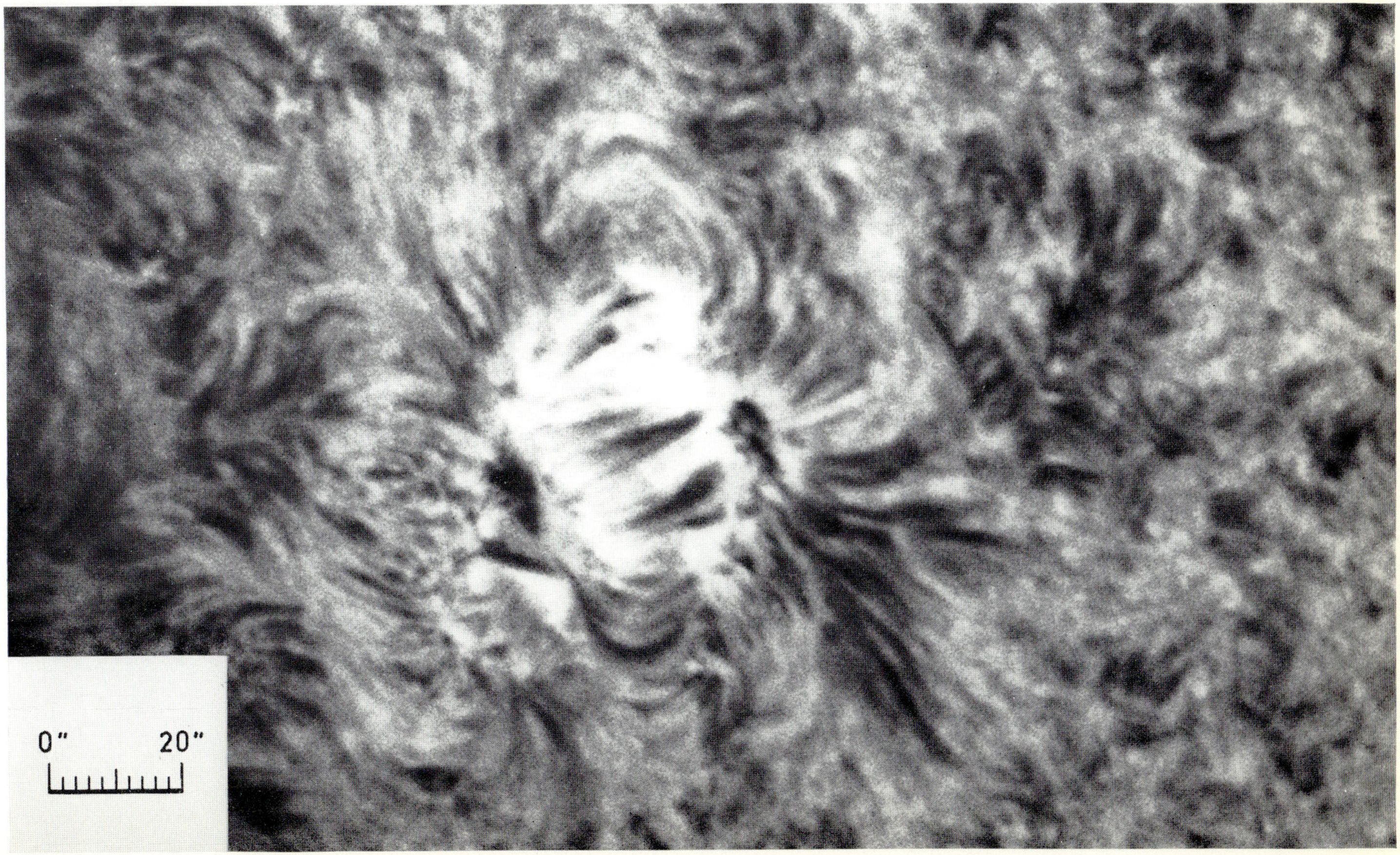

PLATE 5.3. A young active region. It appears to show a typically bipolar magnetic configuration. Note the prominent 'arch' filaments linking the two sides (see text).

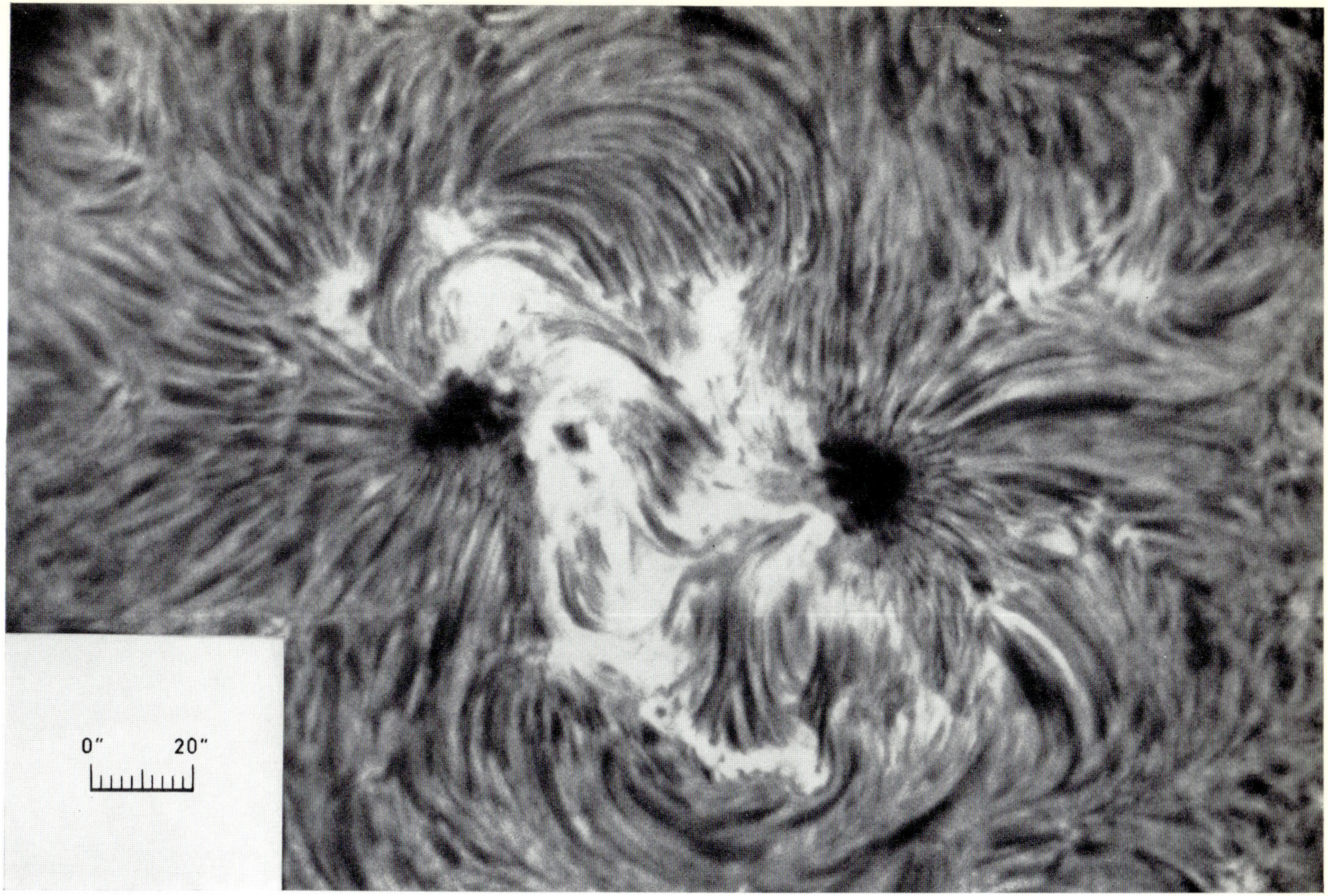

PLATE 5.4. Well-developed active region showing the 'iron filings' pattern of fibrils characteristic of a bipolar sunspot group.

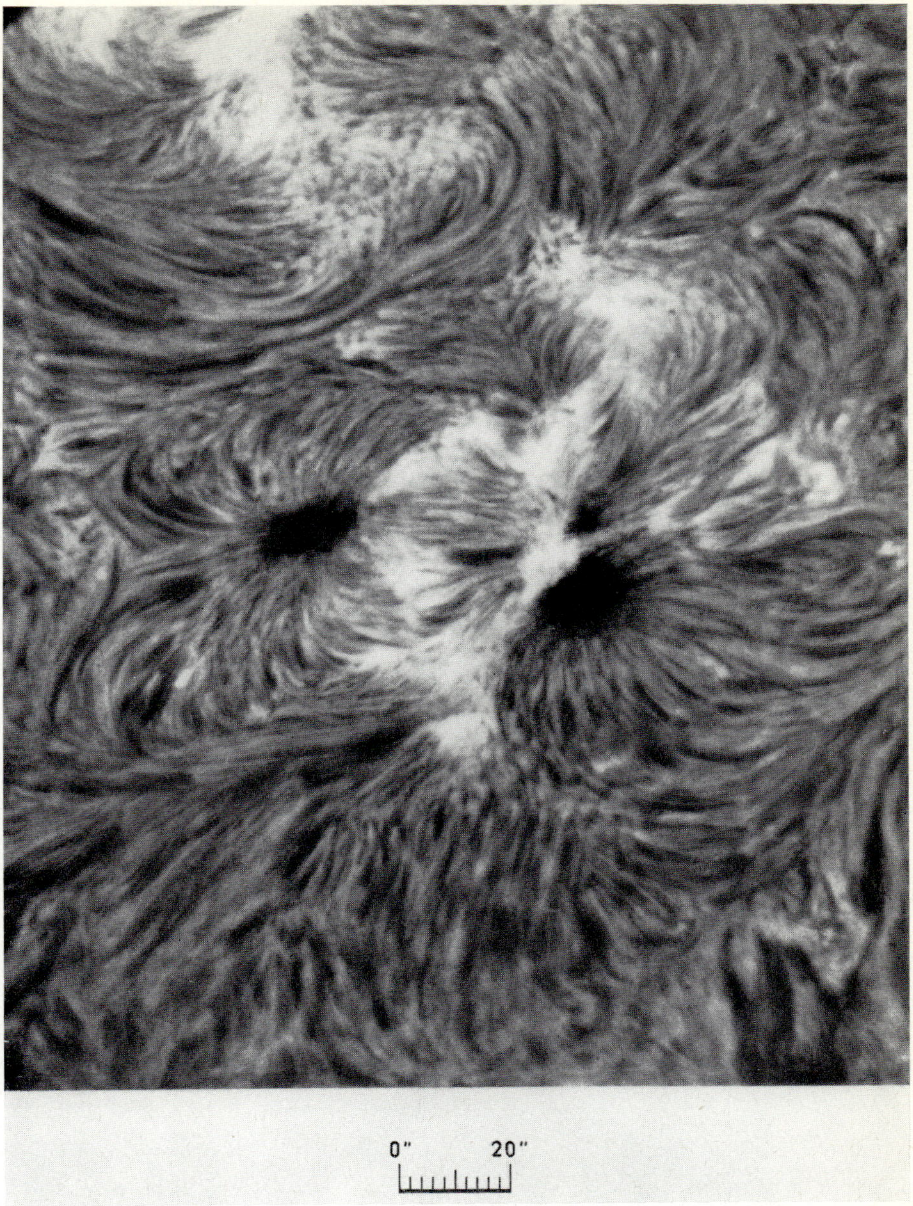

PLATE 5.5. Large bipolar active region showing a very extensive and complicated fibril structure. Note that the arrangement of the individual fibrils is affected not only by the sunspots but also by the faculae present.

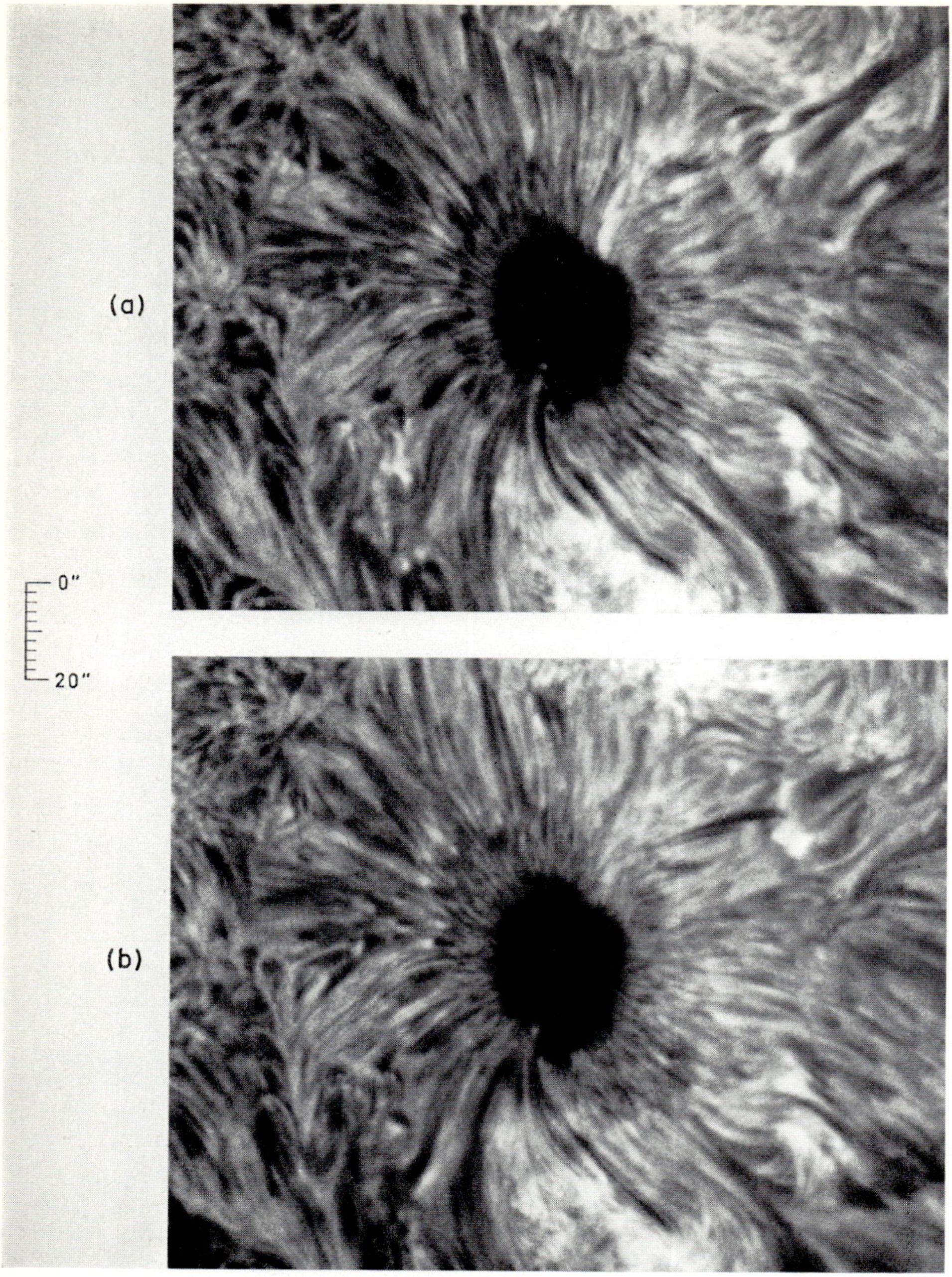

PLATE 5.6. Photographs taken 1·5 h apart showing the dark fibrils around an isolated sunspot. During this time the general pattern of their arrangement and, in particular, their curvature retain a well-marked similarity. However, most of the individual fibrils can no longer be recognized as the same structures.

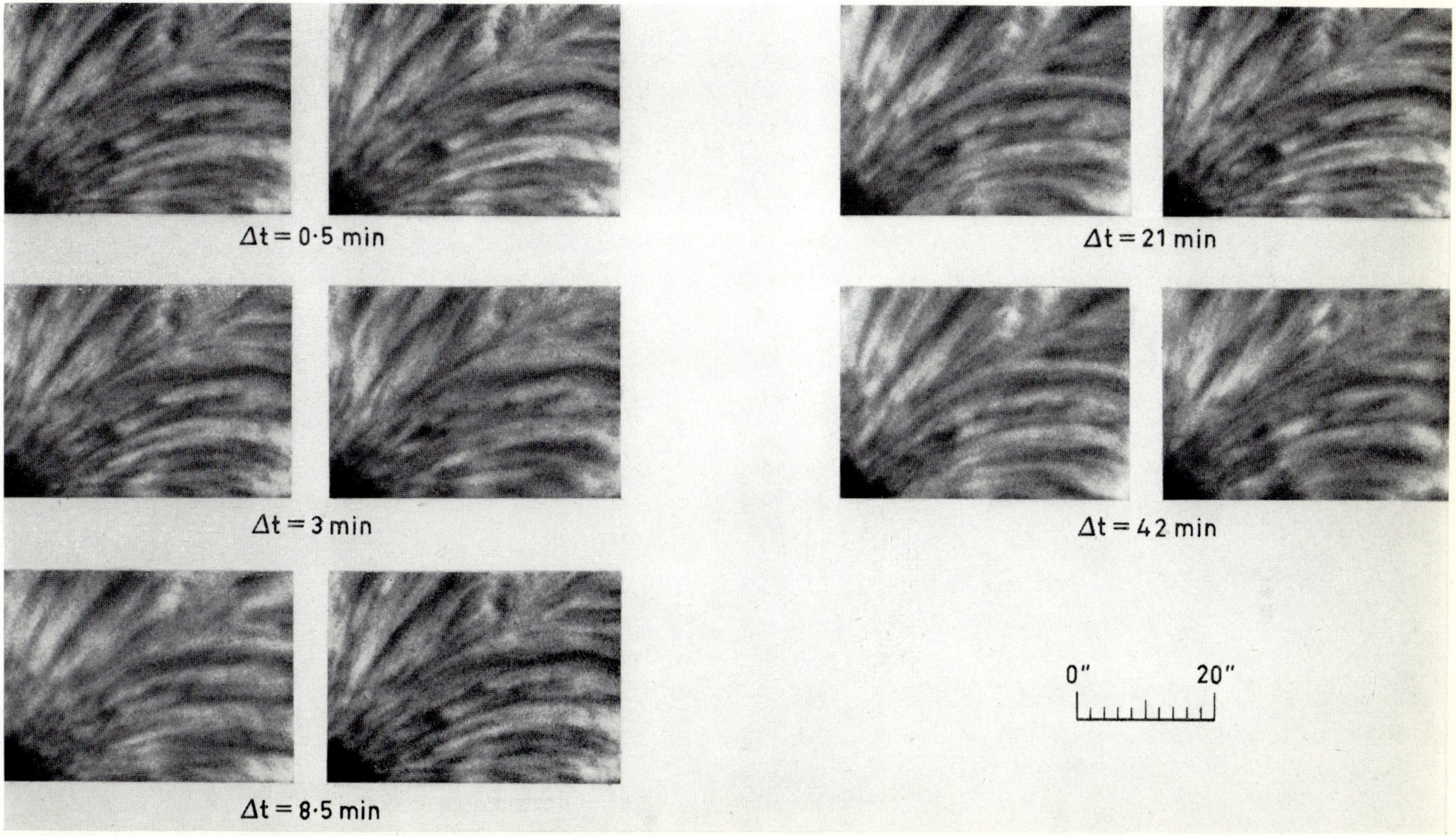

PLATE 5·7. Pairs of photographs of the same region of the superpenumbra of an isolated sunspot showing changes in the individual fibrils over intervals ranging from 0·5 min to 42 min.

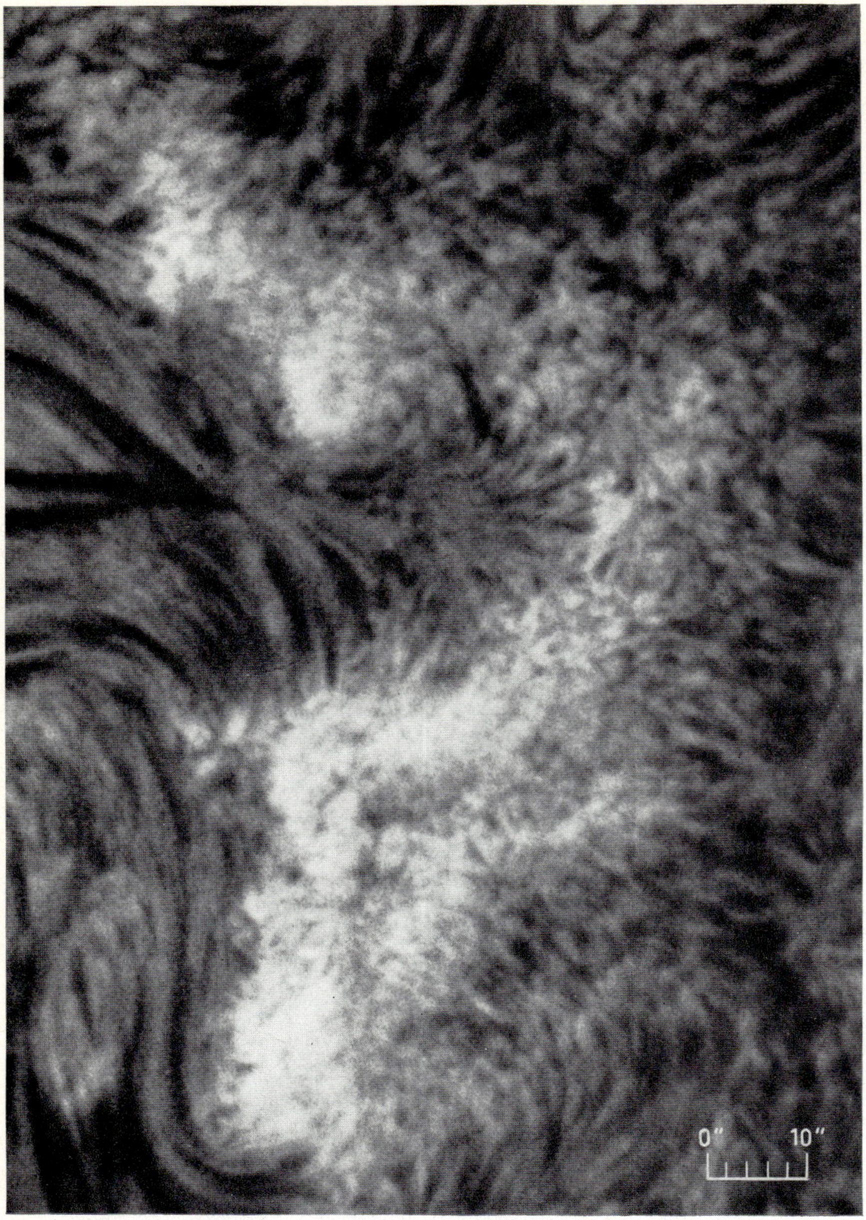

PLATE 5.8. Fine structure of a facular region close to a small isolated sunspot, showing partial resolution into facular granules. The individual granules have apparent diameters of about 1″ of arc, their centres being typically separated by about 1″·5; the smallest separation is about 1″.

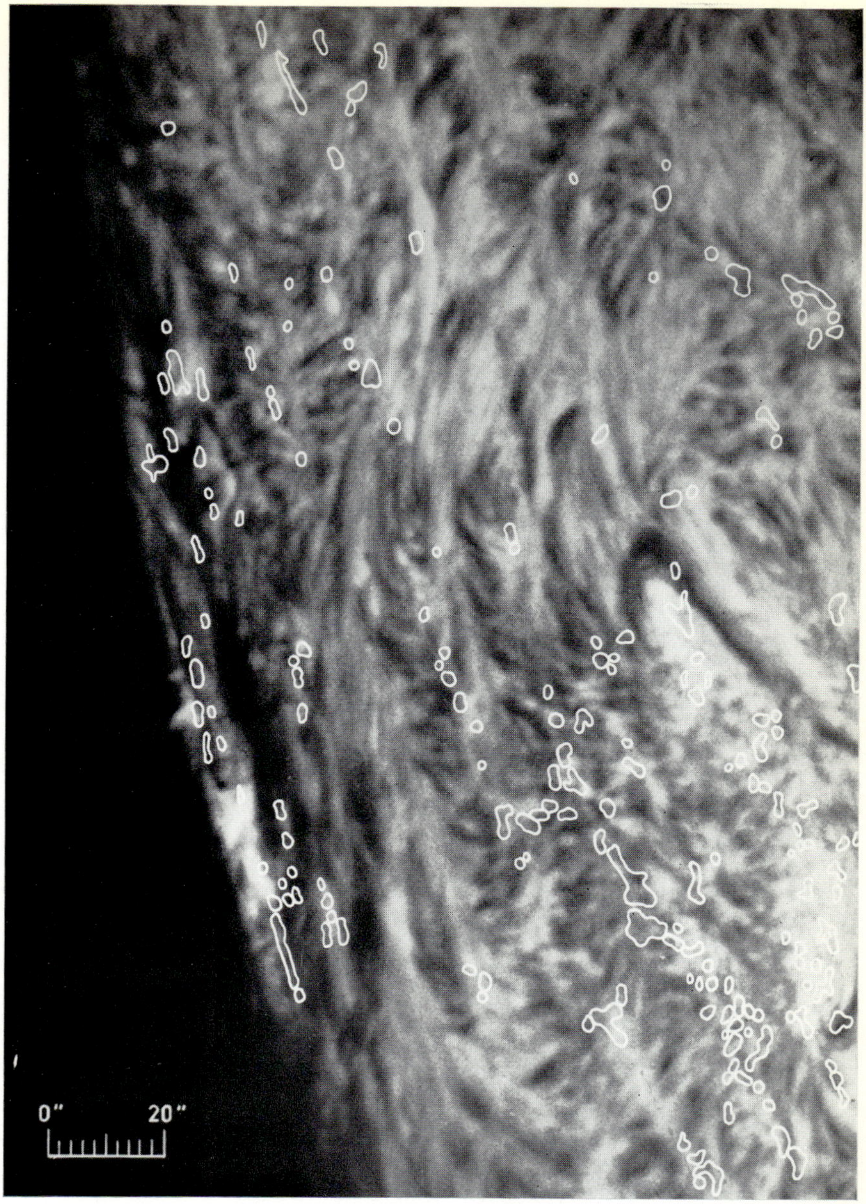

PLATE 5.9. Composite illustration, showing the outlines of the photospheric facular granules (white lines) superimposed on an Hα photograph of a small active region near the east limb taken some 22 min earlier. The photospheric faculae are widely distributed over most of the photograph and appear to be associated indiscriminately with both bright and dark Hα regions. Shading has been used to reduce the effect of limb darkening.

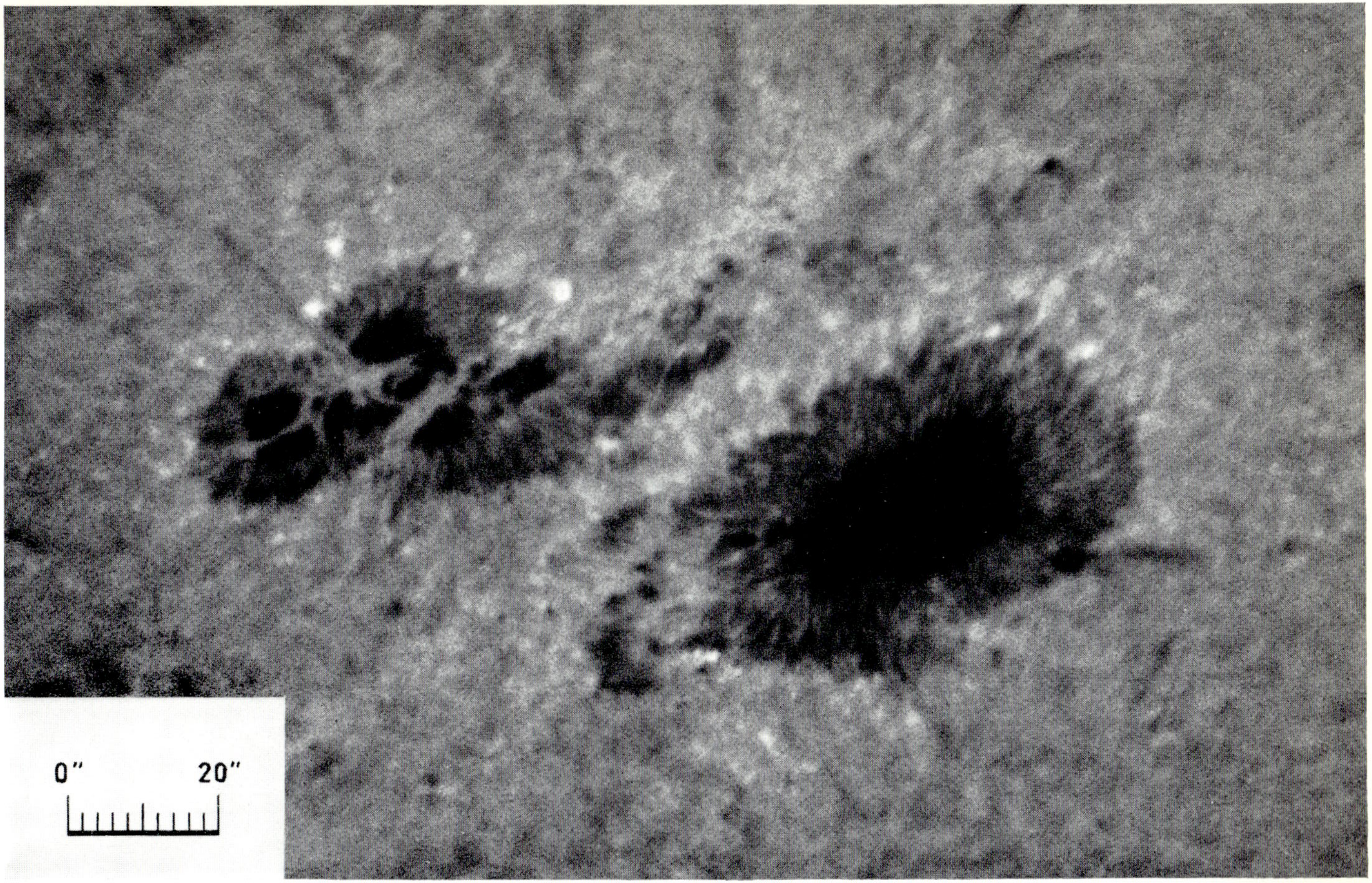

PLATE 5.10. Moustaches in the immediate vicinity of a large sunspot group photographed through a 0·5 Å Halle filter set at Hα + 1·0 Å; at this setting the filter transmits twin 0·5 Å bands, one centred on Hα − 1·0 Å and the other on Hα + 1·0 Å. Note that most of the moustaches are located on or close to the outer borders of the spot penumbrae.

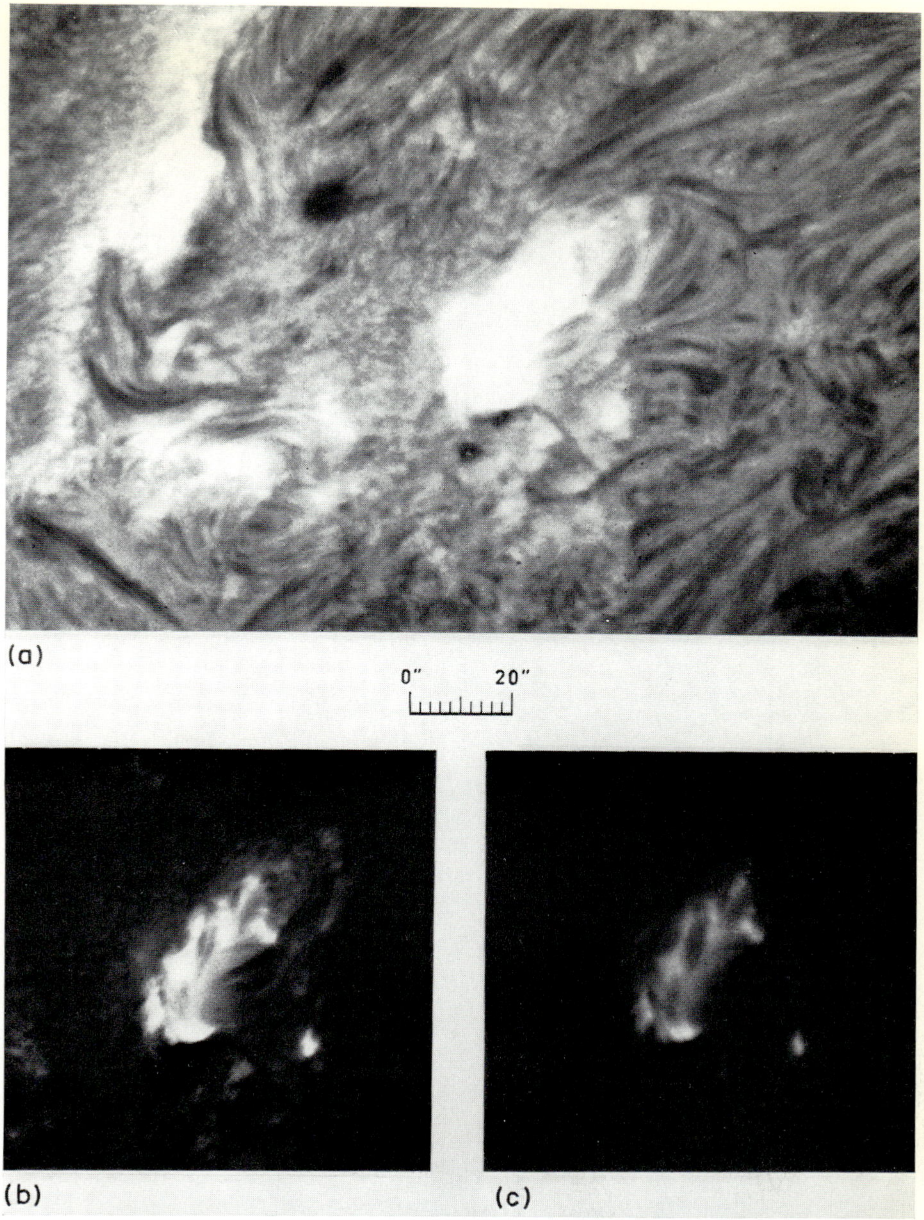

PLATE 5.11. Appearance of a Class Ib flare on photographs taken with different exposures (30 August 1967). (a), (b), and (c) show a normal photograph and two graded under-exposures taken at 00.18.06, 00.18.31, and 00.18.41 UT respectively.

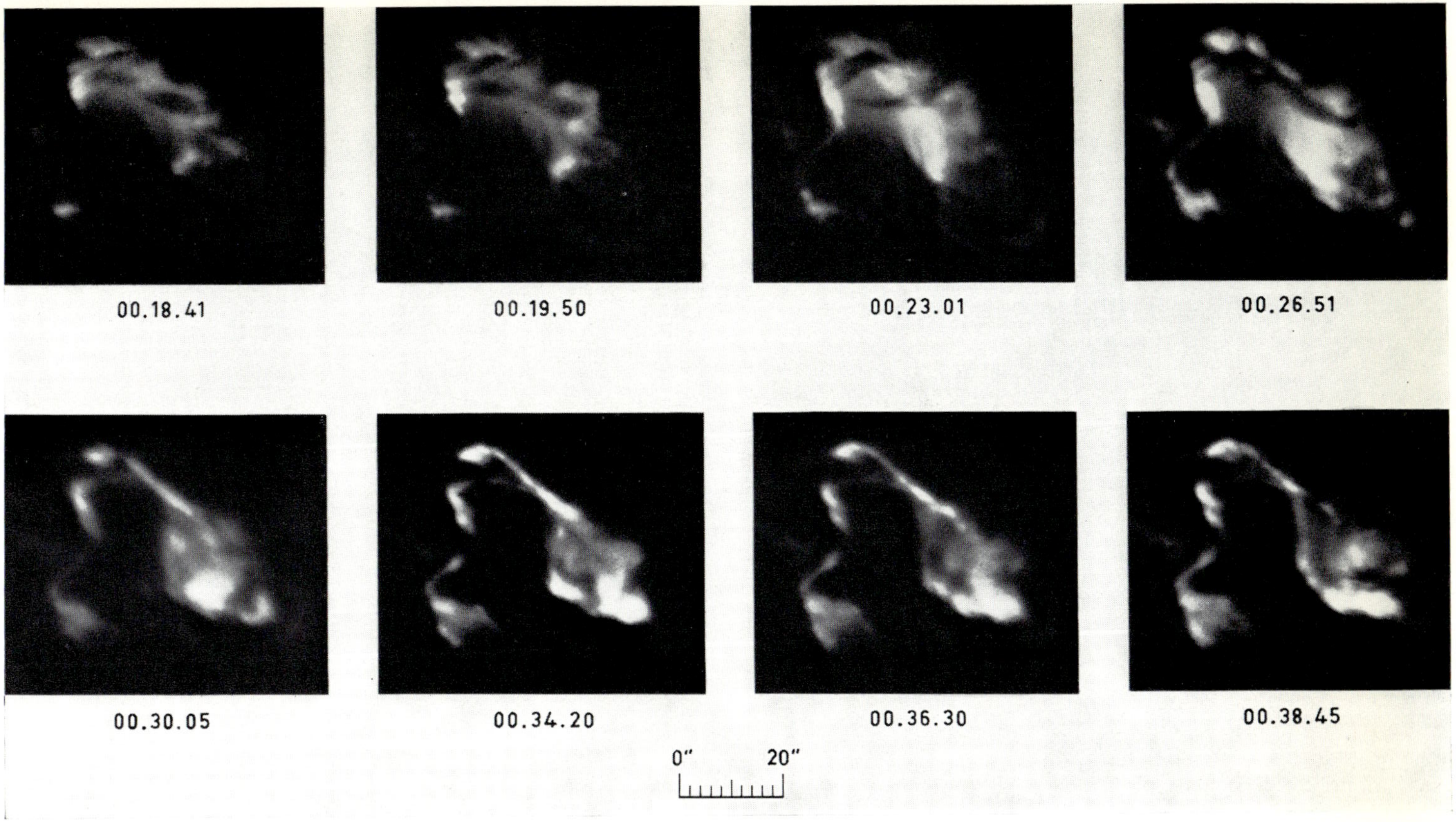

PLATE 5.12. Sequence of underexposures showing the fine structure of the flare of Plate 5.11 and its changes over a period of some 20 min. Flare maximum occurs at 00·31 UT. Note the segmented or waisted appearance of the thin bright filament connecting the left- and right-hand parts of the flare (00.34.20). (The photographs in Plate 5.12 are orientated at right angles to those in Plate 5.11.)

[Frazier, 1972] – for the view that the fibrils map out magnetic lines of force.[7] One of Vrabec's magnetograms, together with a simultaneous photograph taken at the centre of Hα, is reproduced in Plate 3.13.

Finally, it is interesting to note that, in a recent theoretical study, Nakagawa *et al.* [1971] have examined the possibility of explaining the form of the superpenumbra around an isolated sunspot in terms of the configuration of the lines of force of an axisymmetric force-free chromospheric magnetic field centred on the spot. A similar attempt has been made by Raadu and Nakagawa [1971] to explain the characteristic 'iron filings' pattern of fibrils and filaments surrounding a bipolar sunspot group, such as that shown in Plate 5.4.

5.3 Chromospheric Facular Granules

5.3.1 FINE STRUCTURE OF CHROMOSPHERIC FACULAE

Modern high-resolution photographs show that Hα faculae in developed active regions possess a distinct granular structure on a scale of about 1–2″ of arc. A good illustration of this structure is given in Plate 5.8, taken with the 12-inch Culgoora refractor. It shows a facular area close to a small isolated sunspot which can be seen to be partially resolved into a conglomerate of *facular granules*. The individual granules appear to have diameters of about 1″, while the typical distance between the centres of adjacent granules is approximately 1″.5. The smallest separation between any pair of clearly resolved granules in Plate 5.8 is about 1″ – a figure which may be compared with the known effective resolution limit of the telescope, 0″.7. As one tunes out into the wings of the Hα line the contrast of the facular granules decreases until, ultimately, the majority appear dark with respect to the background [Kiepenheuer, 1960a]. Smith [1968] claims that individual Hα facular granules show a relatively rapid appearance and disappearance, but a proper determination of their lifetimes has yet to be made. The same remark applies to their possible velocities and magnetic fields (cf. Section 5.3.3). Other high-quality illustrations of the granular structure of Hα faculae have been given by Smith [1968, Figs. 3 and 4] and Zirin [1971b, Fig. 4].

In a recent paper Bhatnagar and Tanaka [1972] claim to have detected brightness oscillations in individual facular granules on sequences of photographs taken at Hα – 0.5 Å; the observed periods range from 259 to 313 sec. They state

[7] Vrabec also points out that Hα photographs taken at wavelengths equally spaced on either side of the line centre show significant differences, thus suggesting that an adequate knowledge of the velocities in fibrils may be another essential pre-requisite to a final understanding of their origin. The particular question of whether the fibrils around isolated sunspots described in Section 5.2.3 are related in any way to the Evershed motions at the chromospheric level has been discussed by Loughhead [1968a] and Haugen [1969]. Spectroscopic measurements of velocities in the Hα line bearing on this question have been reported by Abdussamatov [1969].

that oscillations can also be observed at the line centre, but with considerably reduced amplitude.

As a general rule the Hα faculae in the close vicinity of sunspots are brighter than those further away and, moreover, the extent and brightness of the former provide a good indication of the activity of the spot group. It has been stated by Kiepenheuer [1957, 1960a,b] that the faculae near spots differ from the rest in that they can no longer be resolved into facular granules but instead show an amorphous or ribbon-like structure. However, failure to resolve bright facular areas may be due simply to the inadvertent overexposure of such areas on normally exposed photographs. Alternatively, it could result from a closer packing of the facular granules in the immediate neighbourhood of sunspots, although no observational study of this question has yet been made.

5.3.2 RELATIONSHIP BETWEEN CHROMOSPHERIC AND PHOTOSPHERIC FACULAE

Although chromospheric faculae can be seen at all heliocentric angles, their photospheric counterparts are visible only near the limb where, as a rule, they are distributed over a large area in the neighbourhood of a sunspot group. On the other hand, photospheric and chromospheric faculae resemble one another in that both show a *granular* structure on a comparable scale. In fact, each photospheric facular area consists of a loose conglomeration of individual bright facular granules superimposed on a background of normal photospheric brightness [Bray and Loughhead, 1967, pp. 55–57, Plate 3.1]. However, not all photospheric facular granules occur in clusters, many appearing as completely isolated bright points. Most photospheric facular granules have diameters in the range 1–2″ of arc and their lifetime is of the order of 2 hours.

In the past it has often been asserted, albeit without direct observational proof, that the chromospheric and photospheric faculae are closely correlated in shape and position. And this supposition, coupled with the rather similar granular appearance exhibited by both phenomena, led to a widely held view that they represent the same structures seen at different heights [see, for example, Kiepenheuer, 1953, pp. 371, 375; de Jager, 1959, p. 173; Smith and Smith, 1963, p. 18; Bray and Loughhead, 1964, pp. 249, 252; Zirin, 1966, p. 380]. Subsequently, however, a careful observational investigation of the problem by Bray [1968] has shown that the correlation between the chromospheric and photospheric faculae is actually weaker, or at least less obvious, than had hitherto been supposed.

This conclusion is well demonstrated by Plate 5.9; it shows an Hα photograph of a small active region near the east limb, superimposed upon which are the outlines of the underlying photospheric facular granules. One can see that the photospheric faculae are widely scattered over most of the photograph and appear to be associated indiscriminately with both bright and dark Hα regions. Note that the brightest Hα region (just to the left of the sunspot) is scarcely associated with any white-light emission at all. A similar comparison for two other active regions showed that, in both cases, photospheric faculae occur in the same general area as the brightest Hα regions but there is no detailed correlation in brightness, size, or position. Other small bright Hα regions present are not associated with photospheric faculae.

Mussayev [1968] has independently reached a similar conclusion to Bray in regard to the relationship between photospheric faculae and the faculae observed in the H and K lines of Ca II [see also Krat, 1965; Stoyanova, 1970].

5.3.3 MAGNETIC FIELDS OF CHROMOSPHERIC FACULAR GRANULES

Like the K faculae [see, for example, Bray and Loughhead, 1964, p. 260], Hα faculae show a definite correlation with the distribution of the longitudinal component of the underlying *photospheric* magnetic field, with the understandable exception of the neighbourhood of 'neutral' lines [Howard and Harvey, 1964; Smith and Ramsey, 1967; Vrabec, 1971; Frazier, 1972]. On the other hand, observations bearing directly on the question of magnetic fields in individual Hα facular granules have yet to be obtained. Some of the difficulties facing attempts to measure chromospheric as opposed to photospheric magnetic fields have been outlined in Section 3.5.4.

5.4 Moustaches and the Fine Structure of Flares

5.4.1 INTRODUCTION

To complete our discussion of the fine structure of the active chromosphere we shall give accounts of the fine detail exhibited by two of the transient phenomena associated with active regions, namely moustaches (Section 5.4.2) and flares (Section 5.4.3). These have been singled out for attention because there are reasons for believing that their study may help to elucidate further the role of magnetic fields in governing the nature and behaviour of the fine structures observed in active regions. Recent years have seen a great increase in our knowledge of the morphological and spectroscopic properties of moustaches, but an up-to-date review of the subject is lacking. For this reason a somewhat more detailed account of the properties of moustaches is given in the next section than might otherwise be justified.

5.4.2 MOUSTACHES

On 21 January 1909 W. M. Mitchell at the Haverford College Observatory in Pennsylvania observed with a spectroscope the transient appearance of two very bright streaks extending outwards for considerable distances from both wings of the Hα line at a point in an active region near the east limb. The subsequent suspension of astronomical work at Haverford prevented Mitchell from following up this discovery and it was left to Ferdinand Ellerman [1917], observing with the 150-foot solar tower telescope and spectrograph at Mt. Wilson, to provide the classical description of the spectral characteristics of the newly observed phenomenon. Typically, it takes the form of a very narrow band of bright emission extending for several Ångströms or more on either side of Hα and the other Balmer lines but not crossing the line centre.[8] Such events occur mainly in the neighbourhood of active sunspot groups, either in the regions between the spots or near the outer borders of their penumbrae (cf. Plate 5.10). Ellerman concluded that they had an average duration of only a few minutes, and this led him to propose the name 'solar hydrogen bomb'. Some later writers used the term 'Ellerman bomb', but this has now been generally supplanted by the rather more aptly descriptive name 'moustache' introduced by Severny [1956a,b]. It is a tribute to Ellerman that, as far as the basic characteristics of moustaches observed in Hα are concerned, the extensive modern observations have added little to his original description [cf. McMath *et al.*, 1960; Severny, 1965, p. 224].

Because of their brightness moustaches are easily photographed through birefringent filters tuned to the wings of Hα and the first photographs were obtained in 1942 by Bernard Lyot in this way. Plate 5.10 shows a number of bright moustaches in the immediate vicinity of a large sunspot group photographed through a 0.5 Å Halle filter set at Hα + 1.0 Å; at this setting (and at Hα − 1.0 Å) the filter transmits twin 0.5 Å bands, one centred on Hα − 1.0 Å and the other on Hα + 1.0 Å [Bruzek, 1968]. Note that most of the moustaches visible are located on or close to the outer borders of the spot penumbrae. Their apparent sizes range from about 3″ of arc to less than 1″, i.e. down to the effective resolution limit of the photograph. Illustrations similar to Plate 5.10 have been published by Beckers [1964, Fig. 9], Bruzek [1967, Fig. 1; 1968, Fig. 1] and, of lesser quality, Severny [1965, Fig. 7].

In 1944 Lyot published the results of his pioneering photographic work on the chromosphere but made no mention of the identity of the small bright features he observed with Ellerman's 'bombs'. Ten years later, A. B. Severny and his collaborators began an extensive study of the spectral properties of moustaches and other related solar phenomena with the newly installed solar tower telescope and spectrograph of the Crimean Astrophysical Observatory [see, for example, Severny, 1957a]. The following year observations of moustaches were also begun with the newly constructed vacuum spectrograph at the McMath–Hulbert Observatory [McMath *et al.*, 1960]. An important refinement in their observing technique was introduced by the McMath–Hulbert

[8] As pointed out below, they also appear in many other Fraunhofer lines, including the H and K lines of Ca II.

workers in 1957: light reflected back from the jaws of the spectrograph slit was sent through a Lyot filter centred on Hα – 0.75 Å or Hα + 0.75 Å and was used as a means of locating individual moustaches and positioning them on the slit. More recently, observations of moustaches have been made at several other observatories and, as a result of all this work, we now have quite a detailed knowledge of their morphological and spectroscopic properties. These may be conveniently summarized under the following headings:

(1) *Basic spectral characteristics.* Moustaches have been observed in all members of the Balmer series up to H_{10} by Severny [1957a] and Koval [1967]. They are conspicuous in the wings of the H and K lines and also occur in a number of strong to weak lines of calcium, iron, and other metals [see, for example, the lists of lines tabulated by Koval, 1967]. According to the latter author moustaches show anomalous behaviour in the sodium D lines and in the magnesium *b* triplet, while Severny [1965] states that they are seen in the helium D_3 line only near the limb. The Soviet observers find that moustaches are usually superimposed on thin bands of continuous emission, which extend over the entire visible spectrum and whose contrast increases towards the violet [see also Engvold and Maltby, 1968, p. 112]. These so-called *grains of continuous emission* seem to be distinct features in their own right and, in fact, are often observed in the absence of moustaches; further information on their properties can be found in the papers by Severny [1957a,b] and Severny and Khokhlova [1959].*

According to Severny [1957b, 1965] moustaches can extend for upwards of 7–10 Å from the centre of Hα and 15 Å from the centres of H and K although, as Koval [1967] has pointed out, in the latter case their extension may be difficult to estimate owing to the presence of associated bands of continuous emission.[9] Actual profiles of the emission in Hα (and other Balmer lines) have been determined by Severny [1957b, 1965, 1968], Koval [1965a], Engvold and Maltby [1968], and Koval and Severny [1970]. These show that the emission is either absent or inconspicuous at the line centre but increases rapidly as one moves out into the wings, reaching a maximum within 0.5–2.0 Å in almost all cases. Thereafter the intensity gradually declines. The question of the actual physical mechanism responsible for the great line broadening observed in moustaches has been discussed by several workers [Severny, 1957b, 1965, 1968; Koval, 1967; Engvold and Maltby, 1968], but a conclusive answer has not yet been reached.

A second interesting feature of the Hα profiles is their apparent *asymmetry*: the majority of moustaches are brighter on the blue side than on the red,

[9] In cases where the continuous emission is absent or weak, Koval finds that the wings of the moustaches in the H and K lines may actually be less extended than in the Balmer lines.

Note added in proof:* see also Bruzek [*Solar Phys.* **26, 94: 1972].

although sometimes the reverse is true. In addition, a few moustaches show symmetrical profiles. There are rare instances where one wing is entirely absent, especially among moustaches observed close to the limb [cf. Severny, 1968, Fig. 7]. As a general rule, however, the asymmetry is independent of position on the disk. Besides noting the difference in intensity, several observers have expressed the opinion that moustaches also tend to extend further on the blue side than on the red [e.g. Severny, 1957a,b, 1965; Engvold and Maltby, 1968], although Koval [1965a] finds that most moustaches extend for equal distances on the two sides of the line. A detailed quantitative investigation of the asymmetry of the Hα and Hβ line profiles of a number of moustaches has recently been undertaken by Koval and Severny [1970], who pay particular attention to the possible presence of associated bands of continuous emission and to the uncertainties in fixing the levels of the continuum both inside and outside the moustaches.

The same asymmetry is seen in the H and K lines and, according to Koval [1967], in many of the other metallic lines exhibiting moustaches (see above). Koval finds that the asymmetry is more pronounced in the weaker lines and, for a given line, is more prominent in the early stage of development of the moustache than later.

Finally, it is interesting to note that rare cases have been observed of moustaches showing an *inclination* to the direction of dispersion [Severny, 1957a,b, 1968]. This effect is similar in character to the tilt of spicule emission streaks described in Section 2.4.4, but its cause is at present unknown.

(2) *Level of formation.* According to Severny [1957a, 1965] the emission of moustaches in the Hα line starts, on the average, at a distance of about 0.75 Å from the line centre and, as we have seen above, can extend outwards for up to 10 Å. This implies that moustaches are basically phenomena of the upper photosphere, a fact consistent with their appearance in many weak metallic lines with Rowland Intensities as low as 2 [Koval, 1967]. On the other hand, well-defined moustaches are occasionally observed in the Hα line *above* the limb [see, for example, Severny, 1968, Figs. 7–9] showing that, in these cases at least, the radiation originates in the low chromosphere. Taken together, these facts indicate that moustaches can occur over a range of heights extending upwards from the photosphere into the low chromosphere.

(3) *Shape and size.* On photographs taken in the wings of the Hα line moustaches generally appear to be circular in shape [Mohler and Dodson, 1958] and to have sizes ranging from about 5$''$ of arc down to a value equal to the resolution limit of the telescope [McMath *et al.*, 1960; Bruzek, 1968]. In the case of the best available observations, such as the photograph given in Plate 5.10, this limit is somewhat less than 1$''$ (cf. Section 3.2.2). According to Severny [1957b] spectroscopic observations indicate that the apparent width of

the wings (perpendicular to the dispersion) invariably decreases with increasing distance from the line centre. However, as Engvold and Maltby [1968] point out, with limited resolution this may result simply from the accompanying decrease in contrast. Severny [1965] claims that moustaches tend to increase in size and become more diffuse during the latter part of their lifetime. Neither Severny nor Engvold and Maltby were able to establish any correlation between apparent size and position on the disk.

(4) *Contrast*. According to McMath *et al*. [1960] the maximum ratio of the intensity of a moustache to that of the undisturbed spectrum at the same wavelength is, near Hα, about 1.3. This value is confirmed by Bruzek [1968] who, from measurements made on calibrated photographs, found values lying in the range 1.15–1.50; he quotes a typical figure of 1.25. The observed intensity ratios are much higher in the vicinity of the K line, where the values determined by the McMath–Hulbert observers were normally 3 or 4 but in several cases ranged as high as 10. Moustaches are therefore much more easily observed spectroscopically in the neighbourhood of the K line than near Hα.

Measurements of the contrast of moustaches in Hα have also been made by Engvold and Maltby [1968, Fig. 2] and in Hα and Hβ by Koval and Severny [1970].

(5) *Location and frequency of occurrence*. From an examination of photographs of some 20 active regions taken at H$\alpha \pm 0.5$ Å and H$\alpha \pm 1.0$ Å, Bruzek [1968] has concluded that moustaches occur throughout all active phases in the lifetime of active regions, but are absent around unipolar sunspots and decaying facular areas. Many moustaches are seen in large complex spot groups where, as illustrated in Plate 5.10, they occur most frequently on or near the outer borders of the penumbrae of the major spots. They are also found near the penumbrae of smaller spots. In addition, moustaches are prominent in the region between the main components of many small and medium-sized bipolar spot groups. Finally, it should be noted that they even occur in the initial stages of the development of a new active region, when the first spots have yet to appear and only a small bright facular area is visible (cf. Section 5.1). Bruzek's conclusions are in good accord with those reported earlier by McMath *et al.* [1960], Koval [1965a,b], and Severny [1965].

Many observers have remarked that moustaches occur very frequently at the ends of small, surge-like, dark filaments [see, for example, McMath *et al.*, 1960; Koval, 1965a; Severny, 1965; Bruzek, 1968].[10] Moreover, in the case of young developing active regions Bruzek [1967, 1968] has found a close and surprising

[10] This association has prompted speculation that the line profile asymmetry described above may be due to the presence above the emitting region of absorbing material with a velocity component directed away from the observer [Koval, 1965a; Engvold and Maltby, 1968], but serious objections to this explanation have been advanced by Severny [1968].

relationship between moustaches and the so-called arch filaments already mentioned in Section 5.2.1. In small bipolar regions of this type the occurrence of moustaches is restricted almost entirely to the area between the leader and follower spots covered or linked by the arch filaments; hardly any moustaches are found outside this area. The moustaches are frequently located near the ends of the arch filaments but are also visible between them. In fact, it seems that moustaches do not occur in young bipolar regions unless arch filaments are present. It is interesting to note that the close relationship between the two phenomena is observed from the very beginning of an active region, even before the first sunspot has appeared [Bruzek, 1968].

Owing to the small size of many moustaches, the number visible in an active region at any given time is highly dependent on the quality of the seeing. Beckers [1964, p. 34] reports that about 50 clearly recognizable moustaches can normally be seen in the vicinity of a spot group, while Bruzek [1968] states that up to 40 medium-sized and up to 10 large-sized moustaches may be present simultaneously in a large spot group. Engvold and Maltby [1968] give a rough estimate based on their best Hα spectra of 100 per spot group. However, it must not be forgotten, as mentioned above, that the number present in any particular case depends on the stage of evolution of the group.

A tendency for moustaches to recur at the same place in an active region has been reported by a number of workers [McMath *et al.*, 1960; Koval, 1965b; Severny, 1965]. However, more recent filter observations made by Bruzek [1968] have shown that this is not entirely correct: new moustaches often appear at locations quite close to but not identical with those of previous ones. Bruzek further remarks that two moustaches are sometimes formed in close proximity to one another.

(6) *Lifetime and evolution.* According to McMath *et al.* [1960] moustaches tend to increase in brightness more rapidly than they subsequently fade away. In addition, Severny [1965] claims that they may increase in size and become more diffuse during the latter part of their lifetime. Several estimates of the actual lifetimes have been given. From spectroscopic observations Severny [1957a,b, 1965] obtained figures ranging from a few minutes up to 20 min, while values found by McMath *et al.* [1960] lie in the range 4.4 min to a few hours, with an average value of about 10 min. On the other hand Bruzek [1968], using filter photographs taken with the 14-inch domeless coudé refractor at Capri, has derived the significantly higher figure of 25 min for the typical lifetime of a moustache, although his extreme values are nevertheless concordant with those given by McMath *et al.*

(7) *Magnetic fields in moustaches.* A number of workers have investigated the interesting question of where moustaches occur in relation to the (photospheric) magnetic field distribution in the active region concerned, but unfortunately

their results are rather contradictory. Howard and Harvey [1964] found that moustaches appear at places where the field has a strength of about 40 gauss or greater, whereas Koval [1965b] concluded that they are mostly seen near the 'neutral' line of the associated sunspot group. However, Bruzek [1968] remarks that his observations fail to confirm Koval's conclusion. Rust [1968] claims that all moustaches originate above what he calls 'satellite sunspots'. These are described as small regions within or near the penumbral border of a large spot which, although invisible in white light, are distinguished by having the opposite polarity to that of the main spot.[11] Finally, we should mention that, according to Koval, some moustaches (16 out of 92 cases) occur in regions with *transverse* fields of 600–1500 gauss; a similar association between moustaches and transverse fields has been noted by Severny [1968]. Taken together, the existing observations fail to provide a consistent picture of the relationship between moustaches and the underlying photospheric magnetic fields.

(8) *Relationship to flares.* Although moustaches are clearly differentiated from flares by their spectroscopic characteristics, the two phenomena nevertheless resemble one another in several important respects [cf. Smith and Smith, 1963, pp. 62–63]. Both are of relatively short duration and tend to rise to maximum brightness more rapidly than they subsequently fade. Furthermore, the frequency of their occurrence is strongly influenced by the configuration and stage of development of the sunspot group with which they are associated. It is interesting, therefore, to enquire whether there is any physical relationship between moustaches and flares.

Unfortunately the answers so far given to this question are contradictory. According to Severny [1957a,b, 1965, 1968] and Koval [1965b] moustaches are closely related to flares: in particular, Severny finds that the broad emission wings observed in the Hα line at the onset of large flares (importance $\geqslant 2$) are produced by highly compact clusters of moustaches of various line widths. In general, however, the locations of moustaches do not coincide with the bright flare emission areas seen at the *centre* of Hα. On the other hand, McMath *et al.* [1960] have found that moustaches are *not* the precursors of flares and, in fact, any association between the times and positions of the two phenomena seems to be one of chance. This latter conclusion is supported by Bruzek [1968], whose filter observations provide no evidence of a direct relationship between flares and moustaches.

5.4.3 FINE STRUCTURE OF FLARES

The observation of the fine structure of chromospheric flares is subject to two basic requirements: (*a*) as illustrated in Plates 5.11 and 5.12, a spatial resolution

[11] With improved resolution Smith [1971] has recently found that both polarities may actually be present in satellite sunspots.

of 1″ of arc or better is needed to reveal fully the fine structure of a flare; and (*b*) in order to avoid loss of detail in the brighter regions of a flare, a graded series of underexposures must be taken down to the level required to show structure in the brightest parts. Only one set of observations satisfying both these criteria has so far been published [Loughhead, 1968b]. They refer to a Class Ib flare photographed on 30 August 1967 with the 12-inch refractor of the Culgoora Solar Observatory; the event began at 00.08 UT and ended at 01.20 UT, the maximum being at 00.31 UT. A few photographs of the flare were first taken with a normal exposure and then successive small groups of photographs were taken with approximately 0.7, 0.5, 0.3, and 0.15 times the normal exposure, the cycle being repeated over and over throughout the lifetime of the flare.

Plate 5.11 illustrates how successive underexposures bring up the fine structure of the flare. No trace of the fine structure is visible in the upper photograph (Plate 5.11a), a normal exposure with good resolution showing – in addition to the flare – neighbouring chromospheric features and a few small sunspots. However, on an intermediate underexposure (Plate 5.11b) the flare is seen to consist of diffuse interconnected sheets of varying brightness, within which a number of small bright condensations are visible. These are more clearly shown in Plate 5.11c, taken with an even greater degree of underexposure. Note also the isolated bright point on the lower right.

Plate 5.12 gives a sequence of underexposures showing changes in the fine structure over a period of some 20 min including flare maximum. In its early stages (00.18.41, 00.19.50 UT) the flare consists of diffuse interconnected sheets or thick filaments containing a number of bright condensations. As the time of maximum (00.31) approaches, the bright areas increase in size and the bright point at the lower left becomes linked to the rest of the flare by a thin curved filament. Some additional tiny points of emission appear on the lower, right-hand side. At 00.23.01 and 00.26.51 small surge-like streaks of dark material, apparently opaque to the underlying emission, become prominent and appear to undergo rapid changes [cf. Svestka, 1968, pp. 22–23].

By the time maximum is reached, all parts of the flare are interconnected, if we accept the gap at the top left as being due to overlying material. A striking feature is the narrow bright filament connecting the left- and right-hand parts of the flare; it shows a number of localized variations in brightness and width (00.34.20–00.38.45). Some of the flare emission, however, is still in the form of vaguely defined diffuse sheets and of bright condensations having diameters of several seconds of arc. It would be a mistake, therefore, to conclude that *all* flare emission occurs in structures of very small dimensions.

Nevertheless, the question of the true nature of the narrow bright filament mentioned above is one of considerable interest. Fig. 5.2 shows the intensity distribution across the filament derived from microphotometry of the frame taken at 00.34.20 UT (see Plate 5.12). Contrary to the impression conveyed by the photograph itself it is apparent from Fig. 5.2 and from normal exposures taken near this time that the filament is not an isolated cylindrical structure completely detached from its surroundings. Rather it represents a sharp localized peak in the brightness distribution, merging on either side into the fainter background of the adjacent flare emission. Within the peak the intensity changes appreciably over a distance of only 1″ of arc.

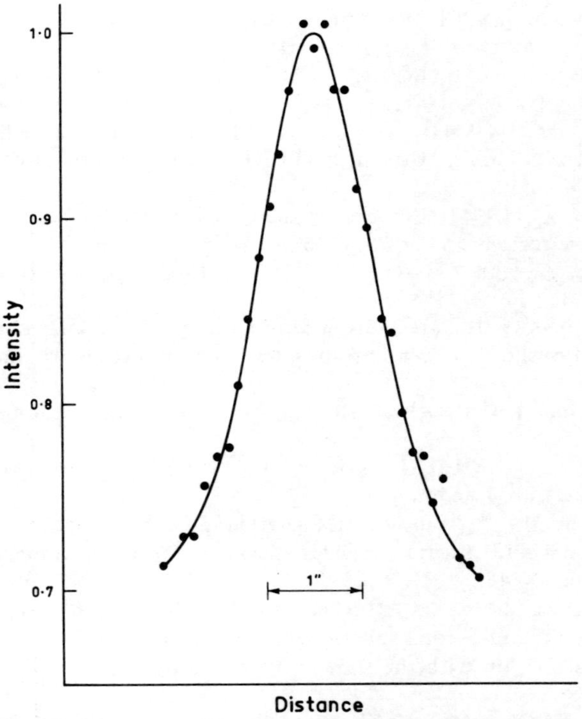

Fig. 5.2. Measured intensity distribution across the flare filament shown in Plate 5.12. The filament is not an isolated cylindrical structure but represents, rather, a sharp localized peak in the general brightness distribution of the flare. Note that the intensity within the peak changes appreciably over a distance of only 1″ of arc. The scanning slit measured 0″.02 by 2″.17, the longer side being aligned parallel to the filament. The intensities are expressed in units of the peak intensity.

Summarizing, we may draw two conclusions from the observations described in this section. In the first place, high-resolution photographs of flares taken with a proper exposure reveal the existence of a wealth of fine structure on a scale extending down to about 1″ of arc. Secondly, the particular Class Ib flare observed shows three distinct types of structural detail: rather vaguely defined diffuse sheets, small bright condensations, and narrow bright filaments. The existence of these types of small-scale detail accords with the general picture of flare structure outlined by Smith and Smith [1963, p. 79 et seq.] in their well-known monograph.

REFERENCES

ABDUSSAMATOV, H. I. [1969] 'Preliminary results of the study of the motion of material in the penumbra of sunspots', *Solnechnye Dannye 1969*, No. 9, p. 101.

D'AZAMBUJA, L. [1930] 'Recherches sur la structure de la chromosphère solaire', *Ann. Obs. Paris-Meudon* **8**, Part 2.

BABCOCK, H. W., and BABCOCK, H. D. [1955] 'The Sun's magnetic field, 1952–1954', *Astrophys. J.* **121**, 349.

BECKERS, J. M. [1964] 'A study of the fine structures in the solar chromosphere', Thesis, Utrecht; AFCRL–Environmental Research Paper No. 49.

BHATNAGAR, A. [1971] 'Fine-scan velocity and magnetic-field measurements in solar active regions', *Solar Phys.* **16**, 40.

BHATNAGAR, A., and TANAKA, K. [1972] 'Intensity oscillation in Hα-fine structure', *Solar Phys.* **24**, 87.

BRAY, R. J. [1968] 'High-resolution photography of the solar chromosphere. II. The relationship between chromospheric and photospheric faculae', *Solar Phys.* **4**, 318.

BRAY, R. J., and LOUGHHEAD, R. E. [1964] *Sunspots,* Chapman and Hall, London.

BRAY, R. J., and LOUGHHEAD, R. E. [1967] *The Solar Granulation,* Chapman and Hall, London.

BRÜCKNER, G. [1968] 'Ein neuer Magnetograph zur simultanen Messung des Longitudinalen und des transversalen Zeemaneffektes im Sonnenspektrum', *Z. Astrophys.* **68**, 48.

BRÜCKNER, G., and WALDMEIER, M. [1969] 'Distribution of magnetic fields in photospheric and chromospheric layers in the active region of 7 July 1966 and their correlation with the flare event of 8 July, 1253–1340 UT', *Annals I.Q.S.Y.* **3**, 24.

BRUZEK, A. [1967] 'On arch-filament systems in spotgroups', *Solar Phys.* **2**, 451.

BRUZEK, A. [1968] 'Bright points (moustaches) and arch filaments in young active regions', *Structure and Development of Solar Active Regions,* ed. K.O. Kiepenheuer, p. 293; Reidel, Dordrecht-Holland.

BRUZEK, A. [1972] 'Properties of solar active regions', *Solar-Terrestrial Physics 1970*, ed. E. R. Dyer, Part I: *The Sun*, p. 49; Reidel, Dordrecht-Holland.

BUMBA, V., and HOWARD, R. [1965] 'A study of the development of active regions on the Sun', *Astrophys. J.* **141**, 1492.

ELLERMAN, F. [1917] 'Solar hydrogen "bombs" ', *Astrophys. J.* **46**, 298.

ENGVOLD, O., and MALTBY, P. [1968] 'A spectroscopic study of the Hα profile of moustaches', *Mass Motions in Solar Flares and Related Phenomena,* Proc. Ninth Nobel Symposium, ed. Y. Öhman, p. 109; Almqvist and Wiksell, Stockholm.

FOUKAL, P. [1971a] 'Morphological relationships in the chromospheric Hα fine structure', *Solar Phys.* **19**, 59.

FOUKAL, P. [1971b] 'Hα fine structure and the chromospheric field', *Solar Phys.* **20**, 298.

FRAZIER, E. N. [1972] 'The relations between chromospheric features and photospheric magnetic fields', *Solar Phys.* **24**, 98.

GIMSE, O. [1967] 'The lifetime of Hα-structures in active regions on the Sun', *Rep. Inst. Theor. Astrophys. Oslo* No. 21.

HAUGEN, E. [1969] 'On the chromospheric velocity field in sunspot regions', *Solar Phys.* **9**, 88.

HESS, W. N. (ed.) [1964] *The Physics of Solar Flares*, Proc. AAS–NASA Symposium, Goddard Space Flight Center, 1963; NASA, Washington.

HOWARD, R. [1959] 'Observations of solar magnetic fields', *Astrophys. J.* **130**, 193.

HOWARD, R., and HARVEY, J. W. [1964] 'Photospheric magnetic fields and chromospheric features', *Astrophys. J.* **139**, 1328.

IOSHPA, B. A., MOGILEVSKY, E. I., and OBRIDKO, W. N. [1964] 'On the force-free magnetic field in the active regions of the solar chromosphere', *Space Research IV,* Proc. Fourth Internat. Space Sci. Symposium, Warsaw, 1963, p. 789; North-Holland, Amsterdam.

JAGER, C. DE [1959] 'Structure and dynamics of the solar atmosphere', *Handbuch der Physik,* ed. S. Flügge, vol. 52, p. 80; Springer, Berlin.

KIEPENHEUER, K. O. [1953] 'Solar activity', *The Sun,* ed. G. P. Kuiper, p. 322; Univ. Chicago Press.

KIEPENHEUER, K. O. [1957] 'Erfahrungen mit einem Lyotschen Hα-filter sowie vorläufige Ergebnisse über die Struktur der Chromosphäre', *Z. Astrophys.* **42**, 209.

KIEPENHEUER, K. O. [1960a] 'Über die Struktur der gestörten Sonnen-atmosphäre', *Z. Astrophys.* **49**, 73.

KIEPENHEUER, K. O. [1960b] 'The optical phenomena forming a solar center of activity', Soc. Ital. Fisica, Rendiconti Scuola Internazionale di Fisica 'Enrico Fermi' (Director, G. Righini), p. 39.

KIEPENHEUER, K. O. [1966] Introduction to *The Fine Structure of the Solar Atmosphere,* ed. K. O. Kiepenheuer, p. 11; Franz Steiner Verlag, Wiesbaden.

KIEPENHEUER, K. O. [1967] 'Morphology of an active region on the Sun', *Solar Physics,* ed. J. N. Xanthakis, p. 383; Interscience, London.

KIEPENHEUER, K. O. (ed.) [1968] *Structure and Development of Solar Active Regions*, I.A.U. Symposium No. 35, Budapest, 1967; Reidel, Dordrecht-Holland.

KLÜBER, H. VON [1968] 'Correlation between local solar magnetic fields and central intensities in some Fraunhofer lines', *Mon. Not. R.A.S.* **141**, 469.

KOVAL, A. N. [1965a] 'An investigation of moustaches and ejections in connection with the development of spot groups', *Izv. Crim. Astrophys. Obs.* **33**, 138.

KOVAL, A. N. [1965b] 'The position of "moustaches" in a spot group relative to the magnetic field', *Izv. Crim. Astrophys. Obs.* **34**, 278.

KOVAL, A. N. [1967] 'An investigation of metallic emission lines in moustaches on the disk', *Izv. Crim. Astrophys. Obs.* **37**, 62.

KOVAL, A. N., and SEVERNY, A. B. [1970] 'On the asymmetry of moustaches', *Solar Phys.* **11**, 276.

KRAT, V. A. [1965] 'Line profiles of the solar chromosphere. II. Facular regions in the H and K lines', *Izv. Pulkovo Astron. Obs.* **24**, No. 178, p. 19.

LIPPINCOTT, S. L. [1955] 'A preliminary investigation of the small changes in the chromospheric structure', *Ann. Astrophys.* **18**, 113.

LOUGHHEAD, R. E. [1968a] 'High-resolution photography of the solar chromosphere. V. The fibrils around isolated sunspots', *Solar Phys.* **5**, 489.

LOUGHHEAD, R. E. [1968b] 'High-resolution photography of the solar chromosphere. III. The fine structure of a Class I flare', *Solar Phys.* **4**, 422.

LYOT, B. [1944] 'Le filtre monochromatique polarisant et ses applications en physique solaire', *Ann. Astrophys.* **7**, 31.

MARTRES, M.-J., MICHARD, R., and SORU-ISCOVICI, I. [1966] 'Étude morphologique de la structure magnétique des régions actives en relation avec les phénomènes chromosphériques et les éruptions solaires. II. Localisation des plages brillantes, filaments et éruptions', *Ann. Astrophys.* **29**, 249.

MARTRES, M.-J., and SORU-ESCAUT, I. [1971] 'Chromospheric absorbing features promising the appearance and the development of an active center', *Solar Phys.* **21**, 137.

McMATH, R. R., MOHLER, O. C., and DODSON, H. W. [1960] 'Solar features associated with Ellerman's "solar hydrogen bombs" ', *Proc. Nat. Acad. Sci.* **46**, 165.

MOHLER, O. C., and DODSON, H. W. [1958] 'Observations of solar "points" ', *Astron. J.* **63**, 309.

MUSSAYEV, M. M. [1968] 'On the connection between calcium flocculi and faculae', *Solnechnye Dannye 1968,* No. 12, p. 73.

NAKAGAWA, Y., RAADU, M. A., BILLINGS, D. E., and McNAMARA, D. [1971] 'On the topology of filaments and chromospheric fibrils near sunspots', *Solar Phys.* **19**, 72.

ÖHMAN, Y. (ed.) [1968] *Mass Motions in Solar Flares and Related Phenomena,* Proc. Ninth Nobel Symposium, Anacapri, 1968; Almqvist and Wiksell, Stockholm.

PIKELNER, S. B. [1971] 'Nature of the fine structure of the middle chromosphere', *Solar Phys.* **20**, 286.

PRATA, S. W. [1971] 'A note on chromospheric fine structure at active region polarity boundaries', *Solar Phys.* **20**, 310.

RAADU, M. A., and NAKAGAWA, Y. [1971] 'The topology of force-free magnetic field near bipolar sunspots', *Solar Phys.* **20**, 64.

RUST, D. M. [1968] 'Chromospheric explosions and satellite sunspots', *Structure and Development of Solar Active Regions,* ed. K. O. Kiepenheuer, p. 77; Reidel, Dordrecht-Holland.

SEVERNY, A. B. [1956a] 'Fine structure in solar spectra', *Observatory* **76**, 241.

SEVERNY, A. B. [1956b] 'The fine structure of the emission of active formations on the Sun', *Astron. J. U.S.S.R.* **33**, 74.

SEVERNY, A. B. [1957a] 'Some results of investigations of nonstationary processes on the Sun', *Astron. J. U.S.S.R.* **34**, 684 (*Sov. Astron. AJ* **1**, 668).

SEVERNY, A. B. [1957b] 'Investigation of the fine structure of the emission in active regions on the Sun', *Izv. Crim. Astrophys. Obs.* **17**, 129.

SEVERNY, A. B. [1965] 'Spectra of solar activity regions', *The Solar Spectrum,* ed. C. de Jager, p. 221; Reidel, Dordrecht-Holland.

SEVERNY, A. B. [1966] 'On the fine structure of solar magnetic fields', *The Fine Structure of the Solar Atmosphere,* ed. K. O. Kiepenheuer, p. 109; Franz Steiner Verlag, Wiesbaden.

SEVERNY, A. B. [1968] 'Mass motions in flares and moustaches indicated by special spectral features', *Mass Motions in Solar Flares and Related Phenomena,* Proc. Ninth Nobel Symposium, ed. Y. Öhman, p. 71; Almqvist and Wiksell, Stockholm.

SEVERNY, A. B., and KHOKHLOVA, V. L. [1959] 'The spectrophotometry of lines in grains of continuous emission', *Izv. Crim. Astrophys. Obs.* **21**, 190.

SMITH, H. J. [1964] 'Solar flares: their structure, development, and motion', *The Physics of Solar Flares*, ed. W. N. Hess, p. 1; NASA, Washington.

SMITH, H. J., and SMITH, E. V. P. [1963] *Solar Flares*, Macmillan, New York.

SMITH, S. F. [1968] 'The formation, structure and changes in filaments in active regions', *Structure and Development of Solar Active Regions,* ed. K. O. Kiepenheuer, p. 267; Reidel, Dordrecht-Holland.

SMITH, S. F. [1971] 'Hα structures and small-scale magnetic field configurations', *Solar Magnetic Fields,* ed. R. Howard, p. 323; Reidel, Dordrecht-Holland.

SMITH, S. F., and RAMSEY, H. E. [1967] 'Flare positions relative to photospheric magnetic fields', *Solar Phys.* **2**, 158.

STOYANOVA, M. N. [1970] 'Vertical structure of plages', *Solar Phys.* **15**, 349.

SVESTKA, Z. [1968] 'Mass motions in flares indicated by line profiles and filtergrams', *Mass Motions in Solar Flares and Related Phenomena,* Proc. Ninth Nobel Symposium, ed. Y. Öhman, p. 17; Almqvist and Wiksell, Stockholm.

TANDBERG-HANSSEN, E. [1967] *Solar Activity,* Blaisdell, Waltham, Massachusetts.

TSAP, T. T. [1970] 'Magnetic fields at different depths of the active regions of solar atmosphere', *Izv. Crim. Astrophys. Obs.* **41–42**, 158.

VEEDER, G. J., and ZIRIN, H. [1970] 'The chromospheric magnetograph', *Solar Phys.* **12**, 391.

VRABEC, D. [1971] 'Magnetic field spectroheliograms from the San Fernando Observatory', *Solar Magnetic Fields*, ed. R. Howard, p. 329; Reidel, Dordrecht-Holland.

WEART, S. R. [1970] 'The birth and growth of sunspot regions', *Astrophys. J.* **162**, 987.

WEART, S. R., and ZIRIN, H. [1969] 'The birth of active regions', *Publ. Astron. Soc. Pac.* **81**, 270.

XANTHAKIS, J. N. (ed.) [1967] *Solar Physics*, Proc. NATO Conf., Athens; Interscience, London.

ZIRIN, H. [1966] *The Solar Atmosphere*, Blaisdell, Waltham, Massachusetts.

ZIRIN, H. [1971a] 'Application of the chromospheric magnetograph to active regions', *Solar Magnetic Fields,* ed. R. Howard, p. 237; Reidel, Dordrecht-Holland.

ZIRIN, H. [1971b] 'The manifold structure of the chromosphere and corona', *Phil. Trans. Roy. Soc.* **A270**, 77.

CHAPTER 6

The Propagation and Dissipation of Waves in a Compressible Gravitationally-Stratified Atmosphere

6.1 Introduction

When we survey our observational knowledge of the chromosphere, outlined in the previous chapters, we find ourselves face-to-face with a number of difficult problems which further observations, by themselves, appear incapable of fully resolving. Foremost amongst these are (*a*) the elucidation of the non-thermal mechanism responsible for the chromospheric temperature rise, and (*b*) the origin of spicules. Fortunately, however, most of these problems are susceptible – at least in part – to a theoretical attack. For example, as we shall see in the next and concluding chapter of the book, the temperature rise can reasonably be accounted for on the basis of dissipation by *shock waves,* although conclusive observational evidence for the existence of such waves has not yet been obtained. These are believed to start their journey outwards from the Sun at heights of a few hundred kilometres, where they develop from acoustic waves which have propagated upwards from the photosphere. Calculation shows that by an appropriate choice of wave parameters the rate of shock dissipation as a function of height can be brought into agreement with that of the known net radiative energy loss from the chromosphere, at least at the lower levels where the radiative loss can be evaluated with reasonable accuracy.

It is appropriate that we should confront theory with the results of observation wherever this is both possible and fruitful, and this task is undertaken in the following chapter. However, the necessary theoretical background for such a confrontation is widely scattered throughout a large number of books and papers dealing with a great diversity of applications spread over several quite separate disciplines. Our aim in the present chapter, accordingly, is to give a systematic introduction, relevant to the theoretical problems of the chromosphere, to the theory of hydrodynamic and magnetohydrodynamic waves. The topics of basic concern to us are the propagation of waves of various types, their degeneration into shock waves, and the subsequent dissipation of the shocks as they proceed upwards through a gravitationally-stratified atmosphere.

We start (Section 6.2.2) with the basic magnetohydrodynamic equations of mass, momentum, and energy which, together with Maxwell's equations, enable us to derive the linearized wave equation for small-amplitude disturbances propagating adiabatically through an inviscid, perfectly conducting gas (Section 6.2.3). General formulae for the wave energy density and energy flux, as well as for the velocity of energy propagation (the group velocity), are also derived (Section 6.2.4). Specializing the discussion to an isothermal atmosphere – a good approximation in the upper photosphere and lower levels of the chromosphere – we next obtain the general dispersion relation governing the behaviour of purely hydrodynamic waves. This relation admits the existence of two very different types of progressive wave, *internal gravity waves* modified by the effect of compressibility and *acoustic waves* modified by the effect of gravity; their properties are outlined in Sections 6.3.2 and 6.3.3. The radiative damping of hydrodynamic waves and, in particular, its dependence on the magnitude of the radiative relaxation time, are discussed in Section 6.3.4.

When a magnetic field is present, the possible modes of wave propagation are naturally much more complex. Nevertheless, at heights in the chromosphere exceeding about 750 km magnetic forces can no longer be ignored even for fields as small as 20 gauss. The dispersion equation in the presence of a magnetic field but without gravity is derived in Section 6.4.2 and used to explore the properties of *generalized Alfvén waves* and of *magneto-acoustic waves.* The full dispersion equation, which takes account of gravity (Section 6.4.3), describes an additional but very complicated class of waves known as *magneto-gravity waves*.

The subject of hydrodynamic shock waves is of fundamental importance to an understanding of the heating of the chromosphere and is considered in some detail in Section 6.5. We start (Section 6.5.1) by describing how finite-amplitude hydrodynamic waves can degenerate into shocks and then proceed to derive the 'jump' conditions across a shock front; these are known as the Rankine–Hugoniot relations and give expression to the discontinuities in density, pressure, and velocity across the front (Section 6.5.2). We next turn to the crucial question of the propagation and dissipation of shock waves, discussing first some underlying thermodynamic concepts related to the increase in the entropy of the gas across a shock front and then outlining the adaption of the classical Brinkley–Kirkwood theory, well-known to fluid dynamicists, to the astrophysically-important case of a radiating gas (Section 6.5.3). In Chapter 7 a modified form of this theory, due to E. Schatzman, is used to compare the shock dissipation and the known net radiative energy loss as functions of height in the low chromosphere, and thereby to deduce both the strength and periods of the shock waves presumed to be responsible for heating this portion of the Sun's atmosphere. The effect of radiative transfer on shock waves is briefly discussed in Section 6.5.4.

In Section 6.6 we turn to the theory of magnetohydrodynamic shock waves, first deriving the relevant jump conditions (Section 6.6.2) and then, with their aid, classifying the main types of magnetohydrodynamic shocks (Section 6.6.3). Finally, we outline the basis of a plausible extension of the Brinkley–Kirkwood theory proposed by D. E. Osterbrock to describe their propagation and dissipation (Section 6.6.4). This theory has been applied to the problem of the heating of the chromosphere both in active regions and above the supergranulation boundaries: the first application lies outside the scope of the present book but the second is discussed in the next chapter.

6.2 Fundamental Equations and Basic Theory

6.2.1 NOTATION AND UNITS

Since the height of the chromosphere is very small compared to the solar radius, we may neglect the effect of curvature and introduce a cartesian reference frame Ox_i ($i = 1, 2, 3$). It is convenient to take the Ox_3 axis to be directed *vertically upwards,* so that the components of the gravitational acceleration vector may be written

$$g_i = (0, 0, -g). \tag{6.1}$$

Extensive use will be made of cartesian tensor notation, but vector notation will occasionally be employed. Readers unfamiliar with cartesian tensors will find the short book by Jeffreys [1952] a helpful introduction to the subject.

The principal symbols employed in this chapter are defined in Table 6.1; c.g.s. units are used throughout.

6.2.2 MAGNETOHYDRODYNAMIC EQUATIONS

Using the notation and units set out in Table 6.1, we may state the magnetohydrodynamic equations for a viscous fluid of finite electrical conductivity as follows.[1]

Conservation of mass:

$$\frac{d\rho}{dt} + \rho \frac{\partial v_i}{\partial x_i} = 0 \tag{6.2}$$

[1] Apart from the use of tensor notation and the inclusion of viscous terms the equations are given here in essentially the same form as in an earlier monograph [Bray and Loughhead, 1964, p. 267], where the reader will find references to detailed discussions of the derivation of the equations and of the assumptions embodied therein. Other sources include the books by Chandrasekhar [1961], Pai [1962], Thompson [1962], Anderson [1963], Jones [1964], Kendall and Plumpton [1964], Kulikovskiy and Lyubimov [1965], Shercliff [1965], Ferraro and Plumpton [1966], Roberts [1967], and Cabannes [1970].

TABLE 6.1

*Principal Symbols Used in Chapter 6**

a	velocity of sound
a^*	velocity of fast-mode magnetohydrodynamic wave normal to field
A	Alfvén velocity
B	Planckian source function
c	velocity of light (2.998×10^{10} cm sec^{-1})
c_v	specific heat of gas at constant volume (per unit mass)
D	energy flux of shock wave
e	internal energy of gas per unit mass
e^*	magnitude of electron charge (4.802×10^{-10} e.s.u.)
E	total energy of gas per unit volume
E_i	electric field vector (in e.s.u.)
F_i	thermal flux vector
g_i	acceleration due to gravity (2.74×10^4 cm sec^{-2} at solar surface)
H_i	magnetic field vector (in e.m.u.)
$\mathcal{H}$	scale height of isothermal atmosphere (see equation 6.73)
I	radiation intensity, integrated over wavelength
j_i	electric current vector (in e.s.u.)
J_i	total energy flux vector
k	Boltzmann constant (1.380×10^{-16} erg deg^{-1}); magnitude of wave propagation vector
k_i	wave propagation vector
m	mass flux across shock front
m_e	electron mass (9.107×10^{-28} gm)
m_f	mass flux across fast magnetohydrodynamic shock
m_H	mass of hydrogen atom (1.673×10^{-24} gm)
m_{ij}	magnetic stress tensor
m_s	mass flux across slow magnetohydrodynamic shock
M	Mach number
M^*	magnetic Mach number
n_i	unit vector normal to shock front
N	Väisälä frequency
N_e	electron number density
p	gas pressure
P_i	electromagnetic Poynting vector
q	heat energy per unit mass (see equation 6.84)
q_e	electric charge density (in e.s.u.)
Q_i	wave energy flux vector
R	Lagrangian coordinate of fluid particle
s	entropy of gas per unit mass
T	gas temperature (kinetic temperature)
t_R	radiative relaxation time
u_i	gas velocity vector relative to shock front
U	velocity of shock wave
v_i	gas velocity vector
V_i	group velocity vector
V_p	phase velocity
w	specific volume of gas ($= 1/\rho$)

* C.g.s. units are used throughout.

TABLE 6.1 *(continued)*

W	wave energy per unit volume
X_i	displacement vector of gas particle
Z	vertical component of particle displacement
γ	ratio of specific heats of gas
δ	shock wave strength (see equation 6.173)
δ_{ij}	Kronecker delta tensor
ϵ_{ijk}	alternating tensor
η	coefficient of viscosity
θ	angle between wave propagation vector and magnetic field; angle between direction of radiation and normal to shock front
κ	coefficient of thermal conductivity; magnitude of (real) wave propagation vector
κ	(linear) absorption coefficient
μ	mean molecular weight of gas
π_{ij}	viscous stress tensor
ρ	gas density
σ	electrical conductivity (in e.s.u.)
σ_R	Stefan constant (5.669×10^{-5} erg cm^{-2} deg^{-4} sec^{-1})
τ	optical depth in continuum
ϕ	gravitational energy of gas per unit mass
Φ	angle between wave propagation vector and vertical
χ_i	complex wave amplitude vector
ψ	angle between group velocity vector and vertical
ω	angular wave frequency
ω_1, ω_2	critical frequencies of gas (no magnetic field)
ω_3, ω_4	critical frequencies of gas (with radiative cooling)
Ω_1, Ω_2	critical frequencies of gas (in magnetic field)

Conservation of momentum:

$$\rho \frac{dv_i}{dt} = \rho g_i + \frac{1}{c} \epsilon_{ijk} j_j H_k - \frac{\partial p}{\partial x_i} + \frac{\partial \pi_{ij}}{\partial x_j} \tag{6.3}$$

Conservation of internal energy:

$$\rho \frac{de}{dt} - \frac{p}{\rho} \frac{d\rho}{dt} = \frac{1}{\sigma} j^2 + \pi_{ij} \frac{\partial v_i}{\partial x_j} - \frac{\partial F_i}{\partial x_i} \tag{6.4}$$

Equation of state:

$$p = \frac{k\rho T}{\mu m_H} \tag{6.5}$$

Current equation:

$$j_i = \sigma\left(E_i + \frac{1}{c} \epsilon_{ijk} v_j H_k\right) \tag{6.6}$$

Maxwell's equations:

$$\frac{\partial H_i}{\partial t} = -c\, \epsilon_{ijk} \frac{\partial E_k}{\partial x_j} \tag{6.7}$$

$$j_i = \frac{c}{4\pi} \epsilon_{ijk} \frac{\partial H_k}{\partial x_j} \tag{6.8}$$

$$\frac{\partial E_i}{\partial x_i} = 4\pi q_e \tag{6.9}$$

$$\frac{\partial H_i}{\partial x_i} = 0. \tag{6.10}$$

In these equations repeated suffixes imply summation, the operator

$$\frac{d}{dt} = \frac{\partial}{\partial t} + v_j \frac{\partial}{\partial x_j}$$

indicates differentiation following the motion of the gas, and the viscous stress tensor is defined as

$$\pi_{ij} = \eta \left(\frac{\partial v_i}{\partial x_j} + \frac{\partial v_j}{\partial x_i} \right) - \frac{2}{3} \eta \, \delta_{ij} \frac{\partial v_k}{\partial x_k}, \tag{6.11}$$

where η is the coefficient of viscosity. It should also be pointed out that the first two terms on the right-hand side of the internal energy equation (6.4) may be written in the forms

$$\frac{1}{\sigma} j^2 = j_i \left(E_i + \frac{1}{c} \epsilon_{ijk} v_j H_k \right)$$

and

$$\pi_{ij} \frac{\partial v_i}{\partial x_j} = \frac{1}{2} \eta \left(\frac{\partial v_i}{\partial x_j} + \frac{\partial v_j}{\partial x_i} - \frac{2}{3} \delta_{ij} \frac{\partial v_k}{\partial x_k} \right)^2 ,$$

which will be recognized as the rates of heat generation per unit volume of the fluid due to ohmic and viscous dissipation respectively.

The thermal flux vector F_i appearing in (6.4) can be expressed as the sum of two terms, one due to the radiation field in the gas and the other due to particle conduction, so that

$$F_i = (F_{rad})_i + (F_{cond})_i. \tag{6.12}$$

As will be discussed in Section 6.5.4, the radiative contribution $(F_{rad})_i$ has to be evaluated by solving the relevant equation of transfer. In the case of a highly ionized gas, such as the upper chromosphere and corona, the particle conduction is dominated by electrons [Chapman, 1954]. The corresponding flux contribution is

$$(F_{cond})_i = -\kappa \frac{\partial T}{\partial x_i}, \tag{6.13}$$

where the coefficient of total conductivity κ is given by the formula

$$\kappa = \kappa_0(T, N_e) \cdot \frac{T^{5/2}}{\ln \Lambda}, \tag{6.14}$$

with

$$\Lambda = \frac{3}{2}\left(\frac{k^3 T^3}{\pi e^{*6} N_e}\right)^{1/2} \tag{6.15}$$

[Delcroix and Lemaire, 1969]. The factor $\kappa_0(T, N_e)$ is a slowly-varying quantity whose magnitude has been conveniently tabulated by Delcroix and Lemaire for the range of values of T and N_e prevailing in the upper chromosphere and corona. Values of the coefficient of thermal conductivity at lower heights, where the chromosphere is only partially ionized (cf. Section 4.4.5), may be obtained from graphs published by Ulmschneider [1970]. His results cover a range of temperature from 3000 to 100 000 K and a range of gas pressure from 10^{-3} to 10^6 dyne cm^{-2}.

The derivation of the formula (6.14) is based on the assumption that there is no magnetic field in the gas. Nevertheless, it remains valid in the presence of a field so long as the direction of the thermal gradient coincides with that of the field. On the other hand, if the thermal gradient is at right angles to the field, the value of the coefficient of thermal conductivity is reduced by a factor whose magnitude increases with the strength of the field. In addition, there is a flow of energy in the direction at right angles to both the magnetic field and the thermal gradient. Explicit formulae for calculating the thermal flux along and perpendicular to the field and in the direction normal to both the field and the thermal gradient are given by Braginskii [1957].

In the absence of a magnetic field the value of the coefficient of viscosity of a highly ionized gas under the conditions prevailing in the upper chromosphere and corona may be obtained from a convenient compilation published by Delcroix and Lemaire [1968]. When a field is present the viscous stresses may be calculated with the aid of formulae given by Braginskii [1957]; see also Spitzer [1962, pp. 146–147].

The internal energy equation (6.4) takes a much simpler form under certain restrictive conditions which will be assumed to apply in some of the discussion that follows. When ionization and other similar forms of internal energy can be neglected, the internal energy of the gas per unit mass is simply

$$e = c_v T = \frac{1}{\gamma - 1}\frac{p}{\rho}, \tag{6.16}$$

where c_v is the specific heat of the gas at constant volume and where the relevant value of γ, the ratio of the specific heats, is that for a monatomic gas, namely 5/3. In addition, when both thermal conduction and ohmic and viscous dissipation can be ignored, equation (6.4) reduces to the familiar relation

$$\frac{dp}{dt} - \frac{\gamma p}{\rho} \cdot \frac{d\rho}{dt} = 0 \tag{6.17}$$

governing variations in the pressure and density of a gas under *adiabatic* conditions, i.e. when there is no heat exchange between different parts of the fluid.

For the purpose of outlining the theory of shock waves (Sections 6.5 and 6.6) it is convenient to express the equations of conservation of mass, momentum, and energy in forms rather different to those given above. The mass equation (6.2) may clearly be written

$$\frac{\partial \rho}{\partial t} + \frac{\partial}{\partial x_j}(\rho v_j) = 0, \tag{6.18}$$

while the momentum equation can be transformed to read

$$\frac{\partial}{\partial t}(\rho v_i) + \frac{\partial}{\partial x_j}(\rho v_i v_j + p\delta_{ij} - \pi_{ij} - m_{ij}) = \rho g_i, \tag{6.19}$$

where

$$m_{ij} = \frac{H_i H_j}{4\pi} - \frac{H^2}{8\pi}\delta_{ij} \tag{6.20}$$

is the magnetic stress tensor (cf. Section 6.6.4). Finally, in place of the internal energy equation one can derive an equation expressing the conservation of the *total* energy of the gas. It may be stated in the form

$$\frac{\partial E}{\partial t} + \frac{\partial J_i}{\partial x_j} = 0, \tag{6.21}$$

where

$$E = \frac{1}{2}\rho v^2 + \rho\phi + \rho e + \frac{H^2}{8\pi} \tag{6.22}$$

is the sum of the kinetic, gravitational, internal, and magnetic energies per unit volume, and

$$J_i = \left(\frac{1}{2}\rho v^2 + \rho\phi + \rho e + p\right) v_i - v_j \pi_{ij} + F_i + P_i \tag{6.23}$$

is the sum of all forms of energy flux. Here ϕ is the gravitational energy per unit mass, so that

$$g_i = -\frac{\partial \phi}{\partial x_i}, \tag{6.24}$$

while

$$P_{\rm i} = \frac{c}{4\pi} \epsilon_{\rm ijk} E_{\rm j} H_{\rm k} \tag{6.25}$$

is the electromagnetic Poynting vector. In the particular case of an infinitely conducting fluid one can show that

$$P_{\rm i} = \frac{1}{4\pi} (H^2 v_{\rm i} - v_{\rm j} H_{\rm j} H_{\rm i}). \tag{6.26}$$

The full import of the total energy equation (6.21) is brought out by integrating it over a given fixed volume V. If S is the surface area of V, then applying Gauss' divergence theorem we have

$$-\frac{\partial}{\partial t} \iiint E \mathrm{d}V = \iint J_{\rm i} \mathrm{d}S_{\rm i}. \tag{6.27}$$

The left-hand side of equation (6.27) represents the time rate of decrease of the total energy within V, while the right-hand side represents the total flux of energy outwards through the surface S. When the expression (6.23) is substituted for $J_{\rm i}$ on the right-hand side of (6.27) it is clear that the first three terms represent the kinetic, gravitational, and internal energy transported out of the volume V by the flow $v_{\rm i}$; the next two terms express the rate at which the pressure and viscous forces acting over the surface S change the energy within V; the next term gives the flux of energy out of V due to particle and radiation conduction; and the final term (the Poynting vector) gives the flux of electromagnetic energy out of V.[2]

Another result, which will be required in the subsequent discussion of the properties of shock waves, is the expression

$$s = c_{\rm v} \ln (p\rho^{-\gamma}) + \text{const.} \tag{6.28}$$

for the *entropy* of the gas per unit mass. By the second law of thermodynamics or, more generally, by Boltzmann's H-theorem, the entropy cannot decrease during the motion of the gas [Chapman and Cowling, 1970, pp. 78–79]. Using (6.28) one can rewrite the equation of adiabatic variation (6.17) in the form

$$\frac{\mathrm{d}s}{\mathrm{d}t} = 0, \tag{6.29}$$

[2] Although the integral $\iint P_{\rm i} \mathrm{d}S_{\rm i}$ gives the rate at which electromagnetic energy flows outwards over the entire closed surface S, it does not necessarily follow that the Poynting vector $P_{\rm i}$ can be identified with the actual flux of electromagnetic energy *at any given point* [see, for example, Stratton, 1941, pp. 131–135; Jones, 1964, pp. 51–53; Shercliff, 1965, pp. 26–28]. However, in calculating the energy flow associated with magnetohydrodynamic waves it is customary, although proof is lacking, to make this identification [cf. Ferraro, 1954].

showing that in the absence of energy dissipation the entropy of any given element of the gas remains constant during its motion. In the presence of dissipation there is an increase in entropy.

Finally, we should mention that to the accuracy of the non-relativistic approximation implicit in the formulation of the magnetohydrodynamic equations the magnetic field H_i may be taken to be invariant under a uniform translation of the frame of reference [see, for example, Ferraro and Plumpton, 1966, p. 15; Ferraro, 1967, p. 86]. This fact, which is of basic importance in an astrophysical context [Alfvén, 1950, p. 6], is made use of in the derivation of the jump conditions across magnetohydrodynamic shock waves in Section 6.6.2.

6.2.3 LINEARIZED WAVE EQUATION

In order to illustrate the essential nature of hydrodynamic and magnetohydrodynamic waves of various types we shall begin our discussion of wave propagation by considering small-amplitude disturbances. Moreover, for the sake of mathematical simplicity, we shall assume the fluid to be perfectly conducting and inviscid, so that $\sigma = \infty$ and $\eta = 0$. Finally, we shall ignore the thermal flux vector F_i and, accordingly, use the equation of adiabatic variation (6.17) in place of the internal energy equation (6.4). (The radiative damping of hydrodynamic waves is discussed later in the chapter.) It is convenient to eliminate the electric field E_i and the current j_i from the basic equations (6.2)–(6.10), which then become

$$\frac{\partial \rho}{\partial t} + v_j \frac{\partial \rho}{\partial x_j} + \rho \frac{\partial v_j}{\partial x_j} = 0 \tag{6.30}$$

$$\rho \frac{\partial v_i}{\partial t} + \rho v_j \frac{\partial v_i}{\partial x_j} = \rho g_i + \frac{1}{4\pi} \epsilon_{ijk}\epsilon_{jlm} \frac{\partial H_m}{\partial x_l} H_k - \frac{\partial p}{\partial x_i} \tag{6.31}$$

$$\rho \frac{\partial p}{\partial t} + \rho v_j \frac{\partial p}{\partial x_j} = \gamma p \left(\frac{\partial \rho}{\partial t} + v_j \frac{\partial \rho}{\partial x_j}\right) \tag{6.32}$$

$$\frac{\partial H_i}{\partial t} = \epsilon_{ijk} \frac{\partial}{\partial x_j} (\epsilon_{klm} v_l H_m). \tag{6.33}$$

Let us now consider small perturbations of the gas from a state of static equilibrium and write

$$\begin{aligned} \rho &= \rho_0 + \rho_1 \\ p &= p_0 + p_1 \\ H_i &= H_{0i} + h_{1i} \\ v_i &= v_{1i}, \end{aligned} \tag{6.34}$$

where the subscript zero refers to the steady state. The steady magnetic field H_{0i} will be taken to be uniform throughout the fluid, whence it follows that the equation of equilibrium of the gas reduces to the familiar hydrostatic relation

$$\frac{\partial p_0}{\partial x_i} = \rho_0 g_i. \tag{6.35}$$

If we now substitute from (6.34) into equations (6.30)–(6.33) and drop all terms of order greater than one in the perturbations $\rho_1, p_1, \ldots$, we obtain the linearized magnetohydrodynamic equations

$$\frac{\partial \rho}{\partial t} + v_j \frac{\partial \rho_0}{\partial x_j} + \rho_0 \frac{\partial v_j}{\partial x_j} = 0 \tag{6.36}$$

$$\rho_0 \frac{\partial v_i}{\partial t} = \rho g_i + \frac{1}{4\pi} \epsilon_{ijk}\epsilon_{jlm} \frac{\partial h_m}{\partial x_l} H_{0k} - \frac{\partial p}{\partial x_i} \tag{6.37}$$

$$\frac{\partial p}{\partial t} = a^2 \frac{\partial \rho}{\partial t} + a^2 v_j \frac{\partial \rho_0}{\partial x_j} - v_j \frac{\partial p_0}{\partial x_j} \tag{6.38}$$

$$\frac{\partial h_i}{\partial t} = \epsilon_{ijk} \frac{\partial}{\partial x_j} (\epsilon_{klm}\, v_l H_{0m}), \tag{6.39}$$

where

$$a = \sqrt{\gamma p_0/\rho_0} \tag{6.40}$$

is the velocity of sound in the medium and the subscripts '1' have, for simplicity, been dropped. In passing we note that a depends only on the temperature T_0 and is therefore constant throughout an isothermal fluid.

To obtain the wave equation one begins by differentiating equation (6.37) with respect to t:

$$\rho_0 \frac{\partial^2 v_i}{\partial t^2} = g_i \frac{\partial \rho}{\partial t} + \frac{1}{4\pi} \epsilon_{ijk}\epsilon_{jlm} \frac{\partial}{\partial x_l}\left(\frac{\partial h_m}{\partial t}\right) H_{0k} - \frac{\partial}{\partial x_i}\left(\frac{\partial p}{\partial t}\right).$$

The next step is to substitute from equations (6.36), (6.38), and (6.39) for the time derivatives on the RHS of this expression. Then, making use of (6.35) and (6.40), one obtains after some algebra the basic equation

$$\frac{\partial^2 v_i}{\partial t^2} = a^2 \frac{\partial}{\partial x_i}\left(\frac{\partial v_j}{\partial x_j}\right) + (\gamma - 1) g_i \frac{\partial v_j}{\partial x_j} + \frac{\partial}{\partial x_i}(g_j v_j) + \frac{H_0^2}{4\pi\rho_0} \frac{\partial}{\partial x_i}\left(\frac{\partial v_j}{\partial x_j}\right)$$
$$- \frac{1}{4\pi\rho_0} H_{0i} \cdot H_{0j} \frac{\partial}{\partial x_j}\left(\frac{\partial v_k}{\partial x_k}\right) - \frac{1}{4\pi\rho_0} H_{0j} \frac{\partial}{\partial x_j}\left(\epsilon_{ilm} H_{0l} \epsilon_{mpq} \frac{\partial v_q}{\partial x_p}\right) \tag{6.41}$$

governing the propagation of waves of all kinds through a compressible, horizontally stratified fluid.

In the sections that follow we shall often find it convenient to express the linearized equations (6.36)–(6.39) and the wave equation (6.41) in terms of the displacement $X_i(x_i,t)$ of a fluid particle at the point x_i at the time t from its equilibrium position x_{0i} instead of in terms of the velocity $v_i(x_i,t)$. Then, to the first order, we have

$$v_i(x_i,t) = \frac{\partial X_i}{\partial t}, \tag{6.42}$$

and so

$$X_i(x_i,t) = \int_0^t v_i(x_i,t')\,dt'. \tag{6.43}$$ [3]

Using these two relations we can transform the linearized equations (6.36)–(6.39) to read

$$\rho = -X_j \frac{\partial \rho_0}{\partial x_j} - \rho_0 \frac{\partial X_j}{\partial x_j} \tag{6.36'}$$

$$\rho_0 \frac{\partial^2 X_i}{\partial t^2} = \rho g_i + \frac{1}{4\pi} \epsilon_{ijk}\epsilon_{jlm} \frac{\partial h_m}{\partial x_l} H_{0k} - \frac{\partial p}{\partial x_i} \tag{6.37'}$$

$$p = a^2 \rho + a^2 X_j \frac{\partial \rho_0}{\partial x_j} - X_j \frac{\partial p_0}{\partial x_j} \tag{6.38'}$$

$$h_i = \epsilon_{ijk} \frac{\partial}{\partial x_j} (\epsilon_{klm} X_l H_{0m}), \tag{6.39'}$$

while the wave equation itself becomes

$$\frac{\partial^2 X_i}{\partial t^2} = a^2 \frac{\partial}{\partial x_i}\left(\frac{\partial X_j}{\partial x_j}\right) + (\gamma - 1) g_i \frac{\partial X_j}{\partial x_j} + \frac{\partial}{\partial x_i}(g_j X_j) + \frac{H_0^2}{4\pi\rho_0} \frac{\partial}{\partial x_i}\left(\frac{\partial X_j}{\partial x_j}\right)$$
$$- \frac{1}{4\pi\rho_0} H_{0i} \cdot H_{0j} \frac{\partial}{\partial x_j}\left(\frac{\partial X_k}{\partial x_k}\right) - \frac{1}{4\pi\rho_0} H_{0j} \frac{\partial}{\partial x_j}\left(\epsilon_{ilm} H_{0l} \epsilon_{mpq} \frac{\partial X_q}{\partial x_p}\right). \tag{6.41'}$$

Finally, eliminating the alternating tensor, we can write the wave equation in the

[3] Strictly speaking, X_i is a function of the equilibrium position x_{0i} of the particle but, provided the displacement remains very small compared with the wavelength, the distinction between x_i and x_{0i} may be disregarded [cf. Schmidt, 1966, pp. 119–120].

useful, alternative form

$$\frac{\partial^2 X_i}{\partial t^2} = a^2 \frac{\partial}{\partial x_i}\left(\frac{\partial X_j}{\partial x_j}\right) + (\gamma - 1) g_i \frac{\partial X_j}{\partial x_j} + \frac{\partial}{\partial x_i}(g_j X_j) + \frac{H_0^2}{4\pi\rho_0}\frac{\partial}{\partial x_i}\left(\frac{\partial X_j}{\partial x_j}\right)$$
$$- \frac{1}{4\pi\rho_0} H_{0i} \cdot H_{0j} \frac{\partial}{\partial x_j}\left(\frac{\partial X_k}{\partial x_k}\right) - \frac{1}{4\pi\rho_0} H_{0j} \frac{\partial}{\partial x_j}\left(H_{0k}\frac{\partial X_k}{\partial x_i} - H_{0k}\frac{\partial X_i}{\partial x_k}\right). \tag{6.44}$$

Derivations of the linearized wave equation in various equivalent forms have been given by a number of authors, including Kuperus [1965], McLellan and Winterberg [1968], and Uchida [1968]. Kuperus makes the assumption that the atmosphere is isothermal, but this restriction is not necessary.

6.2.4 EQUATION OF WAVE ENERGY

We now require formulae for the energy density and energy flux associated with a wave. An elegant method of deriving a wave energy equation within the framework of the first-order equations (6.36)–(6.39) has been formulated by Eckart [1960, p. 52 et seq.]. We multiply equation (6.36) by p/ρ_0, (6.37) scalarly by v_i, (6.38) by $p/\rho_0 a^2$, and (6.39) scalarly by $h_i/4\pi$, and add the results. With the aid of (6.38′) we obtain after some algebra the equation

$$\frac{\partial W}{\partial t} + \frac{\partial Q_i}{\partial x_i} = 0, \tag{6.45}$$

where

$$W = \frac{1}{2}\rho_0 v^2 + \frac{p^2}{2\rho_0 a^2} - \left(\frac{1}{2a^2} g^2 \rho_0 + \frac{1}{2} g \frac{d\rho_0}{dz}\right) Z^2 + \frac{h^2}{8\pi} \tag{6.46}$$

and

$$Q_i = p v_i + \frac{1}{4\pi}(H_{0j} h_j v_i - h_j v_j H_{0i}). \tag{6.47}$$

Here $z \equiv x_3$ is the vertical coordinate and Z is the vertical component of the particle displacement.

The first term on the right-hand side of equation (6.46) is the kinetic energy per unit volume associated with the wave motion, while the second term is the pressure energy familiar from the theory of acoustics [see, for example, Landau and Lifshitz, 1959, Chap. 8], where it is sometimes called the 'elastic' energy. The third term arises from the presence of the gravitational field and is termed by Eckart [loc. cit.] the 'thermobaric' energy. The fourth term represents the magnetic energy per unit volume associated with the wave motion. Within the framework of the linear theory, therefore, it seems physically reasonable to identify W as the overall wave energy density and, similarly, to identify the vector Q_i as the flux of wave energy. With this interpretation equation (6.45) expresses the conservation of *wave* energy and must be distinguished from (6.21) which expresses the conservation of the *total* energy of the gas.

One important result which we shall need later is an expression for the velocity at which the *energy* of the wave motion is propagated through the medium. If we denote this velocity by V_i, then clearly $WV_i = Q_i$ and hence

$$V_i = Q_i/W. \qquad (6.48)$$

It should be emphasized that the direction of energy propagation is not in general the same as that of the wave itself.

In conclusion, we may point out that Moore and Spiegel [1964] have obtained an equation for the wave energy in the absence of a magnetic field which can be shown to be identical to that obtained by omitting the magnetic terms in equation (6.45) above. The same remark applies to the wave energy equation given by Kato [1966a] for an isothermal atmosphere with no magnetic field, provided that one notes that there is a missing factor of ρ_0 in Kato's term for the thermobaric energy. Anderson [1963, pp. 47–50] and Bazer and Hurley [1963], on the other hand, have derived expressions for the energy density and energy flux vector in the case where a magnetic field is present but gravity is ignored ($g = 0$). Bearing in mind the assumptions made in the preceding discussion, namely that the steady velocity of the fluid is zero and the steady magnetic field is uniform, one can easily verify that Bazer and Hurley's results reduce to the expressions given above with $g = 0$. The same holds true for Anderson's formula for W, but his expression for Q_i differs from that given by equation (6.47) by the inclusion of several additional terms. His results are therefore not consistent with the energy equation (6.45). However, it can be shown that the additional terms make no contribution to the energy flux in the case where the wave variables vary harmonically in space and time (cf. Sections 6.3 and 6.4).

6.3 Hydrodynamic Waves in an Isothermal Atmosphere

6.3.1 INTRODUCTION

In the absence of viscous effects any displaced fluid element in a stratified atmosphere with no magnetic field is subject to two distinct forces: (*a*) a pressure force arising from the compressibility of the gas, and (*b*) a buoyancy force due to the gravitational stratification of the atmosphere. So long as the atmosphere is convectively stable, both these forces will act as restoring forces and hence a fluid element, if slightly displaced, will execute a harmonic oscillation about its initial position. Such oscillations permit the propagation of waves through the gas which will be basically *pressure* (i.e. *acoustic*) waves if the pressure force is dominant and internal *gravity* waves if the buoyancy force is dominant. Since the two forces are very different in character, one must expect the characteristics of the two wave modes to differ in a number of important respects, such as the ability to transfer mechanical energy, the direction of propagation, and the phase and amplitude relations between the variations in density, pressure, temperature, and velocity. An understanding of these differences is an essential pre-requisite to evaluating the possible roles which the two types of waves may play in transporting energy upwards through the various levels of the solar atmosphere (cf. Section 7.3.2).

In the interests of mathematical simplicity we shall confine our discussion of hydrodynamic wave propagation to the case of an unbounded *isothermal* atmosphere, in which the velocity of sound is everywhere the same. Even with this simple atmospheric model the conditions of wave propagation are surprisingly varied and complex: in fact, the propagation is both *dispersive* (i.e. the phase velocity varies with wavelength) and *anisotropic* (i.e. there is a preferred direction, defined by gravity). Thus, despite its apparent simplicity, the isothermal atmosphere permits the description of many important aspects of wave propagation. On the other hand, it must be pointed out that the phenomena of reflection and refraction are excluded since all waves propagate along straight lines to infinity. In a non-isothermal atmosphere, the reflection or refraction of waves may limit or even, under some circumstances, prevent the upward transport of wave energy.

6.3.2 DISPERSION EQUATION AND CRITICAL FREQUENCIES: ACOUSTIC AND GRAVITY WAVES

As we have seen in Section 6.2.3, the propagation of waves through a compressible stratified atmosphere is governed by the wave equation (6.44) which, in the absence of a magnetic field, reduces to the form

$$\frac{\partial^2 X_i}{\partial t^2} = a^2 \frac{\partial}{\partial x_i}\left(\frac{\partial X_j}{\partial x_j}\right) + (\gamma - 1)\, g_i \frac{\partial X_j}{\partial x_j} + \frac{\partial}{\partial x_i}(g_j X_j). \tag{6.49}$$

In the case of an isothermal atmosphere a is constant. Hence putting i = 1, 2, 3 in turn in equation (6.49) we obtain three linear equations with constant coefficients, for which, using a complex representation, we seek plane-wave solutions of the form

$$X_i = \chi_i\, e^{i(\omega t - k_j x_j)}. \tag{6.50}$$

Here χ_i is the complex amplitude, ω is the angular frequency, and k_j is the j-th component of the wave propagation vector. Since there is no preferential direction in the (x_1, x_2) plane we may put $k_2 = 0$ without loss of generality. It then follows from the second of equations (6.49) that $X_2 = 0$, while the condition that the remaining equations possess non-trivial solutions becomes simply

$$\begin{vmatrix} a^2k_1^2 - \omega^2 & a^2k_1k_3 - \mathrm{i}gk_1 \\ a^2k_1k_3 - \mathrm{i}(\gamma - 1)gk_1 & a^2k_3^2 - \mathrm{i}\gamma gk_3 - \omega^2 \end{vmatrix} = 0.$$

Expanding the determinant we obtain the general dispersion relation

$$\omega^4 - [a^2(k_1^2 + k_3^2) - \mathrm{i}\gamma gk_3]\,\omega^2 + (\gamma - 1)g^2k_1^2 = 0 \tag{6.51}$$

governing the propagation of hydrodynamic waves through an isothermal atmosphere.

Assuming that ω and k_1 are real we may solve equation (6.51) for the wave vector component k_3, obtaining the result

$$k_3 = \mathrm{i}\cdot\frac{\gamma g}{2a^2} \pm \sqrt{(\omega^2 - \omega_1^2)\frac{1}{a^2} - (\omega^2 - \omega_2^2)\frac{k_1^2}{\omega^2}}, \tag{6.52}$$

where

$$\omega_1 = \frac{\gamma g}{2a} \tag{6.53}$$

and

$$\omega_2 = \frac{\sqrt{\gamma - 1}\cdot g}{a} \tag{6.54}$$

are two critical frequencies whose significance is explained below. Equation (6.52) shows that the wave amplitude increases with height as $\exp[(\gamma g/2a^2)z]$, this being a direct consequence of the decrease in density with height. In order to separate out the amplitude factor it is convenient to re-define the vertical component of the wave vector as

$$k_3' = k_3 - \mathrm{i}\cdot\frac{\gamma g}{2a^2}, \tag{6.55}$$

whence it follows from equation (6.52) that

$$k_3' = \pm\sqrt{(\omega^2 - \omega_1^2)\frac{1}{a^2} - (\omega^2 - \omega_2^2)\frac{k_1^2}{\omega^2}}. \tag{6.56}$$

In the particular case of vertical propagation ($k_1 = 0$) we thus see that

$$k_3' = \pm\sqrt{(\omega^2 - \omega_1^2)\frac{1}{a^2}},$$

which shows that progressive waves are possible only if

$$\omega > \omega_1 \, .$$

For the general case $k_1 \neq 0$, it follows from equation (6.56) that no progressive waves can exist in the frequency range

$$\omega_2 < \omega < \omega_1 \, . \tag{6.57}$$

This is a standard result in the linear theory of wave motions in compressible, stratified fluids in the presence of a gravitational field.[4] Waves with frequencies $\omega < \omega_2$ are essentially *internal gravity waves* modified by the effect of compressibility, whereas waves with frequencies $\omega > \omega_1$ are *acoustic waves* modified by the effect of gravity (cf. Fig. 6.1).

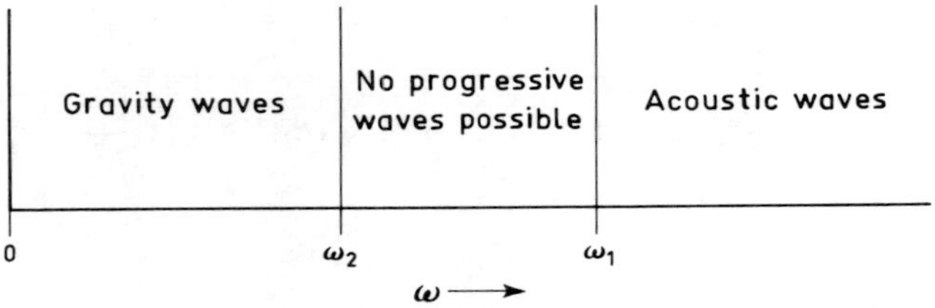

Fig. 6.1. Separation of the domains of gravity and acoustic waves according to the values of the two critical frequencies ω_1 and ω_2 defined in the text. No progressive waves can exist in the range $\omega_2 < \omega < \omega_1$.

Before proceeding to discuss the significance of these critical frequencies it is convenient for later purposes to rewrite the dispersion equation (6.51) with the aid of (6.55) in the form

$$\omega^4 - \omega_1^2 \, \omega^2 - a^2 \kappa^2 \omega^2 + a^2 k_1^2 \, \omega_2^2 = 0, \tag{6.58}$$

where

$$\kappa^2 = k_1^2 + k_3'^2 \, . \tag{6.59}$$

Solving for ω^2 we have

$$\omega^2 = \frac{1}{2} \left[\omega_1^2 + a^2 \kappa^2 \pm \sqrt{(\omega_1^2 + a^2 \kappa^2)^2 - 4a^2 k_1^2 \omega_2^2} \, \right] , \tag{6.60}$$

which shows that equation (6.58) admits of two real positive roots. It can be shown that the root obtained by taking the positive sign in (6.60) is greater than ω_1 and thus corresponds to acoustic waves, while the root taken with the

[4] The theory finds extensive application in studies of wave phenomena in the oceans and atmosphere of the Earth [see, for example, Eckart, 1960; Hines, 1960; Tolstoy, 1963; Murgatroyd, 1970]. An up-to-date review of the theory is given by Lighthill [1967].

negative sign is less than ω_2 and thus corresponds to gravity waves. In the latter case, equation (6.60) yields the result

$$V_p^2 = \frac{1}{2}\left[\frac{\omega_1^2}{\kappa^2} + a^2 - \sqrt{\left(\frac{\omega_1^2}{\kappa^2} + a^2\right)^2 - 4a^2 \sin^2 \Phi \cdot \frac{\omega_2^2}{\kappa^2}}\right], \qquad (6.61)$$

where $V_p = \omega/\kappa$ is the phase velocity of the gravity wave and $\Phi = \sin^{-1}(k_1/\kappa)$ is the angle that the wave vector makes with the vertical. Since $V_p = 0$ when $\Phi = 0$ or $180°$, it is clear that gravity waves cannot propagate vertically. Another important conclusion which follows from a consideration of the roots of equation (6.58) is that the phase velocity of gravity waves is always less than the velocity of sound, whereas the phase velocity of (modified) acoustic waves is greater than the velocity of sound.

The critical frequency ω_1 represents a low-frequency cut-off for the propagation of acoustic waves arising from the presence of a gravitational field. At high frequencies ($\omega \gg \omega_1$) the waves are scarcely affected by gravity, but they become increasingly modified as the frequency decreases until, at $\omega = \omega_1$, propagation is no longer possible and only standing waves can be excited.

On the other hand, the high-frequency cut-off for internal gravity waves, ω_2, corresponds to the basic resonance frequency of the atmosphere [Worrall, 1972]. Its significance is best understood by temporarily dropping the restriction to an isothermal atmosphere and considering the motion of a fluid element displaced vertically from its equilibrium position. Because of the speed with which pressure changes are equalized throughout a gas, we shall suppose that the pressures inside and outside the displaced element are always the same and that, in addition, density changes can occur unimpeded under adiabatic conditions. If z is the height of the element in the equilibrium state, the change in its pressure resulting from a small vertical displacement δz is clearly

$$\delta p = -\rho g \, \delta z.$$

Making use of the fact that the quantity $p\rho^{-\gamma}$ remains constant for a given fluid element under adiabatic conditions (see equation 6.17), we find that the change in density inside the displaced element is

$$(\delta\rho)_{\text{inside}} = \delta p / a^2 ,$$

where $a = \sqrt{\gamma p/\rho}$ is the local velocity of sound. The density just outside the element, which was $\rho(z)$ before the displacement, changes by the amount

$$(\delta\rho)_{\text{outside}} = \frac{d\rho}{dz} \cdot \delta z,$$

and hence the fluid element experiences a buoyancy force

$$g\,[(\delta\rho)_{\text{outside}} - (\delta\rho)_{\text{inside}}] = g \cdot \delta z\left(\frac{\mathrm{d}\rho}{\mathrm{d}z} + \frac{\rho g}{a^2}\right).$$

If the buoyancy force is the only force acting on the element, its equation of motion is accordingly

$$\frac{\partial^2(\delta z)}{\partial t^2} + N^2\,\delta z = 0, \tag{6.62}$$

where

$$N^2 = -g\left[\frac{\mathrm{d}}{\mathrm{d}z}(\ln\rho) + \frac{g}{a^2}\right]. \tag{6.63}$$

Thus, provided the term in brackets is negative and hence N^2 positive, the element will execute simple harmonic oscillations with the frequency N. This is termed the *Väisälä frequency.*

In the present context it is instructive to use the equation of state (6.5) to transform the expression (6.63) to read

$$N^2 = g\left[\frac{(\gamma-1)g}{a^2} + \frac{1}{T}\frac{\mathrm{d}T}{\mathrm{d}z}\right], \tag{6.64}$$

showing that N is equal to the critical frequency ω_2 defined by the relation (6.54) if the atmosphere is isothermal. However, if the temperature decreases outwards (i.e. $\mathrm{d}T/\mathrm{d}z < 0$), the value of N^2 is reduced and may even become negative: when this happens, oscillations can no longer occur. We can conveniently explore the consequences of a negative temperature gradient by introducing the (constant) adiabatic temperature gradient

$$\left(\frac{\mathrm{d}T}{\mathrm{d}z}\right)_{\mathrm{ad}} = -\frac{(\gamma-1)\,gT}{a^2} \tag{6.65}$$

into (6.64), which then becomes

$$N^2 = \frac{g^2(\gamma-1)}{a^2}\left[1 - \left|\frac{\mathrm{d}T}{\mathrm{d}z}\right| \Big/ \left|\frac{\mathrm{d}T}{\mathrm{d}z}\right|_{\mathrm{ad}}\right]. \tag{6.66}$$

In the particular case of a gradient equal to the adiabatic value, we see from (6.66) that $N^2 = 0$, so that gravity waves cannot propagate in an atmosphere in adiabatic equilibrium. If, on the other hand,

$$\left|\frac{\mathrm{d}T}{\mathrm{d}z}\right| > \left|\frac{\mathrm{d}T}{\mathrm{d}z}\right|_{\mathrm{ad}}, \tag{6.67}$$

then $N^2 < 0$ and convective instability may set in: in other words a fluid element, once displaced, does not oscillate about its equilibrium position but continues to move upwards.[5] We are thus led to the important conclusion that gravity waves cannot propagate in the solar convection zone, where (6.67) holds. However, this does not prevent their generation in the overlying atmosphere, where (6.67) no longer holds, by the motions at the top of the convection zone.

The above discussion also makes it clear why a gravity wave cannot propagate in a strictly vertical direction: as pointed out by Schatzman and Souffrin [1967, p. 71], the buoyancy force acting on a displaced element is due to its interaction with other elements of different density located at the same height, but this interaction is precluded by strictly vertical propagation.

The extension of the theory of hydrodynamic wave propagation to the case of a non-isothermal atmosphere has been considered by Moore and Spiegel [1964] in terms of a technique based on the WKB approximation. They find that the critical frequencies ω_1 and ω_2 are now given by the expressions

$$\omega_1^2(z) = \frac{\gamma^2 g^2}{4a^2(z)} \tag{6.68}$$

and

$$\omega_2^2(z) = \frac{g^2(\gamma - 1)}{a^2(z)} \left[1 - \frac{\mathrm{d}T}{\mathrm{d}z} \bigg/ \left(\frac{\mathrm{d}T}{\mathrm{d}z}\right)_{\mathrm{ad}}\right], \tag{6.69}$$

where $a(z)$ is the value of the sound velocity at height z. The relation (6.69) agrees with the formula (6.66) for the Väisälä frequency, derived above by a simple argument.

6.3.3 PROPAGATION OF ENERGY BY ACOUSTIC AND GRAVITY WAVES

Expressions for the total energy density and energy flux of the wave motions governed by the dispersion relation (6.58) are given by equations (6.46) and (6.47) with the magnetic terms omitted. We note, however, that these expressions involve *real* quantities whereas the wave solutions derived in the previous section are expressed in terms of complex functions (cf. equation 6.50). However, taking the properties of these functions into account [see, for example, Stratton, 1941, pp. 135–136] one can easily derive the expressions

$$\overline{W} = \frac{1}{4}\rho_0 v_i v_i^* + \frac{pp^*}{4\rho_0 a^2} + \frac{1}{4a^2} g^2 \rho_0(\gamma - 1) ZZ^* \tag{6.70}$$

$$\overline{Q}_i = \frac{1}{4}(pv_i^* + p^* v_i) \tag{6.71}$$

[5] The inequality (6.67) is only a necessary, but not a sufficient condition for convective instability since the effects of viscosity and thermal conduction have been ignored [Bray and Loughhead, 1967, p. 71].

for the time-averaged values of the energy density and energy flux respectively, the asterisks denoting complex conjugates. The actual calculation of $\overline{W}$ and $\overline{Q}_i$ in terms of the solutions (6.50) requires considerable algebra and we shall merely state the results:

$$\overline{W} = \frac{\rho_0 \omega^2}{2(\omega^2 - a^2 k_1^2)} \cdot (2\omega^2 - \omega_1^2 - a^2 \kappa^2) \cdot ZZ^*$$

$$\overline{Q}_1 = \frac{\rho_0 \omega}{2(\omega^2 - a^2 k_1^2)} \cdot a^2 (\omega^2 - \omega_2^2) k_1 \cdot ZZ^* \qquad (6.72)$$

$$\overline{Q}_3 = \frac{\rho_0 \omega}{2(\omega^2 - a^2 k_1^2)} \cdot a^2 \omega^2 k_3' \cdot ZZ^*.$$

These three quantities are independent of height, as one would expect from the conservation of wave energy. This is easily verified when one recalls that $\rho_0 \propto e^{-z/\mathcal{H}}$ and $ZZ^* \propto e^{z/\mathcal{H}}$, where

$$\mathcal{H} = a^2/\gamma g \qquad (6.73)$$

is the scale height of the (isothermal) atmosphere.

In accordance with equation (6.48) it follows from the expressions (6.72) that the velocity of energy propagation has the components

$$V_1 = \frac{a^2 (\omega^2 - \omega_2^2) k_1}{\omega(2\omega^2 - \omega_1^2 - a^2 \kappa^2)}$$

and (6.74)[6]

$$V_3 = \frac{a^2 \omega^2 k_3'}{\omega(2\omega^2 - \omega_1^2 - a^2 \kappa^2)} .$$

According to equation (6.60) the quantity $(2\omega^2 - \omega_1^2 - a^2\kappa^2)$ is positive for acoustic waves but negative for gravity waves. Hence it follows from the second of the relations (6.74) that, in the case of gravity waves, the vertical component of the group velocity vector is opposite in sign to the vertical component of the

[6] It is well known that the velocity of wave energy propagation in an isotropic medium is the group velocity $V_i = \partial\omega/\partial k_i$, and it can be shown that this is also true for an anisotropic medium, such as the stratified atmosphere being considered. If we write the dispersion relation (6.58) in the form

$$D \equiv \omega^4 - \omega_1^2 \omega^2 - a^2 (k_1^2 + k_3'^2) \omega^2 + a^2 k_1^2 \omega_2^2 = 0,$$

it follows from the rule for differentiating implicit functions that

$$V_i = \partial\omega/\partial k_i = -\frac{\partial D}{\partial k_i} \Big/ \frac{\partial D}{\partial \omega}.$$

Performing the differentiations we recover the expressions (6.74).

wave vector. This means that, for these waves, the upward transfer of energy is accomplished by waves whose phase propagates downwards, and vice versa [cf. Eliassen and Palm, 1954; Hines, 1960].

Following Kato [1966b] it is instructive to express the group velocity and the components of the wave vector in terms of the frequency ω and the angle ψ which the group velocity vector makes with the vertical. Clearly we have

$$V = \frac{V_1}{\sin\psi} = \frac{V_3}{\cos\psi}, \tag{6.75}$$

and hence substituting for V_1 and V_3 from (6.74) we obtain the expression

$$k_1 = \frac{\omega^2 k_3'}{\omega^2 - \omega_2^2} \cdot \tan\psi . \tag{6.76}$$

(It is interesting to note that this relation can be rewritten in the form

$$\tan\Phi = \frac{\omega^2}{\omega^2 - \omega_2^2} \cdot \tan\psi, \tag{6.77}$$

where $\Phi = \tan^{-1}(k_1/k_3')$ is the inclination of the wave vector to the vertical.) Substituting from equation (6.76) in the dispersion relation (6.58) we find that

$$k_1 = \pm\frac{1}{a}\sqrt{\frac{\omega^4(\omega^2-\omega_1^2)\sin^2\psi}{(\omega^2-\omega_2^2)(\omega^2-\omega_2^2\cos^2\psi)}}$$

and

$$k_3' = \pm\frac{1}{a}\sqrt{\frac{(\omega^2-\omega_1^2)(\omega^2-\omega_2^2)\cos^2\psi}{\omega^2-\omega_2^2\cos^2\psi}}. \tag{6.78}$$

Finally, by combining equations (6.74), (6.75), and (6.78), one can derive the formula

$$V^2 = a^2 \cdot \frac{(\omega^2-\omega_2^2)^3\,(\omega^2-\omega_1^2)(\omega^2-\omega_2^2\cos^2\psi)}{\omega^2\left[(\omega^2-\omega_2^2)^2+\omega_2^2(\omega_1^2-\omega_2^2)\sin^2\psi\right]^2} \tag{6.79}$$

for the magnitude of the group velocity vector in terms of ω and ψ. Equation (6.79) shows at once that, in the case of gravity waves (i.e. $\omega < \omega_2$), propagation of energy is possible only for values of ψ exceeding a critical value ψ_c defined by the relation

$$\omega^2 = \omega_2^2\cos^2\psi_c. \tag{6.80}$$

When $\psi = \psi_c$, $V = 0$ and there is no flow of energy. No such restriction applies to the propagation of energy by acoustic waves.

6.3.4 RADIATIVE DAMPING OF WAVES

Damping of hydrodynamic waves due to viscous dissipation and particle conduction is entirely negligible in those layers of the solar atmosphere where small-amplitude disturbances are likely to occur (cf. Section 7.4.2). Radiative damping, on the other hand, may have a marked effect on both the conditions for propagation and the characteristics of the waves, particularly in the upper photosphere, where the time of radiative relaxation is likely to be short compared to the periods of the waves (Table 7.3). As we shall see below, the condition for the propagation of gravity waves, which depend on the existence of buoyancy forces, is more stringent in the presence of radiative damping than in its absence (cf. Section 6.3.2). Equally significant are the changes which result in the phase and amplitude relationships among the perturbations in density, pressure, temperature, and velocity. In fact, accurate observations of such relationships may ultimately prove to be of crucial importance in identifying the wave modes actually present in the solar photosphere and chromosphere. In this connection, Schatzman and Souffrin [1967] have pointed out two other consequences of radiative damping which should not be overlooked. One is the heating of the medium through which the waves propagate, and the other is the retardation of the normal growth of wave amplitude resulting from the upward decrease in density in a stratified atmosphere. It is the growth in wave amplitude that is responsible for the appearance of shock waves at greater heights in the atmosphere (see Section 7.4.2 and Fig. 7.4).

When the time of radiative relaxation is long compared to the wave period, the wave propagation is governed by the familiar adiabatic relation

$$\frac{\mathrm{d}p}{\mathrm{d}t} = \frac{\gamma p}{\rho} \frac{\mathrm{d}\rho}{\mathrm{d}t} \tag{6.81}$$

given in Section 6.2.2. On the other hand, when the relaxation time is very small compared to the wave period, the propagation becomes effectively isothermal and so, in an isothermal atmosphere, we have simply

$$\frac{\mathrm{d}T}{\mathrm{d}t} = 0, \tag{6.82}$$

whence it follows from the equation of state (6.5) that

$$\frac{\mathrm{d}p}{\mathrm{d}t} = \frac{p}{\rho} \frac{\mathrm{d}\rho}{\mathrm{d}t} \,. \tag{6.83}$$

This is *formally* the same as the relation obtained by putting $\gamma = 1$ in equation (6.81). A number of authors have therefore attempted to estimate the effects of radiative damping in intermediate cases by the simple expedient of replacing γ in equation (6.81) by an 'effective gamma' γ_e having a value somewhere in the range $1 < \gamma_e < \gamma$. However, a detailed hydrodynamic treatment of the problem by Souffrin [1966] has shown that this procedure leads to erroneous results, both in the description of the wave properties and in the evaluation of the energy dissipation.

(1) *Equation of internal energy.* In order to give a proper hydrodynamic treatment it is necessary to take account of radiative loss of energy in the equation of conservation of internal energy. Under conditions of local thermodynamic equilibrium the increase de in the internal energy of unit mass of gas resulting from an input of heat energy dq is, by the first law of thermodynamics,

$$\mathrm{d}e = \mathrm{d}q - p \cdot \mathrm{d}\left(\frac{1}{\rho}\right),$$

where the quantity $p \cdot \mathrm{d}\left(\frac{1}{\rho}\right)$ represents the work done by the gas in expanding against its surroundings. Thus we have

$$\mathrm{d}q = \mathrm{d}e - \frac{p}{\rho^2} \cdot \mathrm{d}\rho \tag{6.84}$$

or, dividing by the time interval dt,

$$\frac{\mathrm{d}q}{\mathrm{d}t} = \frac{\mathrm{d}e}{\mathrm{d}t} - \frac{p}{\rho^2} \cdot \frac{\mathrm{d}\rho}{\mathrm{d}t}. \tag{6.85}$$

If we now substitute for e from equation (6.16) we obtain the equation of internal energy in the convenient form

$$\frac{\mathrm{d}p}{\mathrm{d}t} = \frac{\gamma p}{\rho} \frac{\mathrm{d}\rho}{\mathrm{d}t} + \rho(\gamma - 1) \frac{\mathrm{d}q}{\mathrm{d}t} \tag{6.86}$$

given by Souffrin [1966].

The calculation of the term dq/dt poses a difficult problem in the theory of radiative transfer which has been considered by a number of workers. In particular, Spiegel [1957] has shown that in a homogeneous grey atmosphere in local thermodynamic equilibrium the smoothing of temperature fluctuations over *optically thin* regions is governed by Newton's law of cooling

$$\frac{\mathrm{d}q}{\mathrm{d}t} = -\frac{c_V \delta T}{t_R}, \tag{6.87}$$

where δT is the temperature perturbation. The quantity t_R is termed the radiative relaxation time and is given by the formula

$$t_R = \frac{\rho c_V}{16 \bar{\kappa} \sigma_R T^3}, \tag{6.88}$$

$\bar{\kappa}$ being the mean linear absorption coefficient per cm [see also Delache and Froeschlé, 1972]. t_R is independent of the geometrical dimensions of the perturbed region since this is assumed to be optically thin; this assumption is justified for all layers of the solar photosphere and chromosphere having an optical depth $\tau < 1$ at the wavelengths concerned. Strictly speaking, the use of the formula (6.88) in the case of a stratified atmosphere is justified only if the vertical dimension of the perturbation is small compared to the local scale height of the atmosphere [Thomas *et al.*, 1971].

From (6.16) and (6.40) it follows that

$$a^2 = \gamma(\gamma - 1)Tc_v, \tag{6.89}$$

so that (6.87) can be rewritten in the form

$$\frac{dq}{dt} = -\frac{a^2}{\gamma(\gamma - 1)t_R} \cdot \frac{\delta T}{T}. \tag{6.90}$$

When the perturbations are small, we have from the equation of state (6.5)

$$\frac{\delta T}{T} = \frac{\delta p}{p} - \frac{\delta \rho}{\rho}$$

and hence, referring back to equation (6.38), we can express the internal energy equation in the linearized form

$$\frac{\partial p_1}{\partial t} = a^2 \frac{\partial \rho_1}{\partial t} + a^2 v_{1j} \frac{\partial \rho_0}{\partial x_j} - v_{1j} \frac{\partial p_0}{\partial x_j} - \frac{1}{\gamma t_R} a^2 \rho_0 \left(\frac{p_1}{p_0} - \frac{\rho_1}{\rho_0}\right). \tag{6.91}$$

Here ρ_1, p_1, and v_{1j} represent the perturbations in density, pressure, and velocity from the equilibrium values ρ_0, p_0, and $v_{0j} = 0$, respectively. Gathering equation (6.91) and the linearized equations of continuity (6.36) and momentum (6.37) given in Section 6.2.3 and dropping the suffixes '1' we have the full set of hydrodynamic equations governing the propagation of small-amplitude waves through an optically thin, stratified atmosphere in the presence of radiative damping:

$$\frac{\partial \rho}{\partial t} + v_j \frac{\partial \rho_0}{\partial x_j} + \rho_0 \frac{\partial v_j}{\partial x_j} = 0 \tag{6.92}$$

$$\rho_0 \frac{\partial v_i}{\partial t} = \rho g_i - \frac{\partial p}{\partial x_i} \tag{6.93}$$

$$\frac{\partial p}{\partial t} = a^2 \frac{\partial \rho}{\partial t} + a^2 v_j \frac{\partial \rho_0}{\partial x_j} - v_j \frac{\partial p_0}{\partial x_j} - \frac{1}{\gamma t_R} a^2 \rho_0 \left(\frac{p}{p_0} - \frac{\rho}{\rho_0}\right). \tag{6.94}$$

The corresponding generalization of Eckart's equation of wave energy (6.45) is

$$\frac{\partial W}{\partial t} + \frac{\partial Q_{\mathrm{j}}}{\partial x_{\mathrm{j}}} = -\frac{1}{\gamma t_{\mathrm{R}}} p \left(\frac{p}{p_0} - \frac{\rho}{\rho_0} \right), \tag{6.95}$$

where W is the wave energy density given by equation (6.46), Q_{j} is the wave energy flux vector given by equation (6.47), and the term on the RHS represents the rate of energy increase per unit volume due to radiative transfer.

(2) *Gravitational oscillation of a fluid element.* Before particularizing to the case of an isothermal atmosphere it is instructive, as we have done previously in Section 6.3.2, to consider the motion of a fluid element displaced vertically from its equilibrium position. We shall use asterisks to distinguish the values of the physical variables inside the element and shall suppose, as before, that the pressures inside and outside are always the same. Thus

$$p^* = p,$$

whence it follows from the equation of state (6.5) that, for small variations,

$$\frac{T^* - T}{T^*} = -\frac{\rho^* - \rho}{\rho^*}. \tag{6.96}$$

Combining (6.96) with equations (6.86) and (6.90) we find that the equation of internal energy for the element becomes, to the first order in variations from the equilibrium state,

$$\frac{\mathrm{d}p}{\mathrm{d}t} - a^2 \frac{\mathrm{d}\rho^*}{\mathrm{d}t} - \frac{a^2}{\gamma t_{\mathrm{R}}} (\rho^* - \rho) = 0. \tag{6.97}$$

Next we let v^* denote the vertical velocity of the fluid element, so that

$$\frac{\mathrm{d}p}{\mathrm{d}t} = v^* \frac{\mathrm{d}p}{\mathrm{d}z} \quad \text{and} \quad \frac{\mathrm{d}\rho}{\mathrm{d}t} = v^* \frac{\mathrm{d}\rho}{\mathrm{d}z}.$$

Using these relations and the equation of hydrostatic equilibrium (6.35) we can then rewrite (6.97) in the form

$$\frac{\mathrm{d}}{\mathrm{d}t}(\rho^* - \rho) + \frac{1}{\gamma t_{\mathrm{R}}}(\rho^* - \rho) + \rho \left(\frac{1}{\rho} \frac{\mathrm{d}\rho}{\mathrm{d}z} + \frac{g}{a^2} \right) v^* = 0.$$

Finally, if we now substitute for $\rho^* - \rho$ from the equation of motion of the element

$$\rho^* \frac{\mathrm{d}v^*}{\mathrm{d}t} = -(\rho^* - \rho) g,$$

we find that, correct to the first order, the movement of the displaced fluid element is governed by the equation

$$\frac{\partial^2 v^*}{\partial t^2} + \frac{1}{\gamma t_R} \cdot \frac{\partial v^*}{\partial t} + N^2 v^* = 0, \tag{6.98}$$

where

$$N^2 = -g\left[\frac{d}{dz}(\ln \rho) + \frac{g}{a^2}\right] \tag{6.99}$$

gives the Väisälä frequency.

In terms of the particle displacement defined by equation (6.42), which we shall here write as δz, equation (6.98) takes the form

$$\frac{\partial^2(\delta z)}{\partial t^2} + \frac{1}{\gamma t_R} \cdot \frac{\partial(\delta z)}{\partial t} + N^2 \delta z = 0. \tag{6.100}$$

This is a generalization of the corresponding equation (6.62) given in Section 6.3.2 and reduces to it in the limit $t_R \to \infty$. Consideration of the solutions of equation (6.100) shows that, as before, the atmosphere is convectively stable if $N^2 > 0$ but unstable if $N^2 < 0$. The condition for marginal stability, $N^2 = 0$, is independent of the value of t_R.

On the other hand, owing to the damping effect due to radiative relaxation, the stability criterion $N^2 > 0$ is no longer a sufficient condition for the existence of oscillations. In fact, examination of the form of the solutions of equation (6.100) indicates that, even with $N^2 > 0$, the particle motion will be oscillatory only if

$$t_R > \frac{1}{2\gamma N}, \tag{6.101}$$

in agreement with a result of Souffrin [1966]. Generalizing, this analysis indicates that in the presence of radiative damping the propagation of gravity waves is possible only if the radiative relaxation time is sufficiently long.

(3) *Dispersion equation and wave modes in an isothermal atmosphere.* For a more general treatment one has to turn to the solution of the linearized hydrodynamic equations (6.92)–(6.94). These have been solved by Souffrin [1966] for the case of an *isothermal* atmosphere, in which a is a constant. For the sake of mathematical simplicity Souffrin makes the additional assumption that the radiative relaxation time t_R is also constant, although, in reality, it must vary with height. Without loss of generality we may put $v_2 = 0$, since there is no preferential direction in the (x_1, x_2) plane. Then, eliminating v_1, one can replace (6.92)–(6.94) by a set of three equations

$$\frac{\partial P}{\partial t} + \frac{1}{t_R} P + \left(\frac{a^2}{g\mathcal{H}} - 1\right) \frac{\partial U}{\partial t} = \frac{\partial G}{\partial t} + \frac{1}{\gamma t_R} G \tag{6.102}$$

$$\frac{\partial^2 U}{\partial t^2} = -g\left(\frac{\partial}{\partial x_3} - \frac{1}{2\mathcal{H}}\right)P - \frac{g^2}{a^2}G \tag{6.103}$$

$$\frac{\partial^2 G}{\partial t^2} = a^2\frac{\partial^2 P}{\partial x_1^2} - \frac{a^2}{g}\frac{\partial^2}{\partial t^2}\left(\frac{\partial}{\partial x_3} - \frac{1}{2\mathcal{H}}\right)U \tag{6.104}$$

in three 'reduced' variables

$$\left.\begin{aligned} U &= a\rho_0^{1/2}\, v_3 \\ P &= \frac{a}{g\rho_0^{1/2}}\cdot\frac{\partial p}{\partial t} \\ G &= \frac{a^3}{g\rho_0^{1/2}}\cdot\frac{\partial \rho}{\partial t} \end{aligned}\right\}, \tag{6.105}$$

where, by equation (6.73),

$$\mathcal{H} = a^2/\gamma g$$

is the scale height of the atmosphere.

The coefficients of all terms in the reduced equations (6.102)–(6.104) are constants and we may therefore seek plane-wave solutions of the form

$$U, P, G \propto e^{i(\omega t - \kappa_j x_j)} \tag{6.106}$$

where

$$\kappa_j = (k_1, 0, k_3'). \tag{6.107}$$ [7]

Substituting from (6.106) and (6.107) into (6.102)–(6.104) we obtain the relations

$$\begin{aligned} i\omega\left(\frac{a^2}{g\mathcal{H}} - 1\right)U + \left(i\omega + \frac{1}{t_R}\right)P - \left(i\omega + \frac{1}{\gamma t_R}\right)G &= 0 \\ \omega^2 U + g\left(ik_3' + \frac{1}{2\mathcal{H}}\right)P - \frac{g^2}{a^2}G &= 0 \\ \frac{a^2\omega^2}{g}\left(ik_3' + \frac{1}{2\mathcal{H}}\right)U + a^2k_1^2P - \omega^2 G &= 0, \end{aligned} \tag{6.108}$$

and equating the determinant of the coefficients to zero we obtain the dispersion

[7] The vertical component of the wave vector is written as k_3' by analogy with the vertical wave vector component defined by equation (6.55) in Section 6.3.2. The choice of the reduced variables (6.105) is governed by the convenience of separating out amplitude factors resulting from the decrease of density with height.

relation

$$\kappa^2 + \frac{\omega_1^2 - \omega^2}{a^2} - \frac{\omega_2^2}{\omega^2} k_1^2 = \frac{\mathrm{i}}{\omega\gamma t_{\mathrm{R}}}\left(\kappa^2 + \frac{\omega_1^2 - \gamma\omega^2}{a^2}\right) \tag{6.109}$$

Here ω_1 and ω_2 are the two critical frequencies for the case of adiabatic propagation defined by equations (6.53) and (6.54) respectively.

Equation (6.109) is the general dispersion relation governing the propagation of waves through an isothermal radiating atmosphere in which the regions of temperature perturbation are optically thin. As such, it is a generalization of the dispersion equation (6.58) given in Section 6.3.2 for the case of adiabatic propagation and reduces to it in the limit $t_{\mathrm{R}} \to \infty$. However, the vertical wave vector component k_3' is now clearly a complex quantity. In fact, if we write

$$k_3' = \varphi + \mathrm{i}\lambda = (l + \mathrm{i}m)^{1/2}, \tag{6.110}$$

say, where φ, λ, l, and m are all real, then it follows from equation (6.109) that

$$l = \frac{\omega^2 - \omega_1^2}{a^2} + k_1^2\left(\frac{\omega_2^2}{\omega^2} - 1\right) - \frac{1}{1 + \omega^2\gamma^2 t_{\mathrm{R}}{}^2}\frac{\omega_2^2}{\omega^2}\left(k_1^2 - \frac{\omega^4}{g^2}\right)$$

and (6.111)

$$m = \frac{\omega\gamma t_{\mathrm{R}}}{1 + \omega^2\gamma^2 t_{\mathrm{R}}{}^2}\frac{\omega_2^2}{\omega^2}\left(k_1^2 - \frac{\omega^4}{g^2}\right).$$

There is, however, an ambiguity as to the correct sign to allocate to the radical in (6.110), which Souffrin [loc. cit.] resolves by imposing the condition that the significant wave solutions correspond to a positive flow of wave energy in the vertical direction. According to equations (6.47), (6.71), and (6.95) the time-averaged value of the vertical component of the wave energy flux is

$$\bar{Q}_3 = \frac{1}{4}(pv_3^* + p^*v_3),$$

or, in terms of the reduced variables (6.105),

$$\bar{Q}_3 = \frac{g}{4a^2\omega} \cdot \mathrm{i}(P^*U - PU^*), \tag{6.112}$$

the asterisks denoting complex conjugates. Using the second and third of the relations (6.108) one can show that

$$\bar{Q}_3 = \frac{g^2}{2a^2\omega^4\left[\left(\frac{g}{2\mathcal{H}} - g\lambda - \frac{g^2k_1^2}{\omega^2}\right)^2 + g^2\varphi^2\right]} \cdot \omega\varphi(\omega^4 - g^2k_1^2)UU^*, \tag{6.113}$$

and the condition that the vertical flux of wave energy be positive is therefore

$$\omega\varphi(\omega^4 - g^2k_1^2) > 0. \tag{6.114}$$

Furthermore, from (6.110)

$$m = 2\varphi\lambda,$$

so that on multiplying the second of equations (6.111) by $\omega(\omega^4 - g^2k_1^2)$ one finds that

$$\omega\varphi\lambda(\omega^4 - g^2k_1^2) = -\frac{1}{2}\cdot\frac{\omega^2\gamma t_R}{1+\omega^2\gamma^2 t_R^2}\cdot\frac{\omega_2^2}{\omega^2 g^2}(\omega^4 - g^2k_1^2)^2$$

$$\leqslant 0. \tag{6.115}$$

Combining the inequalities (6.114) and (6.115) we thus see that the condition for a positive vertical flux of wave energy is

$$\lambda \leqslant 0 \tag{6.116}$$

for all k_1 and ω. This fixes the physically significant sign of the radical $(l + im)^{1/2}$. It follows also that variables U, P, and G all decrease with height as $e^{-|\lambda|z}$.

The physical significance of the fact that the vertical wave vector component $k_3' = \varphi + i\lambda$ is a complex quantity is that in a radiating atmosphere the wave modes, unlike those in the adiabatic case (Section 6.3.2), are no longer purely progressive or purely stationary. The modes which were progressive in the adiabatic case now experience spatial damping due to the effect of radiative cooling, and the modes which were non-progressive are modified in such a way that they can now carry enough energy to balance the dissipation [Schatzman and Souffrin, 1967, p. 73].

Souffrin [1966] has given a detailed discussion of the various wave modes allowed by the dispersion equation (6.109), and here we need only quote his conclusions. He finds that the permitted modes fall into three separate categories defined in relationship to two critical frequencies

$$\omega_3^2 = \frac{1}{2}\left\{\omega_1^2 - \frac{1}{\gamma t_R^2} + \left[\left(\omega_1^2 - \frac{1}{\gamma t_R^2}\right)^2 + \frac{4\omega_1^2}{\gamma^2 t_R^2}\right]^{1/2}\right\}, \tag{6.117}$$

and

$$\omega_4^2 = \omega_2^2 - \frac{1}{\gamma^2 t_R^2} \quad \text{if} \quad t_R\omega_2 \geqslant \frac{1}{\gamma}$$

or (6.118)

$$\omega_4^2 = 0 \quad \text{if} \quad t_R\omega_2 < \frac{1}{\gamma},$$

where ω_1 and ω_2 are given by (6.53) and (6.54) respectively. Note that $\omega_3 \to \omega_1$ and $\omega_4 \to \omega_2$ in the limit $t_R \to \infty$. Waves with frequencies in the range

$$\omega_4 < \omega < \omega_3$$

are classified by Schatzman and Souffrin [loc. cit.] as 'mainly damped' waves, whereas those outside this range are classified as 'mainly progressive'. The latter are further classified as acoustic-type or gravity-type waves according as $\omega > \omega_3$ or $\omega < \omega_4$, respectively.

Equation (6.118) shows that gravity-type waves can exist only if

$$t_R > \frac{1}{\gamma\omega_2}, \tag{6.119}$$

i.e. provided the time of radiative relaxation is sufficiently long. This result contradicts the condition obtained by the artifice of replacing γ in the equation of adiabatic propagation by an effective value γ_e in the range $1 < \gamma_e < \gamma$. In fact, the latter procedure leads to the erroneous conclusion that gravity-type waves can exist for all non-zero values of t_R.

Finally, it should be noted that the problem of the radiative damping of hydrodynamic waves has recently been re-examined by Stix [1970]. He generalizes the discussion of the dispersion relation (6.109) to admit complex values of the wave frequency ω and, as a result, finds that the condition (6.119) is replaced by the rather less restrictive requirement

$$t_R > \frac{1}{2\gamma\omega_2}. \tag{6.120}$$

Since ω_2 is equal to the Väisälä frequency N in the case of an isothermal atmosphere (cf. equation 6.64), the condition (6.120) is identical to the condition (6.101) derived earlier from our simple analysis of the motion of a displaced fluid element.

6.4 Magnetohydrodynamic Waves in an Isothermal Atmosphere

6.4.1 INTRODUCTION

In the presence of a magnetic field the particles of an ionized fluid are subject to an additional restoring force, and as a result the modes of wave propagation are correspondingly more complex. The relative importance of magnetic and hydrodynamic forces can be crudely assessed by comparing the 'magnetic pressure' $H_0^2/8\pi$ (cf. equation 6.20) with the ordinary gas pressure p_0. Magnetic forces will predominate when

$$\frac{H_0^2}{8\pi} > p_0,$$

which in turn implies that

$$\frac{H_0^2}{4\pi\rho_0} > \frac{2p_0}{\rho_0} \simeq \frac{\gamma p_0}{\rho_0}.$$

Furthermore, introducing the Alfvén wave velocity

$$A = \frac{H_0}{\sqrt{4\pi\rho_0}}, \tag{6.121}$$

we see that the latter inequality is equivalent to the condition

$$A > a. \tag{6.122}$$

In the low solar chromosphere the sound velocity a is roughly constant with height, whereas A increases rapidly outwards as a consequence of the steep decrease in density (cf. Table 7.3). Hence, even with a field as small as 20 gauss, the condition (6.122) holds above a height of some 750 km (see Section 7.4.8).

Our discussion of magnetohydrodynamic waves is divided into two parts. In the first (Section 6.4.2) we consider the situation in which the magnetic field is important but gravity can be ignored, while in the second (Section 6.4.3) we discuss the much more complicated case in which both the magnetic field and gravity are taken into account.

6.4.2 ALFVÉN AND MAGNETO-ACOUSTIC WAVES

When gravity is neglected ($g = 0$), the general wave equation (6.44) takes the form

$$\frac{\partial^2 X_i}{\partial t^2} = (a^2 + A^2)\frac{\partial}{\partial x_i}\left(\frac{\partial X_j}{\partial x_j}\right) - \frac{1}{4\pi\rho_0} H_{0i} \cdot H_{0j}\frac{\partial}{\partial x_j}\left(\frac{\partial X_k}{\partial x_k}\right)$$
$$- \frac{1}{4\pi\rho_0} H_{0j}\frac{\partial}{\partial x_j}\left(H_{0k}\frac{\partial X_k}{\partial x_i} - H_{0k}\frac{\partial X_i}{\partial x_k}\right), \tag{6.123}$$

where A is the Alfvén velocity defined by equation (6.121). In the absence of gravity, the density ρ_0 is uniform throughout the gas and so also, therefore, is A. For convenience, we shall take the x_3-axis to be parallel to the magnetic field, i.e.

$$H_{0i} = (0, 0, H_0). \tag{6.124}$$

As before, we seek plane-wave solutions of the form

$$X_i = \chi_i\, e^{i(\omega t - k_j x_j)}, \tag{6.125}$$

where we may put $k_2 = 0$ without loss of generality since there is symmetry around the x_3-axis.

Substituting from equation (6.125) we find that the equations (6.123) possess non-trivial solutions provided

$$\begin{vmatrix} (a^2+A^2)k_1^2+A^2k_3^2-\omega^2 & 0 & a^2k_1k_3 \\ 0 & A^2k_3^2-\omega^2 & 0 \\ a^2k_1k_3 & 0 & a^2k_3^2-\omega^2 \end{vmatrix} = 0.$$

Expanding the determinant we obtain the condition

$$(A^2k_3^2-\omega^2)\left\{[(a^2+A^2)k_1^2+A^2k_3^2-\omega^2]\,(a^2k_3^2-\omega^2)-a^4k_1^2k_3^2\right\}=0,$$

which is satisfied if either

$$\omega^2 - A^2k_3^2 = 0 \qquad (6.126)$$

or

$$\omega^4-(a^2+A^2)k^2\omega^2+a^2A^2k^2k_3^2=0, \qquad (6.127)$$

where

$$k^2 = k_1^2 + k_3^2.$$

The first relation (6.126) describes the propagation of transverse waves in the direction of the vector $k_i = (k_1, 0, k_3)$; these may be regarded as *generalized Alfvén waves*. It is easily verified that the density variation ρ is zero and hence the fluid behaves as if it were incompressible. The only non-zero wave parameters are the particle displacement X_2, the velocity component v_2, and the magnetic field perturbation h_2. The wave energy density $\overline{W}$ is made up of equal amounts of kinetic and magnetic energies, each of magnitude $\frac{1}{4}\rho_0A^2k_3^2X_2X_2^*$. The only non-vanishing component of the energy flux vector is

$$\overline{Q}_3 = \frac{1}{2}\rho_0A^2\omega k_3X_2X_2^*,$$

whence it follows that the velocity of energy propagation (group velocity) has the single component

$$V_3 = \pm A.$$

Energy is therefore transported parallel to the magnetic field at a speed equal to the Alfvén velocity irrespective of whether the wave is a pure Alfvén wave propagating along the x_3-axis ($k_1 = 0$) or a generalized Alfvén wave propagating at some arbitrary angle to the x_3-axis ($k_1 \neq 0$). From equation (6.126) we see that the phase velocity of the waves is given by

$$V_p^2 = A^2\cos^2\theta, \qquad (6.128)$$

where $\theta = \cos^{-1}(k_3/k)$ is the angle between the wave propagation vector and the magnetic field.

The second relation derived above, namely

$$\omega^4 - (a^2 + A^2)\, k^2 \omega^2 + a^2 A^2 k^2 k_3^2 = 0, \qquad (6.127)$$

is the dispersion equation for *magneto-acoustic waves.* Solving equation (6.127) for ω^2 we obtain the result

$$\omega^2 = \frac{1}{2}(a^2 + A^2)k^2 \pm \frac{1}{2}k^2 \left[(a^2 + A^2)^2 - 4a^2 A^2 \frac{k_3^2}{k^2} \right]^{1/2},$$

whence it follows that

$$V_p^2 = \frac{1}{2}(a^2 + A^2) \pm \frac{1}{2} \left[(a^2 + A^2)^2 - 4a^2 A^2 \cos^2\theta \right]^{1/2}, \qquad (6.129)$$

where $V_p = \omega/k$ is the phase velocity of the waves and $\theta = \cos^{-1}(k_3/k)$ is the angle between the wave propagation vector and the magnetic field. The positive sign in equation (6.129) distinguishes *fast-mode* magneto-acoustic waves and the negative sign *slow-mode* waves. According to equation (6.129) the phase velocity of a fast-mode wave exceeds the greater of a and A, while that of a slow-mode wave is less than the smaller of a and A. The fast-mode wave can propagate in all directions, whereas the slow-mode wave clearly cannot propagate at right angles to the magnetic field. In the presence of a magnetic field, therefore, the medium becomes highly anisotropic.

Expressions for the total energy density and energy flux of magneto-acoustic waves can be derived in the manner outlined in Section 6.3.3. In the present case, with $g = 0$, the time-averaged values of the energy density and energy flux are

$$\overline{W} = \frac{1}{4}\rho_0 v_i v_i^* + \frac{pp^*}{4\rho_0 a^2} + \frac{h_i h_i^*}{16\pi} \qquad (6.130)$$

and

$$\overline{Q}_i = \frac{1}{4}(pv_i^* + p^* v_i) + \frac{1}{16\pi} H_{0j}(h_j v_i^* + h_j^* v_i) - \frac{1}{16\pi}(h_j v_j^* + h_j^* v_j)\, H_{0i}, \qquad (6.131)$$

where the asterisks denote complex conjugates. Once again the derivation of explicit formulae for $\overline{W}$ and $\overline{Q}_i$ in terms of the wave solutions (6.125) requires considerable algebraic manipulation, and we shall merely state the results:

$$\begin{aligned}
\overline{W} &= \frac{\rho_0 \omega^2}{2} \cdot \frac{\omega^2 - a^2 k_3^2}{a^4 k_1^2 k_3^2} \cdot [2\omega^2 - (a^2 + A^2)k^2]\, ZZ^* \\
\overline{Q}_1 &= \frac{\rho_0 \omega}{2k^2} \cdot \frac{\omega^2 - a^2 k_3^2}{a^4 k_1^2 k_3^2} \cdot \omega^4 k_1 ZZ^* \qquad (6.132) \\
\overline{Q}_3 &= \frac{\rho_0 \omega}{2k^2} \cdot \frac{\omega^2 - a^2 k_3^2}{a^4 k_1^2 k_3^2} \cdot (\omega^4 - a^2 A^2 k^4) k_3 ZZ^*.
\end{aligned}$$

From (6.132) it follows that the velocity of energy propagation (group velocity) has the components

$$V_1 = \frac{\omega^4 k_1}{\omega k^2 [2\omega^2 - (a^2 + A^2)k^2]}$$

and (6.133)

$$V_3 = \frac{(\omega^4 - a^2 A^2 k^4)k_3}{\omega k^2 [2\omega^2 - (a^2 + A^2)k^2]}.$$

Finally, we can easily verify that

$$\frac{V_i k_i}{k} = \frac{\omega}{k} \equiv V_p, \tag{6.134}$$

in agreement with a conclusion of Lighthill [1960] that the component of the group velocity in the direction of wave propagation is the phase velocity. It should be noted, however, that this result does not hold in the presence of a gravitational field [cf. Bazer and Hurley, 1963].[8]

6.4.3 MAGNETO-GRAVITY WAVES

It remains now to consider the mathematically much more complicated case of wave propagation in a compressible medium in which account is taken of both a magnetic field and gravity. We start with the full wave equation (6.44), written in the form

$$\frac{\partial^2 X_i}{\partial t^2} = \left(a^2 + \frac{H_0^2}{4\pi\rho_0}\right) \frac{\partial}{\partial x_i}\left(\frac{\partial X_j}{\partial x_j}\right) + (\gamma - 1) g_i \frac{\partial X_j}{\partial x_j} + \frac{\partial}{\partial x_i}(g_j X_j)$$
$$- \frac{1}{4\pi\rho_0} H_{0i} H_{0j} \frac{\partial}{\partial x_j}\left(\frac{\partial X_k}{\partial x_k}\right) - \frac{1}{4\pi\rho_0} H_{0j} \frac{\partial}{\partial x_j}\left(H_{0k}\frac{\partial X_k}{\partial x_i} - H_{0k}\frac{\partial X_i}{\partial x_k}\right). \tag{6.135}$$

In the presence of a gravitational field the equilibrium density ρ_0 is a function of height and hence the coefficients in equation (6.135) are not all constant. However, provided the wavelength $2\pi/k$ of the perturbations is small compared to the scale height of the atmosphere, we may neglect the variation of ρ_0 with height and thus derive a locally-valid dispersion relation for the medium. [A formal justification of this procedure, based on the application of a generalized WKB approximation, is given by McLellan and Winterberg, 1968].

Once again we seek plane-wave solutions of the form

$$X_i = \chi_i \, e^{i(\omega t - k_j x_j)}, \tag{6.136}$$

[8] There is an extensive literature on the subject of magneto-acoustic waves. Useful bibliographies have been given by Lighthill [1960], Jones [1964, p. 749], Ferraro and Plumpton [1966, p. 246], and many other authors.

but this time we cannot put $k_2 = 0$ since the magnetic field may be inclined to the vertical. Substituting from equation (6.136) one finds that the equations (6.135) possess non-trivial solutions provided

$$\begin{vmatrix}
(a^2 + A^2)\,k_1^2 - \dfrac{1}{2\pi\rho_0} H_{0\,1}(H_{0\,j}k_j)\,k_1 + \dfrac{(H_{0\,j}k_j)^2}{4\pi\rho_0} - \omega^2 & (a^2 + A^2)\,k_1 k_2 - \dfrac{1}{4\pi\rho_0}(H_{0\,j}k_j)(H_{0\,1}k_2 + H_{0\,2}k_1) & (a^2 + A^2)\,k_1 k_3 - \dfrac{1}{4\pi\rho_0}(H_{0\,j}k_j)(H_{0\,1}k_3 + H_{0\,3}k_1) - \mathrm{i}gk_1 \\
(a^2 + A^2)\,k_1 k_2 - \dfrac{1}{4\pi\rho_0}(H_{0\,j}k_j)(H_{0\,1}k_2 + H_{0\,2}k_1) & (a^2 + A^2)\,k_2^2 - \dfrac{1}{2\pi\rho_0} H_{0\,2}(H_{0\,j}k_j)\,k_2 + \dfrac{(H_{0\,j}k_j)^2}{4\pi\rho_0} - \omega^2 & (a^2 + A^2)\,k_2 k_3 - \dfrac{1}{4\pi\rho_0}(H_{0\,j}k_j)(H_{0\,2}k_3 + H_{0\,3}k_2) - \mathrm{i}gk_2 \\
(a^2 + A^2)\,k_1 k_3 - \dfrac{1}{4\pi\rho_0}(H_{0\,j}k_j)(H_{0\,1}k_3 + H_{0\,3}k_1) - \mathrm{i}(\gamma - 1)gk_1 & (a^2 + A^2)\,k_2 k_3 - \dfrac{1}{4\pi\rho_0}(H_{0\,j}k_j)(H_{0\,2}k_3 + H_{0\,3}k_2) - \mathrm{i}(\gamma - 1)gk_2 & (a^2 + A^2)\,k_3^2 - \dfrac{1}{2\pi\rho_0} H_{0\,3}(H_{0\,j}k_j)\,k_3 + \dfrac{(H_{0\,j}k_j)^2}{4\pi\rho_0} - \mathrm{i}\gamma gk_3 - \omega^2
\end{vmatrix} = 0,$$

where, as before,

$$A = \frac{H_0}{\sqrt{4\pi\rho_0}} = \left(\frac{H_{0\mathrm{j}}H_{0\mathrm{j}}}{4\pi\rho_0}\right)^{1/2} \tag{6.137}$$

is the Alfvén velocity. Expanding the determinant we obtain the condition

$$\begin{aligned}
\omega^6 - \omega^4 \left[(a^2 + A^2)\,k^2 + \frac{(H_{0\,j}k_j)^2}{4\pi\rho_0} - \mathrm{i}\gamma gk_3 \right] + \omega^2 \Bigg\{ g^2(\gamma - 1)(k^2 - k_3^2) \\
- \mathrm{i}\gamma g \left[\frac{(H_{0\,j}k_j)^2}{4\pi\rho_0} \cdot k_3 + H_{0\,3}\frac{H_{0\,j}k_j}{4\pi\rho_0} \cdot k^2 \right] + 2a^2 \frac{(H_{0\,j}k_j)^2}{4\pi\rho_0} \cdot k^2 \\
+ H_0^2 \frac{(H_{0\,j}k_j)^2}{(4\pi\rho_0)^2} \cdot k^2 \Bigg\} - g^2(\gamma - 1)\frac{(H_{0\,j}k_j)^2}{4\pi\rho_0}(k^2 - k_3^2) \\
+ \mathrm{i}\gamma g H_{0\,3}\frac{(H_{0\,j}k_j)^3}{(4\pi\rho_0)^2} \cdot k^2 - a^2 \frac{(H_{0\,j}k_j)^4}{(4\pi\rho_0)^2} \cdot k^2 = 0,
\end{aligned}$$

which is satisfied if either

$$\omega^2 - \frac{(H_{0j}k_j)^2}{4\pi\rho_0} = 0 \tag{6.138}$$

or

$$\omega^4 - \omega^2\,[(a^2 + A^2)\,k^2 - i\gamma g k_3] + g^2(\gamma - 1)(k^2 - k_3^2) + a^2\,\frac{(H_{0j}k_j)^2}{4\pi\rho_0}\,k^2$$

$$-i\gamma g H_{03}\,\frac{(H_{0j}k_j)}{4\pi\rho_0}\,k^2 = 0. \tag{6.139}$$

The first of these relations governs the propagation of generalized Alfvén waves, whose properties have already been described in the previous section. The second gives the dispersion equation for *magneto-gravity* waves. Because of its mathematical complexity we shall not attempt to explore the consequences of equation (6.139) for the most general type of magneto-gravity wave. However, some insight can be gained by considering the simple case where the magnetic field is horizontal and the wave propagates in a direction which is both at right angles to the field and inclined at an arbitrary angle to the vertical. Then we may take $H_{0i} = (0, H_0, 0)$ and $k_i = (k_1, 0, k_3)$, and the dispersion equation reduces to

$$\omega^4 - [(a^2 + A^2)(k_1^2 + k_3^2) - i\gamma g k_3]\,\omega^2 + (\gamma - 1)g^2 k_1^2 = 0. \tag{6.140}$$

This is formally the same as the dispersion equation (6.51) for purely hydrodynamic waves if a^2 is replaced by $a^2 + A^2$.

Referring back to the results presented in Section 6.3.2 we thus see that, for real ω and k_1,

$$k_3 = i \cdot \frac{\gamma g}{2(a^2 + A^2)} \pm \sqrt{(\omega^2 - \Omega_1^2)\frac{1}{a^2 + A^2} - (\omega^2 - \Omega_2^2)\,\frac{k_1^2}{\omega^2}}\,, \tag{6.141}$$

where

$$\Omega_1 = \frac{\gamma g}{2\sqrt{a^2 + A^2}}$$

and (6.142)

$$\Omega_2 = \frac{\sqrt{\gamma - 1}\cdot g}{\sqrt{a^2 + A^2}}$$

are the critical frequencies of the gas in the presence of a horizontal magnetic field. (It should be noted that Ω_1 and Ω_2 are smaller than the corresponding frequencies ω_1 and ω_2 in the absence of a field – cf. equations 6.53 and 6.54.)

As before, we see that no progressive waves are possible in the frequency range

$$\Omega_2 < \omega < \Omega_1 . \tag{6.143}$$

Bel and Mein [1971] have studied the properties of magneto-gravity waves in the more complicated case of vertical propagation in the presence of a magnetic field inclined in an arbitrary direction. Their results demonstrate the existence under these circumstances of two wave modes, which are the gravity-modified counterparts of the fast- and slow-mode magneto-acoustic waves discussed in Section 6.4.2. For further discussion of magneto-gravity waves the reader is referred to McLellan and Winterberg [1968] and Chiu [1971]; the latter considers the extension of the theory to the case of waves of arbitrary amplitude.

6.5 Hydrodynamic Shock Waves

6.5.1 INTRODUCTION

If we ignore gravity, the propagation of acoustic waves through an initially uniform medium parallel to an axis $O\xi$, say, is governed by the equation

$$\frac{\partial^2 v}{\partial t^2} = a_0^2 \frac{\partial^2 v}{\partial \xi^2}, \tag{6.144}$$

where v is the fluid velocity in the ξ-direction and a_0 is the velocity of sound (cf. equation 6.41). Equation (6.144) admits of a solution of the form

$$v = f(\xi - a_0 t), \tag{6.145}$$

corresponding to a progressive wave whose profile (i.e. the distribution of velocity, density, etc.) moves in the positive ξ-direction with speed a_0, its shape remaining unchanged.

However, equation (6.144) is valid only when the wave amplitude is small and if this is not so, we must revert to the basic non-linear hydrodynamic equations given in Section 6.2.2. For unidimensional flow the equations of continuity, momentum, and adiabatic variation are respectively

$$\frac{\partial \rho}{\partial t} + v \frac{\partial \rho}{\partial \xi} + \rho \frac{\partial v}{\partial \xi} = 0 \tag{6.146}$$

$$\frac{\partial v}{\partial t} + v \frac{\partial v}{\partial \xi} + \frac{1}{\rho} \frac{\partial p}{\partial \xi} = 0 \tag{6.147}$$

$$\left(\frac{\partial}{\partial t} + v \frac{\partial}{\partial \xi}\right) p\rho^{-\gamma} = 0. \tag{6.148}$$

The last of these can be integrated immediately to give

$$p\rho^{-\gamma} = p_0 \rho_0{}^{-\gamma}, \tag{6.149}$$

where p_0 and ρ_0 are the pressure and density in the undisturbed medium. Using (6.149) one can reduce the problem to the solution of the two non-linear equations

$$\frac{\partial \rho}{\partial t} + v \frac{\partial \rho}{\partial \xi} + \rho \frac{\partial v}{\partial \xi} = 0 \qquad (6.150)$$

and

$$\frac{\partial v}{\partial t} + v \frac{\partial v}{\partial \xi} + \frac{a^2}{\rho} \frac{\partial \rho}{\partial \xi} = 0, \qquad (6.151)$$

where

$$a = \sqrt{\frac{\gamma p}{\rho}} \ .$$

It can be shown that equations (6.150) and (6.151) admit of particular solutions corresponding to progressive waves of the form

$$v = f\,[\xi - (a + v)\,t]\,, \qquad (6.152)$$

where

$$a = a_0 + \frac{1}{2}(\gamma - 1)\,v \qquad (6.153)$$

and the velocity is a function only of density [see, for example, Landau and Lifshitz, 1959, pp. 366–370; Pack, 1965, pp. 215–216]. This represents a generalization of the solution (6.145) for the small-amplitude case and reduces to it when $v \ll a_0$.

However, the progressive wave described by (6.152) is essentially different in character from that described by (6.145). The velocity of a point in the wave profile is now

$$u = a + v \qquad (6.154)$$

and so may be regarded as a superposition of the propagation of a disturbance relative to the gas with the (local) velocity of sound and of the movement of the gas itself with velocity v. The velocity u is a function of density and is therefore different at different points in the wave profile; moreover, it can be shown that $\mathrm{d}u/\mathrm{d}\rho > 0$. Therefore, the greater the density at a given point in the wave profile the greater is the velocity of propagation at that point. In regions of compression $\rho > \rho_0$ and $a > a_0$ (i.e. $v > 0$), while in rarefactions $\rho < \rho_0$ and $a < a_0$ (i.e. $v < 0$).

The difference in velocities at different points of the wave profile causes its shape to change as the wave advances, as illustrated in Fig. 6.2. The region of compression A will continually gain on the region of rarefaction B, thus steepening the profile until eventually a discontinuity develops. At this stage the velocity profile has a vertical tangent and the solution of the hydrodynamical equations (6.150) and (6.151) becomes multi-valued and therefore physically meaningless. Heat conduction and viscosity become more and more important as the wave steepens until, finally, a wave of permanent form develops in which a balance is achieved between the energy dissipation and the effects of non-linearity. The transition layer over which this balance takes place has a thickness of the order of a few mean free paths of the gas particles and is called a 'shock layer'. Mathematically the shock layer may be regarded as a surface of discontinuity which, in turn, is termed a shock wave.

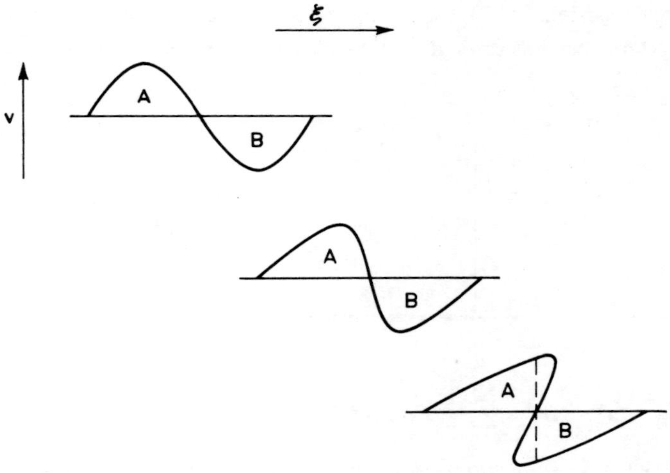

Fig. 6.2. Schematic solution of equations (6.150) and (6.151) representing a wave of finite amplitude propagating in the ξ-direction. Note the progressive steepening of the profile which leads eventually to the formation of a shock (see text).

The preceding discussion makes it clear that shocks must eventually be produced whenever waves propagate into a region of decreasing density, thereby experiencing growth in the amplitudes of their velocities. This must occur, for example, when acoustic waves travel upwards through the solar chromosphere. Moreover, the appearance of shocks results in the dissipation of energy and thus leads to a damping of the wave. This is illustrated in Fig. 6.2: when the discontinuity is formed (dashed line), the highest part of the wave profile is cut off. In the course of time, as the profile is bent over, its height becomes less and the profile is 'smoothed' to one of smaller amplitude, i.e. the wave is damped [cf. Landau and Lifshitz, 1959, p. 369]. The theory of shock wave dissipation is discussed in Section 6.5.3.

To conclude this introduction, we may remark that Chiu [1970] has recently succeeded in obtaining exact solutions which describe the propagation of plane travelling waves of finite amplitude through an isothermal atmosphere pervaded by a uniform gravitational field. These demonstrate analytically the formation of shocks from modified acoustic or gravity waves travelling upwards through the atmosphere.

6.5.2 JUMP CONDITIONS ACROSS A SHOCK FRONT: THE RANKINE–HUGONIOT RELATIONS

The discontinuities in density, pressure, and velocity across a shock wave are determined by the basic equations of conservation of mass, momentum, and total energy (Section 6.2.2), together with the condition that the entropy of the gas increases on passing through the shock front.[9] We shall limit our discussion

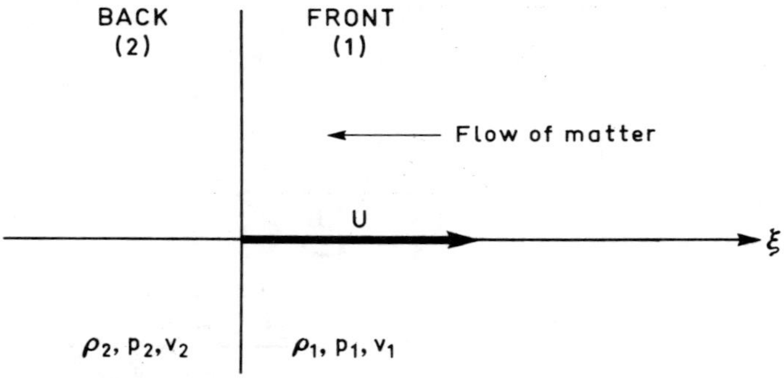

Fig. 6.3. Hydrodynamic shock wave: by convention, the side of the shock from which the gas flows across the shock is referred to as the front side and the other as the back side. The subscripts 1 and 2 are used to distinguish between the values of the density, pressure, and velocity in front of and behind the shock respectively.

to the case of a plane shock wave propagating in, say, the ξ-direction and shall adopt the convention of referring to the side of the shock from which the gas flows across the shock as the front side and the other as the back side. The subscripts 1 and 2 will be used to distinguish between quantities at the front and back of the shock wave (cf. Fig. 6.3).

In the absence of gravity, equations (6.18), (6.19), and (6.21) expressing the conservation of mass, momentum, and energy for a unidimensional flow parallel

[9] The standard texts on the theory of shock waves include those of Courant and Friedrichs [1948], von Mises [1958], and Landau and Lifshitz [1959]. Useful introductions to the subject are given by Cole [1965], Pack [1965], Ferraro and Plumpton [1966], and Becker [1968].

to the ξ-axis take the forms

$$\frac{\partial \rho}{\partial t} + \frac{\partial}{\partial \xi}(\rho v) = 0 \tag{6.155}$$

$$\frac{\partial}{\partial t}(\rho v) + \frac{\partial}{\partial \xi}\left(\rho v^2 + p - \frac{4}{3}\eta \frac{\partial v}{\partial \xi}\right) = 0 \tag{6.156}$$

$$\frac{\partial}{\partial t}\left(\frac{1}{2}\rho v^2 + \frac{1}{\gamma - 1}p\right) + \frac{\partial}{\partial \xi}\left[\rho v\left(\frac{1}{2}v^2 + \frac{\gamma}{\gamma - 1}\frac{p}{\rho}\right) - \frac{4}{3}\eta v \frac{\partial v}{\partial \xi} + F\right] = 0, \tag{6.157}$$

where v is the velocity of the gas. To derive the 'jump' relations across the shock it is convenient to transform to a new set of axes at rest with respect to the shock, which we may take to be located at $\xi = 0$. Equations (6.155)–(6.157) remain unchanged under this transformation provided we replace the velocities of the gas in front of and behind the shock by the corresponding velocities

$$\begin{aligned} u_1 &= v_1 - U \\ u_2 &= v_2 - U \end{aligned} \tag{6.158}$$

relative to the shock front, U being the velocity of the shock in the original frame of reference.

We now integrate equations (6.155)–(6.157) across the shock layer, which we take to extend from $-\epsilon$ to $+\epsilon$, where ϵ is arbitrarily small. Then

$$\int_{-\epsilon}^{\epsilon} \frac{\partial \rho}{\partial t} d\xi + \Big[\rho u\Big]_{-\epsilon}^{\epsilon} = 0$$

$$\int_{-\epsilon}^{\epsilon} \frac{\partial}{\partial t}(\rho u)\, d\xi + \left[\rho u^2 + p - \frac{4}{3}\eta \frac{\partial u}{\partial \xi}\right]_{-\epsilon}^{\epsilon} = 0$$

$$\int_{-\epsilon}^{\epsilon} \frac{\partial}{\partial t}\left(\frac{1}{2}\rho u^2 + \frac{1}{\gamma - 1}p\right) d\xi + \left[\rho u\left(\frac{1}{2}u^2 + \frac{\gamma}{\gamma - 1}\frac{p}{\rho}\right) - \frac{4}{3}\eta u \frac{\partial u}{\partial \xi} + F\right]_{-\epsilon}^{\epsilon} = 0.$$

We next suppose that the effects of viscosity and thermal flux may be neglected and assume that all time derivatives remain bounded [cf. von Mises, 1958, pp. 198–201]. Finally, letting ϵ approach zero, we obtain the classical *Rankine–Hugoniot shock relations*

$$\rho_1 u_1 = \rho_2 u_2 \tag{6.159}$$

$$\rho_1 u_1^2 + p_1 = \rho_2 u_2^2 + p_2 \tag{6.160}$$

$$\frac{1}{2}u_1^2 + \frac{\gamma}{\gamma - 1}\cdot\frac{p_1}{\rho_1} = \frac{1}{2}u_2^2 + \frac{\gamma}{\gamma - 1}\cdot\frac{p_2}{\rho_2}. \tag{6.161}$$

Although these relations have been derived for a plane shock front, they remain valid when the front is curved since the transition layer is taken to be of negligible thickness [Cole, 1965, p. 35]. Moreover, they are unaltered by the presence of a gravitational field [see, for example, Ono *et al.*, 1961; Uchida, 1963]. It should also be noted that in astrophysical contexts it may not be permissible to neglect the thermal flux term F in equation (6.157) since the transfer of energy by radiation, in particular, is often very efficient. In such cases the first two of the jump relations (6.159)–(6.161) remain unaltered, but the third assumes the form

$$\rho_1 u_1 \left(\frac{1}{2} u_1^2 + \frac{\gamma}{\gamma - 1} \cdot \frac{p_1}{\rho_1}\right) + F_1 = \rho_2 u_2 \left(\frac{1}{2} u_2^2 + \frac{\gamma}{\gamma - 1} \cdot \frac{p_2}{\rho_2}\right) + F_2 \tag{6.162}$$

[cf. Ledoux and Walraven, 1958, p. 558]. The effect of radiative transfer on shock waves is discussed in Section 6.5.4.

Given the shock wave velocity U, the Rankine–Hugoniot relations (6.159)–(6.161) suffice to determine the state of the gas on one side of the shock front in terms of the state on the other side. To solve these equations it is convenient to introduce the *Mach number* $M_1 = |u_1|/a_1$ of the gas ahead of the shock. Then it can readily be shown that the ratios of the state variables are given by the expressions

$$\frac{\rho_2}{\rho_1} = \frac{u_1}{u_2} = \frac{(\gamma + 1)M_1^2}{(\gamma - 1)M_1^2 + 2} \tag{6.163}$$

$$\frac{p_2}{p_1} = \frac{2\gamma M_1^2 - (\gamma - 1)}{\gamma + 1} \tag{6.164}$$

$$\frac{T_2}{T_1} = \frac{[2\gamma M_1^2 - (\gamma - 1)]\,[(\gamma - 1)M_1^2 + 2]}{(\gamma + 1)^2 M_1^2} \tag{6.165}$$

[see, for example, Landau and Lifshitz, 1959, pp. 329–331; Lindsay, 1960, pp. 286–289].

Owing to energy dissipation in the shock layer the entropy of the gas increases across the shock, and this in turn places important restrictions on the way in which various quantities can change across the shock. In particular, the Mach numbers M_1 and M_2 are subject to the conditions

$$M_1 > 1 \text{ and } M_2 < 1, \tag{6.166}$$

implying that the gas moves with supersonic speed relative to the shock wave on the front side and with subsonic speed at the back.[10]

[10] The inequalities (6.166) constitute a necessary condition for the existence of a shock wave. For a discussion of the stability of shock waves see Landau and Lifshitz [1959, pp. 325–329].

According to equations (6.164) and (6.165) the ratios p_2/p_1 and T_2/T_1 vary as M_1^2 for large values of the Mach number, showing that the pressure and temperature discontinuities across a shock wave can be arbitrarily great. On the other hand, the density ratio ρ_2/ρ_1 tends to the limiting value $(\gamma + 1)/(\gamma - 1)$; hence in the case of a monatomic gas the compression cannot exceed the limit

$$\rho_2 = 4\rho_1. \tag{6.167}$$

In addition, it follows from equation (6.164) that as $p_2 \to p_1$, $M_1 \to 1$, implying, as one would expect, that a very weak shock behaves as a sound wave.

It is important to note that both the Rankine–Hugoniot relations (6.159)–(6.161) and the formulae (6.163)–(6.165) for the ratios of the state variables can be expressed in a number of alternative forms, which will be useful in our subsequent discussion of shock wave dissipation (Section 6.5.3). From (6.159) and (6.160) one can show that

$$u_1^2 - u_2^2 = (p_2 - p_1)\left(\frac{1}{\rho_1} + \frac{1}{\rho_2}\right), \tag{6.168}$$

and furthermore that

$$u_2 - u_1 = \sqrt{(p_2 - p_1)\left(\frac{1}{\rho_1} - \frac{1}{\rho_2}\right)}. \qquad (6.169)^{11}$$

Moreover, equation (6.168) can be combined with (6.161) to yield the expression

$$\frac{\rho_2}{\rho_1} = \frac{(\gamma + 1)p_2 + (\gamma - 1)p_1}{(\gamma + 1)p_1 + (\gamma - 1)p_2}, \tag{6.170}$$

which in turn can be inverted to give the pressure ratio

$$\frac{p_2}{p_1} = \frac{(\gamma + 1)\rho_2 - (\gamma - 1)\rho_1}{(\gamma + 1)\rho_1 - (\gamma - 1)\rho_2}. \tag{6.171}$$

In addition, one may note that the ratio of the temperatures T_2 and T_1 behind and ahead of the shock is

$$\frac{T_2}{T_1} = \frac{p_2\rho_1}{p_1\rho_2} = \frac{a_2^2}{a_1^2}, \tag{6.172}$$

where a_2 and a_1 are the values of the sound velocity on the respective sides.

The last three relations can be written in even simpler forms by introducing the parameter

$$\delta = (p_2 - p_1)/p_1, \tag{6.173}$$

[11] The sign of the radical on the right-hand side of (6.169) is determined by the fact that we are supposing the shock wave to advance in the direction of increasing ξ, as illustrated in Fig. 6.3. Thus $u_1 < 0$ and $u_2 < 0$, while by (6.168) $|u_2| < |u_1|$; hence $u_2 - u_1 > 0$.

which specifies the *strength* of the shock in terms of the fractional increase in pressure across it. Then we have

$$\frac{\rho_2}{\rho_1} = \frac{2\gamma + (\gamma + 1)\delta}{2\gamma + (\gamma - 1)\delta} \qquad (6.174)$$

$$\frac{p_2}{p_1} = 1 + \delta \qquad (6.175)$$

$$\frac{T_2}{T_1} = (1 + \delta) \cdot \frac{2\gamma + (\gamma - 1)\delta}{2\gamma + (\gamma + 1)\delta} . \qquad (6.176)$$

Finally, we see from equation (6.164) that δ is related to the Mach number M_1 of the shock by the simple expression

$$\delta = \frac{2\gamma(M_1^2 - 1)}{\gamma + 1} , \qquad (6.177)$$

and so $\delta \to 0$ as $M_1 \to 1$.

6.5.3 PROPAGATION AND DISSIPATION OF SHOCK WAVES: THE BRINKLEY–KIRKWOOD THEORY

As we have already remarked in Section 6.1, an understanding of the theory of the propagation and dissipation of shock waves is an essential pre-requisite to any realistic discussion of the energy balance and dynamical processes in the chromosphere. A number of such theories exist, but the one which has so far found most application in the field of astrophysics is that of Brinkley and Kirkwood [1947]. The physical concepts involved in this theory are rather subtle and, from the point of view of readers unfamiliar with the subject, merit an exposition in somewhat greater depth than other topics in this chapter. Further justification for a fairly extended treatment is provided by the absence of any suitable account in book form in the astrophysical literature, although a good exposition of the Brinkley–Kirkwood theory on a purely fluid dynamical level is given by Cole [1965] in his monograph *Underwater Explosions*. However, before turning to the theory itself, we shall first briefly discuss some basic thermodynamic concepts associated with the passage of a shock wave through a gas.

(1) *Some underlying thermodynamic concepts.* The passage of a given mass of gas across a shock front is accompanied by an increase in its entropy, and this in turn is associated with an increase in the heat energy of the gas. According to the formula (6.28) the increase in entropy of unit mass of gas crossing the shock front is

$$\Delta s = s_2 - s_1 = c_v \ln\left(\frac{p_2 \rho_2^{-\gamma}}{p_1 \rho_1^{-\gamma}}\right) , \qquad (6.178)$$

which can be rewritten with the aid of equations (6.174) and (6.175) in the form

$$\Delta s = c_V \left[\ln(1+\delta) - \gamma \ln \frac{2\gamma + (\gamma+1)\delta}{2\gamma + (\gamma-1)\delta} \right]. \tag{6.179}$$

As we shall see in Section 7.4.5, in the low chromosphere we are concerned only with *weak* or *moderately weak* shocks satisfying the condition

$$\delta < 1 . \tag{6.180}$$

This means that one can expand the logarithmic terms in equation (6.179), obtaining the result

$$\Delta s = c_V \frac{\gamma^2 - 1}{12\gamma^2} \delta^3 , \tag{6.181}$$

correct to the third order in the shock strength δ. [An alternative method of deriving (6.181) is given by Lighthill, 1956, p. 291.] This is a standard result in the theory of shock waves and indicates that the increase in entropy across a weak shock depends on the cube of the pressure jump $p_2 - p_1$ [see, for example, Courant and Friedrichs, 1948, p. 142].

The increase in the heat energy of the gas which accompanies the increase in entropy is manifest both in a compression and in a rise in temperature of the gas. Following the passage of the shock a certain amount of this energy is used by the gas in returning to its initial state, while the remainder may appear in the form of radiation or mass motion. In principle, if the atomic properties of the gas are known, it should be possible to specify the exact thermodynamic path followed by the gas on the return cycle. But in practice it has been found necessary to choose simple thermodynamic paths which approximate as closely as possible the actual variation of the physical parameters involved. For example, in discussing the question of chromospheric heating by shock waves Weymann [1960] and Osterbrock [1961] assume that the gas behind the shock cools rapidly by radiation at constant density until the entropy returns to its original value, and then expands adiabatically back to its initial condition [see also Bhatnagar and Kushwaha, 1962]. Since the entropy remains constant during an adiabatic expansion (cf. equation 6.29), the decrease in entropy along the radiative path must equal in magnitude the increase in entropy across the shock front (see Fig. 6.4). Moreover, because the density does not change along this path, it follows from the first law of thermodynamics (6.84) that the heat energy radiated away per unit mass of gas, Δq say, is equal to the decrease in the internal energy of the gas from state 2 to state 3.

In order to calculate Δq we first note that the change in entropy along the radiative path is clearly

$$-\Delta s = s_3 - s_2 = c_v \ln\left(\frac{p_3\rho_3^{-\gamma}}{p_2\rho_2^{-\gamma}}\right). \tag{6.182}$$

Then, using equations (6.178) and (6.182) and remembering that $\rho_3 = \rho_2$ for the particular thermodynamic cycle being considered, we find that the pressure of the gas in state 3 is

$$p_3 = p_1\left(\frac{\rho_2}{\rho_1}\right)^{\gamma}. \tag{6.183}$$

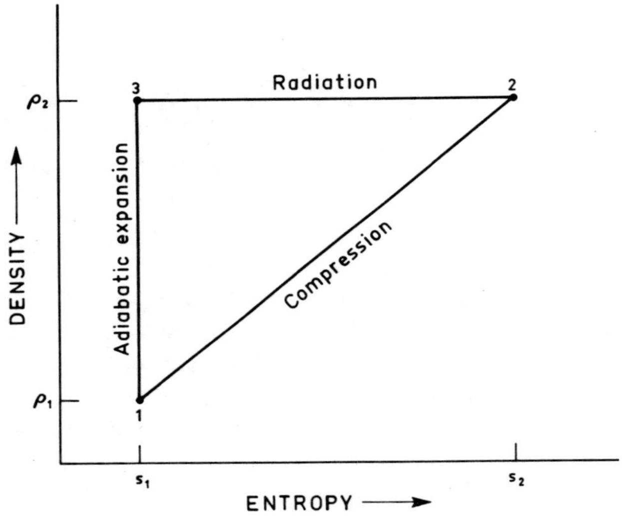

Fig. 6.4. Post shock wave thermodynamic cycle in density–entropy space (schematic), as postulated by Weymann [1960]. The gas is heated and compressed on crossing the shock front (states 1–2), radiates at constant density (2–3), and finally expands adiabatically back to its original condition (3–1).

Furthermore, according to the formula (6.16), the decrease in the internal energy of the gas per unit mass in passing from state 2 to state 3 is given by

$$-\Delta e = \frac{1}{\gamma - 1}\left(\frac{p_3}{\rho_2} - \frac{p_2}{\rho_2}\right). \tag{6.184}$$

Thus, putting $\Delta q = \Delta e$, we obtain with the aid of (6.183) the formula

$$\Delta q = \frac{1}{\gamma - 1} \cdot \frac{p_1}{\rho_1} \cdot \frac{\rho_1}{\rho_2}\left[\frac{p_2}{p_1} - \left(\frac{\rho_2}{\rho_1}\right)^{\gamma}\right] \tag{6.185}$$

for the amount of heat energy dissipated per unit mass of gas for the thermodynamic cycle postulated by Weymann.[12]

Using the jump relations (6.174) and (6.175) we can rewrite the formula (6.185) in terms of the shock strength δ defined by (6.173) and hence show with the aid of the binomial theorem or otherwise that, in the case of weak or moderately weak shocks ($\delta < 1$),

$$\Delta q = \frac{\gamma + 1}{12\gamma^2} \cdot \frac{p_1}{\rho_1} \cdot \delta^3,$$

or (6.186)

$$\Delta q = \frac{\gamma + 1}{12\gamma^2} \cdot \frac{(p_2 - p_1)^3}{\rho_1 p_1^2},$$

correct to the third order in δ. These expressions agree with those derived by Weymann [1960] and Osterbrock [1961] for the case of weak shocks. (In passing we should mention that the formula 6.186 is incorrectly quoted by Billings [1966, p. 191] as giving the energy dissipated per unit *volume* instead of per unit *mass.*)

Since the propagation of a shock wave is accompanied by the dissipation of energy, the energy carried by the wave must evidently decrease with distance. In order to derive an expression for the energy of a shock we shall assume that the wave has a roughly sawtooth shaped profile (cf. Fig. 6.5), in which the quantities ρ, p, and v vary continuously from the peak values ρ_2, p_2, and v_2 immediately behind the shock front to the equilibrium values ρ_1, p_1, and 0 some distance to the rear. (For simplicity we thus suppose the gas to be at rest in the equilibrium state.) Furthermore, it is advantageous at this stage to replace the Eulerian coordinate ξ of a fluid particle by the Lagrangian coordinate R, defined as the initial position of the particle in the undisturbed fluid, and regard the quantities ρ, p, and v as functions of R and t instead of – as hitherto – ξ and t. Following the approach of undimensional Lagrangian hydrodynamics [see, for example, Loughhead, 1957], we now focus our attention on a plane section of fluid

[12] In this particular case the result (6.185) can be derived without calculating Δs explicitly. Since the expansion of the gas from state 3 to state 1 takes place adiabatically, we have

$$p_3 \rho_2^{-\gamma} = p_1 \rho_1^{-\gamma},$$

whence, substituting for p_3 in equation (6.184), we immediately obtain (6.185).

We should mention that Skalafuris [1969] has criticized the procedure of calculating the energy dissipation from the entropy increase across the shock front, pointing out that work is a path function while entropy is a point function. Instead, he derives an expression for the energy dissipation by directly computing the work done along each path of the particular thermodynamic cycle considered.

particles normal to the ξ-axis, specified by the Lagrangian coordinate R. By definition the energy $D(R)$ of a shock wave at R, per unit area of the front, is equal to the work done on the fluid *ahead* of the plane section R as a result of the displacement of unit area of this surface by the excess pressure generated by the shock. In a time $\mathrm{d}t$ the section R moves a distance $v\,\mathrm{d}t$ in the direction of

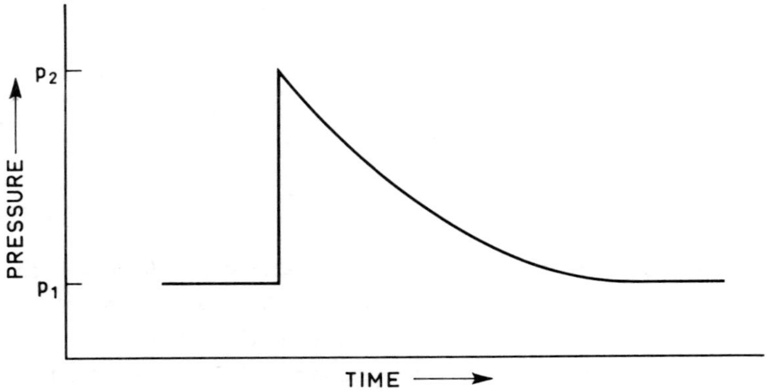

Fig. 6.5. Profile of a shock wave, showing the pressure at a fixed point plotted against time (see text).

the excess pressure force acting at R and hence the total work done by the latter force, per unit area of surface, is clearly

$$D(R) = \int_{t_0(R)}^{\infty} (p - p_1) v \,\mathrm{d}t, \tag{6.187}$$

where $t_0(R)$ is the time of arrival of the shock front at R, $p(R,t)$ and $v(R,t)$ are the values of the pressure and fluid velocity at time t, and the integration is performed at constant R.

In travelling a distance $\mathrm{d}R$ a unit area of the shock front passes through a mass $\rho_1\,\mathrm{d}R$ of the fluid and therefore the shock wave loses an amount of energy

$$\mathrm{d}D = -\rho_1\,\Delta q\,\mathrm{d}R$$

per unit cross section, where Δq is the amount of heat energy dissipated per unit mass of gas (equation 6.185). Hence the equation of attenuation of shock wave energy in an initially uniform medium is

$$\frac{\mathrm{d}D}{\mathrm{d}R} = -\rho_1\,\Delta q, \tag{6.188}$$

where the operator $\mathrm{d}/\mathrm{d}R$ indicates a derivative following the motion of the shock.

(2) *The Brinkley–Kirkwood theory.* Equation (6.188) does not by itself suffice to enable us to calculate D as a function of R. To do this we shall avail ourselves of the theory originally devised by Brinkley and Kirkwood [1947]. One complication, however, is that this theory was developed to describe the propagation of blast waves in water and in air and, accordingly, no account was taken of cooling of the medium behind the shock by radiation. Instead, it was assumed that the fluid returns *adiabatically* to its original pressure (but not density). Under these conditions Brinkley and Kirkwood showed that

$$\frac{dD}{dR} = -\rho_1 \, \Delta h, \qquad (6.189)$$

where Δh is the net increment in the enthalpy

$$h = e + p/\rho \qquad (6.190)$$

of the shocked fluid per unit mass after its return to the original pressure p_1 [see also Odgers and Kushwaha, 1959; Cole, 1965, pp. 135–136]. This result can be

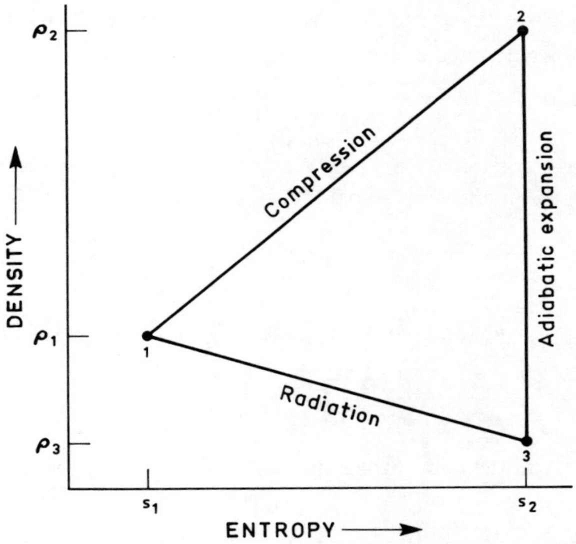

Fig. 6.6. Post shock wave thermodynamic cycle in density–entropy space (schematic), as postulated by Schatzman [1949]. The gas is heated and compressed on crossing the shock front (states 1–2), expands adiabatically (2–3), and finally radiates back to its original condition (3–1).

adapted to the case of a radiating gas if we now suppose that, after the initial adiabatic expansion back to pressure p_1 and density ρ_3, the gas behind the shock cools by radiation and returns to its original density ρ_1, thereby completing the thermodynamic cycle (see Fig. 6.6). It then follows from the first law of thermodynamics (6.84) that the heat energy radiated away in the final stage of the cycle is, per unit mass,

$$\Delta q = \Delta e + p_1 \cdot \Delta \left(\frac{1}{\rho}\right) \equiv \Delta h, \tag{6.191}$$

a result which incidentally explains why the enthalpy h is sometimes called the *heat content* of the fluid [cf. Reiss, 1965, pp. 43–44].

This particular thermodynamic cycle, involving first an adiabatic expansion of the shocked gas back to its original pressure and then a cooling by radiation back to its original density, was first applied to the discussion of shock waves in the solar chromosphere by Schatzman [1949] and has since been widely assumed in astrophysical contexts [see, for example, Odgers and Kushwaha, 1959; Uchida, 1960; Bhatnagar and Kushwaha, 1961; Kogure and Osaki, 1962; Saito, 1964; Schatzman, 1967]. It is often referred to as the 'Schatzman' cycle in order to distinguish it from the 'Weymann' cycle described previously (Fig. 6.4).

In order to calculate the heat dissipation per unit mass of gas, Δq, for the Schatzman cycle we first note that, as a consequence of (6.16), the definition (6.190) can be rewritten in the form

$$h = \frac{\gamma}{\gamma - 1} \cdot \frac{p}{\rho}. \tag{6.192}$$

Hence the increase in enthalpy across the shock front is

$$\frac{\gamma}{\gamma - 1}\left(\frac{p_2}{\rho_2} - \frac{p_1}{\rho_1}\right),$$

while the change during the adiabatic expansion back to the original pressure p_1 is

$$\frac{\gamma}{\gamma - 1}\left(\frac{p_1}{\rho_3} - \frac{p_2}{\rho_2}\right),$$

where the density at the end of the expansion, ρ_3, is determined by the adiabatic relation

$$p_1 \rho_3^{-\gamma} = p_2 \rho_2^{-\gamma}. \tag{6.193}$$

Adding the two contributions we see that the net increase in the enthalpy of the shocked gas after its return to pressure p_1 and density ρ_3 is

$$\Delta h = \frac{\gamma}{\gamma - 1}\left(\frac{p_1}{\rho_3} - \frac{p_1}{\rho_1}\right),$$

and, therefore, taking account of the relations (6.191) and (6.193), we obtain for the heat dissipation per unit mass along the Schatzman path the expression

$$\Delta q = \frac{\gamma}{\gamma - 1} \cdot \frac{p_1}{\rho_1}\left[\frac{\rho_1}{\rho_2}\left(\frac{p_2}{p_1}\right)^{\frac{1}{\gamma}} - 1\right]. \qquad (6.194)^{13}$$

In view of the jump relation (6.170), this result implies that Δq can be regarded as a function of the single variable p_2, the peak pressure behind the shock front.

In the limiting case of weak shocks (strength $\delta < 1$) equation (6.194) reduces to the form

$$\Delta q = \frac{\gamma + 1}{12\gamma^2} \cdot \frac{(p_2 - p_1)^3}{\rho_1 p_1^2} \qquad (6.195)$$

given by Brinkley and Kirkwood [1947], Schatzman [1949, 1967], and others. It is noteworthy that the limiting value of Δq for weak shocks is the same irrespective of whether we choose the Weymann or Schatzman cycle in evaluating the heat dissipation (cf. equation 6.186).

The next step is to derive an equation governing the decrease with distance R of the *peak* excess pressure

$$P_2 = p_2 - p_1 \qquad (6.196)$$

immediately behind the shock front. Briefly, the procedure adopted is to use the equations of conservation of mass and momentum (6.2) and (6.3), together with the equation of adiabatic variation (6.17),[14] to obtain two partial differential equations for the pressure p and velocity v behind the shock front as functions of position R and time t. Specialized to the shock front at $\xi = R$, these then provide two relations among the four derivatives of the peak velocity and peak excess pressure behind the shock front, $\partial v_2/\partial t$, $\partial v_2/\partial R$, $\partial P_2/\partial t$, and $\partial P_2/\partial R$. A third such relation is obtained from the Rankine–Hugoniot condition (6.160) for the conservation of momentum across the front, while together the three conditions (6.159)–(6.161) determine the shock velocity U as a function of the peak excess pressure P_2.

Finally, a fourth relation is obtained from an approximate evaluation of the integral (6.187) determining the shock wave energy $D(R)$. Taken together these four equations enable the derivatives $\partial P_2/\partial t$ and $\partial P_2/\partial R$ to be computed as

[13] This differs by a factor of γ from the result of a similar calculation undertaken by Uchida [1960, equation 30], the origin of the discrepancy being the latter's incorrect identification of Δq with the change in internal energy Δe instead of the change in enthalpy Δh (see equation 6.191).

[14] The introduction of this relation is justified since the assumed first stage of relaxation of the gas behind the shock front entails its adiabatic expansion back to the original pressure p_1. This is not the case, however, for the Weymann cycle, where the initial relaxation involves a cooling of the gas by radiation, a process which is clearly not adiabatic.

functions of P_2 and D. Now, the time derivative following the motion of the shock is clearly

$$\frac{\mathrm{d}}{\mathrm{d}t} = \frac{\partial}{\partial t} + U\frac{\partial}{\partial R}$$

(see Section 6.2.2). But in a time $\mathrm{d}t$ the front moves a distance $\mathrm{d}R = U\mathrm{d}t$ and hence the spatial derivative following the motion of the shock is given by the expression

$$\frac{\mathrm{d}}{\mathrm{d}R} = \frac{\partial}{\partial R} + \frac{1}{U}\frac{\partial}{\partial t}. \tag{6.197}$$

Consequently, knowing $\dfrac{\partial P_2}{\partial t}$ and $\dfrac{\partial P_2}{\partial R}$, we can derive an equation for P_2 of the form

$$\frac{\mathrm{d}P_2}{\mathrm{d}R} = F(P_2, D). \tag{6.198}$$

Adding to this the shock wave energy equation

$$\frac{\mathrm{d}D}{\mathrm{d}R} = -\rho_1 \cdot \Delta q(P_2) \tag{6.199}$$

derived above (cf. equations 6.189, 6.191, and 6.194), we have the two basic relations governing the decrease of the shock wave energy $D(R)$ and peak excess pressure $P_2(R)$ with distance.

To obtain the first two of the four equations in the derivatives $\partial v_2/\partial t$, $\partial v_2/\partial R$, $\partial P_2/\partial t$, and $\partial P_2/\partial R$, we start with the unidimensional equations of conservation of mass and momentum and the equation of adiabatic variation in their usual Eulerian forms

$$\frac{\mathrm{d}\rho}{\mathrm{d}t} + \rho\frac{\partial v}{\partial \xi} = 0 \tag{6.200}$$

$$\rho\frac{\mathrm{d}v}{\mathrm{d}t} = -\frac{\partial p}{\partial \xi} \tag{6.201}$$

$$\frac{\mathrm{d}p}{\mathrm{d}t} = \frac{\gamma p}{\rho} \cdot \frac{\mathrm{d}\rho}{\mathrm{d}t}. \tag{6.202}$$

Now, if ξ is the position at time t of a plane section of fluid particles specified by the Lagrangian coordinate R, conservation of mass evidently requires that $\rho_1\,\mathrm{d}R = \rho\,\mathrm{d}\xi$ and hence

$$\left(\frac{\partial \xi}{\partial R}\right)_t = \frac{\rho_1}{\rho}. \tag{6.203}$$

Furthermore, it follows by definition that the fluid velocity at time t is

$$v = \left(\frac{\partial \xi}{\partial t}\right)_R . \tag{6.204}$$

Using (6.203) and (6.204) and writing $a^2 = \gamma p/\rho$, we can reduce equations (6.200)–(6.202) to the two relations

$$\left(\frac{\partial v}{\partial t}\right)_R + \frac{1}{\rho_1}\left(\frac{\partial p}{\partial R}\right)_t = 0$$

$$\frac{\rho}{\rho_1}\left(\frac{\partial v}{\partial R}\right)_t + \frac{1}{\rho a^2}\left(\frac{\partial p}{\partial t}\right)_R = 0.$$

These are valid at any point behind the shock front during the initial stage of adiabatic relaxation of the shocked gas, and at the front itself, $\xi = R$, they become

$$\frac{\partial v_2}{\partial t} + \frac{1}{\rho_1} \cdot \frac{\partial P_2}{\partial R} = 0 \tag{6.205}$$

$$\frac{\rho_2}{\rho_1} \cdot \frac{\partial v_2}{\partial R} + \frac{1}{\rho_2 a_2^2} \cdot \frac{\partial P_2}{\partial t} = 0, \tag{6.206}$$

where

$$a_2^2 = \frac{\gamma p_2}{\rho_2} \tag{6.207}$$

and, for brevity, the subscripts R and t have been dropped.

Taking account of the transformation formulae (6.158) and remembering that the gas ahead of the shock is stationary (i.e. $v_1 = 0$), one can easily show from the Rankine–Hugoniot relations (6.159) and (6.160) that the peak excess pressure P_2 and shock velocity U are given by the expressions

$$P_2 = \rho_1 U v_2 \tag{6.208}$$

and

$$U^2 = \frac{P_2}{\rho_1(1 - \rho_1/\rho_2)} . \tag{6.209}$$

To these we may add the relation

$$\rho_2 = \rho_1 \frac{(\gamma + 1)P_2 + 2\gamma p_1}{(\gamma - 1)P_2 + 2\gamma p_1} \tag{6.210}$$

already derived in Section 6.5.2 (equation 6.170), from which it follows that ρ_2

can be expressed as a function of P_2 alone, ρ_1 and p_1 being constants. Equations (6.208) and (6.209) then demonstrate that v_2 and U are also functions of P_2 alone. Application of the operator (6.197) to the jump condition (6.208) now yields the required third relation among the four derivatives $\partial v_2/\partial t$, $\partial v_2/\partial R$, $\partial P_2/\partial t$, and $\partial P_2/\partial R$, namely:

$$\frac{\partial v_2}{\partial t} + U\frac{\partial v_2}{\partial R} - \frac{\alpha}{\rho_1 U} \cdot \frac{\partial P_2}{\partial t} - \frac{\alpha}{\rho_1} \cdot \frac{\partial P_2}{\partial R} = 0, \tag{6.211}$$

where the quantity

$$\alpha = 1 - \rho_1 v_2 \frac{\mathrm{d}U}{\mathrm{d}P_2} = 1 - \frac{P_2}{U} \cdot \frac{\mathrm{d}U}{\mathrm{d}P_2} \tag{6.212}$$

is a function of P_2.

The fourth and final relation is obtained by expressing the shock wave energy integral

$$D(R) = \int_{t_0(R)}^{\infty} (p - p_1)v \, \mathrm{d}t \tag{6.187}$$

in the reduced form

$$D(R) = (p_2 - p_1)v_2 \cdot \nu(R) \cdot \mu, \tag{6.213}$$

where

$$\nu(R) = \int_0^{\infty} f(R,\tau) \, \mathrm{d}\tau \tag{6.214}$$

$$f(R,\tau) = \frac{p - p_1}{p_2 - p_1} \cdot \frac{v}{v_2} \tag{6.215}$$

and

$$\tau = (t - t_0)/\mu. \tag{6.216}$$

The reduced time variable τ is chosen to normalize the initial slope of $f(R,\tau)$ to -1 by taking

$$\frac{1}{\mu} = -\left\{ \frac{\partial}{\partial t} \ln \left[(p - p_1)v\right] \right\}_{t=t_0} \tag{6.217}$$

Performing the indicated differentiation we obtain the relation

$$\frac{1}{P_2} \cdot \frac{\partial P_2}{\partial t} + \frac{1}{v_2} \cdot \frac{\partial v_2}{\partial t} = -\frac{1}{\mu},$$

whence substituting for μ from (6.213) we have

$$\frac{1}{P_2} \cdot \frac{\partial P_2}{\partial t} + \frac{1}{v_2} \cdot \frac{\partial v_2}{\partial t} = -\frac{P_2 \cdot v_2 \cdot \nu(R)}{D}. \tag{6.218}$$

Further progress requires an evaluation of the integral $\nu(R)$ in (6.218). Guided by the results of blast wave experiments in air [see, for example, Penney and Pike, 1950, p. 78 et seq.], Brinkley and Kirkwood made the assumption that ν is a constant; it numerical value, ν_0 say, depends on the shape of the wave profile, being equal to 1 for an exponential decay and to 2/3 for a linear decay (sawtooth wave). Equation (6.218) thus becomes finally

$$\frac{1}{P_2}\cdot\frac{\partial P_2}{\partial t}+\frac{1}{v_2}\cdot\frac{\partial v_2}{\partial t}=-\frac{P_2 v_2 \nu_0}{D}. \qquad (6.219)$$

Collecting results, we see that equations (6.205), (6.206), (6.211), and (6.219) provide four relations among the derivatives $\partial v_2/\partial t$, $\partial v_2/\partial R$, $\partial P_2/\partial t$, and $\partial P_2/\partial R$, in which all the coefficients can be expressed as functions of P_2. Solving for $\partial P_2/\partial t$ and $\partial P_2/\partial R$ and making use of (6.197) one obtains after some algebra an equation governing the decrease with distance of the peak excess pressure behind a shock wave in the form

$$\frac{dP_2}{dR}=-\nu_0\frac{P_2^3}{D}\cdot\frac{1}{\rho_1 U^2}\cdot\frac{\beta}{2(1+\alpha)-\beta}, \qquad (6.220)$$

where

$$\alpha=1-\frac{P_2}{U}\cdot\frac{dU}{dP_2} \qquad (6.212)$$

and

$$\beta=1-\left(\frac{\rho_1 U}{\rho_2 a_2}\right)^2. \qquad (6.221)$$

In the limiting case of weak shocks (strength $\delta < 1$) the expression (6.220) becomes simply (taking $\nu_0 = 2/3$)

$$\frac{dP_2}{dR}=-\frac{\gamma+1}{12\gamma^2}\cdot\frac{P_2^4}{Dp_1^2}, \qquad (6.222)$$

while the corresponding limiting equation for the shock wave energy is, by equations (6.195) and (6.199),

$$\frac{dD}{dR}=-\frac{\gamma+1}{12\gamma^2}\cdot\frac{P_2^3}{p_1^2}. \qquad (6.223)$$

These two equations have the solutions

$$D=\frac{\gamma+1}{6\gamma^2 p_1^2}\cdot\frac{P_0^3}{(R-R_0)^{1/2}}$$

$$P_2=\frac{P_0}{(R-R_0)^{1/2}}, \qquad (6.224)$$

where P_0 and R_0 are constants. The relations (6.224) imply that, in the limiting case of weak shocks, both the shock wave energy and the peak excess pressure decrease asymptotically as $R^{-1/2}$ at large distances.

(3) *Extension to non-uniform media.* The Brinkley–Kirkwood theory of shock wave propagation has been extended by Ono *et al.* [1961] to the general case of spherical shock waves propagating through an initially non-uniform medium in the presence of a gravitational field.[15] Their results have been particularized by Saito [1964] to the case of plane shock waves propagating *vertically upwards* through an atmosphere in *hydrostatic equilibrium* under a constant gravitational field. Saito gives the appropriate generalizations of the basic equations (6.220) and (6.199) in the forms

$$\frac{\mathrm{d}}{\mathrm{d}R}\left(\frac{p_2}{p_1}\right) = L\cdot\frac{\nu_0 p_1}{D} + M\cdot\frac{\mathrm{d}(\ln p_1)}{\mathrm{d}R} + N\cdot\frac{\mathrm{d}(\ln \rho_1)}{\mathrm{d}R} \tag{6.225}$$

and

$$\frac{\mathrm{d}D}{\mathrm{d}R} = -\rho_1(\Delta h + \Delta\phi). \tag{6.226}$$

In these equations $\Delta\phi$ represents the change in the gravitational energy of the shocked gas per unit mass after adiabatic expansion back to its original pressure p_1 and the quantities L, M, and N have the values

$$\begin{aligned} L &= \frac{(1-\rho_1/\rho_2)(p_2/p_1-1)^2(1-b)}{2k-b-1} \\ M &= \frac{(p_2/p_1-1)(b+2)}{2k-b-1} \\ N &= -\frac{(p_2/p_1-1)}{2k-b-1}, \end{aligned} \tag{6.227}$$

where

$$k = -\frac{1}{2}\left[1-\frac{(p_2/p_1-1)}{(1-\rho_1/\rho_2)}\cdot\frac{\mathrm{d}(\rho_1/\rho_2)}{\mathrm{d}(p_2/p_1)}\right]$$

and

$$b = \frac{1}{\gamma}\cdot\frac{(\rho_1/\rho_2)(p_2/p_1-1)}{(p_2/p_1)(1-\rho_1/\rho_2)}. \tag{6.228}$$

The first term on the right-hand side of equation (6.225) describes the normal dissipative damping of the peak excess pressure, while the second and third terms take account of the changes in p_2/p_1 resulting from the pressure and density gradients in the undisturbed medium.

[15] A similar extension by Kogure and Osaki [1962] neglected gravity.

In extending the theory to non-uniform media Ono *et al.* [1961] retained the similarity restraint imposed on the energy-time integral $\nu(R)$ defined by equation (6.214). At the same time they pointed out that the validity of this procedure was open to doubt. Subsequently, however, Saito [1964] compared equation (6.225) in certain limiting cases with the corresponding relation derived by Ono *et al.* on the basis of an alternative theory of shock wave propagation formulated by Chisnell [1955]. He found good agreement in these cases and therefore concluded that the use of equations (6.225) and (6.226) is justified to a good degree of approximation.

In conclusion, we should mention that although the Brinkley–Kirkwood theory has been quite successful in predicting the results of blast wave experiments and has been extensively applied in the field of astrophysics, it is only one of a number of theories of shock wave propagation which fluid dynamicists have developed over the years. References to these alternative approaches can be found, amongst other places, in the book by Cole [1965] and in the articles by Ono *et al.* [1961] and Schatzman [1967]. One of these methods, due to Whitham [1958], has been employed by Bird [1964], Ulmschneider [1967], and Skalafuris [1969] to study the behaviour of shock waves in gravitationally-stratified stellar atmospheres. However, the methods of both Chisnell [1955] and Whitham [1958] have been criticized by Schatzman and Souffrin [1967] on the grounds that they do not take proper account of energy dissipation.

6.5.4 EFFECT OF RADIATIVE TRANSFER ON SHOCK WAVES

Owing to its negligible thickness, the transition layer of a shock wave must be optically thin and therefore some of the radiation originating *behind* the shock as a result of energy dissipation will flow forward into the medium in front. This in turn implies the existence of a finite energy flux on both sides of the shock, so that, strictly speaking, one should use the third Rankine–Hugoniot relation in the form

$$\rho_1 u_1\left(\frac{1}{2}u_1^2 + \frac{\gamma}{\gamma-1}\cdot\frac{p_1}{\rho_1}\right) + F_1 = \rho_2 u_2\left(\frac{1}{2}u_2^2 + \frac{\gamma}{\gamma-1}\cdot\frac{p_2}{\rho_2}\right) + F_2$$

given by equation (6.162). In the case of a shock wave propagating upwards through the solar chromosphere the medium ahead is optically thin over much of the wavelength spectrum and therefore there will be a net outward *loss* of energy from the immediate neighbourhood of the shock, i.e. $F_1 - F_2 > 0$. The formulae for the ratios ρ_2/ρ_1 and T_2/T_1 of the densities and temperatures behind and in front of the shock then become

$$\frac{\rho_2}{\rho_1} = \frac{2(\gamma-1)(F_2 - F_1) + [(\gamma+1)p_2 + (\gamma-1)p_1]u_2}{[(\gamma+1)p_1 + (\gamma-1)p_2]u_2} \qquad (6.229)$$

and

$$\frac{T_2}{T_1} = \frac{2(\gamma - 1)(F_1 - F_2) + [(\gamma + 1)p_1 + (\gamma - 1)p_2]u_1}{[(\gamma + 1)p_2 + (\gamma - 1)p_1]u_1} \cdot \frac{p_2}{p_1} \tag{6.230}$$

respectively [Ledoux and Walraven, 1958, p. 561; Uchida, 1961]. Recalling that u_2 and u_1 are both negative (see footnote 11) and taking note of equation (6.170) we see at once that, for a given shock strength $\delta = (p_2 - p_1)/p_1$, the ratio ρ_2/ρ_1 is now greater than in the case $F_1 - F_2 = 0$, whereas T_2/T_1 is smaller. The effect of radiative transfer is thus to lessen the discontinuity in temperature but to increase that in density. It also tends to inhibit the initial formation of the shock [Prasad, 1967].

It follows, therefore, that the transfer of energy by radiation modifies the profiles of the temperature and other properties of the gas on the two sides of a shock discontinuity out to distances which depend, among other factors, on the opacity of the medium. Accordingly, in recent years much attention has been paid to the effect of radiative transfer on the properties of shock waves, both by astrophysicists and by space scientists concerned with problems associated with the re-entry by rockets and space vehicles of the Earth's atmosphere [cf. Penner and Olfe, 1968]. Provided the rate of decay of the shock due to dissipation is not too great, we may assume that the flow behind and ahead of the shock front is approximately steady. Then, putting all time derivatives equal to zero in the conservation equations (6.155)–(6.157), we find that the structure of the shock zone is governed by the equations

$$\frac{\partial}{\partial \xi}(\rho u) = 0 \tag{6.231}$$

$$\frac{\partial}{\partial \xi}(\rho u^2 + p) = 0 \tag{6.232}$$

$$\frac{\partial}{\partial \xi}\left[\rho u\left(\frac{1}{2}u^2 + \frac{\gamma}{\gamma - 1} \cdot \frac{p}{\rho}\right) + F\right] = 0, \tag{6.233}$$

to which we must add an appropriate transfer equation governing the variation of the radiation flux F.

However, most solutions of these equations to date have been derived subject to two basic simplifying assumptions: (*a*) the regions behind and ahead of the shock are optically thick, and (*b*) the emission and absorption of radiation occur under conditions of local thermodynamic equilibrium (LTE). The former leads to the boundary condition that there is no net flux of radiation at infinite distances on either side of the shock, while the latter greatly simplifies the calculation of the radiation source function and hence the flux F. For example, in the simple case of a grey atmosphere the flux at optical thickness τ measured from the shock front is given by the formula

$$F(\tau) = 2\pi \int_0^\pi I(\theta, \tau) \cos\theta \sin\theta \, d\theta, \tag{6.234}$$

where $I(\theta,\tau)$ is the intensity integrated over wavelength in a direction making an angle θ with the positive direction of the ξ-axis and τ is defined by the relation

$$d\tau = \bar{\kappa} d\xi, \tag{6.235}$$

$\bar{\kappa}$ being the mean linear absorption coefficient of the gas [see, for example, Chandrasekhar, 1950: p. 3]. The intensity $I(\theta, \tau)$ in turn is determined by the usual transfer equation

$$\cos\theta \frac{\partial}{\partial\tau} I(\theta, \tau) = B(\tau) - I(\theta, \tau), \tag{6.236}$$

where

$$B = \frac{\sigma_R}{\pi} T^4$$

is the LTE source function, σ_R being the Stefan–Boltzmann radiation constant. The results of such calculations [see, for example, Prasad and Sachdev, 1966; Prasad, 1967; Goulard, 1967, and the references cited therein; Klimishin, 1968; Penner and Olfe, 1968, Section 6-4; Wang, 1970] show that radiative transfer of energy tends to smooth out the profiles of the state variables across the shock region. In particular, the gas *approaching* the shock is heated by the radiation emanating from it and this, in the case of strong shocks, can give rise to a sizeable 'precursor' effect.

Unfortunately, however, the results of such calculations cannot be taken straight over to the case of the solar chromosphere. The latter is certainly not optically thick at all wavelengths and, with the possible exception of the lowest layers, departs widely from a state of local thermodynamic equilibrium (cf. Section 4.3.4). In fact, a fully adequate theory of radiation-modified shocks under the conditions obtaining in stellar atmospheres still awaits development. A critical survey of the present state of the subject is to be found in a review given by Goulard [1967] at the Fifth Symposium on Cosmical Gas Dynamics and in the record of the discussion which follows.

6.6 Magnetohydrodynamic Shock Waves

6.6.1 INTRODUCTION

The process of formation of a shock wave in the presence of a magnetic field is essentially the same as in the absence of a field and, provided the electrical conductivity is high, the physical thickness of the transition layer is again of the order of a few mean free paths. When the magnetic field is parallel to the direction of flow, no electric currents are induced in the fluid and the field has no effect on the shock wave. However, when the magnetic field is inclined to the direction of flow, currents are induced which influence both the momentum and energy balance of the fluid on the two sides of the shock. In the particular case of a field normal to the direction of motion the speed of propagation of compression waves is given by equation (6.129) with $\cos\theta = 0$, viz.

$$a^* = \sqrt{a^2 + A^2}, \tag{6.237}$$

a being the local velocity of sound and A the local Alfvén velocity. In this case, as we shall see later, a shock wave does not develop unless the velocity of the fluid exceeds a^*. Moreover, since a^* increases with the magnetic field strength, the velocity of the gas may be supersonic on both sides of a magnetohydrodynamic shock.[16]

6.6.2 JUMP CONDITIONS ACROSS A MAGNETOHYDRODYNAMIC SHOCK FRONT

To derive the jump conditions across a magnetohydrodynamic shock it is again convenient to transform to a new set of axes at rest with respect to the shock. In this reference frame the velocities of the gas are, in vector notation,

$$\boldsymbol{u}_1 = \boldsymbol{v}_1 - U\boldsymbol{n}$$

and (6.238)

$$\boldsymbol{u}_2 = \boldsymbol{v}_2 - U\boldsymbol{n}$$

ahead of and behind the shock respectively, U being the speed of the shock in the original reference frame and $\boldsymbol{n}$ a unit vector normal to the shock front. For simplicity we shall take the electrical conductivity to be infinite and shall assume that the effects of viscosity, thermal conduction, and radiation may be neglected.

Then, following the procedure outlined in Section 6.5.2, we integrate equations (6.18), (6.19), and (6.21) expressing the conservation of mass, momentum, and total energy across the shock layer and so obtain, with the aid

[16] There is an extensive literature on the theory of magnetohydrodynamic shock waves, references to which are to be found in the books by Pai [1962], Anderson [1963], Jones [1964], Kulikovskiy and Lyubimov [1965], Cabannes [1970], and others. The difficult question of the existence and stability of various theoretically possible types of magnetohydrodynamic shocks is discussed at length by Anderson and by Cabannes [see also the review article by Germain, 1960]. Readers relatively unfamiliar with the subject will find helpful introductions in the article by Chu and Grad [1965] and in the book by Ferraro and Plumpton [1966].

of equation (6.26), the jump relations

$$[\rho u_\mathrm{n}]_1^2 = 0 \tag{6.239}$$

$$\left[\rho u_\mathrm{n}\boldsymbol{u} + \left(p + \frac{H^2}{8\pi}\right)\boldsymbol{n} - \frac{1}{4\pi}H_\mathrm{n}\boldsymbol{H}\right]_1^2 = 0 \tag{6.240}$$

$$\left[u_\mathrm{n}\left(\frac{1}{2}\rho u^2 + \frac{\gamma}{\gamma-1}p + \frac{H^2}{4\pi}\right) - \frac{1}{4\pi}H_\mathrm{n}\,(\boldsymbol{H}\cdot\boldsymbol{u})\right]_1^2 = 0, \tag{6.241}$$

where $[L]_1^2 = L_2 - L_1$ denotes the change in any quantity L in passing from the front (1) to the back (2) of the shock.

Two additional jump conditions are imposed by the electromagnetic field equations (6.6)–(6.10). When the electrical conductivity is infinite it follows from equation (6.6) that, in the original frame of reference,

$$\boldsymbol{E} = -\frac{1}{c}\,\boldsymbol{v} \times \boldsymbol{H}, \tag{6.242}$$

and using this result we can show that the magnetic field is subject to the relation

$$\frac{\partial \boldsymbol{H}}{\partial t} - \operatorname{curl}\,(\boldsymbol{v} \times \boldsymbol{H}) = 0. \tag{6.243}$$

In addition, from (6.10) we have

$$\operatorname{div}\boldsymbol{H} = 0. \tag{6.244}$$

These equations remain invariant under a transformation to a reference frame at rest with respect to the shock provided $\boldsymbol{u}$ is written in place of $\boldsymbol{v}$. Then integrating (6.244) and (6.243) across the shock layer we obtain the remaining jump conditions

$$[H_\mathrm{n}]_1^2 = 0 \tag{6.245}$$

and

$$[\boldsymbol{u}H_\mathrm{n} - \boldsymbol{H}u_\mathrm{n}]_1^2 = 0. \tag{6.246}$$ [17]

Mathematically, the jump conditions (6.239)–(6.241), (6.245), and (6.246) admit the possibility of the existence of many different types of magnetohydrodynamic shock waves, some of which we shall discuss in the next section. However, before we consider the classification of the various possible types, it is instructive to consider how a magnetic field influences the properties of a shock

[17] In passing, we note that there is a current sheet flowing in the surface of discontinuity; the magnitude of the current per unit width of the sheet is, by equation (6.8),

$$j^* = \frac{c}{4\pi}\,\boldsymbol{n} \times [\boldsymbol{H}]_1^2.$$

wave in the simple case where the magnetic field has a single component H, say, parallel to the plane of the shock front and normal to the fluid velocity u. Then the jump relations may be written

$$\rho_1 u_1 = \rho_2 u_2 \tag{6.247}$$

$$H_1 u_1 = H_2 u_2 \tag{6.248}$$

$$\rho_1 u_1^2 + p_1 + \frac{H_1^2}{8\pi} = \rho_2 u_2^2 + p_2 + \frac{H_2^2}{8\pi} \tag{6.249}$$

$$u_1 \left(\frac{1}{2} \rho_1 u_1^2 + \frac{\gamma}{\gamma - 1} p_1 + \frac{H_1^2}{4\pi} \right) = u_2 \left(\frac{1}{2} \rho_2 u_2^2 + \frac{\gamma}{\gamma - 1} p_2 + \frac{H_2^2}{4\pi} \right). \tag{6.250}$$

The solution of these equations is discussed by Ferraro and Plumpton [1966, pp. 99–103] in terms of the two auxiliary variables

$$X = \frac{\rho_2}{\rho_1} = \frac{u_1}{u_2} = \frac{H_2}{H_1}$$

and (6.251)

$$Y = \frac{p_2}{p_1}.$$

Equations (6.247)–(6.250) then reduce to two equations in X and Y, from which it can readily be shown that the condition for a shock to develop is satisfied only if

$$u_1^2 > a_1^{*2} = a_1^2 + A_1^2, \tag{6.252}$$

as previously stated. It is therefore convenient to introduce the *magnetic* Mach number

$$M_1^* = |u_1|/a_1^*, \tag{6.253}$$

so that the condition for the formation of a magnetohydrodynamic shock wave becomes simply

$$M_1^* > 1. \tag{6.254}$$

Since M_1^* is smaller than the ordinary Mach number $M_1 = |u_1|/a_1$, it is evident that a magnetic field tends to inhibit the formation of shocks.

If we now let

$$\Pi = \frac{H_1^2}{8\pi p_1} \tag{6.255}$$

denote the ratio of the 'magnetic pressure' to the gas pressure in front of the shock, it can be shown that the ratio X defined above is given by the positive root, X_0 say, of the quadratic equation

$$\Pi(2-\gamma)X^2 + \left[\gamma(\Pi+1) + \frac{1}{2}(\gamma-1)(2\Pi+\gamma)M_1^{*2}\right]X - \frac{1}{2}(\gamma+1)(2\Pi+\gamma)M_1^{*2} = 0. \tag{6.256}$$

The ratio X_0 thus depends on the parameter Π, the magnetic Mach number M_1^*, and the ratio of the specific heats of the gas γ. Ferraro and Plumpton's calculations show that X_0 decreases with an increase in Π, i.e. with an increasing magnetic field. For very strong shocks ($M_1^* \gg 1$) we have approximately

$$X_0 = (\gamma+1)/(\gamma-1),$$

as in the absence of a magnetic field.

Similarly, we can derive an equation for the ratio $Y = p_2/p_1$, from which it can be shown that the rise in pressure across the shock is less than in the absence of the field.

6.6.3 CLASSIFICATION OF MAGNETOHYDRODYNAMIC SHOCK WAVES

As remarked above, the jump conditions (6.239)–(6.241), (6.245), and (6.246) admit the possible existence of many different types of magnetohydrodynamic shock waves. One convenient way of illustrating some of the various types is to derive an equation for the *mass flux* across the shock front which, by equation (6.239), is

$$m = \rho_1 u_{1n} = \rho_2 u_{2n}. \tag{6.257}$$

Following the treatment of Ferraro and Plumpton [1966] we shall use the symbol

$$w = 1/\rho$$

to denote the specific volume of the gas and the notations

$$[L] = L_2 - L_1$$

$$\bar{L} = \frac{1}{2}(L_1 + L_2)$$

to indicate respectively the jump in any quantity L across the shock front and the mean of the values of L on the two sides. Furthermore, it is convenient to introduce a cartesian reference frame $O\zeta, \eta, \xi$, orientated so that the ξ-axis is parallel to the unit vector $\boldsymbol{n}$ normal to the shock front. Finally, since by equation (6.245) the normal component of the magnetic field is continuous across the shock, we may write

$$H_n = H_{1n} = H_{2n}. \tag{6.258}$$

Resolving the jump relations (6.246) and (6.240) into cartesian components we obtain the equations

$$H_n[u_\xi] - m\{\bar{H}_\xi[w] + \bar{w}[H_\xi]\} = 0 \tag{6.259}$$

$$H_n[u_\eta] - m\{\bar{H}_\eta[w] + \bar{w}[H_\eta]\} = 0 \tag{6.260}$$

$$m[u_\xi] - \frac{H_n}{4\pi}[H_\xi] = 0 \tag{6.261}$$

$$m[u_\eta] - \frac{H_n}{4\pi}[H_\eta] = 0 \tag{6.262}$$

$$\frac{1}{4\pi}\left\{\bar{H}_\xi[H_\xi] + \bar{H}_\eta[H_\eta]\right\} + m^2[w] + [p] = 0, \tag{6.263}$$

whence it follows that, for $m \neq 0$,

$$\left(m^2\bar{w} - \frac{H_n^2}{4\pi}\right)[H_\xi] + m^2\bar{H}_\xi[w] = 0$$

and (6.264)

$$\left(m^2\bar{w} - \frac{H_n^2}{4\pi}\right)[H_\eta] + m^2\bar{H}_\eta[w] = 0.$$

Then, provided that $m^2\bar{w} - \dfrac{H_n^2}{4\pi} \neq 0$,[18] we may substitute for $[H_\xi]$ and $[H_\eta]$ in equation (6.263) and so obtain the quartic equation

$$m^4\bar{w} + m^2\left\{\frac{[p]}{[w]}\bar{w} - \frac{1}{4\pi}(\bar{H})^2\right\} - \frac{[p]}{[w]}\cdot\frac{H_n^2}{4\pi} = 0 \tag{6.265}$$

for the mass flux m.

Before proceeding further, it is instructive to consider the limiting form of equation (6.265) for a disturbance of infinitesimal amplitude. Then we have $\bar{w} \to w \equiv 1/\rho$, $[w] = \delta(1/\rho) = -\delta\rho/\rho^2$, $\bar{\boldsymbol{H}} \to \boldsymbol{H}$, and $[p] = \delta p$. If we suppose that the disturbance is advancing into a medium at rest (i.e. $v_{1\,n} = 0$), it follows from equation (6.238) that

$$m = -\rho U,$$

where U is the speed of the disturbance. In addition, since we have assumed the

[18] The relation

$$m^2\bar{w} - \frac{H_n^2}{4\pi} = 0$$

defines the so-called 'transverse shock'. It is, however, non-dissipative and hence does not concern us here [see, for example, Ferraro and Plumpton, 1966, pp. 114–115].

medium on each side of the shock to be uniform in the present context, we see from equation (6.38′) that

$$\delta p = a^2 \delta \rho,$$

a being the velocity of sound in the gas. Collecting results, we may now rewrite equation (6.265) in the form

$$U^4 - (a^2 + A^2)U^2 + a^2 A^2 \cos^2\theta = 0, \qquad (6.266)$$

where A is the Alfvén velocity defined by (6.121) and θ is the angle between the direction of propagation of the disturbance and the magnetic field $\boldsymbol{H}$. Solving (6.266) for U^2 we obtain the relation

$$U^2 = \frac{1}{2}(a^2 + A^2) \pm \frac{1}{2}\left[(a^2 + A^2)^2 - 4a^2 A^2 \cos^2\theta\right]^{1/2}, \qquad (6.267)$$

which is identical to the expression (6.129) derived in Section 6.4.2 for the phase velocity of magneto-acoustic waves. The positive sign corresponds to the fast-mode wave and the negative sign to the slow-mode wave.

Likewise, in the case of a disturbance of finite amplitude there exist *fast* and *slow* shocks characterized by two values, m_f and m_s say, of the mass flux m. Treating equation (6.265) as a quadratic in m^2 one can show that m_f and m_s obey the inequalities

$$m_\mathrm{s}^2 \leqslant -\frac{[p]}{[w]} \leqslant m_\mathrm{f}^2$$

and (6.268)

$$m_\mathrm{s}^2 \leqslant \frac{H_\mathrm{n}^2}{4\pi\overline{w}} \leqslant m_\mathrm{f}^2.$$

As we shall see in the next chapter, the distinction between fast and slow shocks is involved in discussions of the problem of shock wave heating in the neighbourhood of the chromospheric network (Section 7.4.8). Shocks in which the magnetic field is parallel to the direction of the shock wave (i.e. $\boldsymbol{H} = H_\mathrm{n}\boldsymbol{n}$) are called *parallel* shocks. In this case, as we have already mentioned in Section 6.6.1, the magnetic field drops out of the jump relations and therefore the shock is purely hydrodynamical in character (Section 6.5). However, this is not so for *perpendicular* shocks, where $\boldsymbol{H}$ is normal to the direction of propagation (i.e. $H_\mathrm{n} = 0$). According to equation (6.265) a perpendicular shock is characterized by a single value of m^2, namely

$$m^2 = \frac{(\overline{H})^2}{4\pi\overline{w}} - \frac{[p]}{[w]},$$

corresponding to the fast shock.

All other shocks are termed *oblique* shocks. They are characterized by two values of m^2, corresponding to fast and slow shocks defined by the inequalities (6.268). As one would expect, it can be shown that a fast shock propagates at a speed, relative to the medium ahead of the shock, exceeding the phase velocity of the fast-mode magneto-acoustic wave. A slow shock, on the other hand, propagates at a speed which is greater than the slow-mode wave velocity but less than the fast-mode velocity. The magnitude of the magnetic field increases across a fast shock but decreases across a slow shock. At the same time, the field vector is rotated away from the normal to the front across a fast shock and towards the normal across a slow shock [Ferraro and Plumpton, 1966].

For a more detailed discussion of the classification and properties of magnetohydrodynamic shocks the reader is referred to the paper by Bazer and Ericson [1959].

6.6.4 PROPAGATION AND DISSIPATION OF MAGNETOHYDRODYNAMIC SHOCK WAVES

The problem of calculating the energy of a magnetohydrodynamic shock wave and its attenuation with distance has been discussed by Osterbrock [1961]. His results are based on a plausible extension of the Brinkley–Kirkwood approach outlined in Section 6.5.3, but have yet to receive experimental confirmation. As before, we suppose the fluid ahead of the shock to be at rest and fix our attention on a plane section of particles specified by the Lagrangian coordinate R. (We retain the coordinate system $O\zeta$, η, ξ of the previous section, which is orientated so that the ξ-axis is parallel to the unit vector $\boldsymbol{n}$ normal to the shock front.) By definition, the energy of a magnetohydrodynamic shock wave at R, per unit area of the front, is equal to the work done on the fluid *ahead* of the plane section R as a result of the displacement of unit area of this surface by the excess pressure and magnetic forces generated by the shock. In time dt the section R moves a distance $v_n \, dt$ in the direction of the excess pressure force acting at R and hence the total work done by the latter force, per unit area of surface, is again

$$\int_{t_0(R)}^{\infty} (p - p_1) v_n \, dt,$$

where $t_0(R)$ is the time of arrival of the shock front at R, $p(R, t)$ and $v_n(R, t)$ are the values of the pressure and normal component of the fluid velocity at time t, and the integration is performed at constant R.

Turning now to the work done by the excess magnetic force we first have to note that the magnetic force per unit volume of the fluid (see equation 6.3) can be expressed, with the aid of equations (6.8) and (6.10), in the form

$$\frac{1}{c}\,\epsilon_{ijk}j_jH_k = \frac{\partial}{\partial x_j}\left(\frac{H_iH_j}{4\pi} - \frac{H^2}{8\pi}\,\delta_{ij}\right). \tag{6.269}$$

Then, integrating (6.269) over any specified volume V (of surface area S, say), we have by Gauss' divergence theorem

$$\iiint \frac{1}{c}\,\epsilon_{ijk}j_jH_k\,\mathrm{d}V = \iint\left(\frac{H_iH_j}{4\pi} - \frac{H^2}{8\pi}\,\delta_{ij}\right)\mathrm{d}S_j.$$

This reproduces a standard result of electromagnetic theory, namely that the magnetic force on the fluid volume V is equal to the integral of the components of the magnetic stress tensor

$$m_{ij} = \frac{H_iH_j}{4\pi} - \frac{H^2}{8\pi}\,\delta_{ij} \tag{6.20}$$[19]

over the surface of the volume. Finally, reverting from tensor to vector notation, one can now see that the total work done by the excess magnetic stresses operating over the section R on the fluid ahead is

$$\int_{t_0(R)}^{\infty}\left[\frac{1}{8\pi}(H^2 - H_1^2)v_n - \frac{1}{4\pi}H_n(\boldsymbol{H} - \boldsymbol{H}_1)\cdot\boldsymbol{v}\right]\mathrm{d}t,$$

per unit area of surface. Here $H(R,t)$ is the value of the magnetic field at time t and we have made use of the fact that the normal component H_n of the field is continuous across the shock (equation 6.245).

Combining results we obtain for the total energy of the shock, per unit area of the front, the expression

$$D(R) = \int_{t_0(R)}^{\infty}\left\{\left[(p - p_1) + \frac{1}{8\pi}(H^2 - H_1^2)\right]v_n - \frac{1}{4\pi}H_n(\boldsymbol{H} - \boldsymbol{H}_1)\cdot\boldsymbol{v}\right\}\mathrm{d}t, \tag{6.270}$$

which effectively forms the starting point of Osterbrock's analysis. Limiting his discussion to weak shocks, Osterbrock evaluates D explicitly in terms of the shock strength and the initial values of the physical parameters involved. He then computes the energy dissipated per unit mass of material, Δq, during the return of the shocked gas to its initial state, assuming the thermodynamic relaxation path postulated by Weymann [1960] (see Section 6.5.3). This procedure yields the propagation equation

$$\frac{\mathrm{d}D}{\mathrm{d}R} = -\,\rho_1\,\Delta q$$

[19] The negative sign in front of the term $(H^2/8\pi)\,\delta_{ij}$ on the right-hand side of (6.20) implies that the 'magnetic pressure' acts in the opposite direction to the *outward* normal to the surface being considered.

which, combined with his expression for $D(R)$, enables Osterbrock to obtain an estimate of the mean 'damping length' of a (fast-mode) magnetohydrodynamic shock of given initial strength. The application of Osterbrock's theory to the dissipation of shock waves in magnetic regions of the solar chromosphere is discussed in Section 7.4.8.

The propagation and dissipation of weak magnetohydrodynamic shock waves has also been considered by Dubov [1960] and Pikelner and Livshitz [1964]. In addition, the Brinkley–Kirkwood theory has been generalized by Saito [1964] to take account of the effect of gravity, and hence of the stratification of the fluid, on the propagation of magnetohydrodynamic shocks through the solar chromosphere. However, his treatment is restricted to the case of a horizontal magnetic field, and the same is true of the somewhat earlier investigation undertaken by Uchida [1963]. A similar restriction to a nearly horizontal field applies to the work of Mäckle [1969], which is concerned with the attenuation of fast-mode magnetohydrodynamic shocks propagating vertically upwards through a uniform atmosphere. On the other hand, Mäckle's method of approach to the problem is different: instead of applying the Brinkley–Kirkwood theory, he evaluates directly the rate of energy dissipation within the actual shock layer due to viscosity and thermal conduction. He finds that the effect of the field is to reduce the energy dissipation of fast-mode shocks and hence, presumably, to enable them to travel further upwards through the atmosphere. Other recent work on the propagation and dissipation of fast-mode shocks is discussed in Section 7.4.8.

The need, nevertheless, remains for the development and, if possible, experimental verification of a general theory of the propagation and dissipation of magnetohydrodynamic shock waves in non-uniform media.

REFERENCES

ALFVÉN, H. [1950] *Cosmical Electrodynamics*, Clarendon Press, Oxford.

ANDERSON, J. E. [1963] *Magnetohydrodynamic Shock Waves*, M.I.T. Press, Cambridge, Massachusetts.

BAZER, J., and ERICSON, W. B. [1959] 'Hydromagnetic shocks', *Astrophys. J.* **129**, 758.

BAZER, J., and HURLEY, J. [1963] 'Geometrical hydromagnetics', *J. Geophys. Res.* **68**, 147.

BECKER, E. [1968] *Gas Dynamics*, Academic Press, New York.

BEL, N., and MEIN, P. [1971] 'Propagation of magneto-acoustic waves along the gravitational field in an isothermal atmosphere', *Astron. Astrophys.* **11**, 234.

BHATNAGAR, M. S., and KUSHWAHA, R. S. [1961] 'Propagation of intense shock waves in stellar envelopes', *Ann. Astrophys.* **24**, 211.

BHATNAGAR, M.S., and KUSHWAHA, R. S. [1962] 'Decay of a shock wave in a stellar atmosphere', *Ann. Astrophys.* **25**, 410.

BILLINGS, D. E. [1966] *A Guide to the Solar Corona*, Academic Press, New York.

BIRD, G. A. [1964] 'The behaviour of shock waves in a gravitational atmosphere', *Astrophys. J.* **139**, 675.

BRAGINSKII, S. I. [1957] 'Transport phenomena in a completely ionized two-temperature plasma', *J. Exp. Theor. Phys. U.S.S.R.* **33**, 459 (*Soviet Phys. JETP* **6**, 358).

BRAY, R. J., and LOUGHHEAD, R. E. [1964] *Sunspots,* Chapman and Hall, London.

BRAY, R. J., and LOUGHHEAD, R. E. [1967] *The Solar Granulation,* Chapman and Hall, London.

BRINKLEY, S. R., and KIRKWOOD, J. G. [1947] 'Theory of the propagation of shock waves', *Phys. Rev.* **71**, 606.

CABANNES, H. [1970] *Theoretical Magnetofluiddynamics*, Academic Press, New York.

CHANDRASEKHAR, S. [1950] *Radiative Transfer*, Clarendon Press, Oxford.

CHANDRASEKHAR, S. [1961] *Hydrodynamic and Hydromagnetic Stability,* Clarendon Press, Oxford.

CHAPMAN, S. [1954] 'The viscosity and thermal conductivity of a completely ionized gas', *Astrophys. J.* **120**, 151.

CHAPMAN, S., and COWLING, T. G. [1970] *The Mathematical Theory of Non-Uniform Gases*, 3rd. Ed., Cambridge Univ. Press.

CHISNELL, R. F. [1955] 'The normal motion of a shock wave through a non-uniform one-dimensional medium', *Proc. Roy. Soc.* A **232**, 350.

CHIU, Y. T. [1970] 'Traveling waves of arbitrary amplitude in compressible hydrodynamics under gravity: an exact solution', *Phys. Fluids* **13**, 2950.

CHIU, Y. T. [1971] 'Arbitrary amplitude magnetoacoustic waves under gravity: an exact solution', *Phys. Fluids* **14**, 1717.

CHU, C. K., and GRAD, H. [1965] 'Magnetofluid dynamics', *Research Frontiers in Fluid Dynamics,* ed. R. J. Seeger and G. Temple, p. 308; Interscience, New York.

COLE, R. H. [1965] *Underwater Explosions,* Dover, New York.

COURANT, R., and FRIEDRICHS, K. O. [1948] *Supersonic Flow and Shock Waves,* Interscience, New York.

DELACHE, P., and FROESCHLÉ, C. [1972] 'Time dependent radiative transfer. Damping of a temperature fluctuation. I. Near L.T.E. grey approximation', *Astron. Astrophys.* **16**, 348.

DELCROIX, A., and LEMAIRE, J. [1968] 'Transport coefficients in the chromosphere and coronal gas. I. Viscosity', *Astrophys. J.* **154**, 1155.

DELCROIX, A., and LEMAIRE, A. [1969] 'Transport coefficients in the chromosphere and coronal gas. II. Thermal conductivity', *Astrophys. J.* **156**, 787.

DUBOV, E. E. [1960] 'The propagation and dissipation of waves in the chromosphere', *Izv. Crim. Astrophys. Obs.* **23**, 362.

ECKART, C. [1960] *Hydrodynamics of Oceans and Atmospheres*, Pergamon, Oxford.

ELIASSEN, A., and PALM, E. [1954] 'Energy flux for combined gravitational-sound waves', *Publ. Inst. Theoret. Astrophys.*, Oslo, No. 3.

FERRARO, V. C. A. [1954] 'On the reflection and refraction of Alfvén waves', *Astrophys. J.* **119**, 393.

FERRARO, V. C. A. [1967] 'Magnetohydrodynamics of liquids', *Magnetism and the Cosmos,* ed. W. R. Hindmarsh *et al.,* p. 85; Oliver and Boyd, Edinburgh.

FERRARO, V. C. A., and PLUMPTON, C. [1966] *An Introduction to Magneto-Fluid Mechanics*, 2nd. Ed., Clarendon Press, Oxford.

GERMAIN, P. [1960] 'Shock waves and shock-wave structure in magneto-fluid dynamics', *Rev. Mod. Phys.* **32**, 951.

GOULARD, R. [1967] 'Radiative control of flow problems in aerodynamics. Summary – Introduction', *Aerodynamic Phenomena in Stellar Atmospheres,* ed. R. N. Thomas, p. 245; Academic Press, London.

HINES, C. O. [1960] 'Internal atmospheric gravity waves at ionospheric heights', *Canadian J. Phys.* **38**, 1441.

JEFFREYS, H. [1952] *Cartesian Tensors,* Cambridge Univ. Press.

JONES, D. S. [1964] *The Theory of Electromagnetism*, Pergamon, Oxford.

KATO, S. [1966a] 'The response of an unbounded atmosphere to point disturbances. I. Time-harmonic disturbances', *Astrophys. J.* **143**, 893.

KATO, S. [1966b] 'The response of an unbounded atmosphere to point disturbances. II. Impulsive disturbances', *Astrophys. J.* **144**, 326.

KENDALL, P. C., and PLUMPTON, C. [1964] *Magnetohydrodynamics*, vol. 1, Pergamon, Oxford.

KLIMISHIN, I. A. [1968] 'The structure of a shock wave travelling through a stellar atmosphere', *Astrofizika* **4**, 257 (*Astrophysics* **4**, 94).

KOGURE, T., and OSAKI, T. [1962] 'Propagation of shock waves in inhomogeneous medium', *Publ. Astron. Soc. Japan* **14**, 254.

KULIKOVSKIY, A. G., and LYUBIMOV, G. A. [1965] *Magnetohydrodynamics,* Addison-Wesley, Reading, Massachusetts.

KUPERUS, M. [1965] 'The transfer of mechanical energy in the Sun and the heating of the corona', *Recherches Astron. Obs. Utrecht* **17**, No. 1, p. 1.

LANDAU, L. D., and LIFSHITZ, E. M. [1959] 'Fluid Mechanics', *Course of Theoretical Physics*, vol. 6; Pergamon, London.

LEDOUX, P., and WALRAVEN, T. [1958] 'Variable stars', *Handbuch der Physik*, ed. S. Flügge, vol. 51, p. 353; Springer, Berlin.

LIGHTHILL, M. J. [1956] 'Viscosity effects in sound waves of finite amplitude', *Surveys in Mechanics*, ed. G. K. Batchelor and R. M. Davies, p. 250; Cambridge Univ. Press.

LIGHTHILL, M. J. [1960] 'Studies on magneto-hydrodynamic waves and other anisotropic wave motions', *Phil. Trans. Roy. Soc.* **252A**, 397.

LIGHTHILL, M. J. [1967] 'Predictions on the velocity field coming from acoustic noise and a generalized turbulence in a layer overlaying a convectively unstable atmospheric region', *Aerodynamic Phenomena in Stellar Atmospheres*, ed. R. N. Thomas, p. 429; Academic Press, London.

LINDSAY, R. B. [1960] *Mechanical Radiation*, McGraw-Hill, New York.

LOUGHHEAD, R. E. [1957] 'Solution of flow problems in unidimensional Lagrangian hydromagnetics', *Aust. J. Phys.* **10**, 213.

MÄCKLE, R. [1969] 'On the energy dissipation of fast hydromagnetic shock waves in the solar chromosphere', *Solar Phys.* **10**, 348.

McLELLAN, A., and WINTERBERG, F. [1968] 'Magneto-gravity waves and the heating of the solar corona', *Solar Phys.* **4**, 401.

MISES, R. VON [1958] *Mathematical Theory of Compressible Fluid Flow*, Academic Press, New York.

MOORE, D. W., and SPIEGEL, E. A. [1964] 'The generation and propagation of waves in a compressible atmosphere', *Astrophys. J.* **139**, 48.

MURGATROYD, R. J. [1970] 'The physics and dynamics of the stratosphere and mesosphere', *Rep. Progress Phys.* **33**, 817.

ODGERS, G. J., and KUSHWAHA, R. S. [1959] 'Shock waves in the atmosphere of the β Cephei star BW Vulpeculae', *Publ. Dominion Astrophys. Obs.* **11**, No. 6, p. 185.

ONO, Y., SAKASHITA, S., and OHYAMA, N. [1961] 'On the mechanism of stellar explosion', *Progress Theor. Phys.*, Supp. No. 20, p. 85.

OSTERBROCK, D. E. [1961] 'The heating of the solar chromosphere, plages, and corona by magnetohydrodynamic waves', *Astrophys. J.* **134**, 347.

PACK, D. C. [1965] 'Shockwave phenomena', *Research Frontiers in Fluid Dynamics*, ed. R. J. Seeger and G. Temple, p. 212; Interscience, New York.

PAI, S. [1962] *Magnetogasdynamics and Plasma Dynamics*, Springer, Vienna.

PENNER, S. S., and OLFE, D. B. [1968] *Radiation and Reentry*, Academic Press, New York.

PENNEY, W. G., and PIKE, H. H. M. [1950] 'Shock waves and the propagation of finite pulses in fluids', *Rep. Progress Phys.* **13**, 46.

PIKELNER, S. B., and LIVSHITZ, M. A.[1964] 'Theory of heating of the active and undisturbed chromosphere', *Astron. J. U.S.S.R.* **41**, 1007 (*Sov. Astron. AJ* 8, 808).

PRASAD, P. [1967] 'Non-linear waves in radiation-gas-dynamics', *J. Indian Inst. Sci.* **49**, No. 4, p. 148.

PRASAD, P., and SACHDEV, P. L. [1966] 'Effect of radiation on the structure of a plane stationary shock', *Publ. Astron. Soc. Japan* **18**, 421.

REISS, H. [1965] *Methods of Thermodynamics*, Blaisdell, New York.

ROBERTS, P. H. [1967] *An Introduction to Magnetohydrodynamics*, Longmans, London.

SAITO, M. [1964] 'Transport of mechanical energy in the solar chromosphere: propagation of shock-pulses in stratifying media', *Publ. Astron. Soc. Japan* **16**, 179.

SCHATZMAN, E. [1949] 'The heating of the solar corona and chromosphere', *Ann. Astrophys.* **12**, 203.

SCHATZMAN, E. [1967] 'Shock waves in the solar atmosphere', *Solar Physics*, ed. J. N. Xanthakis, p. 133; Interscience, London.

SCHATZMAN, E., and SOUFFRIN, P. [1967] 'Waves in the solar atmosphere', *Ann. Rev. Astron. Astrophys.* **5**, 67.

SCHMIDT, G. [1966] *Physics of High Temperature Plasmas*, Academic Press, New York.

SHERCLIFF, J. A. [1965] *A Textbook of Magnetohydrodynamics*, Pergamon, Oxford.

SKALAFURIS, A. J. [1969] 'The structure of a shock front in atomic hydrogen. IV. Stability, dissipation, and propagation', *Astrophys. Space Sci.* **3**, 234.

SOUFFRIN, P. [1966] 'Hydrodynamique d'une atmosphère perturbée par une zone convective turbulente sous-jacente', *Ann. Astrophys.* **29**, 55.

SPIEGEL, E. A. [1957] 'The smoothing of temperature fluctuations by radiative transfer', *Astrophys. J.* **126**, 202.

SPITZER, L. [1962] *Physics of Fully Ionized Gases*, 2nd. Ed., Interscience, New York.

STIX, M. [1970] 'On radiative relaxation of chromospheric oscillations', *Astron. Astrophys.* **4**, 189.

STRATTON, J. A. [1941] *Electromagnetic Theory*, McGraw-Hill, New York.

THOMAS, J. H., CLARK, P. A., and CLARK, A. [1971] 'Trapped gravity waves and the five-minute oscillations of the solar atmosphere', *Solar Phys.* **16**, 51.

THOMPSON, W. B. [1962] *An Introduction to Plasma Physics,* Pergamon, Oxford.

TOLSTOY, I. [1963] 'The theory of waves in stratified fluids including the effects of gravity and rotation', *Rev. Mod. Phys.* **35**, 207.

UCHIDA, Y. [1960] 'On the exciters of Type II and Type III solar radio bursts', *Publ. Astron. Soc. Japan* **12**, 376.

UCHIDA, Y. [1961] 'On the formation of solar chromospheric spicules and flare-surges', *Publ. Astron. Soc. Japan* **13**, 321.

UCHIDA, Y. [1963] 'An effect of the magnetic field in the shock wave heating theory of the solar corona', *Publ. Astron. Soc. Japan* **15**, 376.

UCHIDA, Y. [1968] 'Propagation of hydromagnetic disturbances in the solar corona and Moreton's wave phenomenon', *Solar Phys.* **4**, 30.

ULMSCHNEIDER, P. [1967] 'The structure of the outer atmosphere of the Sun and of cool stars', *Z. Astrophys.* **67**, 193.

ULMSCHNEIDER, P. [1970] 'Thermal conductivity in stellar atmospheres. I. Without magnetic field', *Astron. Astrophys.* **4**,144.

WANG, C. P. [1970] 'Effect of thermal radiation on the propagation of plane shock waves', *Phys. Fluids* **13**, 1717.

WEYMANN, R. [1960] 'Heating of stellar chromospheres by shock waves', *Astrophys. J.* **132**, 452.

WHITHAM, G. B. [1958] 'On the propagation of shock waves through regions of non-uniform area or flow', *J. Fluid Mech.* **4**, 337.

WORRALL, G. [1972] 'Oscillations in an isothermal atmosphere: the solar five-minute oscillations', *Astrophys. J.* **172**, 749.

CHAPTER 7

Theories of the Heating of the Chromosphere and of the Origin of Spicules

7.1 Introduction

In the previous chapter we have discussed those aspects of the theory of wave propagation and dissipation which are particularly relevant to the theoretical problems of the chromosphere. Foremost amongst the problems now confronting us are (*a*) the elucidation of the non-thermal mechanism responsible for heating the chromosphere, and (*b*) the origin of spicules. An essential pre-requisite to any attack on the first problem is a knowledge of the net radiative energy loss from the chromosphere as a function of height. Accordingly, we begin our final chapter by evaluating the dominant sources of line and continuum radiation and thereby estimating both the overall radiative loss from the chromosphere as a whole and the net loss as a function of height (Section 7.2).

For a number of years it has been widely accepted that the heating of the chromosphere is due to the dissipation of upward-travelling shock waves formed in the chromosphere from hydrodynamic waves which start their upward propagation at the photospheric level. After a brief outline of the origin and subsequent development of this hypothesis (Section 7.3.1), we then proceed to consider in Section 7.3.2 the crucial question of the critical periods and allowable modes of wave propagation in the upper photosphere and low chromosphere, paying particular attention to the important role played by radiative damping. These topics are discussed separately for the non-magnetic regions overlying the interiors of the supergranule cells and the magnetic regions located along their boundaries. In both cases, the available evidence strongly supports the view that the waves ultimately responsible for the chromospheric heating are gravity-modified *acoustic* waves with periods below a certain theoretical upper limit, ~195 sec. Modified *gravity* waves, on the other hand, are unlikely to play any significant part in heating the chromosphere although, as explained in Section 7.3.4, they may possibly be associated with the 5-minute oscillations described in Section 3.5.1.

Details of the transformation of the upward-propagating acoustic waves into shocks are considered in Section 7.4.2: it appears that waves with periods of about 100 sec or less are transformed into shocks within the first 1000 km or less of the base of the chromosphere. The weak shocks so formed are believed to be responsible for heating the lower layers of the chromosphere. Estimates of their initial strengths and of their rates of dissipation as a function of height have been obtained by several authors, using a modified form of the Brinkley–Kirkwood theory of shock propagation first formulated by E. Schatzman in 1949. An up-dated account of this theory is given in Section 7.4.3, while its overall validity is briefly examined in Section 7.4.4. Following P. Ulmschneider, we then compare computed curves showing the shock dissipation and the net radiative energy loss as functions of height (Section 7.4.5): this comparison leads to the conclusion that the shocks involved have periods of only some tens of seconds. The short-period acoustic waves presumed to give rise to these shocks have so far escaped detection, for reasons which are briefly considered in Section 7.4.6.

Acoustic waves with periods exceeding about 100 sec reach greater heights than those of shorter period before developing into shocks; such waves, accordingly, are likely candidates for heating the middle and upper regions of the chromosphere. On the other hand, our present knowledge of the heating of these layers is much less reliable than that of the low chromosphere; the existing information is summarized in Section 7.4.7.

Regions of the low chromosphere coinciding with the chromospheric network are pervaded by magnetic fields whose strengths – measured at the photospheric level – range from 10 to about 50 gauss. As we shall see in Section 7.4.8, the effect of the field is to reduce the height up to which shock heating is operative, at the same time *increasing* the dissipation at lower levels. This provides a plausible explanation for the enhanced heating implied by observations of the network.

Existing discussions of shock heating of the chromosphere embody the implicit assumption that the chromosphere is *convectively stable* within the meaning of the classical Schwarzschild criterion. This criterion, however, is valid only when all changes in the gas occur *adiabatically*, an assumption which is not necessarily applicable under the conditions obtaining in the chromosphere. Accordingly, in Section 7.5, we follow the procedure of R. J. Defouw to examine in a more fundamental manner the stability of an optically thin, radiating gas characterized by a (simplified) 'heat-loss function' involving only the gas density and temperature. Depending on the form of this function, it turns out that the gas may be thermally unstable even in the absence of gravity. Furthermore, in the presence of gravity a thermally unstable atmosphere is also *convectively* unstable. Unfortunately, existing studies of this potentially important problem have not yet advanced to the stage of giving a definite answer to the question of whether, at any given height, the chromosphere is stable or unstable.

Finally, we turn to the origin of spicules (Section 7.6). This problem has attracted the attention of a number of authors who have proposed a variety of physical mechanisms. These fall naturally into three classes, discussed briefly in Sections 7.6.2–7.6.4. At the present time the last of these appears the most promising: spicules are explained in terms of the upward expulsion of gas in the neighbourhood of neutral sheets in the weak magnetic fields surrounding the supergranules. In no instance, however, can the suggested explanation really be regarded as constituting a fully elaborated theory of the spicule phenomenon as a whole. The important question of whether spicules play a significant role in maintaining either the mass or energy balance of the corona is considered in Section 7.6.5.

7.2 Net Radiative Energy Loss From the Chromosphere

7.2.1 INTRODUCTION

An essential preliminary to any discussion of the heating of the chromosphere is the evaluation as a function of height of the net radiative energy loss per unit volume. Referring back to Section 6.2.2, we see that in a steady state the general equation of energy conservation (6.21) takes the form

$$\frac{\partial J_i}{\partial x_i} = 0, \tag{7.1}$$

where J_i is the i-th component of the sum of all forms of energy flux. In the present context we are evidently concerned only with mechanical, radiative, and conductive energy fluxes, so that

$$J_i = (F_{\mathrm{mech}})_i + (F_{\mathrm{rad}})_i + (F_{\mathrm{cond}})_i. \tag{7.2}$$

Ignoring spatial fluctuations in a horizontal direction, we obtain from (7.1) and (7.2) the basic equation of energy balance at each point in the atmosphere

$$-\frac{\mathrm{d}F_{\mathrm{mech}}}{\mathrm{d}h} - \frac{\mathrm{d}F_{\mathrm{cond}}}{\mathrm{d}h} = \frac{\mathrm{d}F_{\mathrm{rad}}}{\mathrm{d}h}, \tag{7.3}$$

where h is the vertical height, measured from the base of the chromosphere ($\tau_{5000} \simeq 0.01$). The two terms on the left-hand side of (7.3) represent the *gain* of energy per unit volume and time due to mechanical dissipation and (particle) conduction respectively, while the term on the right-hand side represents the net *loss* of energy due to radiation.[1] Note that the quantity $\mathrm{d}F_{\mathrm{rad}}$ represents the net

[1] With the h-axis taken to be directed vertically upwards, the quantity $\mathrm{d}F_{\mathrm{mech}}/\mathrm{d}h$, for example, must necessarily be *negative* if mechanical energy is to be liberated and hence $(-\mathrm{d}F_{\mathrm{mech}}/\mathrm{d}h)$ is *positive*. Similarly, $\mathrm{d}F_{\mathrm{rad}}/\mathrm{d}h$ is necessarily a positive quantity in the present context.

radiative energy loss per unit cross section from a chromospheric shell of thickness dh located at a height h, while the integral $\int(\mathrm{d}F_{\mathrm{rad}}/\mathrm{d}h)\mathrm{d}h$ gives the overall net radiative loss over the total height of the chromosphere.

Anticipating a later result (Section 7.2.6), we may neglect the conductive flux in the lower regions of the chromosphere. Where this is valid (7.3) reduces to

$$-\frac{\mathrm{d}F_{\mathrm{mech}}}{\mathrm{d}h}=\frac{\mathrm{d}F_{\mathrm{rad}}}{\mathrm{d}h}. \tag{7.4}$$

In the upper chromosphere, on the other hand, downward conduction of heat from the hot corona can no longer be ignored and equation (7.3) must be used (Section 7.2.6).

At the lower levels (7.4) provides a means of determining the mechanical energy dissipation from a knowledge of the net radiation loss at each height. However, it is important to realize that the latter quantity is not simply equal to the radiation observed to emerge from the chromosphere, a quantity we shall term the *total* radiative loss. The reason for this is that the total loss from any given layer dh contains a contribution – which may be large – originating in scattering of radiation from the photosphere and from other parts of the chromosphere. Evidently this contribution must be subtracted from the total radiation loss to derive the portion which owes its origin to mechanical heating.

The distinction between total and net radiative loss may be clarified by appealing to the equation of transfer, which for a given frequency ν in a line transition k → l takes the form

$$\mu\frac{\partial I_\nu}{\partial h}=j_\nu-\kappa_\nu I_\nu, \tag{7.5}$$

where $I_\nu(h,\theta)$ is the specific intensity in a direction making an angle θ with the vertical and $\mu=\cos\theta$. In addition, the equation for the radiation flux in the line, F_ν, can be written as

$$F_\nu=2\pi\int_{-1}^{+1}\mu I_\nu\,\mathrm{d}\mu \tag{7.6}$$

(cf. equation 6.234). Differentiating (7.6) with respect to h and substituting from (7.5) we find that

$$\frac{1}{4\pi}\frac{\mathrm{d}F_\nu}{\mathrm{d}h}=j_\nu-\kappa_\nu\times\frac{1}{2}\int_{-1}^{+1}I_\nu\mathrm{d}\mu=j_\nu-\kappa_\nu J_\nu. \tag{7.7}$$

If we now put

$$F_{\mathrm{kl}}=\int_0^\infty F_\nu\mathrm{d}\nu,\qquad j_{\mathrm{kl}}=\int_0^\infty j_\nu\mathrm{d}\nu,\quad\text{and}\quad\kappa_\nu=\kappa_{\mathrm{kl}}\phi_\nu, \tag{7.8}$$

where ϕ_ν is the normalized absorption profile, and integrate (7.7) over the transition $k \to 1$ we obtain the relation

$$\frac{dF_{kl}}{dh} = 4\pi\left(j_{kl} - \kappa_{kl}\int_0^\infty J_\nu\phi_\nu\,d\nu\right). \tag{7.9}$$

This equation is easy to interpret: the first term on the right-hand side, $4\pi j_{kl}$, is the total radiation emitted in the transition $k \to 1$ over all solid angles and is equal to $h'\nu_{kl}A_{kl}N_k$, where A_{kl} is the transition probability and N_k is the population of the upper level. The second term on the right-hand side, which depends on the radiation field, represents the radiation (e.g., from the photosphere), *absorbed* by the layer. Accordingly, following Athay [1966], we take the term on the left-hand side as representing the portion of the total or observed radiation which owes its origin to mechanical heating so that, in the absence of conduction

$$\frac{dF_{kl}}{dh} = \left(\frac{dF_{rad}}{dh}\right)_{kl} = -\left(\frac{dF_{mech}}{dh}\right)_{kl}. \tag{7.10}$$

Equation (7.9) can be expressed in a more convenient form by putting $S_{kl} = j_{kl}/\kappa_{kl}$; then

$$\frac{dF_{kl}}{dh} = 4\pi j_{kl}\rho_{kl} = h'\nu_{kl}A_{kl}N_k\rho_{kl}, \tag{7.11}$$

where

$$\rho_{kl} = 1 - \frac{\int_0^\infty J_\nu\phi_\nu\,d\nu}{S_{kl}}. \qquad (7.12)^2$$

From (7.10) and (7.11) we see that the problem of determining dF_{mech}/dh resolves itself essentially into determining the quantity ρ_{kl} for all spectral lines (and continua) contributing significantly to the net radiation loss from the chromosphere. However, as pointed out by Athay, the number of cases for which ρ_{kl} has actually been calculated falls far short of what is needed. Moreover, the above analysis shows that a knowledge of ρ_{kl} is also required for a determination of dF_{kl}/dh from *observation*: for let the observed flux from the layer dh be represented by dF^*_{kl}. Then for an optically thin layer

$$\frac{dF^*_{kl}}{dh} = 4\pi j_{kl}\,.$$

[2] Our derivation of (7.11) follows that of Athay [1966], our quantity dF_{kl}/dh being the same as Athay's $4\pi\Delta H_{kl}/\Delta h$. The factor ρ_{kl} is equal to the so-called 'net radiative bracket' [Mihalas, 1970, p. 140].

Hence, from (7.10) and (7.11)

$$-\left(\frac{dF_{mech}}{dh}\right)_{kl} = \frac{dF_{kl}}{dh} = \rho_{kl}\frac{dF^*_{kl}}{dh}.$$

Thus $|\,dF_{mech}/dh\,|_{kl} < dF^*_{kl}/dh$, since ρ_{kl} lies between 0 and 1. Only when $\rho_{kl} = 1$ is equality achieved.

7.2.2 NET RADIATIVE LOSS IN THE BALMER AND LYMAN SERIES

Athay [1966] has shown that for a *series* of lines such as the Balmer or Lyman series of hydrogen, the impractibility of determining the ρ_{kl}'s separately for each line can be circumvented by summing over the series. From (7.11) we then have

$$\sum_k \frac{dF_{kl}}{dh} = \sum_k h'\nu_{kl} A_{kl} N_k \rho_{kl}$$
$$\simeq \langle h'\nu_{kl}\rangle \sum_k A_{kl} N_k \rho_{kl}\,. \tag{7.13}$$

By introducing the equations of steady-state equilibrium (Section 4.3.4), one can transform the sum on the right-hand side to a dependence on collisional transition rates C_{lm} and LTE departure coefficients b_l and b_m instead of on A_{kl} and ρ_{kl}. In this way the contribution of the whole series to the net radiative loss can be approximately evaluated. Similar calculations have been carried out by Jordan [1969] and Ulmschneider [1970, 1971a] and, for spicules only, by Nguyen-Ngan [1970].

According to Athay, the net loss in the Balmer series comes from the height range $h = 500$–1000 km; the net loss in Lyman α, which dominates the loss in the Lyman series, comes from $h \cong 1500$ km. Ulmschneider [1970, Fig. 1) has plotted the Balmer and Lyman losses as a function of h for a number of chromospheric models, including the BCA (Section 4.2.5). For all of these models except the BCA, maximum loss for both series occurs at $h \simeq 600$–700 km, but for the BCA the region of maximum loss in these series is shifted to lower heights.

7.2.3 NET RADIATIVE LOSS FROM HELIUM AND HEAVIER ELEMENTS

Athay [1966] remarks that the total energy radiated in the resonance series of neutral and ionized helium is only ~5 per cent of that radiated in Lyman α. Since the net loss must be smaller than the total radiative loss (Section 7.2.1), it is evident that the net loss from He can be neglected. Excluding H and He lines, Athay estimates that the total flux $\int (dF^*/dh)dh$ radiated in ultraviolet and X-ray lines from heavier elements in the range 1–1325 Å, originating in the upper chromosphere and corona, is 2×10^5 erg cm^{-2} sec^{-1}; this figure is comparable with the observed Lyman α flux. For most of these lines the net loss can be equated to the total energy radiated, i.e. $\rho_{kl} = 1$. Ulmschneider [1971a] quotes the somewhat smaller figure of 7.4×10^4 erg cm^{-2} sec^{-1} for XUV lines lying below 1206 Å. Athay arbitrarily sets the mean level of origin of the radiation from the heavier elements at $h = 7500$ km; this point is included in Fig. 7.1.

A detailed calculation of the height dependence of $\mathrm{d}F_{\mathrm{rad}}/\mathrm{d}h$ throughout the range, say, $h = 2000$–$20\,000$ km, is not feasible at the present time. On the other hand, if existing calculations are reliable, this region makes only a relatively small contribution to the overall net loss from the whole chromosphere, $\int(\mathrm{d}F_{\mathrm{rad}}/\mathrm{d}h)\mathrm{d}h$.

7.2.4 NET RADIATIVE LOSS FROM H$^-$

We turn now to the dominant contribution to the net radiative loss from the chromosphere namely, emission from the negative hydrogen ion. According to Athay [1966] and Jordan [1969], this originates in the region $h \simeq 0$–500 km, i.e. just below the region in which Balmer and Lyman series losses become important. Most authors have followed Osterbrock [1961] in quoting an approximate expression for the net loss from the H$^-$ layer; this expression is proportional to $\Delta\tau$, the optical thickness of the layer, and to $T_0^3(T - T_0)$, where T is the local electron or kinetic temperature and T_0 is the 'boundary' temperature of the Sun [Athay, 1966, 1970; Ulmschneider, 1970, 1971a]. However, as pointed out by Athay [1970] the specification of both $\Delta\tau$ and $T - T_0$ is subject to much uncertainty.

Only one calculation based, in effect, on an evaluation of ρ_{H^-} and the relevant departure coefficient b_{H^-} is known to the authors [Jordan, 1969, pp. 70–74, 86–88], and uncertainty in the magnitude of the H$^-$ loss represents the major source of error in current estimates of the overall net radiative loss from the chromosphere. The difficulties inherent in a more exact treatment of the problem have been discussed by Praderie and Thomas [1972].

7.2.5 NET RADIATIVE LOSS AS A FUNCTION OF HEIGHT

Figure 7.1 shows the net radiative loss from all sources, $\mathrm{d}F_{\mathrm{rad}}/\mathrm{d}h$, as a function of height according to the calculations of Athay [1966] and Ulmschneider [1970]. The latter's results are based on the BCA model of the low chromosphere (Section 4.2.5) and at low heights, e.g. $h < 500$ km, are in good agreement with Athay's. However, for $h > 1000$ km there are serious differences amounting to two orders of magnitude. These differences give some idea of the possible uncertainties in our present knowledge of $\mathrm{d}F_{\mathrm{rad}}/\mathrm{d}h$. Ulmschneider's curve is much steeper than Athay's, implying a much more rapid shedding of mechanical energy in the low chromosphere.

Another quantity of fundamental importance in discussions of the heating of

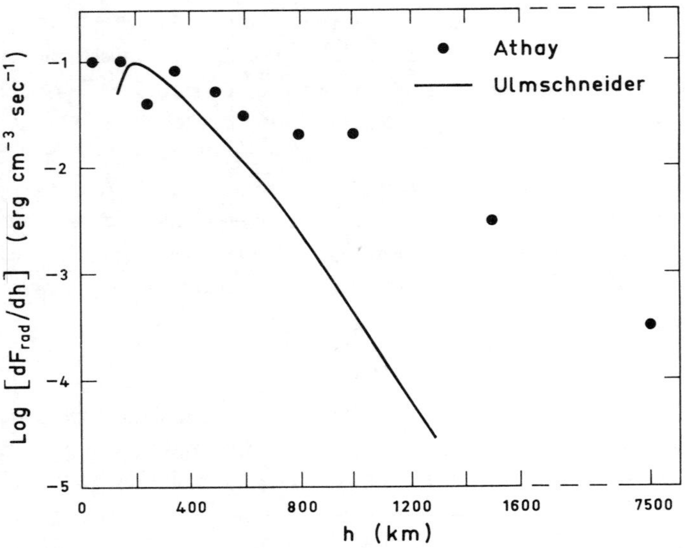

Fig. 7.1. Net radiative energy loss dF_{rad}/dh as a function of height according to the calculations of Athay [1966, Table 6] and Ulmschneider [1970, Fig. 1]. The zero point of the height scale corresponds to the base of the chromosphere, $\tau_{5000} \simeq 0.01$; Ulmschneider's height scale has been made consistent with Athay's by subtracting 300 km. The isolated point on the far right represents an estimate of the contribution of ultraviolet and X-ray lines from elements heavier than H and He (see Section 7.2.3).

the chromosphere is the value of the integral $\int(dF_{rad}/dh)dh$ (cf. Section 7.3.1), the integration being performed over the total height range of the chromosphere. Various estimates of this integral are given in Table 7.1. Differences among these estimates can be attributed to the use of different chromospheric models and to variations in the procedures employed in calculating the contributions to dF_{rad}/dh from the various sources outlined in Sections 7.2.2–7.2.4. A major

TABLE 7.1

Net Radiative Energy Loss from Chromosphere Integrated Over Height

REFERENCE	$\int(dF_{rad}/dh)dh$ (erg cm^{-2} sec^{-1})
Osterbrock [1961]	4×10^7
Athay [1966]	5.6×10^6
Athay [1970]	2×10^6
Ulmschneider [1970]	2.2×10^6

source of possible error lies in the evaluation of the net loss due to H^-, as pointed out above. The determinations of the radiative loss referred to above all ignore the inhomogeneous structure of the chromosphere, but Nguyen-Ngan [1970] has attempted to calculate the loss specifically from spicules.

It is of interest to compare the net radiative energy loss from the chromosphere with the net energy loss from the corona, the solar wind, and the photosphere. This comparison is shown in Table 7.2. We see that the chromospheric loss is an order of magnitude greater than the coronal loss but is nevertheless very small in comparison to the enormous radiation output of the photospheric layers.

TABLE 7.2

Comparison Between Net Radiative Loss From Chromosphere and Net Energy Loss from Other Regions

REGION	NET LOSS (erg cm^{-2} sec^{-1})	REFERENCE
Chromosphere	$2\text{–}6 \times 10^6$	Table 7.1
Corona	1.5×10^4	Osterbrock [1961]
	3×10^5	Athay [1966]
Solar wind: heat flux	2×10^4	Ulmschneider [1971a]
kinetic energy	$< 2 \times 10^4$	
Photosphere	6.5×10^{10}	Jordan [1969]

7.2.6 DOWNWARD CONDUCTIVE FLUX IN THE UPPER CHROMOSPHERE

Hitherto in our discussion of the net radiative loss from the chromosphere we have ignored the conductive term dF_{cond}/dh in (7.3), thus enabling us to equate $|dF_{mech}/dh|$ to dF_{rad}/dh. The downward conductive flux may be determined from the formula

$$F_{cond} = -\kappa \frac{dT}{dh}, \tag{6.13}$$

where κ is the coefficient of conductivity; references to the evaluation of κ under the physical conditions prevailing in the chromosphere and corona are given in Section 6.2.2. Calculations by a number of authors have demonstrated that in the *low* chromosphere $|F_{cond}|$ is indeed quite negligible in comparison with $\int(dF_{rad}/dh)dh$ [see, for example, Athay, 1966; Ulmschneider, 1970, 1971a; Jordan, 1969]. Thus Ulmschneider [1970, Table III] has calculated that for heights less than $h \simeq 2000$ km, $|F_{cond}|$ ranges from only 0.4 to 6 erg cm^{-2} sec^{-1}, i.e. approximately six orders of magnitude smaller than $\int(dF_{rad}/dh)dh$

(cf. Table 7.1). In the low chromosphere, therefore, our neglect of the conductive term in (7.3) is justified.

In the upper chromosphere, on the other hand, the situation is quite different. In this region the ionization is almost complete (Table 4.18) and (6.13) can be written in the form

$$F_{\text{cond}} = -KT^{5/2} \frac{\mathrm{d}T}{\mathrm{d}h}, \tag{7.14}$$

where K is a slowly-varying function of T and N_e (see Section 6.2.2). The form of (7.14) suggests that as the transition to the corona is approached – where both T and $\mathrm{d}T/\mathrm{d}h$ become large – $|F_{\text{cond}}|$ can be expected to rise to values much higher than in the low chromosphere. Calculations confirm this expectation, yielding a value of $\sim 5 \times 10^5$ erg cm^{-2} sec^{-1} for the flux conducted downward into the upper chromosphere from the corona [Kopp and Kuperus, 1968; Ulmschneider, 1971a].

With the aid of a model of the middle and upper chromosphere it would be a comparatively simple matter, using equation (7.14), for one to calculate the conductive term $\mathrm{d}F_{\text{cond}}/\mathrm{d}h$ appearing in (7.3) as a function of h. If we also knew $\mathrm{d}F_{\text{rad}}/\mathrm{d}h$ as a function of h, we could then proceed to apply (7.3) to establishing the run of $\mathrm{d}F_{\text{mech}}/\mathrm{d}h$. At the present time, however, our knowledge of $\mathrm{d}F_{\text{rad}}/\mathrm{d}h$ in these layers is not adequate to allow us to adopt this procedure. The alternative – but less satisfactory – procedure of attempting to derive by calculation a 'self-consistent' model of the middle and upper chromosphere and its shock heating is described in Section 7.4.7.

7.3 Wave Motions in the Upper Photosphere and Low Chromosphere

7.3.1 INTRODUCTION

The idea that the heating of the chromosphere might be due to the dissipation of the energy of upward-propagating shock waves was first put forward by Biermann [1946, 1948]. In simple terms the basic physical picture underlying his suggestion may be described as follows: according to modern knowledge of the photosphere, there exists above the level of the photospheric granulation seen in white light (optical depth $\tau_{5000} \simeq 1$) a *convectively stable* region some hundreds of kilometres thick extending up into the low chromosphere [see, for example, Bray and Loughhead, 1967, p. 80]. The bottom of this layer is continually buffeted by rising granules and, indeed, there is direct observational evidence that the granules actually penetrate into it at least up to a height corresponding to $\tau_{5000} \simeq 0.1$ [op. cit., pp. 31–33]. This is only a lower limit to the height of penetration and, in fact, indirect evidence derived from velocity observations indicates that convective overshoot may persist right up to the height of the temperature minimum [Frazier, 1968]. Biermann argued that the resulting disturbance of the stable layer gives rise to the excitation of *acoustic waves* (Section 6.3.2), which then propagate upwards, carrying with them (in the form of acoustic wave energy) some portion of the original kinetic energy of the granules. As they proceed upwards, their velocity amplitude increases rapidly as a consequence of the marked decrease in gas density with height in the chromosphere: within a comparatively short distance the amplitude becomes comparable to the velocity of sound and the acoustic waves develop into *shocks* (Section 6.5). The subsequent dissipation of these shocks at higher levels is responsible for the heating of the chromosphere.[3]

In the year that Biermann's second paper appeared, the suggestion that the heating of the chromosphere (and also the corona) might be due to the dissipation of the energy of a train of acoustic waves generated by the granulation was advanced independently by Schwarzschild [1948]. However, unlike Biermann, he did not stress the indispensable role played by shock waves in the actual dissipation of the wave energy. Nevertheless, it is instructive to outline his analysis of another basic and related problem, namely the question of whether the kinetic energy transported up by the granules is sufficient to make good the combined net energy loss from the chromosphere and corona. The kinetic energy carried by a single rising granule can be estimated from the formula

$$E_{\text{granule}} = \frac{1}{2} \rho A d v^2,$$

where $\rho \simeq 3 \times 10^{-7}$ gm cm^{-3} is the mean density at the top of the convection zone ($\tau_{5000} \simeq 1$), $A = 2.8 \times 10^{16}$ cm^2 is the average area of a granule, $d \simeq 3.5 \times 10^7$ cm is the effective thickness of the convecting layer, and $v \simeq 1 \times 10^5$ cm sec^{-1} is the mean upward velocity of a granule. Furthermore, if $N = 2.2 \times 10^6$ is the total number of granules on the Sun and $t = 600$ sec their mean lifetime,[4] it follows that the total kinetic energy brought up by the granules to the top of the convection zone over the entire surface of the Sun per unit time is

$$L_{\text{granule}} = \frac{N E_{\text{granule}}}{t} = 5.4 \times 10^{30} \text{ erg sec}^{-1}.$$

On the other hand, we can deduce from Table 7.2 that the net outflow of energy per second from the chromosphere and corona, including that carried by the

[3] Historically, it is interesting to note that at about the time that Biermann put forward his suggestion, Thomas [1948] attributed the heating of the chromosphere to a system of 'superthermic' (i.e. supersonic) jets, which he identified with spicules.

[4] All numerical values are taken from Bray and Loughhead [1967] except that of the granule velocity, where a higher value has been adopted in the light of more recent measurements [see, for example, Beckers, 1968].

solar wind is, at most,

$$L = 3.9 \times 10^{29} \text{ erg sec}^{-1}.$$

Comparing the two figures we see that the amount of kinetic energy transported upwards by the granules exceeds by an order of magnitude the combined net energy loss from the chromosphere and corona.

Of course, one would not expect *all* the granule kinetic energy to be transformed into acoustic wave energy, nor is there any a priori reason why the region of wave generation should necessarily be confined to the stable layer immediately overlying the solar convection zone. In fact, many later writers have discussed the possibility that the waves are excited within the convection zone itself by a regime of *turbulent motions.* However, as we shall emphasize in Section 7.3.3, there is no observational evidence to substantiate the assumption that the motions in this region constitute fully developed turbulence in the strict aerodynamic sense. At the present time, accordingly, a full understanding of the detailed mechanism of acoustic wave generation in the photosphere is still lacking. An ab initio calculation of the amount of wave energy transported up into the chromosphere is therefore impossible and an estimate of this quantity has to be obtained by other means (see Section 7.4.5).

Besides acoustic waves, there is also the possibility that *internal gravity waves* may be generated in the stable layer lying above the convection zone, the properties of which have been discussed in Sections 6.3.2 and 6.3.3. This idea was proposed by Whitaker [1963] who, following earlier suggestions of M. Krook and C. O. Hines, attempted to explain the heating of the corona in terms of the dissipation of gravity waves by thermal conduction. However, no account was taken of radiative damping of the waves but this effect, as we shall see later, has since proved of crucial importance.

We shall begin our discussion of wave motions in the upper photosphere and low chromosphere by evaluating the critical frequencies and periods for the propagation of acoustic and gravity waves, first in the adiabatic approximation and then in the presence of radiative damping (Section 7.3.2). In the case of gravity waves the latter turns out to be of great significance at very low heights: in fact, no waves of this type can propagate in the upper photosphere and only those with periods exceeding ~260 sec can travel further upwards from the base of the chromosphere. On the other hand, provided their period is less than ~195 sec, there is no obstacle to the upward propagation of acoustic-type waves, apart from some energy loss due to radiative damping at the lowest levels. Consequently, the balance of the available evidence continues to support Biermann's original identification of acoustic waves as the ultimate agency responsible for heating the chromosphere.

After a brief discussion of possible mechanisms for the generation of waves

(Section 7.3.3), we turn to the important question of whether the 5-minute oscillations observed in the upper photosphere and low chromosphere (described in Section 3.5.1) are to be interpreted as direct manifestations of the waves which carry up the energy to heat the chromosphere. The short answer at the present time appears to be 'no', and other possible interpretations of the 5-minute oscillations are briefly referred to.

Fuller accounts of the historical developments underlying modern shock wave theories of chromospheric and coronal heating are to be found in papers by Kuperus [1965, 1969] and Jordan [1969].

7.3.2 CRITICAL PERIODS AND ALLOWABLE WAVE MODES

In this section we shall use the theory outlined in Chapter 6 to evaluate the critical periods and other relevant parameters pertaining to wave propagation in the upper photosphere and low chromosphere and, with the aid of this data, attempt to identify the particular type of wave actually responsible for the upward transmission of energy in this part of the Sun's atmosphere. All calculations are made on the basis of the BCA model given in Table 4.6, with the zero of the height scale shifted to correspond to the base of the chromosphere (i.e. $\tau_{5000} \simeq 0.01$). On this scale the region of interest is taken to extend from $h = -200$ km up to $h = 1000$ km. The lower level lies slightly below the height to which the photospheric granules are known, from direct observations, to extend (Section 7.3.1). The upper level, as we shall see in Section 7.4.2, represents an inferred upper limit to the height of shock formation in the chromosphere for short-period waves. Following Ulmschneider [1967], we shall adopt the values $\gamma = 5/3$ and $\mu = 1.44$ for the ratio of the specific heats and mean molecular weight respectively over the whole height range.

(1) *Some important parameters pertaining to wave propagation.* As a first step we present in Table 7.3 the values at 100-km intervals of the sound velocity

$$a = \sqrt{\frac{\gamma p}{\rho}}\,; \qquad (6.40)$$

the Alfvén velocity

$$A = \frac{H}{\sqrt{4\pi\rho}} \qquad (6.121)$$

for representative field strengths of 2, 20, and 50 gauss; and the radiative relaxation time

$$t_R = \frac{\rho c_v}{16\bar{\kappa}\sigma_R T^3}\,. \qquad (6.88)$$

The field strengths of 2 and 20 gauss selected in the calculation of A are typical of those observed – at the photospheric level – in the interiors and boundaries respectively of the supergranule cells; the higher figure of 50 gauss is representative of the stronger fields sometimes found at the vertices of the cells (Section 3.4.4). The values of t_R tabulated are those calculated by Ulmschneider [1971b] on the basis of the Harvard-Smithsonian Reference Atmosphere (cf. Section 4.2.5). Values of t_R up to a height of approximately 500 km calculated with the aid of the BCA model are given by Stix [1970, Fig. 1] and Souffrin [1972].

TABLE 7.3

Sound Velocity, Alfvén Velocity, and Radiative Relaxation Time

h (km)	a (km sec^{-1})	A (km sec^{-1}) 2 gauss	20 gauss	50 gauss	t_R (sec)
−200	7.72	0.01	0.12	0.29	3.4
−100	7.29	0.02	0.16	0.41	11.6
0	7.05	0.02	0.25	0.62	29.4
100	7.04	0.04	0.39	0.98	76.7
200	7.05	0.06	0.62	1.56	207
300	7.10	0.10	1.00	2.49	499
400	7.20	0.16	1.57	3.94	741
500	7.30	0.25	2.47	6.18	504
600	7.41	0.38	3.83	9.58	429
700	7.52	0.59	5.88	14.7	382
800	7.60	0.88	8.77	21.9	352
900	7.67	1.34	13.4	33.6	323
1000	7.74	1.94	19.4	48.5	294

Note: the values of a and A are calculated on the basis of the BCA model, t_R on the basis of the HSRA (see text).

Three important conclusions follow at once from an examination of Table 7.3:

(i) over the entire height range the value of the sound velocity a does not differ from the mean value of ~7.4 km sec^{-1} by more than about 5 per cent. To this degree of approximation, therefore, this portion of the solar atmosphere may be regarded as effectively isothermal and, accordingly, we are justified in applying the theory of wave propagation in an isothermal gravitationally-stratified atmosphere;

(ii) at the base of the chromosphere the value of the Alfvén velocity A for all the assumed field strengths is much smaller than the sound velocity a. However, A increases with height much more rapidly than a and at h = 800 km exceeds a when H = 20 gauss and at h = 600 km when H = 50 gauss;

(iii) the value of the radiative relaxation time t_R is only about 7 sec in the upper photosphere but increases by some two orders of magnitude within a distance of a few hundreds of kilometres.

(2) *Critical periods in the adiabatic approximation.* Using the values of a given in Table 7.3 we next compute the two critical frequencies

$$\omega_1 = \frac{\gamma g}{2a} \tag{6.53}$$

and

$$\omega_2 = \frac{\sqrt{\gamma - 1} \cdot g}{a} \tag{6.54}$$

arising in the theory of adiabatic wave propagation in an isothermal or quasi-isothermal atmosphere (Section 6.3.2). However, instead of tabulating the frequencies as functions of height we shall find it more convenient to tabulate the corresponding critical periods $P_1 = 2\pi/\omega_1$ and $P_2 = 2\pi/\omega_2$. These are given in the second and third columns of Table 7.4.

TABLE 7.4

Critical Periods as Functions of Height

h (km)	P_1 (sec)	P_2 (sec)	P_3 (sec)	P_4 (sec)	P_N (sec)
−200	213	217	274	∞	341
−100	201	205	249	∞	241
0	194	197	215	257	209
100	193	197	198	203	196
200	194	197	195	198	194
300	195	199	195	199	193
400	198	202	198	202	194
500	201	204	201	204	196
600	203	208	204	208	198
700	207	210	208	210	205
800	209	212	209	212	203
900	211	215	212	215	209
1000	212	217	213	218	207

As we have seen in Section 6.3.2, waves with periods $P < P_1$ are essentially acoustic waves modified by gravity, whereas waves with periods $P > P_2$ are gravity waves modified by compressibility. Table 7.4 shows that the minimum value of P_1 is 193 sec and this implies that only acoustic waves with periods less than 193 sec are capable of propagating from the photosphere up into the low chromosphere. Similarly, the maximum value of P_2 at heights up to 1000 km is 217 sec and hence only gravity waves with periods greater than 217 sec can propagate up to this height in the chromosphere.[5] However, as we shall see below, this conclusion has to be modified when radiative damping is taken into account. At any given height only *standing* waves can exist in the period range $P_1 < P < P_2$.

An error arises in the calculation of the cut-off period P_2 for gravity waves due to the neglect of the finite temperature gradient in the assumed atmospheric model. This error can be estimated by calculating the Väisälä frequency N of purely gravitational oscillations (see Section 6.3.2) from the formula

$$N^2 = \omega_2^2 + g\frac{\mathrm{d}(\ln T)}{\mathrm{d}h}, \tag{7.15}$$

which is obtained by combining equations (6.54) and (6.64). The corresponding values of the cut-off period $P_N = 2\pi/N$ are given in the final column of Table 7.4 [see also Thomas *et al.*, 1971, Fig. 1b]. One sees that the value of P_N at each height does not differ very greatly from that of P_2 except in the upper photosphere, where the Väisälä period is significantly larger. However, owing to the neglect of radiative damping up to this point in our discussion, it turns out that neither P_2 nor P_N gives a realistic estimate of the cut-off period for gravity waves in the upper photosphere, as we shall now show.

(3) *Critical periods in the presence of radiative damping.* We must next consider how the foregoing conclusions are changed when we drop the assumption of adiabatic propagation and take the effect of radiative damping into account (cf. Section 6.3.4). Comparison of the values of the radiative relaxation time t_R in Table 7.3 with those of the cut-off period P_2 in Table 7.4 shows at once that gravity waves will be strongly damped in the upper photosphere and near the base of the chromosphere. The only significant effect on acoustic waves will also occur at these low heights, but will evidently decrease with the value of the wave period (see also Section 7.4.2). Quantitative studies of the spatial or temporal damping of modified acoustic waves by radiation have been made by Noyes and Leighton [1963], Stix [1970], and Souffrin [1972].

According to the approximate theory advanced by Souffrin [1966] and outlined in Section 6.3.4, in the presence of radiative damping the critical periods P_1 and P_2 are replaced by the periods $P_3 = 2\pi/\omega_3$ and $P_4 = 2\pi/\omega_4$ respectively, where the corresponding frequencies ω_3 and ω_4 are defined by equations (6.117) and (6.118). The computed values of P_3 and P_4 from

[5] More precisely, the periods of gravity waves capable of propagating energy are limited to the range

$$P_2 |\cos\psi|^{-1} > P > P_2,$$

where ψ is the angle the group velocity vector makes with the vertical (see Section 6.3.3: equation 6.79).

$h = -200$ km up to $h = 1000$ km are given in the fourth and fifth columns of Table 7.4 and are also plotted in Fig. 7.2. Note that the minimum value of P_3 (195 sec) differs inappreciably from that of P_1 (193 sec) and hence our conclusion regarding the propagation of acoustic-type waves remains valid. The situation regarding gravity-type waves, however, is very different: no gravity

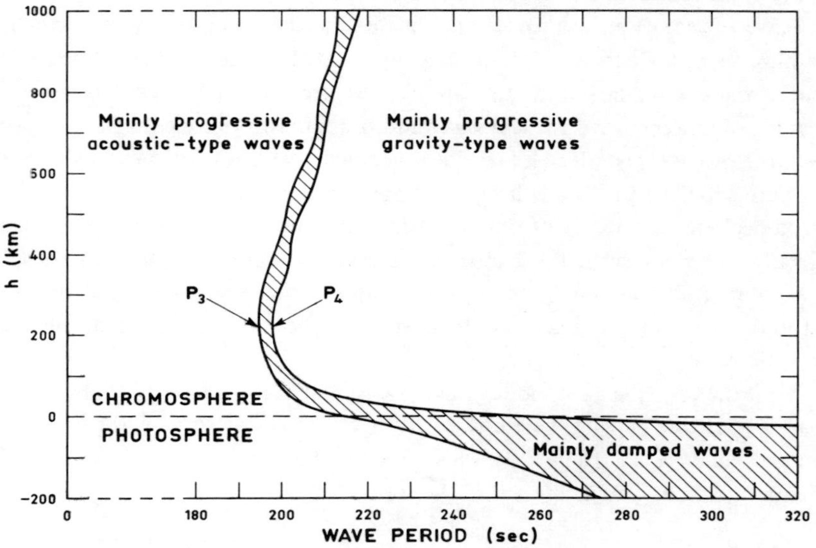

Fig. 7.2. Critical periods P_3 and P_4 for wave propagation in the upper photosphere and low chromosphere, with radiative damping taken into account. For a detailed explanation, see text.

wave can propagate in the upper photosphere [see also Frazier, 1968] and only waves generated *above* this region with periods exceeding 257 sec can propagate from the base of the chromosphere into the atmosphere above. Moreover, waves with periods in the range $P_3 < P < P_4$ are no longer purely stationary but are modified in such a way that they can now carry enough energy to balance the radiative dissipation (see Section 6.3.4).

Armed with all the foregoing information, we can now attempt to answer the crucial question of which particular type of wave, acoustic or gravity, is responsible for carrying upwards from the photosphere the energy needed to heat the chromosphere and corona. The short answer is that the balance of the available evidence seems to lean heavily on the side of *acoustic* waves. Basically the reason for this conclusion is that, apart from a certain amount of radiative damping in the lowest layers, there is no obstacle to the upward propagation of acoustic waves, provided their periods are less than ~195 sec. Gravity waves, on the other hand, cannot propagate at all in the upper photosphere and, owing to their greater period ($\gtrsim$210 sec), are more heavily damped in the very low chromosphere than acoustic waves. A priori, therefore, gravity waves cannot be regarded as a promising means of transmitting energy upwards from the photosphere into the overlying chromosphere. This view accords with the conclusion reached by Kuperus [1969].

(4) *Wave propagation in magnetic regions.* Finally, we must comment briefly on wave propagation in magnetic regions of the quiet chromosphere, which, as we have seen in Chapter 3, are principally located in the neighbourhood of the supergranule boundaries. In the absence of gravity, three types of wave are possible (Section 6.4.2): Alfvén waves and fast- and slow-mode magneto-acoustic waves. However, Osterbrock [1961] has shown that both Alfvén and slow-mode magneto-acoustic waves are heavily damped in the upper photosphere and low chromosphere as a result of ohmic dissipation unless the field strength exceeds 50 gauss or so. Since the fields in the supergranule boundaries generally lie below this value (Section 3.4.4), it follows that in magnetic regions of the quiet chromosphere we are likely to be concerned only with fast-mode magneto-acoustic waves.

Nevertheless, the fast-mode waves will still be mainly hydrodynamic in character at low heights, where the Alfvén velocity A is small compared to the sound velocity a (cf. Table 7.3). This can easily be seen by writing the dispersion equation (6.129) for such waves in the form

$$\omega^2 = \frac{1}{2}a^2k^2\left[\left(1+\frac{A^2}{a^2}\right)+\sqrt{\left(1+\frac{A^2}{a^2}\right)^2-4\frac{A^2}{a^2}\cos^2\theta}\,\right] \tag{7.16}$$

which for $A \ll a$, approximates closely to the familiar relation

$$\omega^2 = a^2k^2$$

valid for purely acoustic waves (Section 6.3.2), regardless of the value of θ, the angle between the wave propagation vector and the magnetic field. On the other hand, at a certain height in the chromosphere A will become comparable to a and the magnetic character of the waves can no longer be ignored. Before this height is reached, however, there are good reasons for concluding that the fast-mode waves believed to be responsible for heating the chromosphere above supergranule boundaries will already have developed into shocks; this topic is discussed in Section 7.4.8.

The whole problem becomes much more complicated when proper account is taken of both gravity and magnetic field (Section 6.4.3). In this case both Alfvén and magneto-gravity waves should be considered, but the possible role played by the latter in heating the chromosphere remains largely unexplored. In general one must expect that such waves will eventually steepen into shocks owing to the decrease in density as they propagate upwards. However, McLellan and Winterberg [1968] have pointed out that growth of the wave amplitude will not occur for certain ranges of the wave number and for certain configurations of the wave propagation, gravitational, and magnetic field vectors: under these particular circumstances shock waves will not eventuate.

7.3.3 SOME REMARKS ON THE MECHANISM OF WAVE GENERATION IN THE PHOTOSPHERE

The arguments presented in the previous section indicate that the waves ultimately responsible for heating the chromosphere are very probably of the acoustic type. Moreover, as we have remarked in Section 7.3.1, the waves are presumably generated by motions in the upper part of the solar convection zone or by convection currents penetrating into the stable layer immediately above. As emphasized by Frazier [1968], the actual state of motion in the upper photosphere must be very complicated involving, amongst other things, detailed interaction between convective and oscillatory modes. Nevertheless, *given a knowledge of the exact mechanism of wave generation* one could, in principle, calculate the upward flux of acoustic energy and compare the result with the available estimates of the total net outward flux of energy from the chromosphere and corona (cf. Table 7.2). A number of authors have attempted to carry out such a procedure, using the theory of acoustic noise generation in a field of homogeneous, isotropic turbulence originally developed by M. J. Lighthill in connection with the problem of jet aircraft noise. Their results purport to show that, in particular, almost all of the acoustic noise is generated in a thin layer less than 100 km thick located at the very top of the convection zone [see, for example, Kuperus, 1969, Fig. 2]. Other authors have considered the effect on the rate of noise production of factors such as the stratification of the atmosphere under gravity and the presence of a magnetic field.

However, as the present authors have pointed out on a number of occasions, there is absolutely no evidence to substantiate the original underlying assumption that motions at the top of the convection zone constitute a regime of fully developed turbulence in the strict sense normally used by aerodynamicists. Accordingly, quantitative estimates of the upward acoustic energy flux based on such a shaky foundation fail to inspire much confidence. Instead, as we shall see in Section 7.4.5, a more objective procedure is to treat the incident flux of wave energy as a parameter in the theory of shock wave heating; initially unknown, its value is determined by comparing computed curves of the mechanical dissipation as a function of height with the known variation of the net radiative energy loss.

Unlike acoustic waves, gravity waves cannot be generated in the solar convection zone (Section 6.3.2). Nor, as we have seen in the previous section, are they able to propagate in the stable part of the photosphere lying immediately above the convection zone, owing to the overwhelming effect of radiative damping at these heights. On the other hand, examination of the values given in Table 7.4 of the critical period P_4 for waves of this type shows that progressive gravity waves can be excited in the low chromosphere from the level $h = 0$ upwards. We now turn to the possible identification of such waves with the observed 5-minute oscillations described in Chapter 3.

7.3.4 INTERPRETATION OF THE OBSERVED 5-MINUTE VELOCITY OSCILLATIONS

In Section 3.5.1 we have summarized the modern observations of the (predominantly) vertical oscillations, having average periods of about 5 min, which are found to occur in the upper photosphere and low chromosphere. The obvious question now arises of whether these oscillations are to be interpreted as direct evidence for the existence of the waves ultimately responsible for heating the chromosphere and corona. Examination of Fig. 7.2 shows at once that in the upper photosphere oscillations with periods of about 300 sec will have the character of 'mainly damped' waves in the sense defined in Section 6.3.4. It is true that a certain amount of energy is carried upwards by these waves; however, this is used to balance the loss resulting from radiative cooling. In the low chromosphere, on the other hand, the oscillations have the character of 'mainly progressive' gravity-type waves. This conclusion accords with the observational evidence obtained from spectral lines of widely different strengths which, although meagre, suggests that the character of the 5-minute oscillations changes from that of standing waves in the upper photosphere to that of progressive waves in the low chromosphere.

The magnitude of the radiative relaxation time t_R increases rapidly with height so that, for example, when the level $h = 300$ km is reached t_R has become significantly larger than the typical oscillation period of 300 sec (cf. Table 7.3). Consequently, at these heights we can expect the properties of any gravity-type waves to approximate to those of purely adiabatic gravity waves, discussed in Sections 6.3.2 and 6.3.3. However, as we have already remarked in Section 7.3.2, it is unlikely that waves of this kind are responsible for the heating of the higher levels.

In summary, one has to conclude that the observed 5-minute oscillations in the upper photosphere and low chromosphere are *not* manifestations of the wave motions ultimately responsible for heating the chromosphere and corona (see also Section 7.4.6).[6]

How then are the observed 5-minute oscillations to be interpreted? It has

[6] The opposite view has been expressed by de Jager [1971].

been suggested by Lighthill [1967, p. 446 et seq.] that the generation of gravity waves might constitute a very efficient process for damping the convection currents which penetrate beyond the top of the solar convection zone into the stable layer above.[7] With this in mind one is tempted to speculate that the rising convective columns expend some of their kinetic energy in exciting standing oscillations of some kind in the upper photosphere and that these, in turn, generate gravity waves above the level $h \simeq 0$.

On the other hand, there is clear observational evidence that there is no simple one-to-one correlation between the ascent of a single photospheric granule and the onset of an oscillation in the overlying atmosphere. Direct observational proof of this fact has been obtained by Frazier [1968], and there are other important pieces of information pointing to the same conclusion. For example, as we have pointed out in Section 3.5.1, a typical 'burst' of 5-minute oscillations lasts some 30–45 min, whereas the average lifetime of a granule is only about 10 min. Furthermore, the oscillating elements are much larger than the granules, most of which have apparent diameters in the range 1–2″ of arc (725–1450 km). By contrast, the phase of the velocity oscillations can be sensibly constant over regions extending up to some tens of thousand kilometres in diameter, although the area over which the amplitude is coherent is considerably smaller. It seems reasonable, therefore, to picture the observed oscillations as the response of the overlying stable region to the *statistically-averaged* effect, both in space and time, of many randomly-occurring granule impacts.

Attempts to calculate the response of the solar atmosphere to such excitation have been made by a number of workers, including Schmidt and Zirker [1963], Meyer and Schmidt [1967], and Stix [1970]. Evidently, a complete explanation of the origin of the 5-minute oscillations would require (*a*) a detailed understanding of the mechanism whereby the observed motions are generated by the granules; (*b*) a full knowledge of the resonance and propagation properties of the atmosphere; and (*c*) an accurate treatment of the effect of radiative damping under the conditions prevailing in the upper photosphere and low chromosphere. With respect to (*c*), the theory outlined in Section 6.3.4 can only be regarded as a rather crude treatment of what is, after all, a complex problem in radiative transfer theory [cf. Ulrich, 1970; Thomas *et al.*, 1971; Ulmschneider, 1971b]. The present theory contains, amongst other limitations, the dubious assumption of LTE at all heights concerned and neglects, in solving the wave equations, the rapid variation of t_R with height. (Strictly speaking, the use of the formula 6.88 for this quantity is valid only for a homogeneous atmosphere, as pointed out in Section 6.3.4.) It would not be surprising if future more exact treatments were

[7] Problems of this kind are also familiar in the field of fluid dynamics [see, for example, Warren, 1960; Townsend, 1966; Deardorff, 1969].

to yield somewhat different numerical values for the critical periods P_3 and P_4 given in Table 7.4 and Fig. 7.2.

Besides the papers cited above, there is an extensive literature dealing with various aspects of the problem of interpreting the 5-minute oscillations. For further discussion of this topic the interested reader is referred to papers by Bahng and Schwarzschild [1963], Noyes and Leighton [1963], Moore and Spiegel [1964], Uchida [1965, 1967], Ulmschneider [1968], Jones [1969], Stein and Leibacher [1969], Leibacher and Stein [1971], and McKenzie [1971].

7.4 Heating of the Chromosphere by Shock Waves

7.4.1 INTRODUCTION

As we have seen in the previous section, there is good reason to believe that the waves ultimately responsible for carrying the energy needed to heat the chromosphere upwards from the photosphere are modified *acoustic* waves whose periods, moreover, must lie below a certain theoretical upper limit. The exact value of the cut-off period is still subject to some uncertainty, but in Section 7.3.2 it has been estimated to be about 195 sec. Taking the effect of radiative damping and other factors into account, we investigate in Section 7.4.2 the transformation of these waves into shocks; we conclude that for waves with periods of the order of 100 sec or less the process should be completed within the first 1000 km of the base of the chromosphere.

We next turn our attention to the fundamental problems of estimating (*a*) the initial strengths of the shocks, and (*b*) their rate of dissipation as they continue to travel upwards through the chromospheric layers. To date, most attempts to solve these problems have been based on a modified form of the Brinkley–Kirkwood theory of shock propagation first formulated by E. Schatzman in 1949. An up-dated account of this theory in a form suitable for our purposes is given in Section 7.4.3, while the question of its overall validity is briefly discussed in Section 7.4.4. Using values of the net radiative energy loss as a function of height given in Section 7.2.5, we find in Section 7.4.5 that the shocks responsible for heating the low chromosphere are *weak* shocks with periods which are probably of the order of some tens of seconds. The short-period acoustic waves presumed to give rise to such shocks have so far escaped detection for reasons which are briefly considered in Section 7.4.6.

Present knowledge of the shock wave heating of the higher levels of the chromosphere rests on less secure foundations, partly because available estimates of the net radiative energy loss in the middle and upper chromosphere may well be seriously in error and partly because current chromospheric models at these heights are themselves subject to considerable uncertainty. Without an accurate model it is not possible to properly evaluate the *conductive* flux, which in these regions can no longer be ignored. Existing work on the heating of the upper chromosphere is briefly summarized in Section 7.4.7.

Finally, in Section 7.4.8 we turn our attention to the question of the heating of the low chromosphere in regions coinciding with the chromospheric network. These regions are pervaded by magnetic fields whose strengths – measured at the photospheric level – range from 10 to about 50 gauss, or perhaps somewhat higher. The effect of the field is to reduce the height up to which shock heating is operative, as a consequence of 'shock refraction'. At the same time, however, there is enhanced dissipation at lower levels and this offers an explanation for the increased heating implied by the presence of the network.

7.4.2 FORMATION OF SHOCKS IN THE LOW CHROMOSPHERE

From the results presented in Section 6.3.2 it can easily be shown that the velocity amplitude of a modified acoustic wave travelling vertically upwards through an isothermal atmosphere increases with height h as $\exp(h/2\mathcal{H})$, where $\mathcal{H}$ is the density scale height. Calculating on the basis of the BCA model given in Table 4.6, one finds that the value of $\mathcal{H}$ within the height range 0–1000 km is about 110–120 km.[8] Thus, in the absence of dissipation, the velocity amplitude of a wave entering the chromosphere with an initial amplitude of only, say, 0.2 km sec^{-1} (cf. Section 3.5.1) would nevertheless grow to a value comparable to the local sound velocity (~7.5 km sec^{-1} according to Table 7.3) over a height of about 900 km. A priori, therefore, it follows that waves entering the chromosphere from below will quickly degenerate into shocks as they proceed upwards.

However, the above argument is not adequate to determine the actual height of shock formation, which is influenced by a number of factors including (*a*) the height of origin of the wave; (*b*) its initial amplitude; (*c*) its period; and (*d*) its damping by dissipative processes. There is no need to elaborate on the effects of the first two factors, which are self-evident. The influence of period can readily be illustrated by considering the case of pure compression waves, for which it can be shown that, for a given initial flux of wave energy, the (longitudinal) velocity perturbation v is independent of period. According to the results presented in Section 6.5.1, the wave profiles will therefore distort over a given time interval by an amount determined, irrespective of period, by the formula

$$u = a + \frac{1}{2}(\gamma + 1)v, \tag{7.17}$$

[8] For a non-isothermal atmosphere the *local* value of the density scale height may be computed from the formula

$$\frac{1}{\mathcal{H}} = -\frac{d(\ln \rho)}{dh}$$

[see, for example, Bray and Loughhead, 1967, p. 72].

where u is the velocity of a point in the profile and a is the (local) velocity of sound. The effect of such an equal distortion on two initially sinusoidal waves of different period is depicted in Fig. 7.3, which is taken from Ulmschneider [1971a]. The profile of the short-period (and hence short-wavelength) wave evidently distorts much more rapidly than the long-period wave and consequently develops into a shock wave in a much smaller vertical distance.[9] This result has led Ulmschneider to suggest that the lower levels of the chromosphere are heated by shocks formed from short-period waves, whereas the higher layers and the overlying corona are heated by shocks originating from waves of longer period (cf. Sections 7.4.5 and 7.4.7).

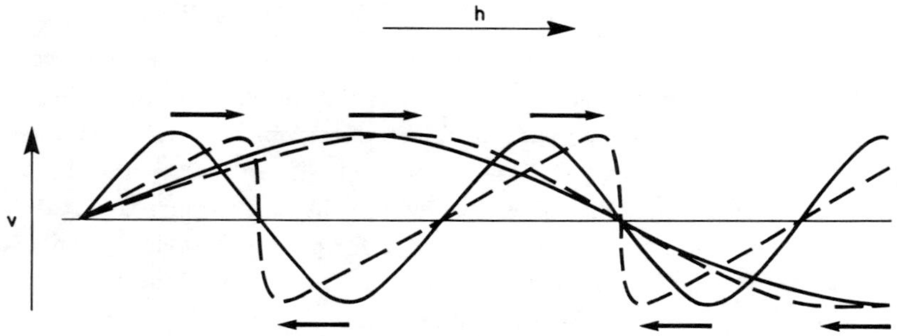

Fig. 7.3. Distortions due to non-linear effects of the profiles of advancing compression waves which start off with equal velocity amplitudes but have different wavelengths (and hence periods) [Ulmschneider, 1971a]. Note that over a given time interval the relative distortion of the short-period (i.e. short-wavelength) wave is much greater than that of the long-period wave.

The final factor to be taken into consideration in discussing the transformation of acoustic waves into shocks is damping arising from viscous dissipation, particle conduction, and radiative transfer. It can be shown that damping due to the first two mechanisms is quite negligible, at least up to a height of about 1000 km [Biermann, 1948; Schirmer, 1950; Osterbrock, 1961; Jordan, 1969]. On the other hand, the loss of wave energy due to radiative damping is by no means negligible in the upper photosphere, although it decreases as the wave moves up into the low chromosphere. Above optical depth $\tau_{5000} = 1$ the effect of radiative damping is determined by the magnitude of the radiative relaxation time

$$t_R = \frac{\rho c_V}{16\bar{\kappa}\sigma_R T^3} \tag{6.88}$$

[9] For a *uniform* gas the distance of shock formation is directly proportional to the period [Bird, 1964]. Detailed calculations by Bird show that in the absence of dissipation this result remains true for a gravitationally-stratified atmosphere having a vertical temperature gradient comparable to that of the chromosphere.

compared to the period of the wave (see Section 6.3.4). Values for t_R in the upper photosphere and low chromosphere computed by Ulmschneider [1971b] on the basis of the Harvard-Smithsonian Reference Atmosphere (Section 4.2.5) are given in Table 7.3. These values demonstrate that only waves with a period of about one second or less can propagate from the upper photosphere into the low chromosphere without suffering radiation damping. On the other hand, the waves currently believed to be responsible for the heating of the chromosphere have periods of the order of tens of seconds, as we shall see in Section 7.4.5.

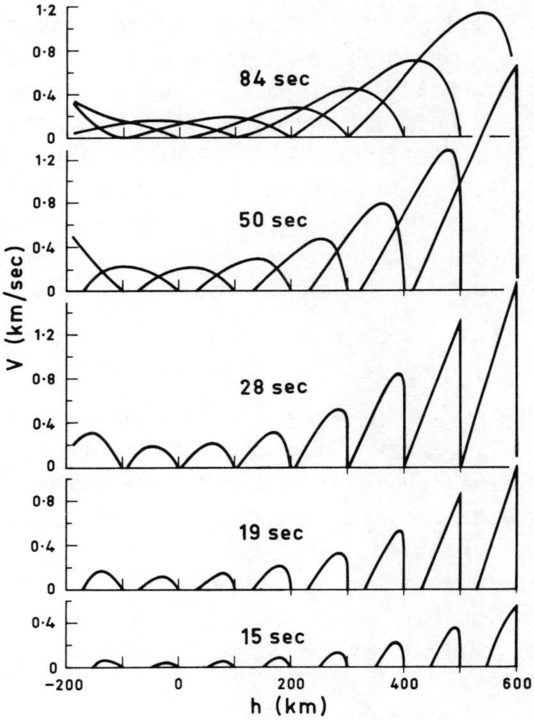

Fig. 7.4. Growth of the velocity amplitudes of acoustic waves propagating from the upper photosphere into the low chromosphere, with radiative damping taken into account [Ulmschneider, 1971b]. The profiles of the advancing waves are plotted at 100-km intervals. They show heavy damping at heights below about 200 km, but thereafter grow rapidly with height and develop into shocks. Figures above the profiles represent the assumed periods.

Taking radiation damping into account Ulmschneider [1971b] has computed the growth of the velocity amplitudes of acoustic waves for a wide range of periods and initial amplitudes. Some typical results of his calculations are shown in Fig. 7.4, where the profiles of the advancing waves are plotted at 100-km intervals. The individual waves are all assumed to originate at an arbitrarily chosen 'critical' height of $h_c = -157$ km, below which the atmosphere is taken to be optically thick. The heavy damping of the waves at low heights ($h \lesssim 200$ km) due to radiation losses is clearly evident in Fig. 7.4 [see also Stein, 1969]. At greater heights, however, the losses are very much reduced, thus allowing the waves to gain rapidly in amplitude and finally to develop into shocks. In fact, Ulmschneider's published graphs (see his Figs. 4 and 5) imply that acoustic waves with periods of about 100 sec or less propagating through the upper photosphere into the chromosphere will be transformed into shocks within the first 1000 km or less of the chromosphere. (In any particular case, of course, the exact height will depend on the assumed height of origin and initial amplitude of the wave.) The same general conclusion has been reached by a number of authors, including Dubov [1960], de Jager and Kuperus [1961], Osterbrock [1961], Bird [1964], Saito [1964], Kuperus [1965, 1969], and Schatzman [1967]. It is significant that the height range 500–1000 km defines the region where, according to the BCA model, the temperature first begins to rise above the value at the chromospheric temperature minimum (see Fig. 4.3), presumably owing to the onset of shock wave heating at these heights.

7.4.3 PROPAGATION OF SHOCKS THROUGH THE LOW CHROMOSPHERE: SCHATZMAN'S THEORY

In the preceding section we have outlined the physical processes by which acoustic waves with periods of the order of 100 sec or less originating in the photosphere give rise to shock waves within the lowest 1000 km of the overlying chromosphere. We turn now to the fundamental problems of estimating (*a*) the initial strengths of the shocks, and (*b*) their rate of dissipation as they continue to travel upwards through the chromospheric layers. To date most attempts to solve these problems have been based on a modified form of the Brinkley–Kirkwood theory of shock wave propagation (Section 6.5.3) first formulated by Schatzman [1949].

When we focus our attention on the real chromosphere we have to bear in mind that we are concerned not with isolated acoustic waves and their transformation into individual shocks, but rather with *trains* of upward-propagating waves. Each train will give rise to a succession of shocks, whose frequency ν (defined as the number passing a given point in unit time) is given by the relation

$$\nu = \frac{1}{P}, \qquad (7.18)$$

where P is the period of the original wave. For mathematical simplicity it is customary to assume that these waves can be characterized by a single unique frequency but, as emphasized by Ulmschneider [1971b], this assumption is not physically realistic. If we were to imagine ourselves observing over a sufficiently large area or during a sufficiently long time interval, then we might expect to see the emergence of trains of acoustic waves having a marked dispersion in periods (and, incidentally, in initial amplitudes). Despite this difficulty, it is clear that some estimate of the value of ν for strictly monochromatic waves or, more generally, of the spectrum of ν values, is mandatory for any calculation of the heating effect of the shocks.

The starting point of most discussions is the expression

$$D(h) = \int_0^\infty (p - p_1) v \mathrm{d}t \tag{7.19}$$

for the energy of a plane shock wave per unit area of the front derived in Section 6.5.3, the wave being here supposed to propagate vertically upwards and to reach height h at time $t = 0$. Retaining as far as possible the notation of Chapter 6, we let ρ_2, p_2, and v_2 denote the peak values of the density, pressure, and gas velocity immediately behind the shock front and ρ_1, p_1, and 0 the equilibrium values of these quantities in the undisturbed fluid. Anticipating later procedures, we shall rewrite (7.19) in the reduced form

$$D(h) = (p_2 - p_1) v_2 t_0, \tag{7.20}$$

where

$$t_0 = \int_0^\infty \frac{p - p_1}{p_2 - p_1} \frac{v}{v_2} \mathrm{d}t. \tag{7.21}$$

In accordance with the similarity restraint embodied in the Brinkley–Kirkwood theory, the quantity t_0 is taken to be independent of height and, in fact, may be regarded as a characteristic time interval measuring the duration of each individual shock.

As an example, let us consider the simple case of a sawtooth shock of duration τ (assumed $\leqslant P$) specified by the relations

$$\left.\begin{aligned} p &= p_2 - \frac{p_2 - p_1}{\tau} t \\ v &= v_2 - \frac{v_2}{\tau} t \end{aligned}\right\} \quad \text{for } t \leqslant \tau$$

and

$$\left.\begin{aligned} p &= p_1 \\ v &= 0 \end{aligned}\right\} \quad \text{for } \tau < t \leqslant P$$

(cf. Fig. 7.5). Then we have

$$t_0 = \int_0^\tau \left(1 - \frac{t}{\tau}\right)^2 \mathrm{d}t = \frac{1}{3}\tau \tag{7.22}$$

and, for the particular case $\tau = P$,

$$t_0 = \frac{1}{3}P = \frac{1}{3\nu}. \tag{7.23}$$

Since the magnitude of the numerical factor on the right-hand side of equation (7.23) depends on the particular assumptions underlying its derivation, it is not surprising that the adopted value varies from author to author. Thus it is taken as 1/8 by Osterbrock [1961] and 1/12 by Ulmschneider [1970, 1971a].

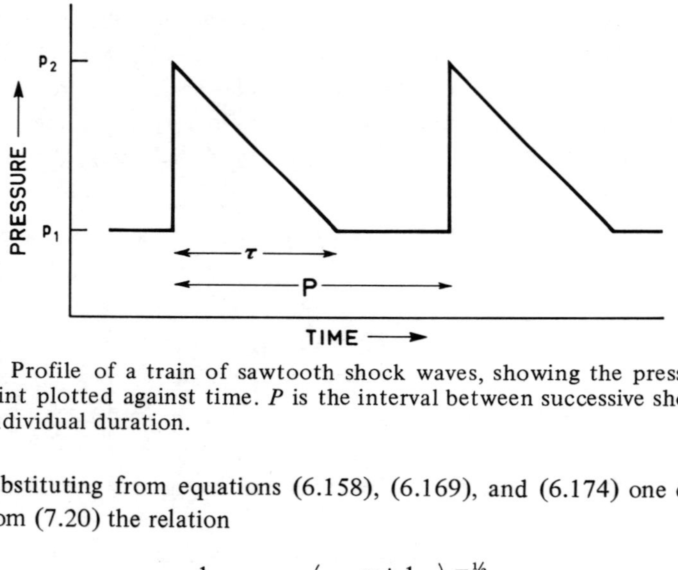

Fig. 7.5. Profile of a train of sawtooth shock waves, showing the pressure at a fixed point plotted against time. P is the interval between successive shocks and τ their individual duration.

By substituting from equations (6.158), (6.169), and (6.174) one can now derive from (7.20) the relation

$$D(h) = \frac{1}{\gamma^2}\rho_1 a_1^3 \delta^2 \left(1 + \frac{\gamma + 1}{2\gamma}\delta\right)^{-\frac{1}{2}} t_0 \tag{7.24}$$

expressing the energy of a shock in terms of the shock strength

$$\delta = (p_2 - p_1)/p_1 \tag{7.25}$$

introduced in Section 6.5.2. As we shall see below (Section 7.4.5), the shocks currently believed to be responsible for the heating of the low chromosphere are *weak* shocks (i.e. $\delta < 1$); in this case (7.24) becomes simply

$$D(h) = \frac{1}{\gamma^2}\rho_1 a_1^3 \delta^2 t_0. \tag{7.26}$$

From this it follows at once that the total flux of energy at height h due to the passage of a train of weak shocks of strength δ and frequency ν is

$$F_{\mathrm{mech}} \equiv \nu D(h) = \nu \frac{1}{\gamma^2} \rho_1 a_1^3 \delta^2 t_0. \tag{7.27}$$

It should be noted that some authors prefer to measure the strength of a shock in terms not of δ but of the fractional compression

$$\eta = (\rho_2 - \rho_1)/\rho_1. \tag{7.28}$$

In the case of weak shocks, one readily finds from the shock condition (6.174) that

$$\eta = \delta/\gamma, \tag{7.29}$$

hence enabling us to rewrite the formula (7.27) in the form

$$F_{\mathrm{mech}} = \nu \rho_1 a_1^3 \eta^2 t_0. \tag{7.30}$$

Over the range of heights for which the condition $\delta < 1$ is satisfied, the Brinkley–Kirkwood theory yields an expression for the height variation of the energy of an individual shock wave in the form

$$\frac{\mathrm{d}D}{\mathrm{d}h} = -\frac{1}{12} \gamma(\gamma + 1) p_1 \eta^3, \tag{7.31}$$

whence it follows that the total energy flux carried by a train of shocks of frequency ν obeys the relation

$$\frac{\mathrm{d}F_{\mathrm{mech}}}{\mathrm{d}h} = -\frac{1}{12} \gamma(\gamma + 1) p_1 \eta^3 \nu. \tag{7.32}$$

The quantity $(-\mathrm{d}F_{\mathrm{mech}}/\mathrm{d}h)$ represents the amount of shock energy dissipated per unit volume and time (cf. footnote 1) and, in the lower portions of the chromosphere – where particle conduction can be neglected – can be equated to the net radiative energy loss $\mathrm{d}F_{\mathrm{rad}}/\mathrm{d}h$ evaluated in Section 7.2. In principle, therefore, one can use (7.32) to determine the shock strength η as a function of height provided, of course, some independent estimate of ν is available. Actual values of η obtained in this way are given in Section 7.4.5. Note also that since $\nu = 1/P$ the shock dissipation decreases as the period of the waves increases provided, of course, that η is kept constant. This is consistent with the hypothesis, mentioned in Section 7.4.2, that the lower levels of the chromosphere are heated by short-period shocks whereas the higher layers are heated by shocks of longer period.

It should be pointed out, however, that as it stands (7.32) takes no account of the decrease with height of the energy flux F_{mech} arising from *refraction* of the shock waves away from the vertical direction. This is caused by the increase of temperature, and therefore of sound velocity, with height in the chromosphere. Arguing on the basis of an analogy with geometrical optics originally proposed by Schatzman [1949], Ulmschneider [1970, 1971a] has concluded that the energy loss due to refraction necessitates the addition of the term

$$-\frac{1}{a_1^2}\frac{da_1^2}{dh}F_{mech}$$

to the right-hand side of (7.32) [see also Osterbrock, 1961]. Thus when both energy dissipation and refraction are taken into account, the change in the flux of the shock wave energy with height obeys the equation

$$\frac{dF_{mech}}{dh} = -\frac{1}{a_1^2}\frac{da_1^2}{dh}F_{mech} - \frac{1}{12}\gamma(\gamma+1)p_1\eta^3\nu. \tag{7.33}$$

The use of the geometrical optics approximation in the derivation of (7.33) is strictly valid only if the 'wavelength' of the shocks is small compared to the local values of the density and pressure scale heights [cf. Osterbrock, 1961; Pikelner and Livshitz, 1964; Kuperus, 1969]. In practice the effect of refraction is comparatively small in the low chromosphere owing to the relatively slow increase of temperature with height [cf. Ulmschneider, 1970, Fig. 2; 1971a, Fig. 3].

Substituting for F_{mech} from (7.30) we obtain after some algebra the equation

$$\frac{d\eta}{dh} = -\frac{\eta}{2}\left[\frac{1}{\gamma}\frac{d\gamma}{dh} + \frac{1}{p_1}\frac{dp_1}{dh} + \frac{3}{2a_1^2}\frac{da_1^2}{dh} + \frac{1}{12t_0}\frac{(\gamma+1)\eta}{a_1}\right] \tag{7.34}$$

governing the variation of the shock strength η with height.[10] In the context of the chromosphere, the first term on the right-hand side determines the change in η resulting from the variation of γ with height; the second specifies the growth of η due to the change of pressure with height; the third specifies the decrease in η arising from the increase in temperature with height; and the final term represents the effect of dissipation.

In the particular case of an atmosphere in hydrostatic equilibrium, we have

$$\frac{dp_1}{dh} = -\rho_1 g,$$

and (7.34) reduces to

$$\frac{d\eta}{dh} = -\frac{\eta}{2}\left[\frac{1}{\gamma}\frac{d\gamma}{dh} - \frac{\gamma g}{a_1^2} + \frac{3}{2a_1^2}\frac{da_1^2}{dh} + \frac{1}{12t_0}\frac{(\gamma+1)\eta}{a_1}\right]. \tag{7.35}$$

[10] Modified forms of (7.34) have been given by Schatzman [1949] and Osterbrock [1961] for the more complicated case of an isotropic field of shock waves propagating upwards at all angles to the vertical.

This equation agrees with an equation obtained by Ulmschneider [1970, 1971a] for shocks whose profile is such that $t_0 = 1/(12\nu)$.

Given an initial value of η one can integrate (7.34) or (7.35) numerically with the aid of an appropriate chromospheric model and so evaluate η as a function of height. The corresponding values for the amount of shock wave energy dissipated per unit volume, $(-\mathrm{d}F_{\mathrm{mech}}/\mathrm{d}h)$, are then obtained directly from (7.32). The results of actual calculations of this kind are presented in Section 7.4.5.

7.4.4 COMMENTS ON THE VALIDITY OF SCHATZMAN'S THEORY; MORE EXACT EQUATIONS

Despite its practical utility, equation (7.34) above, based on Schatzman's approach, unfortunately leads to results which differ from those derived from the rigorous form of the Brinkley–Kirkwood theory outlined in Section 6.5.3. This can be demonstrated by considering the simple case of a uniform atmosphere, for which (7.34) reduces to

$$\frac{\mathrm{d}\eta}{\mathrm{d}h} = -\frac{\gamma + 1}{24 a_1 t_0}\eta^2.$$

The solution of this equation takes the form

$$\eta = \frac{24 a_1 t_0}{\gamma + 1} \cdot \frac{1}{h - b},$$

where b is a constant, implying that the strength of a weak shock in a uniform atmosphere decreases asymptotically as h^{-1} at large distances. In contradiction to this result, the corresponding solution given by the Brinkley–Kirkwood theory (see equation 6.224) yields an asymptotic decrease following an $h^{-1/2}$ law!

Further light on the inadequacy of the Schatzman approach can be obtained by turning to the generalizations of the basic equations of the Brinkley–Kirkwood theory given by Saito [1964] for the case of plane shock waves of arbitrary strength propagating vertically upwards through an atmosphere in hydrostatic equilibrium under gravity. His results have already been reproduced in Section 6.5.3 as equations (6.225)–(6.228) which, in the limiting case of *weak* shocks, can be shown with the aid of Saito's equation (8) to reduce to the relations

$$\frac{\mathrm{d}\eta}{\mathrm{d}h} = -\frac{\eta}{2}\left[\frac{-\gamma g}{a_1^2} + \frac{1}{2a_1^2}\frac{\mathrm{d}a_1^2}{\mathrm{d}h} + \nu_0\frac{\gamma(\gamma+1)}{4} \cdot \frac{p_1 \eta^3}{D}\right] \tag{7.36}$$

and

$$\frac{\mathrm{d}D}{\mathrm{d}h} = -\frac{\gamma(\gamma+1)}{12} p_1 \eta^3 - \frac{g}{2a_1^2} \cdot \frac{D}{\eta}. \qquad (7.37)^{11}$$

[11] Equivalent limiting forms of Saito's equations have also been given by Schatzman [1967], except that he neglects the first term on the right-hand side of (7.37). However, in the case of a uniform atmosphere (i.e., one with $g = 0$), this evidently leads to the erroneous result, D = const.

In these equations no attention is paid to refraction, γ is taken to be constant with height, and ν_0 is a constant of order unity. It should be noted that equation (7.37) implies that the upward-propagating shock loses energy not only because of dissipation but also as a result of gravitational damping. As a consequence, shocks having a strength less than a certain limiting value will not be capable of traversing the chromosphere [Saito, 1964].

With the same assumptions the Schatzman approach yields the single equation (cf. equation 7.35 above)

$$\frac{d\eta}{dh} = -\frac{\eta}{2}\left[\frac{-\gamma g}{a_1^2} + \frac{1}{2a_1^2}\frac{da_1^2}{dh} + \frac{1}{12t_0}\frac{(\gamma+1)\eta}{a_1}\right]. \qquad (7.38)$$

Thus it gives results differing from Saito's theory in two respects: (*a*) the outright omission of (7.37), and (*b*) the substitution of the approximate formula (7.26) for D in the last term on the right-hand side of (7.36). The magnitude of the errors resulting from these approximations over a height range comparable to that encountered in the low chromosphere is unknown but clearly merits detailed examination.

7.4.5 DISSIPATION OF SHOCK WAVE ENERGY IN THE LOW CHROMOSPHERE

We have seen in Section 7.4.3 that the strength η of the shocks in a train of frequency ν can be estimated from equation (7.32) by equating the magnitude of the shock wave energy dissipation to the net radiative energy loss dF_{rad}/dh evaluated in Section 7.2. For this purpose it is convenient to rewrite (7.32) in the form

$$\eta = \left[\frac{12P}{\gamma(\gamma+1)p_1} \cdot \frac{dF_{rad}}{dh}\right]^{1/3}, \qquad (7.39)$$

where P (= $1/\nu$) is the assumed period of the waves. At the time of writing the best available computation of the net radiative energy loss as a function of height is probably that of Ulmschneider [1970]. His values of dF_{rad}/dh, calculated on the basis of the BCA model of the low chromosphere (Table 4.6), cover the height range from $h \simeq 200$ km to $h = 1300$ km and are plotted in Fig. 7.1. Using these results we have presented in Table 7.5 values of η at four selected heights for values of P ranging from 10 to 200 sec. In all cases γ was assigned the value 5/3 appropriate at these heights (Section 7.3.2).

Examination of Table 7.5 shows that, for any given period, the shock strength is almost constant with height. On the other hand, the value of η at any given height gradually increases with period. Nevertheless, in all cases η satisfies the inequality $\eta < 1$, thus justifying the assertion in Section 7.4.3 that the shocks responsible for heating the low chromosphere are *weak* shocks. Of course, the validity of this conclusion depends on the reliability of the adopted values of the gas pressure p_1 and of the estimates of dF_{rad}/dh or, strictly speaking, on the ratio $(dF_{rad}/dh)/p_1$. Both these quantities are influenced in turn by the particular choice of chromospheric model. However, thanks to the presence of a cube root sign on the right-hand side of (7.39), η is fairly insensitive to changes in the two parameters. In fact, in the case of waves with periods of 25 sec or less, the value of the quantity $(dF_{rad}/dh)/p_1$ has to increase by some two orders of magnitude before the condition $\eta < 1$ is violated; the factor is smaller for waves of longer period. The conclusion that the shocks responsible for the heating of the low chromosphere are weak in the sense that $\eta < 1$ has either been inferred or assumed by a number of writers on the subject, including Schatzman [1949], Osterbrock [1961], and Ulmschneider [1970, 1971a].

TABLE 7.5

Values of Shock Strength η in the Low Chromosphere

PERIOD (sec)	HEIGHT (km) 400	600	800	1000
10	0.14	0.16	0.16	0.15
25	0.19	0.22	0.22	0.21
50	0.24	0.27	0.28	0.26
75	0.27	0.31	0.32	0.30
100	0.30	0.34	0.35	0.33
200	0.38	0.43	0.44	0.41

Given an initial value of η at some specified height in the low chromosphere one can numerically integrate equation (7.34), or some variant thereof, on the basis of an adopted chromospheric model and so evaluate η as a function of height.[12] The corresponding values for the shock wave energy dissipation, $(-dF_{mech}/dh)$, may then be obtained directly via (7.32) provided, of course, the shock frequency ν is known or can be assumed. The most recent calculations of this kind are those of Ulmschneider [1970, 1971a], who considered a number of chromospheric models.

In his first paper Ulmschneider adopted the procedure of treating the wave period P and the initial flux of shock wave energy $F_{mech}(0)$ as free parameters and, for each pair of chosen values, computed the shock dissipation $(-dF_{mech}/dh)$ as a function of height. (The corresponding initial values of η

[12] As an alternative to numerical integration the equation can be transformed into an explicit expression for η [see Pikelner and Livshitz, 1964; Jordan, 1970].

were calculated via equation 7.30 on the assumption that $t_0 = 1/(12\nu)$.) Fig. 7.6 gives the shock dissipation over the height range 200–900 km obtained by Ulmschneider on the basis of the BCA model and, for comparison, a plot of the net radiative energy loss as a function of height calculated on the basis of the same model (cf. Fig. 7.1). An examination of Fig. 7.6 leads to the important

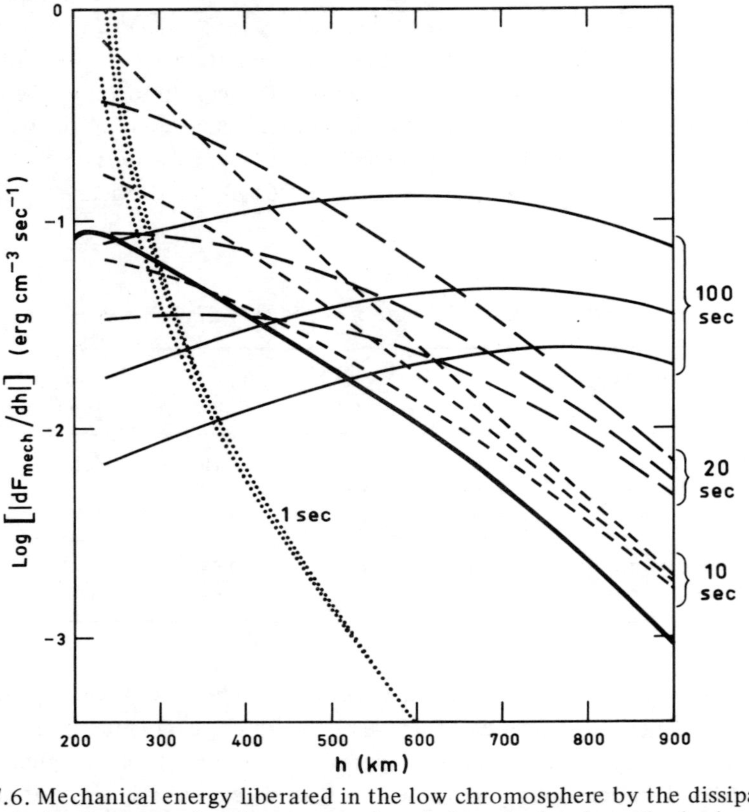

Fig. 7.6. Mechanical energy liberated in the low chromosphere by the dissipation of weak shock waves of various assumed periods, calculated on the basis of the BCA model [Ulmschneider, 1970]. The three curves for each period correspond to different assumed values of the initial mechanical energy flux. For comparison, the thick curve shows the net radiative energy loss over the same height range taken from Fig. 7.1.

conclusion that, regardless of the assumed initial value of the shock wave energy flux, the curves of shock dissipation and net radiative energy loss can be brought into reasonable coincidence only if the wave period is close to 10 sec. Moreover, Ulmschneider found that this conclusion remained valid when the calculations were repeated for four alternative chromospheric models.

However, in the second paper mentioned above Ulmschneider obtained somewhat higher estimates for the shock wave period, namely 25 and 30 sec; for these calculations he used models proposed by Noyes and Kalkofen and the Harvard-Smithsonian Reference Atmosphere. Figure 3 of this paper shows, amongst other things, the fairly close agreement achieved over the height range 500 to 900 km between the curves of shock dissipation and net radiative energy loss when computed on the basis of the Noyes–Kalkofen model with a shock period of 25 sec. The same diagram also illustrates the variation in the shock strength η over the same height range: the maximum value is less than 0.5. In a third paper Ulmschneider [1971b] extended his discussion to the more complicated case of a *spectrum* of exciting acoustic waves as opposed to a single wave of unique period and initial amplitude (cf. Section 7.4.3). Once again the dissipation at a height of 500 km turned out to be dominated by shocks with periods of about 28 sec.

Finally, reference should be made to the earlier computations of Osterbrock [1961] who used essentially the same simplified theory of weak shock propagation as Ulmschneider.[13] However, unlike Ulmschneider, Osterbrock did not attempt to derive a value for the period but, instead, assumed a figure of 83 sec. His numerical results cover the height range $h = 0$ to $h = 6000$ km [see Osterbrock's Table 8, part 1], but the model on which they are based has since been supplanted by the more recent models presented in Section 4.4 [cf. Jordan, 1970]. Certain shortcomings of Osterbrock's approach have been discussed by Uchida [1963].

7.4.6 SOME COMMENTS ON THE OBSERVATIONAL IMPLICATIONS OF ULMSCHNEIDER'S THEORY

Ulmschneider's theory, described in the previous section, attributes the heating of the chromosphere at low heights ($h \lesssim 1000$ km) to short-period acoustic waves which propagate up from the photosphere and develop into shocks within a few hundred kilometres or so of the base of the chromosphere. To obtain agreement between the curves of mechanical dissipation and net radiative energy loss as functions of height it is necessary to postulate that the periods of the original acoustic waves are of the order of 10–30 sec. However, no waves with periods as short as this have so far been detected either in the upper photosphere or low chromosphere. On the contrary, the observed oscillatory motions described in Section 3.5.1 have periods ranging from about 150 to 400 sec, the average value being close to 300 sec.

[13] Osterbrock's analysis actually refers to weak *magnetohydrodynamic* shocks (see Section 6.6.4) but, as Jordan [1970] has pointed out, comparison between his results for the case $H = 0.5$ gauss and $H = 2$ gauss indicates that for the former case the results (for low heights) are virtually the same as in the absence of a field.

A way out of this apparent dilemma has been suggested by Ulmschneider [1970], who has pointed out that the question of whether a wave makes an observable contribution to the Doppler shift of a given spectral line depends on the magnitude of its wavelength λ compared to the depth of formation of the line. From Section 6.3.2 the wavelength of a gravity-modified acoustic wave of angular frequency $\omega = 2\pi/P$ is easily shown to be

$$\lambda = \frac{2\pi a}{(\omega^2 - \omega_1^2)^{1/2}}, \tag{7.40}$$

where a is the velocity of sound and $\omega_1 = \gamma g/2a$ is the cut-off frequency for acoustic waves. At the level corresponding to, say, $\tau_{5000} = 0.1$, where $a = 7.45$ km sec^{-1} and $\omega_1 = 0.03$ sec^{-1}, this gives a value for λ of only 187 km for a wave with a period of 25 sec. At $\tau_{5000} = 0.01$, corresponding to the base of the chromosphere (i.e. $h = 0$), we have $\lambda = 178$ km. Both these figures are evidently comparable to the thickness of the region of formation of a typical Fraunhofer line; accordingly the assumed wave will fail to give rise to any measurable Doppler shift, although it will of course contribute to broadening of the line.[14] The fact that waves with periods of the order of 10–30 sec have escaped detection is therefore not surprising and cannot, at this stage, be regarded as constituting a decisive objection to Ulmschneider's theory.

A second possible point of contact between observation and theory is concerned with the magnitude of the *velocity amplitudes* of the assumed gravity-modified acoustic waves in the upper photosphere. The theory of Sections 6.3.3 and 7.4.3 allows us to derive a lower limit for this quantity: for vertical propagation the flux of acoustic wave energy is given by the formula

$$F_{\text{acoustic}} = \frac{1}{2}\rho V^2 \cdot \frac{a^2 P}{\lambda}, \tag{7.41}$$

where ρ is the gas density and V the velocity amplitude. To derive a lower limit to F_{acoustic} we neglect the loss of wave energy due to radiative cooling below $h \simeq 200$ km (Section 7.4.2); this lower limit must be equal to the flux of mechanical energy F_{mech} at the height of shock formation. Taking $t_0 = (1/12)P$ in conformity with Ulmschneider [1970], we have from equation (7.30)

$$F_{\text{mech}} = \frac{1}{12}\rho a^3 \eta^2, \tag{7.42}$$

η being the shock strength given by Table 7.5. For the sake of argument let us assume a height of shock formation of 400 km: then for a wave period of 25 sec,

[14] The effect of acoustic waves on spectral line profiles has been discussed theoretically by a number of authors, including Kulander and Jefferies [1966], Olson [1966], and Eriksen and Maltby [1967].

$\eta = 0.19$, and hence $F_{mech} = 1.4 \times 10^6$ erg cm^{-2} sec^{-1}. By equation (7.41) it follows that

$$F_{acoustic} = \frac{1}{2}\rho V^2 \cdot \frac{a^2 P}{\lambda} > 1.4 \times 10^6 \text{ erg cm}^{-2} \text{ sec}^{-1}.$$

Accordingly, at the photospheric level $\tau_{5000} = 0.1$, for example,

$$V > 0.05 \text{ km sec}^{-1}.$$

This is a barely detectable velocity with current techniques; even if we take the initial acoustic flux to be, say, four times greater in order to allow for loss of energy by radiative cooling, the velocity amplitude at $\tau_{5000} = 0.1$ only increases to about 0.1 km sec^{-1}. As we have already seen in Section 3.4.3, velocities of this magnitude are hard to detect against the background of the oscillatory and granulation velocity fields in the photosphere. This further reduces the chance of observing the 10–30 second acoustic waves in the upper photosphere postulated by Ulmschneider.

7.4.7 SHOCK WAVE HEATING OF THE MIDDLE AND UPPER CHROMOSPHERE

As we have seen in Section 7.4.5, the heating of the chromosphere at low heights can be satisfactorily accounted for in terms of short-period (~10–30 sec) acoustic waves which propagate up from the photosphere and develop into shocks within a few hundred kilometres or so of the base of the chromosphere. Below a height of about 200 km the original acoustic waves are damped by radiative loss and it is therefore not surprising that the resulting shock waves turn out to be *weak*, in the sense that $\eta < 1$. On the other hand, theory indicates that acoustic waves of longer period, say 100 sec or greater, should reach greater heights before developing into shocks. Furthermore, since the effect of radiative damping is much diminished at heights exceeding $h \simeq 200$ km (cf. Section 7.4.2), the long-period waves evidently have a greater opportunity to grow in amplitude. Consequently, there is good reason to expect that shocks reaching the higher levels of the chromosphere or the corona are *strong* shocks ($\eta > 1$) and, indeed, this expectation has been borne out by the calculations of Kuperus [1965] and Ulmschneider [1971a], discussed below.

At these heights the weak-shock theory developed in Section 7.4.3 is no longer applicable and the mechanical dissipation must now be calculated by a method valid for shocks of arbitrary strength. There are a number of different approaches to this problem, among which those of the following authors are worthy of note:

(1) Saito [1964] has used the generalized form of the Brinkley–Kirkwood theory given in Section 6.5.3(3), which governs the vertical propagation of

shock waves through a gravitationally-stratified atmosphere in hydrostatic equilibrium;

(2) Kuperus [1965] has employed the hydrodynamical theory of the formation of shocks from large amplitude acoustic waves given by Landau and Lifshitz [1959]; and

(3) Ulmschneider [1967, 1971a] has used a reformulated version of Whitham's theory of shock propagation (Section 6.5.3(3)) which takes due account of dissipation following the procedures adopted in the classical Brinkley–Kirkwood theory.

In the higher levels of the chromosphere it is no longer permissible to neglect the conductive energy term in the basic equation of energy balance

$$-\frac{\mathrm{d}F_{\mathrm{mech}}}{\mathrm{d}h}-\frac{\mathrm{d}F_{\mathrm{cond}}}{\mathrm{d}h}=\frac{\mathrm{d}F_{\mathrm{rad}}}{\mathrm{d}h}, \tag{7.3}$$

as we have pointed out in Section 7.2.6. Accordingly, it is not possible to follow the procedure of Section 7.4.5 of equating the mechanical dissipation directly to the radiative energy loss. Ideally, given a reliable model of the upper chromospheric layers, one would be able to calculate $\mathrm{d}F_{\mathrm{cond}}/\mathrm{d}h$ at each height by means of the formula (7.14). If, in addition, the net radiative loss was accurately known, one could plot the quantity $\mathrm{d}F_{\mathrm{rad}}/\mathrm{d}h + \mathrm{d}F_{\mathrm{cond}}/\mathrm{d}h$ as a function of height. This curve could then be matched against a family of curves depicting the mechanical dissipation $(-\mathrm{d}F_{\mathrm{mech}}/\mathrm{d}h)$ calculated for a range of assumed values of the initial mechanical flux $F_{\mathrm{mech}}(0)$ and of the shock period P. At the present time, however, such a procedure is not feasible since (*a*) the available estimates of $\mathrm{d}F_{\mathrm{rad}}/\mathrm{d}h$ in the middle and upper chromosphere may well be in error by an order of magnitude or more; and (*b*) current models of the chromosphere at these heights are themselves subject to considerable uncertainty.

Nevertheless, Kuperus and Ulmschneider attempted to answer the fundamental question of whether it is possible, at least in principle, to explain the heating of these layers by the mechanism of shock heating by deriving self-consistent, shock heated models of the middle and upper chromosphere and of the overlying corona. Apart from the use of different shock wave theories to evaluate $(-\mathrm{d}F_{\mathrm{mech}}/\mathrm{d}h)$, their procedures are basically similar; at each height a computation is made of the net radiative energy loss $\mathrm{d}F_{\mathrm{rad}}/\mathrm{d}h$ on the basis of a discussion of the various emission processes involved. Using the formula (7.14) for the conductive flux they then determine the value of the temperature which, at the point in question, accurately satisfies the equation of energy balance (7.3). The corresponding density or pressure at each point can then be calculated with the aid of the equation of hydrostatic equilibrium.

Kuperus extended his integrations downwards from the level of the coronal temperature maximum until a point was reached at which the temperature had dropped to ~10 000 K. He assumed a shock period of 300 sec, a value which seems unrealistically high in the light of the discussion in Section 7.3.2.

Ulmschneider [1971a], on the other hand, introduced fully developed shock waves at arbitrarily chosen heights of 5000 and 20 000 km and carried his computations upwards from the lower level far out into the corona; the calculations were carried out for assumed periods of 100 and 200 sec.

In the future we can no doubt look forward to rocket and satellite observations in the extreme ultraviolet leading both to improved models of the middle and upper chromosphere and to more accurate estimates of the net radiative energy loss as a function of height. This knowledge will, in turn, enable a more refined attack to be made on the problem of the heating of these layers, perhaps along the lines of the ideal procedure suggested above.

7.4.8 SHOCK WAVE HEATING IN THE NEIGHBOURHOOD OF THE CHROMOSPHERIC NETWORK

In discussing the problem of chromospheric heating we have, up to this point, ignored any possible influence of magnetic fields. This is presumably allowable in regions of the chromosphere overlying the *interiors* of the supergranule cells where, as we have seen in Section 3.4.4, the observed field strengths (at the photospheric level) are only about 2 gauss. However, the situation is different at the cell *boundaries*, which are closely correlated with much larger fields and with a bright network visible in Hα, the K line of Ca II, certain weaker Fraunhofer lines, and (very weakly) the photospheric continuum (Sections 3.3.2, 3.3.3, 3.4.2, 3.4.4). As far as our present knowledge extends, these fields are predominantly longitudinal and their strengths generally range from 10 to about 50 gauss or somewhat higher; at the vertices of the supergranule cells, fields as large as 100 gauss may occur. Most workers believe that these fields are responsible for *local excess heating* in the upper photosphere and in the overlying chromosphere, and that the chromospheric and photospheric networks are a manifestation of this heating. Accordingly, we now turn our attention to the possible influence of such fields, located along the supergranule boundaries, on the dissipation of shock wave energy in the overlying chromosphere.[15]

Unfortunately, no direct observational information is available as to just how far the photospheric fields do indeed extend up into the chromosphere (cf. Section 3.5.4). However, this problem has been discussed theoretically by Clark [1968], who finds that the upward extension depends critically on the magnitude of the *vertical* motions at the photospheric level. In the light of the

[15] The role played by magnetic fields in heating the *active* chromosphere lies beyond the scope of this book; the interested reader is referred to papers by Osterbrock [1961], Pikelner and Livshitz [1964], and Pikelner [1965].

subsequent confirmation by Frazier and other observers of the existence of downflows of the order of 0.1 km sec^{-1} at the supergranule boundaries (Section 3.4.3), Clark's results are consistent with a penetration of the field up to heights of 1000 km or more with only a fairly moderate diminution in the strength of the longitudinal component. The physical picture underlying Clark's analysis is simple and instructive: consider a bunch of vertical field lines being pushed towards the network boundary by the horizontal motions in the interior of a supergranule cell. At the photospheric level, where hydrodynamic forces are dominant (see Section 3.4.5), this process results in a concentration of the field at the cell boundaries. However, whether or not the continuations of the field lines into the overlying chromosphere remain closely bunched evidently depends on whether or not the matter originally 'frozen' into the field can flow away (downwards) along the lines of force. The observations mentioned above imply that it does. On the other hand, it must be emphasized that if the field is to exert any influence on vertically propagating waves of the longitudinal type, such as the fast-mode magneto-acoustic waves discussed below, then it has to diverge to some extent with increasing height.

If we neglect the presence of gravity, three modes of wave propagation are possible in the presence of a magnetic field: generalized Alfvén waves and fast- and slow-mode magneto-acoustic waves (Section 6.4.2). However, for the reasons explained in Section 7.3.2, it is usually assumed that the waves responsible for heating the low chromosphere in magnetic regions are magneto-acoustic waves of the fast-mode type. Furthermore, in the photosphere the value of the Alfvén velocity

$$A = \frac{H}{\sqrt{4\pi\rho}} \tag{7.43}$$

is small compared to the velocity of sound a for the field strengths concerned (cf. Table 7.3), and hence the fast-mode waves are basically hydrodynamic in character. On the other hand, at a certain height in the chromosphere A will become comparable with a and the magnetic character of the waves can no longer be ignored. Assuming a typical field strength in the chromospheric network of, say, 20 gauss, one finds that the equality $A = a$ is satisfied when

$$p = \frac{H^2}{4\pi\gamma} = 19.1 \text{ dyne cm}^{-2},$$

corresponding on the BCA model (Table 4.6) to a height of $h \simeq 750$ km. Consequently, bearing in mind the discussions of Sections 7.4.2 and 7.4.5, one may draw the important conclusion that the fast-mode magneto-acoustic waves are likely to develop into shocks *before* their magnetic characteristics become important provided, of course, their periods are comparable with those of the acoustic waves present in non-magnetic regions. From the point where the shock finds itself in the region in which the Alfvén velocity is comparable to the sound velocity, one evidently has to consider how the propagation and dissipation are altered by the presence of the field.

As we have seen in Section 7.4.3, hydrodynamic shocks propagating upwards through the low chromosphere are gradually refracted away from the vertical as a consequence of the increase in the sound velocity with height. When a magnetic field is present, the operative parameter becomes the wave velocity of the fast-mode magneto-acoustic waves, which is given by equation (6.129). This quantity increases much more rapidly with height than the ordinary sound velocity as a consequence of the growth in the Alfvén velocity (7.43) resulting from the decrease in the gas density. Accordingly, one must expect *enhanced shock refraction* in the presence of a field, and this in turn will reduce the height up to which shock heating is effective. This question has been quantitatively investigated by Yamazaki *et al.* [1969] who, using a generalized form of Chisnell's theory (cf. Section 6.5.3), studied the upward propagation of fast-mode magnetohydrodynamic shocks through an isothermal, gravitationally-stratified atmosphere pervaded by a uniform magnetic field. They found that the shocks are rapidly refracted away from the vertical irrespective of the direction of the field and, moreover, the amount of refraction increases with the strength of the field. Yamazaki *et al.* therefore arrive at the important result that a magnetic field does not, in fact, exert any *guiding* influence on a fast-mode shock (cf. Section 7.6.2).

Subsequently, Milkey [1970] has used the method of geometrical optics to study the transport and dissipation of energy by weak fast-mode magnetohydrodynamic shocks in an inhomogeneous, anisotropic medium and has applied his results to the problem of the heating of the low chromosphere in the magnetic regions overlying the supergranule boundaries. Milkey's computations are based on Osterbrock's [1961] extension of the Brinkley–Kirkwood theory of shock wave propagation described in Section 6.6.4, and give the amount of mechanical energy dissipated as a function of height over the range 200 to ~1200 km for five separate models in hydrostatic equilibrium. We shall designate these models by the letters α, β, γ, δ, and ϵ.

Model α has a uniform vertical field of strength 2 gauss; the temperature is taken to follow the BCA model up to the top of the photosphere but to remain constant above this height. Model β differs from α only in that the temperature follows the BCA model throughout. The remaining models γ, δ, and ϵ have the BCA temperature distribution and a force-free configuration of diverging magnetic field lines with axial field strengths of 50, 100, and 100 gauss respectively; ϵ differs from δ in that allowance is made for a horizontal variation in temperature designed to simulate the assumed rise which occurs in the network. The period of the shock waves was (arbitrarily) taken to be 12.6 sec.

Milkey's results for the four models β, γ, δ, and ϵ are presented in Fig. 7.7, which gives for each model the energy dissipation as a function of height expressed as a ratio of that calculated for the 'standard' model α. These curves illustrate the important conclusion that an increase in the field strength from 2

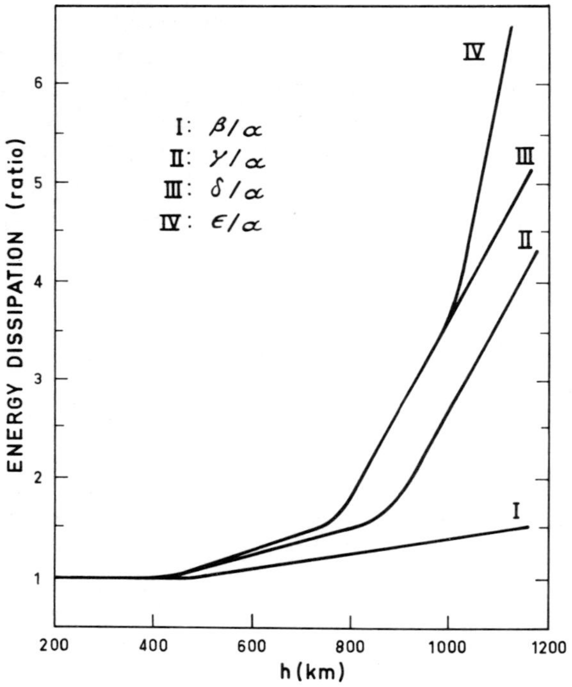

Fig. 7.7. Shock wave heating of the low chromosphere above supergranule boundaries under the influence of the magnetic field [Milkey, 1970]. Curves I–IV give the mechanical dissipation for four separate models, expressed as a ratio of that calculated for a certain 'standard' model (for details, see text).

to 50 or 100 gauss significantly *enhances* the mechanical energy dissipation above the level where the equality $A = a$ is achieved. On the other hand, the heating *below* this level is not greatly altered by the presence of such stronger fields. This result accordingly offers an explanation for the enhanced heating implied by observations of the chromospheric network.[16]

As pointed out by Milkey, it is interesting to note that curves II (50-gauss field) and III (100-gauss field) are nearly parallel: the only apparent difference is

[16] Since shock dissipation does not begin until the original magneto-acoustic waves have penetrated sufficiently high into the chromosphere, it is evident, as Durrant and Michalitsanos [1971] have emphasized, that some other mechanism is required to explain the temperature enhancement observed in the *photospheric* network (Section 3.4.2).

a shift in height due to the fact that for model δ the equality $A = a$ is reached at a lower height than for model γ. According to Milkey the results are not greatly dependent on the assumed input flux of energy from the underlying photosphere. In fact, an increase in the input flux simply results in an increased dissipation within the first few hundred kilometres of shock propagation without significantly changing the dissipation at greater heights.

This explanation of the heating of the chromospheric network, if correct, must also explain why the Hα bright mottles are observed only over a fairly localized height range near the base of the chromosphere, a typical mottle height (measured at its centre) being about 2000 km (cf. Table 3.1). Above this height, therefore, there ought to be a sharp drop in the dissipation of the corresponding magnetohydrodynamic shock. Unfortunately, it is not possible to test this prediction on the basis of Milkey's results: above the level where $A = a$ there is a rapid increase in the 'wavelength' of the shocks due to the decrease in density with height and, as a consequence, the condition for the validity of an approach based on geometrical optics is violated within a few scale heights (see Section 7.4.3).

Another question left unanswered by Milkey's investigation is that of the effect of the magnetic field on the longer-period waves ($P > 100$ sec) which, outside magnetic regions, are believed to be responsible for the heating of the middle and upper chromosphere (Section 7.4.7). At these heights we come into the domain of the spicules and it is noteworthy that, if their identification with dark Hα mottles is correct (Section 3.2.6), they also are located in the neighbourhood of the chromospheric network and, therefore, of its magnetic field. This raises the important question of whether magnetohydrodynamic shock waves may also play a role in the formation of spicules, a topic discussed in Section 7.6.2.

7.5 Thermal-Convective Stability of the Chromosphere

In our foregoing discussion of shock wave heating of the chromosphere we have implicitly assumed that the atmosphere is convectively stable, at least to the extent implied by the well-known Schwarzschild criterion which, in the particular case where the gradients $\mathrm{d}T/\mathrm{d}h$ and $(\mathrm{d}T/\mathrm{d}h)_{\mathrm{ad}}$ are both positive, takes the form

$$\frac{\mathrm{d}T}{\mathrm{d}h} > \left(\frac{\mathrm{d}T}{\mathrm{d}h}\right)_{\mathrm{ad}} \qquad (7.44)^{17}$$

[17] It must be remembered that we have taken the h-axis to be directed vertically upwards, whereas in the literature the Schwarzschild criterion is usually quoted for the case where both gradients are positive with respect to an axis, Oξ say, pointing vertically downwards. Under these circumstances the criterion for stability assumes the more familiar form

$$\frac{\mathrm{d}T}{\mathrm{d}\xi} < \left(\frac{\mathrm{d}T}{\mathrm{d}\xi}\right)_{\mathrm{ad}}$$

[see, for example, Bray and Loughhead, 1967, p. 71].

As we have seen in Section 6.3.2, the derivation of (7.44) follows from a simple consideration of the behaviour of a fluid element subjected to a small vertical displacement under adiabatic conditions, i.e. ignoring heat exchange with its surroundings. However, heat exchange can no longer be ignored in the case of a gas – such as the chromosphere – whose temperature at every point is maintained by a balance between mechanical (and conductive) dissipation and radiative cooling. Under such circumstances a much more sophisticated analysis of the conditions for convective stability is evidently required.

Following Defouw [1970a] we can obtain valuable physical insight into this problem by the simple expedient of considering the behaviour of a small fluid element displaced from its equilibrium position. To do this it is convenient to introduce the so-called 'heat-loss function'

$$L \equiv \frac{1}{\rho}\left[\frac{\mathrm{d}F_{\mathrm{rad}}}{\mathrm{d}h} - \left(-\frac{\mathrm{d}F_{\mathrm{mech}}}{\mathrm{d}h}\right)\right]$$

which at each point represents, per unit mass of gas, the net (time) rate of radiative energy loss minus the rate of gain due to mechanical dissipation; by convention, L does not take account of energy transfer by conduction, but in a rigorous hydrodynamic analysis this term may be included separately in the energy equation [see, for example, Defouw, loc. cit.]. In terms of L the condition (7.4) of energy balance in the *low* chromosphere becomes simply

$$L(\rho, T) = 0 \tag{7.45}$$

where, for simplicity, we shall assume that L depends only on the gas density and temperature.[18] Furthermore, owing to the speed with which pressure changes are equalized throughout a gas, we shall also suppose that the fluid element always remains in pressure equilibrium with its surroundings, so that

$$p^* = p,$$

where p^* is the pressure inside the element and p the pressure outside.

The equation of internal energy (6.86) for the element takes the form

$$\frac{\mathrm{d}p^*}{\mathrm{d}t} - \frac{\gamma p^*}{\rho^*}\frac{\mathrm{d}\rho^*}{\mathrm{d}t} + \rho^*(\gamma - 1)\cdot L(\rho^*, T^*) = 0, \tag{7.46}$$

where for small deviations from an initial state of equilibrium specified by (7.45) we have

$$L(\rho^*, T^*) = \frac{\partial L}{\partial \rho}(\rho^* - \rho) + \frac{\partial L}{\partial T}(T^* - T).$$

[18] According to Defouw [loc. cit.] this is strictly true only in the case of free-free emission from an optically thin, fully ionized gas.

We now rewrite this result with the aid of (6.96) in the form

$$L(\rho^*,T^*) = -\frac{T^*}{\rho^*}\left(\frac{\partial L}{\partial T} - \frac{\rho^*}{T^*}\frac{\partial L}{\partial \rho}\right)(\rho^* - \rho)$$

and introduce the equation of motion of the element

$$\rho^*\frac{\mathrm{d}v^*}{\mathrm{d}t} = -(\rho^* - \rho)g,$$

in which v^* denotes its vertical velocity. Then, following the method used in Section 6.3.4(2) one can show that, correct to the first order, the movement of the element is governed by the equation

$$\frac{\partial^2 v^*}{\partial t^2} + \frac{1}{c_\mathrm{p}}\left(\frac{\partial L}{\partial T} - \frac{\rho}{T}\frac{\partial L}{\partial \rho}\right)\frac{\partial v^*}{\partial t} + \frac{g}{T}\left[\frac{\mathrm{d}T}{\mathrm{d}h} - \left(\frac{\mathrm{d}T}{\mathrm{d}h}\right)_\mathrm{ad}\right]v^* = 0, \tag{7.47}$$

where

$$\left(\frac{\mathrm{d}T}{\mathrm{d}h}\right)_\mathrm{ad} = \frac{\gamma - 1}{\gamma}\frac{T}{p}\frac{\mathrm{d}p}{\mathrm{d}h} \tag{7.48}$$

is the adiabatic temperature gradient in the surrounding atmosphere. Under adiabatic conditions (i.e. $L \equiv 0$), (7.47) reduces to the relation (6.62) previously derived in Section 6.3.2. We should point out that, as a consequence of (7.48), the adiabatic gradient $(\mathrm{d}T/\mathrm{d}h)_\mathrm{ad}$ will be positive or negative according as the pressure increases or decreases outwards in the region concerned. It is interesting to note that the adopted homogeneous model of Section 4.4.6 – taken at its face value – shows both cases: out to a height of 7000 km there is a steady drop in pressure, but beyond this level there is a slow *rise* to the top of the chromosphere.

Equation (7.47) admits solutions of the form $v^* \propto e^{nt}$, where the growth rate n takes either of the values

$$n = -\frac{1}{2c_\mathrm{p}}\left(\frac{\partial L}{\partial T} - \frac{\rho}{T}\frac{\partial L}{\partial \rho}\right) \pm \left\{\frac{1}{4c_\mathrm{p}^2}\left(\frac{\partial L}{\partial T} - \frac{\rho}{T}\frac{\partial L}{\partial \rho}\right)^2 - \frac{g}{T}\left[\frac{\mathrm{d}T}{\mathrm{d}h} - \left(\frac{\mathrm{d}T}{\mathrm{d}h}\right)_\mathrm{ad}\right]\right\}^{1/2}. \tag{7.49}$$

This relation shows that even in the case of a uniform medium with no gravity ($g = 0$), where convection as such cannot occur, some form of fluid instability can nevertheless exist. Under such circumstances the condition for instability becomes simply

$$\frac{1}{c_\mathrm{p}}\left(\frac{\partial L}{\partial T} - \frac{\rho}{T}\frac{\partial L}{\partial \rho}\right) < 0. \tag{7.50}$$

This type of instability, occurring in the absence of gravity, is referred to as 'thermal instability'. It was first described by E. N. Parker in 1953 and has since been discussed in relation to the chromosphere by a number of authors, including Weymann [1960], Thomas and Athay [1961, Chapter 5], Field [1965], and Jordan [1969, pp. 22–23]. The general subject of thermal instability has been discussed by Defouw [1970b] and Getling *et al.* [1971].

In the presence of gravity, on the other hand, (7.49) yields the two values

$$n = \pm\left\{-\frac{g}{T}\left[\frac{\mathrm{d}T}{\mathrm{d}h} - \left(\frac{\mathrm{d}T}{\mathrm{d}h}\right)_\mathrm{ad}\right]\right\}^{1/2} \tag{7.51}$$

in the particular case of purely adiabatic motion. In the chromosphere the value of the temperature gradient $\mathrm{d}T/\mathrm{d}h$ is positive (i.e. the temperature increases outwards) at all heights above the temperature minimum (cf. Figs. 4.3 and 4.12). If moreover $(\mathrm{d}T/\mathrm{d}h)_\mathrm{ad}$ is negative, as is the case, for example, under hydrostatic equilibrium (see equation 6.65), then both values of n given by (7.51) are imaginary and the atmosphere is therefore convectively stable. Under these circumstances $n = \mathrm{i}N$, where N is simply the Väisälä frequency for gravitational *oscillations* whose significance has been discussed in Section 6.3.2. But if on the other hand $(\mathrm{d}T/\mathrm{d}h)_\mathrm{ad}$ is positive, one of the values of n will be real and positive and hence *convection* will occur, provided the atmosphere satisfies the Schwarzschild criterion for convective instability

$$\frac{\mathrm{d}T}{\mathrm{d}h} < \left(\frac{\mathrm{d}T}{\mathrm{d}h}\right)_\mathrm{ad}.$$

Finally, if we drop the restriction to adiabatic conditions, the relation (7.49) leads at once to the very important conclusion that a *thermally* unstable atmosphere, i.e. one in which the inequality (7.50) holds, is also *convectively* unstable. For this reason it is customary to describe this type of instability as 'thermal-convective' instability. There is thus a close connection between the thermal and convective instabilities of a fluid, the nature of which is analysed in some detail by Defouw [1970a]. Moreover, it should be emphasized that the above conclusion is completely independent of the atmospheric temperature gradient, $\mathrm{d}T/\mathrm{d}h$. Consequently, the question of whether the chromosphere is convectively stable or not at any particular height hinges essentially on the functional dependence of the heat-loss function L. In general this is quite complicated and therefore it cannot be assumed, a priori, that a chromospheric model constructed to satisfy the equation of energy balance (7.3) is necessarily stable. If not, then evidently we must expect the presence of dynamical motions (cf. Section 7.6.3) and the model will have to be modified accordingly.

Our theoretical discussion above has ignored the effects of damping due to viscosity and thermal conduction. To explore these effects Defouw [1970a] has used the basic equations of hydrodynamics in the Boussinesq approximation to re-examine the stability of a thin layer of fluid characterized by a heat-loss function $L(\rho, T)$. This treatment can be regarded as an extension of the classical Rayleigh problem [see, for example, Bray and Loughhead, 1967, pp. 67–70]. Defouw seeks periodic solutions proportional to $e^{nt}e^{i(k_x x + k_y y)} \sin k_z z$ and finds that the condition for the onset of thermal-convective instability now takes the form

$$\frac{1}{c_p}\left(\frac{\partial L}{\partial T} - \frac{\rho}{T}\frac{\partial L}{\partial \rho}\right) + \left(1 + \frac{\nu}{\chi}\right)\chi k^2 < 0, \tag{7.52}$$

where ν is the kinematic viscosity, χ the coefficient of thermal diffusion, and $k = \sqrt{k_x^2 + k_y^2 + k_z^2}$ is the wave-number of the perturbation. Comparison of (7.52) with the earlier condition (7.50) shows that the effect of conduction and viscosity is to stabilize perturbations corresponding to large values of k and hence to small physical dimensions. On the other hand, thermal-convective instability *must* ensue if the condition (7.50) is satisfied and the perturbation dimension is sufficiently large. It may be remarked that Defouw has further generalized the analysis to allow for the presence of a vertical magnetic field; for details of the results the interested reader is referred to his [1970a] paper [see also Bhatia, 1971].

In a subsequent paper Defouw [1970c] has applied his stability analysis to a simple hydrogen model of the chromosphere, taking proper account of the part played by hydrogen ionization. For simplicity, the mechanical dissipation per unit mass of gas is taken to be constant with height. Assuming that T increases with height, Defouw concludes that regions of his model chromosphere where $T < 17\,500$ K are stable, whereas regions where $T > 17\,500$ K are unstable if either the temperature gradient or the density is sufficiently large. He also examines the effect of a magnetic field. However, despite its considerable interest, Defouw's analysis contains many simplifications and idealizations and, accordingly, caution must be exercised in applying his results to the real chromosphere. In conclusion, it should be pointed out that analyses of this type serve primarily to establish the conditions required for the *onset* of instability and provide no description of the resulting motion of the gas. However, an exploratory attack on this latter problem, albeit for a highly idealized and simplified situation, has been made by Bessey and Kuperus [1970].

7.6 Theories of the Origin of Spicules

7.6.1 INTRODUCTION

As we have seen in Chapters 2 and 4, a vast amount of effort has been devoted to direct photographic and spectroscopic observations of spicules and much is now known about their properties and physical conditions. This information, together with the intrinsic interest of the phenomenon, has stimulated many authors to propose a variety of physical mechanisms which might conceivably explain their origin. To be fully successful a theory of spicules must explain a large number of diverse observational facts and, in particular, must answer the following questions:

(1) why is a spicule invariably a long, narrow object?

(2) why is the density of a spicule greater than its surroundings, and why, above about 10 000 km, is its temperature less?

(3) why are spicules localized in the neighbourhood of supergranule boundaries and, therefore, in regions of weak magnetic field?

(4) what is the driving force which keeps a spicule moving upwards at a more or less constant velocity despite the retarding effect of gravity?

(5) why do changes in the observed Doppler velocity of a spicule appear to be propagated along its length at a speed comparable to that of an Alfvén wave?

However, explanations of spicules advanced to date tend to concentrate on relatively isolated aspects of the spicule problem. They fall naturally into one of three classes: (*a*) those which envisage spicules as the direct manifestation of hydrodynamic or magnetohydrodynamic shock waves travelling vertically upward (Section 7.6.2); (*b*) those which attribute the origin of spicules to dynamical motions resulting from thermal or thermal-convective instability at some level in the low chromosphere (Section 7.6.3); and (*c*) those which attempt to explain spicules in terms of the expulsion of gas upwards in the neighbourhood of neutral lines or sheets in the pattern of weak magnetic fields surrounding the supergranules (Section 7.6.4). This last mechanism appears to show considerable promise, but in no instance can the suggested explanation really be regarded as constituting a fully-fledged theory of the spicule phenomenon as a whole. Accordingly, we shall confine ourselves to briefly outlining some of the suggested hypotheses, together with some of the major criticisms which have been raised against them.

Whatever their mode of origin, spicules must evidently carry both kinetic energy and mass from the low chromosphere up to the base of the corona. Consequently, one is led to ask whether they play a significant role in maintaining either the mass or energy balance of the corona; these questions are discussed in Section 7.6.5.

7.6.2 HYDRODYNAMIC OR MAGNETOHYDRODYNAMIC SHOCKS

The idea that a spicule may be a direct manifestation of the passage of a hydrodynamic shock wave or train of shock waves was first proposed by Unno and Kawabata [1955] and was subsequently elaborated by Uchida [1961]. The latter used the Rankine–Hugoniot shock relations to compare the physical conditions inside and outside a spicule, paying particular attention to the possible loss of energy from the back side of the shock front due to radiation. However, it is now recognized that at the heights concerned such radiation loss can be neglected, owing to the magnitude of the radiative relaxation time [cf. Wentzel and Solinger, 1967].

Uchida realized that his hypothesis provided no reason why spicules should appear as long narrow structures, and he therefore suggested that the original disturbance producing the spicule is guided along magnetic lines of force, presumed to be vertical or near-vertical. Quite independently, Osterbrock [1961] advanced an explanation of spicules in terms of magnetohydrodynamic shocks. Osterbrock's mechanism is complicated and involves both fast- and slow-mode shocks. The fast-mode shocks are supposed to originate near the base of the chromosphere in regions of weak magnetic field, i.e. in the neighbourhood of the supergranule boundaries. These shocks propagate outwards until they reach the level at which the Alfvén velocity becomes equal to the sound velocity. At this level, by a number of processes described only qualitatively by Osterbrock, a certain fraction of the fast-mode shocks are supposed to be transformed into slow-mode shocks. The character of these latter shocks is such that they behave as purely hydrodynamic shocks propagating only in the direction of the field. Osterbrock identifies the spicules with the chromospheric material carried upwards by these slow-mode shocks. One objection to Osterbrock's mechanism is that owing to the rapid increase of the Alfvén velocity with height (Table 7.3), fast-mode shocks will be rapidly refracted away from the vertical, regardless of the angle which their original direction of propagation makes with the direction of the field (see Section 7.4.8). Accordingly, refraction will tend to seriously reduce the amount of energy available for generating vertically propagating slow-mode shocks.

Later authors who have also attempted to explain spicules in terms of magnetohydrodynamic shocks include Parker [1964], Lüst and Scholer [1966], and Wentzel and Solinger [1967]. However, these papers, in common with those of Uchida and Osterbrock, concentrate on more or less isolated aspects of spicule behaviour and therefore hardly constitute adequate theories of the phenomenon as a whole.

An interesting point in connection with shock wave theories of spicules is the question of whether the observed height to which a spicule ascends, nearly 12 000 km on average, can be accounted for by a *single* shock or whether a *train* of shocks must be invoked. In principle, this question can be resolved by calculation, since behind the shock front the motion of the gas is governed by the ordinary hydrodynamic equations and hence depends on the pressure gradient and on the initial velocity imparted to the gas by the shock. On this basis, Pikelner [1971] finds that a single shock would raise a spicule only to a height of 1500 km [see also Osterbrock, 1961], which evidently implies that a train of shocks is needed, whose period is short compared to the spicule's time of ascent. Accurate measurements of a spicule's height as a function of time might throw additional light on this question, but it is doubtful whether existing observations have the necessary precision [cf. Athay and Thomas, 1957].

7.6.3 FLUID INSTABILITY

It has been suggested by Defouw [1970c] that spicules may owe their origin to thermal-convective instability of the kind described in Section 7.5, occurring in the low chromosphere in the neighbourhood of supergranule boundaries and their associated magnetic fields. However, Defouw's analysis is limited to an investigation of the conditions necessary for the *onset* of instability and provides no description of the resulting (developed) motions: at this stage, therefore, we have no proper basis for identifying any such motions with spicules.

Earlier, Kuperus and Athay [1967] and Kopp and Kuperus [1968] had attempted to identify spicules with motions resulting from thermally-induced instability in the low chromosphere. The instability was attributed to the liberation of a large amount of energy in a very thin layer in the low chromosphere, conducted down to this level from the corona as the result of a steep temperature gradient. However, they made no attempt to give a proper hydrodynamical description of the developed motions nor of the effect of such motions on the assumed steep temperature gradient [see, however, Bessey and Kuperus, 1970].

7.6.4 MAGNETIC EXPULSION

Pikelner [1969, 1971] and Uchida [1969] have proposed mechanisms involving the *annihilation* of part of the longitudinal magnetic field at a point in the neighbourhood of a supergranule boundary, a process accompanied both by the liberation of heat and by the *upward expulsion* of material in the form of a spicule. Both mechanisms are based on the observation that along the supergranulation boundaries the field has a complex fine structure involving changes in polarity over quite small distances (see Section 3.4.4 and the caption to Fig. 3.6). Consequently, at various places along these boundaries and particularly, perhaps, at the vertices of the supergranule cells, one may expect to find 'neutral points' or 'neutral lines'.

At such locations oppositely-directed lines of force will be pressed together either by the opposing horizontal motions of the gas in adjacent supergranules (Pikelner) or by motions associated with the photospheric granulation or with the 5-minute oscillations described in Section 3.5.1 (Uchida). The effect of this compression will be to hasten the annihilation of the field along the neutral lines (or more realistically, sheets), thereby causing lines of force in the chromosphere to sever their connection with the underlying photosphere, a process pictured in Fig. 7.8. These severed lines of force then reconnect and move upwards. At the same time, gas is squeezed out in a vertical direction from the neighbourhood of such a neutral sheet: the ejected gas is carried up by the receding, reconnected field lines and is observed as a spicule. According to Pikelner, his mechanism implies that spicules cannot be formed at levels below about 1000–2000 km, since at such levels the gas pressure is large enough to prevent the attainment of the necessary degree of compression.

The foregoing description may be regarded as a somewhat simplified picture of the basic mechanism of field annihilation in the neighbourhood of a neutral point. This process is of fundamental importance in a number of astrophysical problems, for example that of the origin of solar flares, and has been extensively studied.[19] However, in the case of spicules the mechanism described, as it stands, is not sufficiently rapid to accomplish the necessary field annihilation within the lifetime of a spicule (~5 min). To achieve this it is necessary to invoke other processes operating in conjunction with the first.

Such processes usually involve either the onset of some form of magnetohydrodynamic instability in the neutral sheet or the generation of wave motions transforming magnetic energy into wave energy. Pikelner invoked the production of ionic acoustic waves, whereas Uchida chose to rely on a mechanism first proposed by Petschek [1964]. This would entail the establishment of a pattern of standing magnetohydrodynamic waves, but subsequent studies have shown that this mechanism is incapable of realization in the form originally proposed [Green and Sweet, 1967; Petschek and Thorne, 1967]. In the case of the spicule problem, the presence of gravity by itself helps to increase the rate of field annihilation: as remarked by Pikelner, when the field lines are forced together, the gas is compressed, cools by radiation, and then slides down the lines of force into the photosphere. This enables the field lines to come still closer together, and the cycle is repeated until the field is annihilated. The process envisaged by Pikelner is illustrated in Fig. 7.8, taken from his [1971] paper.

Evidently, we do not yet have a very clear understanding of the particular mechanisms that may be involved. Nevertheless, the Pikelner–Uchida proposal seems to have a number of important points in its favour. Firstly, it exploits the

[19] See, for example, the reviews by Akasofu [1966, p. 126 et seq.]; Stern [1966, pp. 166–170]; Piddington [1969, pp. 50–55].

presence of the magnetic field to obtain the energy needed to accelerate the spicules and maintain their upward motion. Secondly, as pointed out by Uchida, the heat liberated in the process of field annihilation may be responsible for producing the chromospheric network, which is coincident with the supergranule boundaries. Finally, as remarked by Pikelner, it may provide an explanation of the intriguing observation that the time lags between *changes* in the Doppler velocity of a spicule at different heights are consistent with a disturbance being propagated along its length with a phase velocity comparable to the likely value of the Alfvén velocity (see Section 2.4.4).

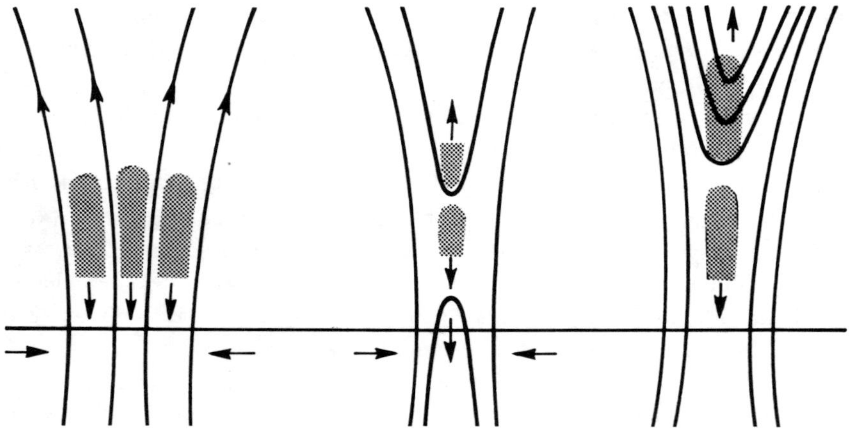

Fig. 7.8 The annihilation and reconnection of magnetic field lines in the neighbourhood of a neutral sheet [Pikelner, 1971]. As the field lines are squeezed together (left), the compressed gas trapped between them cools and flows downwards under gravity. At the same time material is carried upwards by the receding, reconnected field lines, thus giving rise to a spicule (centre and right).

On the other hand, Beckers [1972] has raised an important objection: existing observations of the photospheric magnetic fields along the supergranule boundaries fail to reveal sufficient small-scale polarity changes to account for the observed number of spicules. However, most of these observations have been made with comparatively modest spatial resolution and it is possible that improved observations may radically change our present rather limited knowledge of the fine structure of the field.

7.6.5 ROLE OF SPICULES IN THE MASS AND ENERGY BALANCE OF THE CORONA

As we have previously remarked in Section 2.4.4, only a minority of spicules appear to ultimately reverse their upward motion and descend back to the low chromosphere. This raises the important question of whether they may play a significant role in maintaining either the mass or energy balance of the corona. Let us assume that at any given moment approximately 50 per cent are moving upwards, 20 per cent are stationary, and the remaining 30 per cent are descending (cf. Section 2.4.1); then the *net* upward flow of mass in the spicules is simply $0.2\,\rho VQ$, where ρ and Q represent respectively the spicule gas density and the fractional area occupied by spicules at the height concerned and V is the magnitude of their mean upward or downward velocity. Calculating on the basis of the spicule model given in Table 4.23 and taking $V = 25$ km sec^{-1} (Section 2.4.2), one finds that the net upward mass flux amounts to 5×10^{-11} gm cm^{-2} sec^{-1} at a height of 10 000 km, a figure chosen because it is close to the value given in Section 2.3.5 for the average *true* height of the spicules. This value for the outward mass flux is comparable to the figure of 3×10^{-11} gm cm^{-2} sec^{-1} recently quoted by Ulmschneider [1971a] for the net outward flow of mass from the corona due to the *solar wind.* Accordingly, if we accept these figures at their face values, we must conclude that the mass lost from the corona due to the solar wind is made good by the injection of spicules from below since, evidently, there can be no *net* loss or gain of mass from the corona [see also Beckers, 1972].

Similarly, the net upward flux of kinetic energy carried by the spicules can readily be shown to be

$$0.2\,(\tfrac{1}{2}\rho V^2)VQ = 1.6 \times 10^2 \text{ erg cm}^{-2}\text{ sec}^{-1}$$

at a height of 10 000 km. By comparison, the overall rate of energy loss from the corona due to radiation and solar wind, but ignoring conduction of heat downward to the chromosphere, is variously estimated to be about 5.5×10^4 to 3.5×10^5 erg cm^{-2} sec^{-1} (see Table 7.2) and is therefore vastly greater. Consequently, in agreement with Beckers [loc. cit.] we can conclude that the spicules appear to be of no significance in maintaining the energy balance, as opposed to the mass balance, of the corona.

REFERENCES

AKASOFU, S.-I. [1966] 'Electrodynamics of the magnetosphere: geomagnetic storms', *Space Science Rev.* **6**, 21.

ATHAY, R. G. [1966] 'Radiative energy loss from the solar chromosphere and corona', *Astrophys. J.* **146**, 223.

ATHAY, R. G. [1970] 'A non-LTE line-blanketed solar model', *Astrophys. J.* **161**, 713.

ATHAY, R. G., and THOMAS, R. N. [1957] 'The numbers and motions of solar spicules', *Astrophys. J.* **125**, 804.

BAHNG, J., and SCHWARZSCHILD, M. [1963] 'Hydrodynamic oscillations in the solar chromosphere', *Astrophys. J.* **137**, 901.

BECKERS, J. M. [1968] 'High-resolution measurements of photosphere and sunspot velocity and magnetic fields using a narrow-band birefringent filter', *Solar Phys.* **3**, 258.

BECKERS, J. M. [1972] 'Solar spicules', *Ann. Rev. Astron. Astrophys.* **10**, 73.

BESSEY, R. J., and KUPERUS, M. [1970] 'Thermally driven motions in a gravitational atmosphere', *Solar Phys.* **12**, 216.

BHATIA, P. K. [1971] 'On thermal-convective instability in a stellar atmosphere', *Publ. Astron. Soc. Japan* **23**, 181.

BIERMANN, L. [1946] 'Zur Deutung der chromosphärischen Turbulenz und des Exzesses der UV-Strahlung der Sonne', *Naturwiss.* **33**, 118.

BIERMANN, L. [1948] 'Über die Ursache der chromosphärischen Turbulenz und des UV-Exzesses der Sonnenstrahlung', *Z. Astrophys.* **25**, 161.

BIRD, G. A. [1964] 'The propagation of acoustic waves through the solar chromosphere', *Astrophys. J.* **140**, 288.

BRAY, R. J., and LOUGHHEAD, R. E. [1967] *The Solar Granulation,* Chapman and Hall, London.

CLARK, A. [1968] 'Chromospheric magnetic fields associated with supergranulation', *Solar Phys.* **4**, 386.

DEARDORFF, J. W. [1969] 'Numerical study of heat transport by internal gravity waves above a growing unstable layer', *Phys. Fluids* **12**, Supp. II, p. 184.

DEFOUW, R. J. [1970a] 'Thermal-convective instability', *Astrophys. J.* **160**, 659.

DEFOUW, R. J. [1970b] 'Thermal instability of a model hydrogen plasma', *Astrophys. J.* **161**, 55.

DEFOUW, R. J. [1970c] 'Convective instability of a model chromosphere', *Solar Phys.* **14**, 42.

DUBOV, E. E. [1960] 'On the energy balance and structure of the chromosphere', *Izv. Crim. Astrophys. Obs.* **22**, 101.

DURRANT, C. J., and MICHALITSANOS, A. G. [1971] 'Shock wave dissipation in magnetically active regions', *Solar Phys.* **18**, 60.

ERIKSEN, G., and MALTBY, P. [1967] 'The effect of progressive sound waves on the profiles of stellar spectral lines', *Astrophys. J.* **148**, 833.

FIELD, G. B. [1965] 'Thermal instability', *Astrophys. J.* **142**, 531.

FRAZIER, E. N. [1968] 'A spatio-temporal analysis of velocity fields in the solar photosphere', *Z. Astrophys.* **68**, 345.

GETLING, A. V., KUZMIN, V. V., and TVERSKOI, B. A. [1971] 'Criteria of thermal (overheat) instability of transparent radiating gas', *Doklady Acad. Nauk U.S.S.R.* **196**, 71 (*Sov. Phys.–Doklady* **16**, 15).

GREEN, R. M., and SWEET, P. A. [1967] 'Note on the dissipation of magnetic fields at neutral lines', *Astrophys. J.* **147**, 1153.

JAGER, C. DE [1971] 'The production of solar and stellar chromospheres and coronae', *Phil. Trans. Roy. Soc.* A **270**, 175.

JAGER, C. DE, and KUPERUS, M. [1961] 'The acoustic energy flux of the Sun and the formation of the corona', *Bull. Astron. Inst. Netherlands* **16**, 71.

JONES, W. L. [1969] 'Non-divergent oscillations in the solar atmosphere', *Solar Phys.* **7**, 204.

JORDAN, S. D. [1969] 'The temperature distribution in the solar chromosphere', NASA Tech. Rep. TR/R-291.

JORDAN, S. [1970] 'The effect of the density scale height on shock-wave heating in the low solar chromosphere', *Astrophys. J.* **161**, 1189.

KOPP, R. A., and KUPERUS, M. [1968] 'Magnetic fields and the temperature structure of the chromosphere–corona interface', *Solar Phys.* **4**, 212.

KULANDER, J. L., and JEFFERIES, J. T. [1966] 'Inference of velocities from line asymmetries', *Astrophys. J.* **146**, 194.

KUPERUS, M. [1965] 'The transfer of mechanical energy in the Sun and the heating of the corona', *Recherches Astron. Obs. Utrecht* **17**, No. 1.

KUPERUS, M. [1969] 'The heating of the solar corona', *Space Sci. Rev.* **9**, 713.

KUPERUS, M., and ATHAY, R. G. [1967] 'On the origin of spicules in the chromosphere–corona transition region', *Solar Phys.* **1**, 361.

LANDAU, L. D., and LIFSHITZ, E. M. [1959] 'Fluid Mechanics', *Course of Theoretical Physics*, vol. 6; Pergamon, London.

LEIBACHER, J., and STEIN, R. F. [1971] 'A new description of the solar five-minute oscillation', *Astrophys. Letters* **7**, 191.

LIGHTHILL, M. J. [1967] 'Predictions on the velocity field coming from acoustic noise and a generalized turbulence in a layer overlaying a convectively unstable atmospheric region', *Aerodynamic Phenomena in Stellar Atmospheres*, ed. R. N. Thomas, p. 429; Academic Press, London.

LÜST, R., and SCHOLER, M. [1966] 'Kompressionswellen in einer isothermen Atmosphäre mit vertikalem Magnetfeld', *Z. Naturforschung* **21A**, 1098.

McKENZIE, J. F. [1971] 'Natural modes of the acoustic-gravity type in the solar atmosphere', *Astron. Astrophys.* **15**, 450.

McLELLAN, A., and WINTERBERG, F. [1968] 'Magneto-gravity waves and the heating of the solar corona', *Solar Phys.* **4**, 401.

MEYER, F., and SCHMIDT, H. U. [1967] 'Die Erzeugung von Schwingungen in der Sonnenatmosphäre durch einzelne Granula', *Z. Astrophys.* **65**, 274.

MIHALAS, D. [1970] *Stellar Atmospheres*, Freeman, San Francisco.

MILKEY, R. W. [1970] 'Chromospheric heating above supergranular boundaries', *Solar Phys.* **14**, 62.

MOORE, D. W., and SPIEGEL, E. A. [1964] 'The generation and propagation of waves in a compressible atmosphere', *Astrophys. J.* **139**, 48.

NGUYEN-NGAN [1970] 'Energy balance in spicules', *Astron. J. U.S.S.R.* **47**, 91 (*Sov. Astron. AJ* **14**, 72).

NOYES, R. W., and LEIGHTON, R. B. [1963] 'Velocity fields in the solar atmosphere. II. The oscillatory field', *Astrophys. J.* **138**, 631.

OLSON, E. C. [1966] 'The effect of acoustic waves and convection on solar line asymmetries', *Astrophys. J.* **143**, 904.

OSTERBROCK, D. E. [1961] 'The heating of the solar chromosphere, plages, and corona by magnetohydrodynamic waves', *Astrophys. J.* **134**, 347.

PARKER, E. N. [1964] 'A mechanism for magnetic enhancement of sound-wave generation and the dynamical origin of spicules', *Astrophys. J.* **140**, 1170.

PETSCHEK, H. E. [1964] 'Magnetic field annihilation', *The Physics of Solar Flares*, ed. W. N. Hess, p. 425; NASA, Washington, D.C.

PETSCHEK, H. E., and THORNE, R. M. [1967] 'The existence of intermediate waves in neutral sheets', *Astrophys. J.* **147**, 1157.

PIDDINGTON, J. H. [1969] *Cosmic Electrodynamics*, Wiley, New York.

PIKELNER, S. B. [1965] 'The nature of the active regions of the Sun', *Vistas in Astronomy*, ed. A. Beer, vol. 6, p. 131; Pergamon, Oxford.

PIKELNER, S. B. [1969] 'A mechanism for the formation of chromospheric spicules', *Astron. J. U.S.S.R.* **46**, 328 (*Sov. Astron. AJ* **13**, 259).

PIKELNER, S. B. [1971] 'Theory of spicules and chromospheric structure', *Comments Astrophys. Space Phys.* **3**, 33.

PIKELNER, S. B., and LIVSHITZ, M. A. [1964] 'Theory of heating of the active and undisturbed chromosphere', *Astron. J. U.S.S.R.* **41**, 1007 (*Sov. Astron. AJ* **8**, 808).

PRADERIE, F., and THOMAS, R. N. [1972] 'On estimates of the nonradiative energy input to the solar chromosphere from the H^- emission', *Astrophys. J.* **172**, 485.

SAITO, M. [1964] 'Transport of mechanical energy in the solar chromosphere: propagation of shock-pulses in stratifying media', *Publ. Astron. Soc. Japan* **16**, 179.

SCHATZMAN, E. [1949] 'The heating of the solar corona and chromosphere', *Ann. Astrophys.* **12**, 203.

SCHATZMAN, E. [1967] 'Shock waves in the solar atmosphere', *Solar Physics*, ed. J. N. Xanthakis, p. 133; Interscience, London.

SCHIRMER, H. [1950] Über die Ausbreitung von Stosswellen in der Sonnen-atmosphäre', *Z. Astrophys.* **27**, 132.

SCHMIDT, H. U., and ZIRKER, J. B. [1963] 'On the oscillations of the solar atmosphere', *Astrophys. J.* **138**, 1310.

SCHWARZSCHILD, M. [1948] 'On noise arising from the solar granulation', *Astrophys. J.* **107**, 1.

SOUFFRIN, P. [1966] 'Hydrodynamique d'une atmosphère perturbée par une zone convective turbulente sous-jacente', *Ann. Astrophys.* **29**, 55.

SOUFFRIN, P. [1972] 'Radiative relaxation of sound waves in an optically thin isothermal atmosphere', *Astron. Astrophys.* **17**, 458.

STEIN, R. F. [1969] 'Chromospheric and coronal heating by shock waves', *Chromosphere–Corona Transition Region*, p. 171; National Center for Atmospheric Research, Boulder, Colorado.

STEIN, R. F., and LEIBACHER, J. [1969] 'On the five-minute oscillation of the solar atmosphere', *Astrophys. Letters* **3**, 95.

STERN, D. P. [1966] 'The motion of magnetic field lines', *Space Science Rev.* **6**, 147.

STIX, M. [1970] 'On radiative relaxation of chromospheric oscillations', *Astron. Astrophys.* **4**, 189.

THOMAS, J. H., CLARK, P. A., and CLARK, A. [1971] 'Trapped gravity waves and the five-minute oscillations of the solar atmosphere', *Solar Phys.* **16**, 51.

THOMAS, R. N. [1948] 'Superthermic phenomena in stellar atmospheres. I. Spicules and the solar chromosphere', *Astrophys. J.* **108**, 130.

THOMAS, R. N., and ATHAY, R. G. [1961] *Physics of the Solar Chromosphere*, Interscience, New York.

TOWNSEND, A. A. [1966] 'Internal waves produced by a convective layer', *J. Fluid Mech.* **24**, 307.

UCHIDA, Y. [1961] 'On the formation of solar chromospheric spicules and flare-surges', *Publ. Astron. Soc. Japan* **13**, 321.

UCHIDA, Y. [1963] 'An effect of the magnetic field in the shock wave heating theory of the solar corona', *Publ. Astron. Soc. Japan* **15**, 376.

UCHIDA, Y. [1965] 'Standing mode of compressional-body gravity wave in the solar chromosphere', *Astrophys. J.* **142**, 335.

UCHIDA, Y. [1967] 'Resonant responses of the solar atmosphere to the gravitational-hydrodynamic waves', *Astrophys. J.* **147**, 181.

UCHIDA, Y. [1969] 'A mechanism for the acceleration of solar chromospheric spicules', *Publ. Astron. Soc. Japan* **21**, 128.

ULMSCHNEIDER, P. [1967] 'The structure of the outer atmosphere of the Sun and of cool stars', *Z. Astrophys.* **67**, 193.

ULMSCHNEIDER, P. H. [1968] 'Possible explanation of the 300-second type oscillation in the solar chromosphere', *Astrophys. J.* **152**, 349.

ULMSCHNEIDER, P. [1970] 'On frequency and strength of shock waves in the solar atmosphere', *Solar Phys.* **12**, 403.

ULMSCHNEIDER, P. [1971a] 'On the computation of shock heated models for the solar chromosphere and corona', *Astron. Astrophys.* **12**, 297.

ULMSCHNEIDER, P. [1971b] 'On the propagation of a spectrum of acoustic waves in the solar atmosphere', *Astron. Astrophys.* **14**, 275.

ULRICH, R. K. [1970] 'The five-minute oscillations on the solar surface', *Astrophys. J.* **162**, 993.

UNNO, W., and KAWABATA, K. [1955] 'On the acoustic noise in the solar atmosphere', *Publ. Astron. Soc. Japan* **7**, 21.

WARREN, F. W. G. [1960] 'Wave resistance to vertical motion in a stratified fluid', *J. Fluid Mech.* **7**, 209.

WENTZEL, D. G., and SOLINGER, A. B. [1967] 'On the motion of shocks through the solar chromosphere', *Astrophys. J.* **148**, 877.

WEYMANN, R. [1960] 'Heating of stellar chromospheres by shock waves', *Astrophys. J.* **132**, 452.

WHITAKER, W. A. [1963] 'Heating of the solar corona by gravity waves', *Astrophys. J.* **137**, 914.

YAMAZAKI, H., HASHIMOTO, Y., and ONO, Y. [1969] 'Propagation of shock waves in inhomogeneous gases. VI. Oblique hydromagnetic shock propagation', *Progress Theoret. Phys.* **41**, 367.

Name Index

Subject Index